WITHDRAWN
WRIGHT STATE UNIVERSITY LIBRARIES

SIGMA RECEPTORS
Chemistry, Cell Biology and Clinical Implications

SIGMA RECEPTORS
Chemistry, Cell Biology and Clinical Implications

edited by

Rae R. Matsumoto
University of Mississippi, University, MS

Wayne D. Bowen
Brown University, Providence, RI

and

Tsung-Ping Su
National Institute on Drug Abuse, Bethesda, MD

FORDHAM
QY
55.7
S54
2007

Library of Congress Control Number: 2006928837

ISBN-10: 0-387-36512-5   e-ISBN-10: 0-387-36514-1
ISBN-13: 978-0-387-36512-1

Printed on acid-free paper.

© 2007 Springer Science+Business Media, LLC
All rights reserved. This work may not be translated or copied in whole or in part without the written permission of the publisher (Springer Science+Business Media, LLC, 233 Spring Street, New York, NY 10013, USA), except for brief excerpts in connection with reviews or scholarly analysis. Use in connection with any form of information storage and retrieval, electronic adaptation, computer software, or by similar or dissimilar methodology now known or hereafter developed is forbidden.
The use in this publication of trade names, trademarks, service marks and similar terms, even if they are not identified as such, is not to be taken as an expression of opinion as to whether or not they are subject to proprietary rights.

9 8 7 6 5 4 3 2 1

springer.com

Contents

Contributing Authors		vii
Preface		xi
1.	σ Receptors: Historical Perspective and Background *Rae R. Matsumoto*	1
2.	Medicinal Chemistry: New Chemical Classes and Subtype-Selective Ligands *Amy Hauck Newman and Andrew Coop*	25
3.	Irreversible σ Compounds *Giuseppe Ronsisvalle and Orazio Prezzavento*	45
4.	Pharmacophore Models for σ_1 Receptor Binding *Seth Y. Ablordeppey and Richard A. Glennon*	71
5.	Cloning of σ_1 Receptor and Structural Analysis of its Gene and Promoter Region *Vadivel Ganapathy, Malliga E. Ganapathy and Katsuhisa Inoue*	99
6.	Site-Directed Mutagenesis *Hideko Yamamoto, Toshifumi Yamamoto, Keiko Shinohara Tanaka, Mitsunobu Yoshii, Shigeru Okuyama and Toshihide Nukada*	113

7.	σ Receptor Modulation of Ion Channels *Chris P. Palmer, Ebru Aydar and Meyer B. Jackson*	127
8.	Subcellular Localization and Intracellular Dynamics of σ_1 Receptors *Teruo Hayashi and Tsung-Ping Su*	151
9.	Intracellular Signaling and Synaptic Plasticity *Francois P. Monnet*	165
10.	Modulation of Classical Neurotransmitter Systems by σ Receptors *Linda L. Werling, Alicia E. Derbez and Samer J. Nuwayhid*	195
11.	σ_2 Receptors: Regulation of Cell Growth and Implications for Cancer Diagnosis and Therapeutics *Wayne D. Bowen*	215
12.	Cognitive Effects of σ Receptor Ligands *Tangui Maurice*	237
13.	σ Receptors and Schizophrenia *Xavier Guitart*	273
14.	Potential Role of σ Ligands and Neurosteroids in Major Depression *Guy Debonnel, Malika Robichaud and Jordanna Bermack*	293
15.	σ Receptors and Drug Abuse *Yun Liu, Yongxin Yu, Jamaluddin Shaikh, Buddy Pouw, AnTawan Daniels, Guang-Di Chen and Rae R. Matsumoto*	315
16.	σ_1 Receptors and the Modulation of Opiate Analgesics *Gavril W. Pasternak*	337
17.	σ Receptors in the Immune System: Implications for Potential Therapeutic Intervention - an Overview *Sylvaine Galiegue, Hubert Vidal and Pierre Casellas*	351
18.	σ Receptors and Gastrointestinal Function *Francois J. Roman, Maria Chovet and Lionel Bueno*	371
Appendix		393
Index		405

Contributing Authors

Seth Y. Ablordeppey, Florida A&M University, College of Pharmacy & Pharmaceutical Sciences, Tallahassee, FL 32307, USA

Ebru Aydar, University of Wisconsin Medical School, Department of Physiology, Madison, WI 53706, USA

Jordanna Bermack, McGill University, Department of Psychiatry, Neurobiological Psychiatry Unit, Montreal QC H3A 1A1, Canada

Wayne D. Bowen, Brown University, Department of Molecular Pharmacology, Physiology and Biotechnology, Division of Biology and Medicine, Providence, RI 02912, USA

Lionel Bueno, Institut National de la Recherche Agronomique, 180 Chemin de Tournefeuille, 31932 Toulouse Cedex, France

Pierre Casellas, Sanofi-Aventis, F-34184 Montpellier Cedex 04, France

Guang-Di Chen, University of Oklahoma Health Sciences Center, Department of Pharmaceutical Sciences, Oklahoma City, OK 73190, USA

Maria Chovet, Pfizer Inc., Research Technologies, 94265 Fresnes Cedex, France

Andrew Coop, University of Maryland School of Pharmacy, Department of Pharmaceutical Sciences, Baltimore, MD 21201, USA

AnTawan Daniels, University of Oklahoma Health Sciences Center, Department of Pharmaceutical Sciences, Oklahoma City, OK, 73190

Guy Debonnel, McGill University, Department of Psychiatry, Neurobiological Psychiatry Unit, Montreal QC H3A 1A1, Canada

Alicia E. Derbez, The George Washington University Medical Center, Department of Pharmacology & Physiology, Washington, DC 20037, USA

Malliga E. Ganapathy, Medical College of Georgia, Department of Medicine, Augusta, GA 30912, USA

Vadivel Ganapathy, Medical College of Georgia, Department of Biochemistry and Molecular Biology, Augusta, GA 30912, USA

Sylvaine Galiegue, Sanofi-Aventis, F-34184 Montpellier Cedex 04, France

Richard A. Glennon, Virginia Commonwealth University, Department of Medicinal Chemistry, Richmond, VA 23298, USA

Xavier Guitart, Prous Science, Provenza 388, Barcelona 08025, Spain

Teruo Hayashi, National Institutes of Health, National Institute on Drug Abuse, Intramural Research Program, Baltimore, MD 21224, USA

Katsuhisa Inoue, Medical College of Georgia, Department of Biochemistry and Molecular Biology, Augusta, GA 30912, USA

Meyer B. Jackson, University of Wisconsin Medical School, Department of Physiology, Madison, WI 53706, USA

Yun Liu, University of Oklahoma Health Sciences Center, Department of Pharmaceutical Sciences, Oklahoma City, OK 73190, USA

Rae R. Matsumoto, University of Mississippi, Department of Pharmacology, University, MS 38677, USA

Tangui Maurice, University of Montpellier II, Cerebral Plasticity Laboratory, 3409 Montpellier Cedex 5, France

Francois Monnet, Institut National de la Sante et de la Recherche Medicale Unite 705-CNRS UMR 7157, Hopital Fernand Widal, 75475 Paris Cedex 10, France

Amy Hauck Newman, National Institute on Drug Abuse, Intramural Research Program, Department of Medicinal Chemistry, Baltimore, MD 21224, USA

Toshihide Nukada, Tokyo Institute of Psychiatry, Department of Neuronal Signaling, Tokyo 156-8585, Japan

Samer J. Nuwayhid, The George Washington University Medical Center, Department of Pharmacology & Physiology, Washington, DC 20037, USA

Shigeru Okuyama, Taisho Pharmaceutical, The 1st Laboratory, Medicinal Research Laboratory, Saitama 330-8530, Japan

Chris P. Palmer, University of Wisconsin Medical School, Department of Physiology, Madison, WI 53706, USA

Gavril W. Pasternak, Memorial Sloan Kettering Cancer Center, Department of Neurology, New York, NY 10021, USA

Buddy Pouw, University of Oklahoma Health Sciences Center, Department of Pharmaceutical Sciences, Oklahoma City, OK 73190 USA

Orazio Pressanvento, University of Catania, Department of Pharmaceutical Sciences, 95125 Catania, Italy

Malika Robichaud, McGill University, Department of Psychiatry, Neurobiological Psychiatry Unit, Montreal QC H3A 1A1, Canada

Giuseppe Ronsisvalle, University of Catania, Department of Pharmaceutical Sciences, 95125 Catania, Italy

Francois Roman, Pfizer Inc., Research Technologies, 94265 Fresnes Cedex, France

Jamaluddin Shaikh, University of Mississippi, Department of Pharmacology, University, MS 38677, USA

Keiko Shinohara Tanaka, Tokyo Institute of Psychiatry, Department of Molecular Psychiatry, Tokyo 156-8585, Japan

Tsung-Ping Su, National Institutes of Health, National Institute on Drug Abuse, Intramural Research Program, Baltimore, MD 21224, USA

Hubert Vidal, Sanofi-Aventis, F-34184 Montpellier Cedex 04, France

Linda L. Werling, The George Washington University Medical Center, Department of Pharmacology & Physiology, Washington, DC 20037, USA

Hideko Yamamoto, Department of Molecular Psychiatry, Tokyo Institute of Psychiatry, Tokyo 156-8585, Japan

Toshifumi Yamamoto, Yokohama City University, Department of Molecular Recognition, Yokohama 236-0027, Japan

Mitsunobu Yoshii, Tokyo Institute of Psychiatry, Department of Neural Plasticity, Tokyo 156-8585, Japan

Yongxin Yu, University of Oklahoma Health Sciences Center, Department of Pharmaceutical Sciences, Oklahoma City, OK 73190, USA

Preface

Over the last 30 years, our understanding of σ receptors has undergone a colossal evolution. They began as theoretical entities, then progressed to enigmatic receptors, and finally to identified proteins with important biological functions.

Since the first book on σ receptors was published in 1994, there have been many significant advances in the field. We now know that σ receptors subserve many critical functions in the body and recent studies indicate that they are promising drug development targets for a host of neurological, psychiatric, cardiovascular, ophthalmological, immunological, and gastrointestinal disorders. This book provides a timely update on the medicinal chemistry, cell biology, and clinical implications of σ receptors. It puts the information in a historical perspective to help new comers to the field successfully navigate the confusing early history surrounding these proteins, and it provides a launching point from which future studies and research directions can easily be developed.

The full impact of σ receptors on biological function has yet to be determined. The existing gaps in our knowledge base offer untold opportunities for future research. It is our hope that the information contained in this book will stimulate new, exciting research on σ receptors and ultimately lead to innovative insights into basic biological mechanisms and novel therapeutic advances.

Rae R. Matsumoto
Wayne D. Bowen
Tsung-Ping Su

Chapter 1

σ RECEPTORS: HISTORICAL PERSPECTIVE AND BACKGROUND

Rae R. Matsumoto
Department of Pharmacology, University of Mississippi, Oxford, MS 38677, USA

1. HISTORICAL PERSPECTIVE

σ Receptors were first proposed in 1976 by Martin and coworkers based on the actions of SKF-10,047 (N-allylnormetazocine) and related benzomorphans (1). The name σ was in fact derived from the first letter "S" from SKF-10,047 which was thought to be the prototypic ligand for these receptors. Over the next 10 years, a series of studies determined that SKF-10,047 interacts with a number of distinct binding sites (Figure 1-1), leading to much confusion about the true identity and nature of σ receptors during its early history.

σ Receptors were originally thought to be a type of opioid receptor. This belief stemmed from a historic study by Martin and colleagues who evaluated SKF-10,047 and other benzomorphans in morphine-dependent and non-dependent chronic spinal dogs (1). In this groundbreaking study, Martin and colleagues discovered that the physiological actions of the tested compounds fell into three distinct groups. They hypothesized that the differences between the groups stemmed from interactions with different subtypes of opioid receptors (1). Martin and his colleagues proposed a μ subtype which mediated the actions of morphine and related compounds, a κ subtype based on the actions of ketocyclazocine and its grouping, and a σ subtype which was characterized by SKF-10,047 and related compounds. Martin's study employed the use of racemic benzomorphans, a mixture of the (+)- and (-)-isomers of the compounds. Therefore, in later studies, when the isomers of benzomorphans were evaluated separately, it was determined that the (-)-isomers accounted for the vast majority of opioid-mediated effects. In the case of SKF-10,047, the (+)-isomer was determined in subsequent studies to produce actions that were insensitive to opioid antagonists (2-4), while the (-)-isomer was responsive to opioid antagonists (5,6). It is now accepted that the opioid-mediated actions of (-)-SKF-10,047 are relayed primarily through μ and κ opioid receptors.

During the 1980s, renewed interest in the (+)-isomer of SKF-10,047 occurred when it was determined that it possessed phencyclidine (PCP)-like properties. During this period, the term σ/PCP made its appearance in the literature and many investigators believed that the σ and PCP sites were identical. There was conclusive evidence that (+)-SKF-10,047 interacted with the PCP binding site, which was ultimately determined to be within the ionophore of the N-methyl-D-aspartate (NMDA) receptor (7-11). However, as selective ligands for the NMDA receptor were identified, it became apparent that [^{3}H](+)-SKF-10,047 binding could only be partially displaced using selective NMDA receptor ligands (11). Therefore, it appeared that (+)-SKF-10,047 bound to another site in addition to the ionophore of the NMDA receptor. This other binding site was ultimately identified as the entity that today retains the designation of the σ receptor.

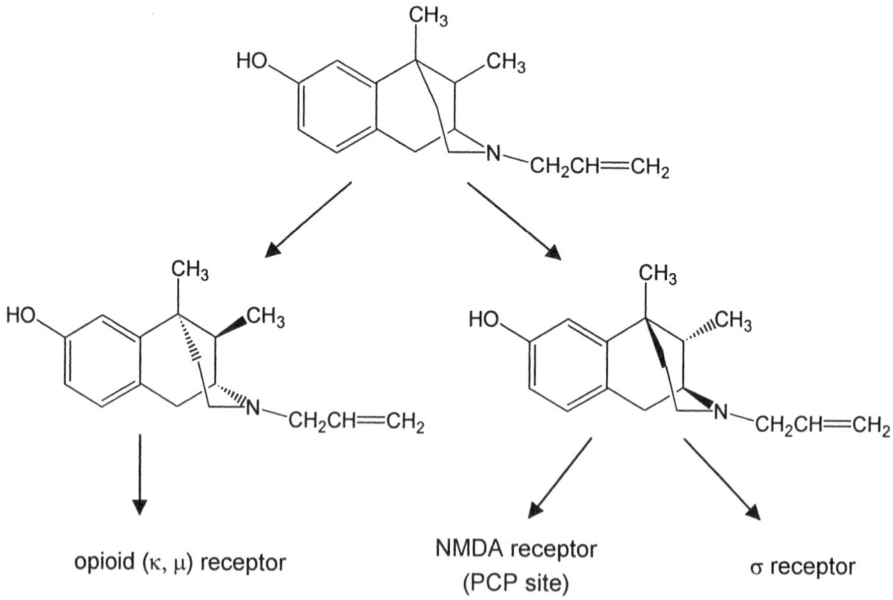

Figure 1-1. Association between different forms of SKF-10,047 and multiple binding sites

1. Sigma receptors

During the early 1980s, pioneering studies by Tsung-Ping Su had already begun shedding light on this additional component of [^{3}H]SKF-10,047 binding. These studies ultimately led to the identification of a unique drug selectivity pattern that characterized σ receptors from other known receptors (Table 1-1). Using guinea pig brain homogenates, Su characterized [^{3}H](±)-SKF-10,047 binding sites that were inaccessible to the opioid etorphine (12,13). Radioligand binding to these sites, which are today recognized as σ receptors, could be inhibited by a number of factors including phospholipase C and divalent cations, and exhibited a drug binding profile that was unlike anything characterized at that time (13). In these classic studies, Su demonstrated that σ receptors displayed high affinity for several (+)-benzomorphans including (+)-pentazocine, dextrallorphan, and (+)-cyclazocine (13). These binding sites were distinct from classical opioid receptors because in addition to having reverse stereoselectivity for benzomorphans (i.e. opioid receptors preferentially bind the (-)-isomer), a number of established opiates and opioid peptides failed to display significant affinities for these sites (13). In addition, these etorphine-inaccessible sites also bound neuroleptics such as haloperidol, the antidepressant imipramine, the β-adrenergic blocker propranolol, and the dissociative anesthetic PCP (13). Together, the data collected by Su clearly indicated the existence of a new and previously uncharacterized receptor, which is today recognized as the σ receptor.

Table 1-1. Drug selectivity profile of select compounds for σ receptors

Opioid-Related:		DA-Related:		Other:	
(+)-SKF-10,047	+++	Haloperidol	++++	PCP	++
(-)-SKF-10,047	+	Fluphenazine	+++	MK-801	---
(+)-Pentazocine	++++	Perphenazine	+++	Propranolol	++
(-)-Pentazocine	++	Chlorpromazine	++	Atropine	---
Dextrallorphan	+++	Pimozide	++	Clonidine	---
(+)-Cyclazocine	++	Molindone	++	Imipramine	++
(+)-EKC	+++	(+)-3PPP	++++	Pyrilamine	+++
Morphine	---	(-)-Butaclamol	+++	Chlorpheniramine	++
Naloxone	---	Clozapine	---	Promethazine	++
β-endorphin	---	Dopamine	---	Cimetidine	---
Leu-enkephalin	---	Apomorphine	---	Histamine	---
Dynorphin (1-13)	---	Amphetamine	---	DTG	+++

Relative affinities based on competition binding studies. ++++ <10 nM; +++ 11-100 nM; ++ 101-1000 nM; + 1001-10,000 nM; --- >10,000. EKC = ethylketocyclazocine. Adapted from refs. (13,14,15,18,19).

The unique pattern of binding that was initially reported by Su was subsequently corroborated and extended in William Tam's laboratory, first using [^{3}H]($\pm$)-ethylketocyclazocine and [^{3}H]SKF-10,047 to bind naloxone-inaccessible sites in the rat central nervous system, and then using [^{3}H](+)-SKF-10,047 and [^{3}H]haloperidol in the guinea pig brain (14,15). Tam confirmed that σ receptors bound a number psychotomimetic opioids (e.g. SKF-10,047, ethylketocylazocine, pentazocine, cyclazocine, bremazocine), the β-blocker propranolol, and the dissociative anesthetic PCP (14,15). In addition, Tam greatly expanded the list of neuroleptics that were shown to bind σ receptors with nanomolar affinity (e.g. haloperidol, perphenazine, fluphenazine, molindone, pimozide, thioridazine, chlorpromazine), and revealed that H1 antihistamines also interacted with these sites (e.g. pyrilamine, promethazine, chlorpheniramine) (14,15). Moreover, Tam demonstrated that the binding profile of drugs at σ receptors differed from the pattern of binding when using [^{3}H]PCP and [^{3}H]spiperone to label NMDA and dopamine receptors (14,15). Together, the studies of Su and Tam identified a unique drug selectivity pattern which characterized the binding sites that are now recognized as σ receptors.

Soon thereafter, more selective radioligands were identified for σ receptors. [^{3}H](+)-3-(3-hydroxyphenyl)-N-(1-propyl)piperidine (3PPP) was successfully used by a number of groups to discriminate σ receptors in binding studies from interactions with opioid, NMDA or dopamine receptors, which were problematic for historic radioligands such as [^{3}H]SKF-10,047 and [^{3}H]haloperidol (16). However, (+)-3PPP was also a presynaptic dopamine autoreceptor agonist (17), necessitating the search for even more selective compounds which could be used as radiolabeled probes to study σ receptors. A crucial breakthrough occurred with the introduction of [^{3}H]di-o-tolylguanidine (DTG) by Eckard Weber's group, the first truly selective radioligand for σ receptors (18). Subsequently, [^{3}H](+)-pentazocine was identified as another selective radioligand for σ receptors; this probe selectively binds to the σ_1 subtype (see below for additional information about σ receptor subtypes) (19,20). The availability of selective radioligands for σ receptors represented a major milestone for the field, and firmly established σ receptors as a viable topic for research.

In contrast to the early history of σ receptor research which was defined by pharmacological studies, the revolution in molecular and cell biology has greatly altered the way in which science is approached. Although its impact on the σ receptor field has been relatively slower than in some other areas, significant advances have been made. Foremost among these achievements was the cloning of the first σ receptor (σ_1 subtype), which is described in more detail in subsequent chapters of this book. This information led to insights about the structure and function of the receptor, and its relationship

1. Sigma receptors

to other known proteins. In addition, the development of cell and molecular biological-based probes allowed investigators to further elucidate σ receptor function; other chapters in this volume detail our current knowledge in these areas.

Table 1-2. Characteristics of σ_1 and σ_2 receptors

Feature	σ_1 Receptor	σ_2 Receptor	Reference
Physical Characteristics:			
Size	25-29 kDa	18-22 kDa	
Sequence [a]	AF030199 (mouse)	n.d.	
	AF004218 (rat)		
	U75283 (human)		
Tissue Expression:			
Brain	High	High	
Heart	High	Low	
Liver	High	High	
Spleen	High	Low	
GI tract	High	High	
Putative Agonists (Ki in nM):			
DTG	74 ± 15	61 ± 13	20
CB-184	7,436 ± 308	13 ± 2	49
Igmesine	n.d.	n.d.	
(+)-Pentazocine	7 ± 1	1361 ± 134	20
PRE-084	n.d.	n.d.	
Pregnenolone	n.d.	n.d.	
SA4503	17 ± 2 [b]	1784 ± 314 [b]	121
(+)-SKF-10,047	29 ± 3	33,654 ± 9,409	20
Putative Antagonists (Ki in nM):			
BD1047	0.9 ± 0.1	47 ± 0.6	33
BD1063	9 ± 1	449 ± 11	33
BMY 14802	60 [b]	230 [b]	50
Lu 28-179	17 [b]	0.12 [b]	50
NE-100	2 ± 0.3 [b]	85 ± 33 [b]	32
Panamesine	n.d.	n.d.	
Progesterone	n.d.	n.d.	
(±)-SM 21	>1000	67 ± 8	122
SR 31742A	n.d.	n.d.	

See Appendix A for chemical names of compounds. [a] Accession numbers for representative sequences. [b] IC_{50} in nM. n.d. = affinities for specific subtypes not determined; existing affinity information based on binding to both subtypes.

2. σ RECEPTOR SUBTYPES AND SPLICE VARIANTS

There are two well established subtypes of σ receptors, which have been designated σ_1 and σ_2. These receptor subtypes can be distinguished from one another based on their molecular weights, tissue distribution, and drug selectivity patterns. Select features of these two subtypes and compounds that are commonly used as agonists and antagonists at σ receptors are summarized in Table 1-2.

The σ_1 subtype has been cloned from a number of species including mouse, rat, guinea pig, and human (21-25; see Chapter 5). This subtype is predicted to be a 223 amino acid protein with at least one transmembrane-spanning region (26,27). It is widely expressed in a number of tissues, including heart and spleen where the expression of the σ_1 subtype appears to predominate over the σ_2 subtype (28,29). σ_1 Receptors appear to translocate during signaling and are linked to the modulation or production of intracellular second messengers (see Chapter 8). In addition, σ_1 receptors can associate with other proteins, including ankyrin B, heat shock protein 70 (hsp70), heat shock conjugate protein (hsc 70), glucose-related protein (GRP78/BiP), and potassium channels (26,30,31). To study their function, (+)-pentazocine is commonly used as a selective agonist at σ_1 receptors, and selective antagonists such as 1-[2-(3,4-dichlorophenyl)ethyl]-4-methyl piperazine (BD1063) and N,N-dipropyl-2-[4-methoxy-3-(2-phenylethoxy) phenyl]ethylamine (NE-100) are also available (32,33). In addition, sequence-specific antibodies, antisense oligodeoxynucleotides, and a σ_1 knockout mouse have been developed to further delineate the functions of this receptor subtype (34-46). Many of these functions are described further in the chapters that follow.

The σ_2 subtype appears to be a distinct physical entity from the σ_1 receptor. Comparisons of their sizes based on affinity labeling studies indicated that the σ_2 subtype is slightly smaller than the σ_1 receptor (47,48). The sequence of the σ_2 receptor has not yet been determined, although considerable progress has been made in this area in recent years. In contrast to σ_1 receptors that readily translocate, σ_2 receptors appear to be lipid raft proteins that affect calcium signaling via sphingolipid products (see Chapter 11). Unfortunately, there are no truly selective σ_2 receptor agonists and antagonists. (+)–1R,5R-(E)-8-(3,4-dichlorobenzylidene)-5-(3-hydroxy-phenyl)-2-methylmorphan-7-one (CB-184), one of the more selective σ_2 agonists, also interacts with μ opioid receptors (49). 1'-[4-[1-(4-fluorophenyl)-1H-indol-3-yl]-1-butyl]spiro[isobenzofuran-1(3H),4'-piperidine (Lu 29-179), one of the more selective σ_2 antagonists, interacts with dopamine D_2 and α-adrenergic receptors (50), while (±)-tropanyl 2-(4-

chlorophenoxy)butanoate [(±)-SM-21] interacts with dopamine transporters (unpublished data). Therefore, to study σ_2 receptor function, nondiscriminating σ ligands such DTG have been used in systems that are enriched in the σ_2 subtype; alternatively, nonselective σ_2 compounds have been used in systems that are enriched in σ receptors. The development of truly selective experimental tools with which to manipulate σ_2 receptors will greatly enhance understanding of their function.

In addition to σ_1 and σ_2 receptors, numerous papers have cited evidence in support of additional subtypes (e.g. 29,51-53). However, these putative subtypes have not yet been well characterized and will therefore not be described here in detail.

In addition to subtypes of σ receptors, there is evidence for splice variants. Thus far, only the σ_1 subtype has been sequenced. Therefore, information about splice variants is currently limited to this subtype. There are at least two truncated versions of the σ_1 receptor (54,55), and the extent to which these splice variants affect physiological functions are as yet unknown. However, studies examining σ_1 receptor polymorphisms in disease states have begun (56-61). The results have been mixed, but available nascent data support the possibility that these polymorphisms have functional consequences (58).

3. ENDOGENOUS LIGAND(S)

The conclusive identification of an endogenous ligand for σ receptors has yet to be achieved. The following sections summarize data supporting the existence of an endogenous ligand for these receptors, and raise the possibility of multiple such compounds. This would be consistent with the structural diversity of synthetic ligands that are known to interact with σ receptors.

3.1 Evidence from binding studies

Receptor binding studies to identify known endogenous ligands with significant affinity for σ receptors have been employed by a number of investigators. Although the vast majority of known endogenous compounds exhibit low to negligible affinities for σ receptors (see Appendix B), some activity has been reported. These possible candidates are described below in further detail.

Su and coworkers were the first to suggest that some neurosteroids serve as endogenous ligands for σ receptors. In particular, progesterone was shown to exhibit nanomolar affinity for σ receptors in guinea pig brain and spleen (62). The interaction of progesterone with brain σ receptors was competitive in nature (increase in K_d, but not B_{max} of $[^3H](+)$-SKF-10,047 binding), suggesting that progesterone binds to the same portion of the receptor as classical σ ligands (62). To further confirm that progesterone interacts with σ receptors, competition binding studies using $[^3H]$progesterone revealed a drug selectivity pattern that was consistent with σ receptors (63,64). The ability of progesterone to bind to σ receptors was subsequently confirmed in a number of laboratories (21,54,65-67). Other neurosteroids with micromolar affinities for σ receptors have also been reported (21,62,64,67), but it is unclear whether all of these steroids produce physiological actions through σ receptors. The limited functional studies that are available nevertheless indicate that some of them act as agonists at σ receptors (e.g. pregnenolone), while others act as antagonists (e.g. progesterone) (68).

Neuropeptide Y has been reported to have significant affinity for σ receptors (69). However, subsequent efforts to confirm this interaction have been unsuccessful (see Appendix B). It therefore does not appear that neuropeptide Y is an endogenous ligand for σ receptors.

A number of investigators have shown that divalent cations significantly inhibit radioligand binding to σ receptors. These divalent cations include magnesium, calcium, manganese, zinc, cadmium, copper (13,70,71). Some of these cations appear to preferentially target the σ_1 subtype while others target the σ_2 subtype. The effects of zinc on σ_2 receptors is particularly noteworthy because binding studies were also performed under physiological conditions in which zinc was released from hippocampal slices by depolarization and shown to displace $[^3H]$DTG, but not $[^3H](+)$-pentazocine binding in the slice, suggesting that it may be an endogenous ligand for σ_2 receptors (71). Further studies are needed to fully evaluate the conditions under which these candidates may serve as endogenous ligands for σ receptors.

3.2 Evidence from fractionation studies

A classical strategy for identifying an endogenous ligand is to extract it from the tissues in which it acts. Su and colleagues were among the first to report putative endogenous ligands for σ receptors, which they collectively named sigmaphins (72). These sigmaphins were isolated from guinea pig brain extracts and the partially purified fractions were shown to displace

1. Sigma receptors 9

binding to σ receptors in a competitive manner (72). Since there was a loss of binding after trypsin digestion, the compounds were thought to be peptides (72). However, to date, the active compounds have not been fully purified and identified.

Soon afterward, O'Donoghue and his colleagues also reported a putative endogenous ligand for σ receptors, which they isolated from extracts of porcine brain (73). The active material was also believed to be a peptide because pronase, a general proteolytic enzyme, could abolish its binding (73). In addition, its absorbance spectrum suggested that it contained phenylalanine residues (73). Additional studies to further purify and characterize the material have not been reported.

The high densities of σ receptors in the liver provided the impetus to search for an endogenous σ ligand in this tissue. A substance was extracted from porcine liver that binds to σ receptors (74). In contrast to the brain-derived compounds, this substance did not appear to be a peptide since it was resistant to pronase digestion (74). In addition, the liver-derived substance was thermostable, soluble in both water and organic solvents, and had a molecular weight of less than 1000 Da (74). However, full purification was not achieved.

In summary, fractionation studies demonstrated the existence of multiple endogenous extracts that bind to σ receptors, although none of them have been fully purified. Since these earlier efforts, there have been significant advances in the development of selective tools to label σ receptors and improved nuclear magnetic resonance and mass spectroscopy technologies to facilitate renewed efforts to discover endogenous σ ligands.

3.3 Evidence from physiological studies

In an elegant series of studies, Chavkin and coworkers demonstrated the release of endogenous ligands with σ-binding properties from hippocampal slices under physiologically relevant conditions. In these studies, fresh hippocampal slices were preloaded with a radioligand to occupy σ receptors. When the brain sections were depolarized using potassium chloride or veratridine, the radioligand that was bound to σ receptors was displaced in a time- and calcium-dependent manner, suggesting that depolarization caused the release of endogenous σ ligands (75). Electrical stimulation of the perforant path and/or mossy fibers, but not other tested regions, of the hippocampus produced similar effects (76), indicating that endogenous σ ligands could be released from specific circuits. Together, the data indicate the existence of σ-binding substances in the brain that can be liberated under conditions that cause neurotransmitter release.

Table 1-3. Representative imaging studies of σ receptors in the nervous system

Species	Tissue	Probe	Reference
Guinea pig	Brain	[^{3}H](+)-3PPP	16, 78, 105
	Brain	[^{3}H]DTG	18, 79, 139
	Brain	[^{3}H]Dextromethorphan	123
	Brain	[^{3}H](+)-Pentazocine (σ_1)	136, 131
	Brain	[^{3}H]DuP 734	126
	Brain	[^{3}H]NE-100 (σ_1)	132
	Brain	In situ (σ_1)	37
	Spinal cord	[^{3}H](+)-3PPP	78
Rat	Brain	[^{3}H](+)-3PPP	78
	Brain	[^{3}H](+)-SKF-10,047	133
	Brain	[^{3}H]DTG	18
	Brain	[^{3}H]DTG (σ_2)	77
	Brain	[^{3}H]Lu 28-179 (σ_2)	134
	Brain	[^{3}H](+)-Pentazocine (σ_1)	77
	Brain	[^{3}H]Nemonapride	135
	Brain	Antibody (143-162) (σ_1)	34
	Brain	Antibody (138-157) (σ_1)	43
	Brain	[^{11}C]SA6298	130
	Brain	[^{11}C]SA4503	94
	Brain	[^{18}F]fluoroethyl SA4503	124
	Pineal gland	[^{3}H]DTG	127
	Spinal cord	[^{3}H](+)-3PPP	78, 90
Mouse	Brain	[^{3}H](+)-SKF-10,047	133
	Brain	[^{3}H](+)-Pentazocine (σ_1)	131
	Brain	[^{11}C]SA6298	130
	Brain	[^{18}F]fluoroethyl SA4503	124
	Brain	In situ (σ_1)	37
Cat	Substantia nigra	[^{3}H]DTG	125
	Brain	[^{11}C]SA6298	130
Primate	Brain	[^{3}H](+)-3PPP	87
	Brain	[^{18}F]fluoroethyl SA4503	124
Human	Brain	[^{3}H]DTG	128
	Brain	[^{3}H]Lu 28-179 (σ_2)	134
	Cerebellum	[^{3}H]DTG	88
	Hippocampus	[^{3}H]DTG	129

4. ANATOMICAL DISTRIBUTION

σ Receptors are present throughout the body and knowledge about their localization can provide clues about their physiological functions. This section summarizes the distribution of σ receptors and their possible implications.

4.1 Nervous system

Following the first reports of σ receptors in the brain, numerous research groups have mapped their distribution in the central nervous system (Table 1-3). σ Receptors are found in the brain and spinal cord, where they subserve a variety of physiological functions.

The highest concentrations of σ receptors in the brain are found in brainstem motor nuclei. Cranial nerves such as the facial, hypoglossal, and trigeminal contain particularly high levels of σ receptors (77-79). Other constituents of brainstem motor circuits including the cerebellum, red nucleus, and inferior olive are also enriched in σ receptors (77-79). This pattern of distribution provided compelling evidence for a role for σ receptors in motor function, which was confirmed in early functional studies (80,81). The basal ganglia also contain moderate levels of σ receptors (77-79). Lesion studies showed that σ receptors are localized on substantia nigra pars compact neurons (78), and this distribution is consistent with the ability of σ receptor agonists to stimulate motor behavior via nigrostriatal dopaminergic pathways (82-85). Consistent with the enrichment of σ_2 receptors in the substantia nigra (77), these receptors were the first subtype to be implicated in motor function (85). Over time, accumulated data from anatomical and functional studies have supported the involvement of both σ_1 and σ_2 subtypes in motor function (77,84-86).

Significant levels of σ receptors are also found in limbic regions of the brain. The localization of σ receptors in the dentate gyrus and pyramidal cell layer of the hippocampus (77-79) are supportive of their role in learning and memory which are described in additional detail in Chapters 9 and 12. Moreover, the hippocampus, and particularly the dentate gyrus, is enriched in the σ_1 subtype (77), which has been implicated in the modulation of cognitive behaviors (Chapter 12). The presence of σ receptors in the olfactory bulb and other limbic and paralimbic areas such as the frontal cortex, cingulate, hippocampus, and amygdala further suggests that they may modulate affective states (79,87). This is consistent with their apparent role in depression and other mood disorders, which are described in further detail in Chapter 14.

Neuroendocrine areas are also enriched in σ receptors. The supraoptic and paraventricular areas of the hypothalamus, which send projections to the pituitary, contain significant densities of σ receptors, as does the adenohypophysis (78,79,88). Other hypothalamic areas also contain significant concentrations of σ receptors (79), and this region of the brain is particularly enriched in the σ_1 subtype (77). Although the anatomical distribution of σ receptors is highly suggestive of a role for σ receptors in the release of hormones from the pituitary, systematic functional studies to address this question have not been conducted. However, the ability of SKF-10,047 to raise plasma corticosterone levels in a naloxone-independent manner (89) supports this possibility. The additional presence of σ receptors in endocrine organs (see below), further indicates that this may be a fertile area for future research.

In contrast to the negligible levels of σ receptors in most sensory regions of the brain, several regions of the visual system contain significant densities of σ receptors (79). These regions include the lateral geniculate and superior colliculus (79). Together with recent reports of σ receptors in the eye (see below), additional studies to further examine the role of σ receptors in visual function are also needed.

The gray matter of the spinal cord contains extremely high densities of σ receptors (78). The motor subserving ventral horn of the spinal cord is especially enriched in these receptors (78,90), which is consistent with a role for these receptors in motor control. In addition, a sensory role for these receptors is suggested by their expression in dorsal root ganglion (78,90). Since the central gray in the midbrain also contains high densities of σ receptors (78,79), it is conceivable that σ receptors modulate sensory pain transmission. The role of σ receptors in pain is described in further detail in Chapter 16.

4.2 Peripheral organs

In addition to their presence in the nervous system, σ receptors are found in a variety of peripheral organs. The early evidence for the existence of σ receptors in the periphery came from binding studies in tissue homogenates, which were sometimes followed by autoradiographic studies to determine discrete localization in tissue slices. More recently, evidence for the existence of σ receptors in peripheral organs has also come from imaging studies in live organisms and Northern blot analysis against transcripts for the σ_1 subtype. The widespread distribution of σ receptors in the body suggests that they perform an essential physiological function.

The heart contains significant levels of σ receptors. Homogenate binding studies indicate that over 80% of the σ receptors in the heart are of the σ_1 subtype (28). These receptors are present on both the parasympathetic neurons that innervate the heart and the cardiac myocytes themselves (28,91-93). In myocytes, σ receptors influence contractility, calcium influx, and beating rate (28,93). In intracardiac neurons, the σ_1 and σ_2 subtypes affect neuronal excitability by modulating calcium and potassium channels, respectively (91,92). The overall effects of these influences on physiological parameters of cardiovascular function are still unclear.

There are several reports of σ_1 receptors in the lung (94-97). It is unclear whether σ_2 receptors are also present. The role of σ receptors in the lung has thus far been unexplored.

The highest levels of σ receptors in the body have been reported in the liver. Both the σ_1 and σ_2 subtypes are present (24,48,97). However, the function of σ receptors in the liver is currently unknown. Early studies hypothesized that they might have a cytochrome P450-like role, but this was not supported by experimental data (98-100).

The kidney contains both σ_1 and σ_2 receptors (24,48,94,97). The function of σ receptors in the kidney has yet to be determined.

Reproductive organs such as the testis, ovaries, vas deferens, and placenta contain σ receptors (67,101-103). The specific subtypes that are present within these tissues are unclear because the studies were performed under conditions where both σ_1 and σ_2 receptors were labeled (103). Autoradiographic studies to localize receptor distribution were performed in some of these tissues. In the testis, σ binding was highest in the ductuli efferentes and ductus epididymis, with lower levels of binding in the seminiferous tubules (103). In the ovaries, the highest densities of σ receptors were present in maturing follicles (103).

The adrenal gland contains σ receptors. The presence of the σ_1 subtype has been confirmed (103); the extent to which σ_2 receptors are expressed is unclear. Although the function of σ receptors in the adrenal gland has not been studied systematically, they may have a role in the modulation of neurosecretory processes (104).

Similar to the heart, the spleen is enriched in σ_1 receptors (29,94,105). Autoradiographic studies revealed that σ_1 receptors are most densely concentration in T cell zones (29). Together with the presence of σ receptors on immune cells (see below) and the ability of steroids to bind to these receptors (62), the data are supportive of a role for these receptors in immune function (see Chapter 17 for additional details).

The gastrointestinal tract contains significant levels of σ receptors, of both σ_1 and σ_2 subtypes (106). Within the gastrointestinal tract, autoradiographic studies revealed high concentrations of σ receptors in the mucosa and submucosal plexus (107,108). Labeling was especially dense at the level of the fundus and duodenum (108). The functional relevance of σ receptors in the gastrointestinal tract is described in detail in Chapter 18.

σ_1 Receptors have recently been reported in the eye (109,110). They are found in the iris-ciliary body and retina, including the projecting terminals of the retina to the superior colliculus (110). Specific cell types that contain σ_1 receptor mRNA and protein include: retinal ganglion cells, photoreceptors, and retinal pigment epithelial cells (109). In addition, they are associated with cells in the inner nuclear layer (109). Investigations into the physiological and therapeutic significance of σ receptors on visual function have only just begun. Data thus far indicate that σ receptor agonists can reduce ocular pressure and protect against retinal cell death (111-113).

4.3 Cell types

σ Receptors are found in a variety of cell types that are not components of organs. Naturally-occurring cells such as blood cells and tumor cells contain significant levels of σ receptors. Blood cells that express σ receptors include: peripheral blood leukocytes, granulocytes, lymphocytes, natural killer cells (52,114,115). The functional role of σ receptors on these cells and their implications for treating a variety of immune disorders are described in detail in Chapter 17. Tumor cells also contain high densities of σ receptors, and recent studies report that they are expressed in especially high densities in proliferating tumors (116-120). The implications of σ receptors in tumors are discussed further in Chapters 11 and 17.

In addition to their expression in cells *in situ*, σ receptors have been reported in many different cell lines. These cell types and the subtype(s) of σ receptor that they express are summarized in Appendix C. Many of these cell types have been valuable experimental tools for delineating σ receptor function, and were used in the studies described in subsequent chapters of this book.

5. SUMMARY

The early history of σ receptors is characterized by classical pharmacological approaches which succeeded in defining a unique drug selectivity pattern and anatomical distribution for these proteins. σ Receptors are widely distributed in the body, where they mediate a variety of physiological functions. The chemical diversity of compounds that interact with σ receptors is vast and includes therapeutically relevant entities including psychotomimetic opiates, neuroleptics, antihistamines, and antidepressants. The recent revolution in molecular biology has provided additional information about σ receptors, including the sequence of one of its major subtypes and a host of experimental tools to aid in selectively deciphering its functions. We now know that σ receptors have important implications for a number of disease states and mounting evidence indicates that they are viable therapeutic targets for medication development. The remaining chapters in this book summarize our current knowledge regarding the medicinal chemistry, cell biological and clinical implications of σ receptors. It is hoped that this information will lay the groundwork for innovative future studies to stimulate new insights into the physiological and therapeutic relevance of σ receptors.

REFERENCES

1. Martin WR, Eades CE, Thompson JA, Huppler RE. The effects of morphine and nalorphine-like drugs in the non-dependent and morphine-dependent chronic spinal dog. J Pharmacol Exp Ther 1976, 197:517-532.
2. Iwamoto ET. Locomotor activity and antinociception after putative mu, kappa and sigma opioid receptor agonists in the rat: influence of dopaminergic agonists and antagonists. J Pharmacol Exp Ther 1981, 217:451-460.
3. Vaupel DB. Naltrexone fails to antagonize the sigma effects of PCP and SKF 10,047 in the dog. Eur J Pharmacol 1983, 92:269-274.
4. Young GA, Khazan N. Differential neuropharmacological effects of mu, kappa and sigma opioid agonists on cortical EEG power spectra in the rat. Stereospecificity and naloxone antagonism. Neuropharmacology 1984, 23:1161-1165.
5. Berzetei-Gurske IP, Toll L. The mu-opioid activity of kappa-opioid receptor agonist compounds in the guinea pig ileum. Eur J Pharmacol 1992, 212:283-286.
6. Khazan N, Young GA, El-Fakany EE, Hong O, Caliigaro D. Sigma receptors mediated the psychotomimetic effects of N-allylnormetazocine (SKF-10,047), but not its opioid agonistic-antagonistic properties. Neuropharmacology 1984, 23:983-987.
7. Mendelsohn LG, Kalra V, Johnson BG, Kerchner GA. Sigma opioid receptor: characterization and co-identity with the phencyclidine receptor. J Pharmacol Exp Ther 1985, 233:597-602.
8. Quirion R, Hammer P Jr, Herkenham M, Pert CB. The phencyclidine (angel dust)/sigma 'opiate' receptor: Its visualization by tritium-sensitive film. Proc Natl Acad Sci 1981, 78:5881-5885.
9. Zukin SR, Brady KT, Slifer BL, Balster RL. Behavioral and biochemical stereoselectivity of sigma opiate/PCP receptors. Brain Res 1984, 174-177.
10. Sircar R, Nichtenhauser R, Ieni JR, Zukin SR. Characterization and autoradiographic visualization of (+)-[^{3}H]SKF10,047 binding in rat and mouse brain: further evidence for phencyclidine/"sigma opiate" receptor commonality. J Pharmacol Exp Ther 1986, 257:681-688.

11. Wong EH, Knight AR, Woodruff GN. [^{3}H]MK-801 labels a site on the N-methyl-D-aspartate receptor channel complex in rat brain membranes. J Neurochem 1988, 50:274-281.
12. Su T-P. Psychotomimetic opioid binding: Specific binding of [^{3}H]-SKF-10047 to etorphine-inaccessible sites in guinea-pig brain. Eur J Pharmacol 1981, 75:81-82.
13. Su T-P. Evidence for sigma opioid receptor: Binding of [^{3}H]SKF-10047 to etorphine-inaccessible sites in guinea-pig brain. J Pharmacol Exp Ther 1982, 223:284-290.
14. Tam SW. Naloxone-inaccessible σ receptor in rat central nervous system. Proc Natl Acad Sci USA 1983, 80:6703-6707.
15. Tam SW, Cook L. σ Opiates and certain antipsychotic drugs mutually inhibit (+)-[^{3}H]SKF 10,047 and [^{3}H]haloperidol binding in guinea pig brain membranes. Proc Natl Acad Sci USA 81:5618-5621.
16. Largent BL, Gundlach AL, Snyder SH. Psychotomimetic opiate receptors labeled and visualized with (+)-[^{3}H]3-(3-hydroxyphenyl)-N-(1-propyl)piperidine. Proc Natl Acad Sci 1984, 81:4983-4987.
17. Hjorth SA, Carlsson A, Wikstrom H, Lindberg P, Sanchez D, Hacksell U, Arvidsson LE, Svensson U, Nilsson JLG. 3-PPP, a new centrally acting dopamine receptor agonist with selectivity for autoreceptors. Life Sci 1981, 28:1225-1238.
18. Weber E, Sonders M, Quarum M, McLean S, Pou S, Keana JFW. 1,3-Di(2-[5-^{3}H]tolyl)guanidine: A selective ligand that labels σ-type receptors for psychotomimetic opiates and antipsychotic drugs. Proc Natl Acad Sci USA 1986, 83:8784-8788.
19. de Costa BR, Bowen WD, Hellewell SB, Walker JM, Thurkauf A, Jacobson AE, Rice KC. Synthesis and evaluation of optically pure [^{3}H]-(+)-pentazocine, a highly potent and selective radioligand for σ receptors. FEBS Lett 1989, 251:53-58.
20. Bowen WD, de Costa BR, Hellewell SB, Walker JM, Rice KC. [^{3}H]-(+)-Pentazocine: a potent and highly selective benzomorphan-based probe for sigma-1 receptors. Mol Neuropharmacol 1993, 3:117-126.
21. Hanner M, Moebius FF, Flandorfer A, Knaus H-G, Striessnig J, Kempner E, Glossmann H. Purification, molecular cloning, and expression of the mammalian σ$_1$ binding site. Proc Natl Acad Sci USA 1996, 93:8072-8077.
22. Mei J and Pasternak GW. Molecular cloning and pharmacological characterization of the rat σ$_1$ receptor. Biochem Pharmacol 2001, 62:349-355.
23. Pan Y-X, Mei J, Xu J, Wan B-L, Zuckerman A, Pasternak GW. Cloning and characterization of a mouse σ$_1$ receptor. J Neurochem 1998, 70:2279-2285.
24. Seth P, Fei YJ, Li HW, Huang W, Leibach FH, Ganapathy V. Cloning and functional characterization of a σ receptor from rat brain. J Neurochem 1998, 70:922-931.
25. Seth P, Leibach FH, Ganapathy V. Cloning and structural analysis of the cDNA and the gene encoding the murine type 1 σ receptor. Biochem Biophys Res Commun 1997, 41:535-540.
26. Aydar E, Palmer CP, Klyachko VA, Jackson MB. The σ receptor as a ligand-regulated auxiliary potassium channel subunit. Neuron 2002, 34:399-410.
27. Jbilo O, Vidal H, Paul R, De Nys N, Bensaid M, Silve S, Carayon P, Davi D, Galiegue S, Bourrie B, Guillemot J-C, Ferrara P, Loison G, Maffrand J-P, Le Fur G, Casellas P. Purification and characterization of the human SR 31747A-binding protein. A nuclear membrane protein related to yeast sterol isomerase. J Biol Chem 1997, 272:27107-27115.
28. Novakova M, Ela C, Barg J, Vogel E, Hasin Y, Eilam Y. Ionotropic action of sigma receptor ligands in isolated cardiac myocytes from adult rats. Eur J Pharmacol 1995, 286:19-30.

29. Wolfe SA Jr, Ha BK, Whitlock BB, Saini P. Differential localization of three distinct binding sites for sigma receptor ligands in rat spleen. J Neuroimmunol 1997, 72:45-58.
30. Hayashi T, Su T-P. Regulating ankyrin dynamics: Roles of sigma-1 receptors. Proc Natl Acad Sci 2001, 98:491-496.
31. Yamamoto H, Kametani F, Namiki Y, Yamamoto T, Karasawa J, Shen H, Ikeda K, Hagino Y, Kobayashi H, Sora I, Nukuda T. Identification of GRP78 as a type-1 sigma receptor (SigmaR1)-associated protein. Soc Neurosci Abst 2002, program #833.9.
32. Chaki S, Tanaka M, Muramatsu M, Otomo S. NE-100, a novel potent sigma ligand, preferentially binds to σ_1 binding sites in guinea pig brain. Eur J Pharmacol 1994, 251:R1-2.
33. Matsumoto RR, Bowen WD, Tom MA, Vo VN, Truong DD, de Costa BR. Characterization of two novel σ receptor ligands: antidystonic effects in rats suggest σ receptor antagonism. Eur J Pharmacol 1995, 280:301-310.
34. Alonso G, Phan V, Guillemain I, Saunier M, Legrant A, Anoal M, Maurice T. Immonocytochemical localization of the sigma$_1$ receptor in the adult rat central nervous system. Neuroscience 2000, 97:155-170.
35. Hayashi T, Maurice T, Su TP. Ca^{2+} signaling via σ_1 receptors: novel regulatory mechanism affecting intracellular Ca^{2+} concentration. J Pharmacol Exp Ther 2000, 293:788-798.
36. Hiramatsu M, Hoshino T. Involvement of kappa-opioid receptors and sigma receptors in memory function demonstrated using an antisense strategy. Brain Res 2004, 1030:247-255.
37. Kitaichi K, Chabot JG, Moebius FF, Flandorfer A, Glossman H, Quirion R. Expression of the purported sigma$_1$ (σ_1) receptor in the mammalian brain and its possible relevance in deficits induced by antagonism of the NMDA receptor complex as revealed using an antisense strategy. J Chem Neuroanat 2000, 20:375-387.
38. Langa F, Codony X, Tovar V, Lavado A, Gimenez E, Cozar P, Cantero M, Dordal A, Hernandez E, Perez R, Monroy X, Zamanillo D, Guitart X, Montoliu L. Generation and phenotypic analysis of sigma receptor type I (σ_1) knockout mice. Eur J Neurosci 2003, 18:301-310.
39. Matsumoto RR, McCracken KA, Friedman MJ, Pouw B, de Costa BR, Bowen WD. Conformationally restricted analogs of BD1008 and an antisense oligodeoxynucleotide targeting σ_1 receptors produce anti-cocaine effects in mice. Eur J Pharmacol 2001, 419:163-174.
40. Matsumoto RR, McCracken KA, Pouw B, Zhang Y, Bowen WD. Involvement of sigma receptors in the behavioral effects of cocaine: evidence from novel ligands and antisense oligodeoxynucleotides. Neuropharmacology 2002, 42:1043-1055.
41. Maurice T, Phan VL, Urani A, Guillemain I. Differential involvement of the sigma$_1$ (σ_1) receptor in the anti-amnesic effect of neuroactive steroids, as demonstrated using an in vivo antisense strategy in the mouse. Br J Pharmacol 2001, 134:1731-1741.
42. Mei J, Pasternak GW. σ_1 receptor modulation of opioid analgesia in the mouse. J Pharmacol Exp Ther 2002, 300:1070-1074.
43. Palacios G, Muro A, Vela JM, Molina-Holgado E, Guitart X, Ovalle S, Zamanillo D. Immunohistochemical localization of the sigma-1 receptor in oligodendrocytes in the rat central nervous system. Brain Res 2003, 961:92-99.
44. Romieu P, Martin-Fardon R, Maurice T. Involvement of the σ1 receptor in the cocaine-induced conditioned place preference. Neuroreport 2000, 11:2885-2888.
45. Romieu P, Meunier J, Garcia D, Zozime N, Martin-Fardon R, Bowen WD, Maurice T. The sigma$_1$ (σ_1) receptor activation is a key step for the reactivation of cocaine conditioned place preference by drug priming. Psychopharmacology 2004, 175:154-162.

46. Takebayashi M, Hayashi T, Su TP. Nerve growth factor-induced neurite sprouting in PC12 cells involves σ-1 receptors: implications for antidepressants. J Pharmacol Exp Ther 2002, 303:1227-1237.
47. Hellewell SB, Bowen WD. A sigma-like binding site in rat pheochromocytoma (PC12) cells: decreased affinity for (+)-benzomorphans and lower molecular weight suggest a different sigma receptor form from that of guinea pig brain. Brain Res 1990, 527:244-253.
48. Hellewell SB, Bruce A, Feinstein G, Orringer J, Williams W, Bowen WD. Rat liver and kidney contain high densities of sigma-1 and sigma-2 receptors: characterization by ligand binding and photoaffinity labeling. Eur J Pharmacol Mol Pharmacol Sect 1994, 268:9-18.
49. Bowen WD, Bertha CM, Vilner BJ, Rice KC. CB-64D and CB-184: ligands with high σ_2 receptor affinity and subtype selectivity. Eur J Pharmacol 1995, 278:257-260.
50. Perregaard J, Moltzen EK, Meier E, Sanchez C. Sigma ligands with subnanomolar affinity and preference for the sigma$_2$ binding site. 1. 3-(omega-aminoalkyl)-1H-indoles. J Med Chem 1995, 38:1998-2008.
51. Booth RG, Wyrick SD, Baldessarini RJ, Kula NS, Myers AM, Mailman RB. New σ-like receptor recognized by novel phenylaminotetralins: ligand binding and functional studies. Mol Pharmacol 1993, 4:1232-1239.
52. Carr DJ, De Costa BR, Radesca L, Blalock JE. Functional assessment and partial characterization of [^{3}H](+)-pentazocine binding sites on cells of the immune system. J Neuroimmunol 1991, 35:153-166.
53. Enomoto R, Ogita K, Yoneda Y. Multiplicity of [^{3}H]1,3-di-o-tolylguanidine binding sites with low affinity for haloperidol in rat brain. Biol Pharm Bull 1993, 16:989-996.
54. Ganapathy ME, Prasad PD, Huang W, Seth P, Leibach FH, Ganapathy V. Molecular and ligand-binding characterization of the σ receptor in the Jurkat human lymphocyte cell line. J Pharmacol Exp Ther 1999, 289:251-260.
55. Zamanillo D, Romero G, Dordal A, Perez P, Vincent L, Mendez R, Andreu F, Hernandez E, Perez R, Monroy X, Ovalle S, Guitart X. Increase of forskolin-stimulated adenylyl cyclase and AP-1 activities by sigma1 receptor expression. FENS Absr 2002, A04617.
56. Inada T, Iijima Y, Uchida N, Maeda T, Iwashita S, Ozaki N, Harano M, Komiyama T, Yamada M, Sekine Y, Iyo M, Sora I, Ujike H. No association found between the type 1 sigma receptor gene polymorphisms and methamphetamine abuse in the Japanese population: a collaborative study by the Japanese Genetics Initiative for Drug Abuse. Ann NY Acad Sci 2004, 1025:27-33.
57. Ishiguro H, Ohtsuki T, Toru M, Itokawa M, Aoki J, Shibuya H, Kurumaji A, Okubo Y, Iwawaki A, Ota K, Shimizu H, Hamaguchi H, Arinami T. Association between polymorphisms in the type 1 sigma receptor gene and schizophrenia. Neurosci Lett 1998, 257:45-48.
58. Miyatake R, Furukawa A, Matsushita S, Higuchi S, Suwaki H. Functional polymorphisms in the sigma$_1$ receptor gene associated with alcoholism. Bio Psychiatr 2004, 55:85-90.
59. Ohmori O, Shinkai T, Suzuki T, Okano C, Kojima H, Terao T, Nakamura J. Polymorphisms of the sigma$_1$ receptor gene in schizophrenia: An association study. Am J Med Genet 2000, 96:118-122.
60. Satoh F, Miyatake R, Furukawa A, Suwaki H. Lack of association between sigma$_1$ receptor gene variants and schizophrenia. Psychiatr Clin Neurosci 2004, 58:359-363.
61. Uchida N, Ujike H, Nakata K, Takaki M, Nomura A, Katsu T, Tanaka Y, Imamura T, Sakai A, Kuroda S. No association between the sigma receptor type 1 gene and

schizophrenia: results of analysis and meta-analysis of case-control studies. BMC Pscyhiatr 2003, 3:13.
62. Su T-P, London ED, Jaffe JH. Steroid binding at σ receptors suggest a link between endocrine, nervous and immune systems. Science 1988, 240:219-221.
63. McCann DJ, Su T-P. Solubilization and characterization of haloperidol-sensitive (+)-[^{3}H]SKF-10,047 binding sites (sigma sites) from rat liver membranes. J Pharmacol Exp Ther 1991, 257:547-554.
64. Yamada M, Nishigami T, Nakasho K, Nishimoto Y, Miyaji H. Relationship between sigma-like site and progesterone-binding site of adult male rat liver microsomes. Hepatology 1994, 20:1271-1280.
65. Klein M, Cooper TB, Musacchio JM. Effects of haloperidol and reduced haloperidol on binding to σ sites. Eur J Pharmacol 1994, 254:239-248.
66. Maurice T, Roman FJ, Privat A. Modulation by neurosteroids of the in vivo (+)-[^{3}H]SKF-10,047 binding to sigma$_1$ receptors in the mouse forebrain. J Neurosci Res 1996, 46:734-743.
67. Ramamoorthy JD, Ramamoorthy S, Mahesh VB, Leibach FH, Ganapathy V. Cocaine-sensitive σ-receptor and its interaction with steroid hormones in the human placental syncytiotrophoblast and in choriocarcinoma cells. Endocrinology 1995, 136:924-932.
68. Maurice T, Urani A, Phan V-L, Romieu P. The interaction between neuroactive steroids and the σ$_1$ receptor function: behavioral consequences and therapeutic opportunities. Brain Res Rev 2001, 37:116-132.
69. Roman F, Pascaud X, Duffy O, Vauche D, Martin B, Junien JL. Neuropeptide Y and peptide YY interact with brain sigma and PCP binding sites. Eur J Pharmacol 1989, 174:301-302.
70. Basile AS, Paul IA, Mirchevich A, Kuijpers G, De Costa B. Modulation of (+)-[^{3}H]pentazocine binding to guinea pig cerebellum by divalent cations. Mol Pharmacol 1992, 42:882-889.
71. Connor MA, Chavkin C. Ionic zinc may function as an endogenous ligand for the haloperidol-sensitive σ$_2$ receptor in rat brain. Mol Pharmacol 1992, 42:471-479.
72. Su T-P, Weissman AD, Yeh S-Y. Endogenous ligands for sigma opioid receptors in the brain ("sigmaphin"): evidence from binding assays. Life Sci 1986, 38:2199-2210.
73. Contreras PC, DiMaggio DA, O'Donohue TL. An endogenous ligand for the sigma opioid binding site. Synapse 1987, 1:57-61.
74. Nagornaia LV, Samovilvo NN, Korobov NV, Vinogradov VA. Partial purification of endogenous inhibitors of (+)-[^{3}H]SKF-10047 binding with sigma opioid receptors of the liver. Biull Eksp Biol Med 1988, 106:314-317.
75. Neumaier JF, Chavkin C. Calcium-dependent displacement of haloperidol-sensitive σ receptor binding in rat hippocampal slices following tissue depolarization. Brain Res 1989, 500:215-222.
76. Connor MA, Chavkin C. Focal stimulation of specific pathways in the rat hippocampus causes a reduction in radioligand binding to the haloperidol-sensitive sigma receptor. Exp Brain Res 1991, 85:528-536.
77. Bouchard P, Quirion R. [^{3}H]1,3-Di(2-tolyl)guanidine and [^{3}H](+)pentazocine binding sites in the rat brain: autoradiographic visualization of the putative sigma$_1$ and sigma$_2$ receptor subtypes. Neuroscience 1997, 76:467-477.
78. Gundlach AL, Largent BL, Snyder SH. Autoradiographic localization of sigma receptor binding sites in guinea pig and rat central nervous system with (+)^{3}H-3-(3-hydroxyphenyl)-N-(1-propyl)piperidine. J Neurosci 1986, 6:1757-1770.
79. McLean S, Weber E. Autoradiographic visualization of haloperidol-sensitive sigma recepotors in guinea-pig brain. Neuroscience 1988, 25:259-269.

80. Tran TT, de Costa BR, Matsumoto RR. Microinjection of sigma ligands into cranial nerve nuclei produces vacuous chewing in rats. Psychopharmacology 1998, 137:191-200.
81. Walker JM, Matsumoto RR, Bowen WD, Gans DL, Jones KD, Walker FO. Evidence for a role of haloperidol-sensitive σ-'opiate' receptors in the motor effects of antipsychotic drugs. Neurology 1988, 38:961-965.
82. Bastianetto S, Rouquier L, Perrault G, Sanger DJ. DTG-induced circling behaviour in rats may involve the interaction between σ sites and nigro-striatal dopaminergic pathways. Neuropharmacology 1995, 34:281-287.
83. Goldstein SR, Matsumoto RR, Thompson TL, Patrick RL, Bowen WD, Walker JM. Motor effects of two sigma ligands mediated by nigrostriatal dopamine neurons. Synapse 1989, 4:254-258.
84. Patrick SL, Walker JM, Perkel JM, Lockwood M, Patrick RL. Increases in rat striatal extracellular dopamine and vacuous chewing produced by two σ receptor ligands. Eur J Pharmacol 1993, 32:243-249.
85. Walker JM, Bowen WD, Patrick SL, William WE, Mascarella SW, Bai X, Carroll FI. A comparison of (-)-deoxybenzomorphans devoid of opiate activity with their dextrorotatory phenolic counterparts suggest role of σ_2 receptors in motor function. Eur J Pharmacol 1993, 231:61-68.
86. Matsumoto RR, Pouw B. Correlation between neuroleptic binding to σ_1 and σ_2 receptors and acute dystonic reactions. Eur J Pharmacol 2000, 401:155-160.
87. Mash DC, Zabetian CP. Sigma receptors are associated with cortical limbic areas in the primate brain. Synapse 1992, 12:195-205.
88. Jansen KLR, Faull RLM, Dragunow M, Leslie RA. Autoradiographic distribution of sigma receptors in human neocortex, hippocampus, basal ganglia, cerebellum, pineal and pituitary glands. Brain Res 1991b, 559:172-177.
89. Eisenberg RM. Plasma corticosterone changes in response to central or peripheral administration of kappa or sigma opiate agonists. J Pharmacol Exp Ther 1985, 223:863-869.
90. Aanonsen LM, Seybold VS. Phencyclidine and sigma receptors in rat spinal cord: Binding characterization and quantitative autoradiography. Synapse 1989, 4:1-10.
91. Zhang H, Cuevas J. Sigma receptors inhibit high-voltage-activated calcium channels in rat sympathetic and parasympathetic neurons. J Neurophysiol 2002, 87:2867-2879.
92. Zhang H, Cuevas J. σ Receptor activation blocks potassium channels and depresses neuroexcitability in rat intracardiac neurons. J Pharmacol Exp Ther 2005, 313:1387-1396.
93. Ela C, Barg J, Vogel Z, Hasin Y, Eilam Y. Sigma receptor ligands modulate contractility, Ca++ influx and beating rate in cultured cardiac myocytes. J Pharmacol Exp Ther 1994, 269:1300-1309.
94. Kawamura K, Ishiwata K, Tajima H, Ishii S, Matsuno S, Homma Y, Senda M. In vivo evaluation of [^{11}C]SA4503 as a PET ligand for mapping CNS sigma$_1$ receptors. Nucl Med Biol 2000, 27:255-261.
95. Moebius FF, Burrows GG, hanner M, Schmid E, Striessnig J, Glossman H. Identification of a 27-kDa high affinity phenylalkylamine-binding polypeptide as the sigma$_1$ binding site by photoaffinity labeling and ligand-directed antibodies. Mol Pharmacol 1993a, 44:966-971.
96. Moebius FF, Burrows GG, Striessnig J, Glossman H. Biochemical characterization of a 22-kDa high affinity antiischemic drug-binding polypeptide in the endoplasmic reticulum of guinea pig liver: potential common target for antiischemic drug action. Mol Pharmacol 1993b, 43:139-148.

97. van Waarde A, Buursma AR, Hospers GA, Kawamura K, Kobayashi T, Ishii K, Oda K, Ishiwata K, Vaalburg W, Elsinga PH. Tumor imagining with two sigma receptor ligands, ^{18}F-FE-SA5845 and ^{11}C-SA4503: a feasibility study. J Nucl Med 2004, 45:1939-1945.
98. Basile AS, Paul IA, de Costa B. Differential effects of cytochrome P-450 induction on ligand binding to σ receptors. Eur J Pharmacol 1992, 227:95-98.
99. Jewell A, Wedlund P, Dwoskin L. Strain differences in rat brain and liver σ binding: lack of cytochrome P450-2D1 involvement. Eur J Pharmacol 1993, 243:249-254.
100. Monnet FP, Debonnel G, de Montigny C. The cytochromes P-450 are not involved in the modulation of the N-methyl-D-aspartate response by sigma ligands in the rat CA3 dorsal hippocampus. Synapse 1993, 13:30-38.
101. Kennedy C, Henderson G. An examination of the putative sigma-receptor in the mouse isolated vas deferens. Br J Pharmacol 1989, 98:429-436.
102. SuT-P, Wu XZ. Guinea pig vas deferens contains sigma but not phencyclidine receptors. Neurosci Lett 1990, 108:341-345.
103. Wolfe SA Jr, Culp SG, De Souza EB. Sigma receptors in endocrine organs: identification, characterization, and autoradiographic localization in rat pituitary, adrenal, testis, and ovary. Endocrinology 1989, 124:1160-1172.
104. Paul IA, Basile AS, Rojas E, Youdim MBH, De Costa B, Skolnick P, Pollard HB, Kuijpers GAJ. Sigma receptors modulate nicotinic receptor function in adrenal chromaffin cells. FASEB J 1993, 7:1171-1178.
105. Brent PJ. Kappa opioid receptor agonists inhibit sigma-1 (σ1) receptor binding in guinea-pig brain, liver and spleen: autoradiographical evidence. Brain Res 1996, 725:155-165.
106. Harada Y, Hara H, Sukamoto T, Characterization of specific (+)-[^{3}H]N-allylnormetazocine and [^{3}H]1,3 di(2-tolyl)guanidine binding sites in porcine gastric fundic mucosa. J Pharmacol Exp Ther 1994, 269:905-910.
107. Hara H, Tanaka K, Harada Y, Sukamoto T. Sigma receptor-mediated effects of a new antiulcer agent, KB-5492, on experimental gastric mucosal lesions and gastric alkaline secretion in rats. J Pharmacol Exp Ther 1994, 269:799-805.
108. Roman F, Pascaud X, Chomette G, Bueno L, Junien JL. Autoradiographic localization of sigma opioid receptors in the gastrointestinal tract of the guinea pig. Gastroenterology 1989, 97:76-82.
109. Shamsul Ola M, Moore P, El-Sherbeny A, Roon P, Agarwal N, Sarthy VP, Casellas P, Ganapathy V, Smith SB. Expression pattern of sigma receptor 1 mRNA and protein in mammalian retina. Brain Res Mol Brain Res 2001, 95:86-95.
110. Wang WF, Ishiwata K, Kiyosawa M, Kawamura K, Oda K, Kobayashi T, Matsuno K, Mochizuki M. Visualization of sigma$_1$ receptors in eyes by ex vivo autoradiography and in vivo positron emission tomography. Exp Eye Res 2002, 75:723-730.
111. Campana G, Bucolo C, Murari G, Spampinato S. Ocular hypotensive action of topical flunarizine in the rabbit: role of σ$_1$ recognition sites. J Pharmacol Exp Ther 2002, 303:1086-1094.
112. Martin PM, Shamsul Ola M, Agarwal N, Ganapathy V, Smith SB. The sigma receptor ligand (+)-pentazocine prevents apoptotic retinal ganglion cell death induced in vitro by homocysteine and glutamate. Brain Res Mol Brain Res 2004, 123:66-75.
113. Senda T, Mita S, Kaneda K, Kikuchi M, Akaike A. Effect of SA4503, a novel σ$_1$ receptor agonist, against glutamate neurotoxicity in cultured rat retinal neurons. Eur J Pharmacol 1998, 342:105-111.
114. Paul R, Lavastre S, Floutard D, Floutard R, Canat X, Casellas P, Le Fur G, Breliere JC. Allosteric modulation of peripheral sigma binding sites by a new selective ligand: SR 31747. J Neuroimmunol 1994, 52:183-192.

115. Wolfe SA Jr, Kulsakdinun C, Battaglia G, Jaffe JH, De Souza EB. Initial identification and characterization of σ receptors on human peripheral blood leukocytes. J Pharmacol Exp Ther 1988, 247:1114-1119.
116. Bem WT, Thomas GE, Mamone JY, Homan SM, Levy BK, Johnson FE, Coscia CJ. Overexpression of sigma receptors in nonneuronal human tumors. Cancer Res 1991, 51:6558-6562.
117. Mach RH, Smith CR, al-Nabulsi I, Whirrett BR, Childers SR, Wheeler KT. Sigma-2 receptors as potential biomarkers of proliferation in breast cancer. Cancer Res 1997, 57:156-161.
118. Thomas GE, Szucs M, Mamone JY, Bem WT, Rush MD, Johnson FE, Coscia CJ. Sigma and opioid receptors in human brain tumors. Life Sci 1990, 46:1279-1286.
119. Wang B, Rouzier R, Albarracin CT, Sahin A, Wagner P, Yang Y, Smith TL, Meric-Bernstam F, Marcelo AC, Horobagyi GN, Pusztai L. Expression for sigma 1 receptor in human breast cancer. Breast Cancer Res Treat 2004, 87:205-214.
120. Wheeler KT, Wang LM, Wallen CA, Childers SR, Cline JM, Keng PC, Mach RH. Sigma-2 receptors as a biomarker of proliferation in solid tumours. Br J Cancer 2000, 82:1223-1232.
121. Matsuno K, Nakazawa M, Okamoto K, Kawashima Y, Mita S. Binding properties of SA4503, a novel and selective σ_1 receptor agonist. Eur J Pharmacol 1996, 306:271-179.
122. Mach RH, Wu L, West T, Whirrett BR, Childers SR. The analgesic tropane analogue (±)-SM 21 has high affinity for σ_2 receptors. Life Sci 1999, 64:PL131-137.
123. Canoll PD, Smith PR, Gottesman S, Musacchio JM. Autoradiographic localization of [^{3}H]dextromethorphan in guinea pig brain: allosteric enhancement by ropizine. J Neurosci Res 1989, 24:31-328.
124. Elsinga PH, Kawamura K, Kobayashi T, Tsukada H, Senda M, Vaalburg W, Ishiwata K. Synthesis and evaluation of [^{18}F]fluoroethyl SA4503 as a PET ligand for the sigma receptor. Synapse 2002, 43:259-267.
125. Graybiel AM, Besson M-J, Weber E. Neuroleptic-sensitive binding sites in the nigrostriatal system: Evidence for differential distribution of sigma sites in the substantia nigra, pars compacta of the cat. J Neurosci 1989, 9:326-338.
126. Heroux JA, Tam SW, De Souza EB. Autoradiographic identification and characterization of σ receptors in guinea pig brain using [^{3}H]1(cyclopropylmethyl)-4-(2'-(4"-fluorophenyl)-2'-oxoethyl)piperidine ([^{3}H]DuP 734), a novel σ receptor ligand. Brain Res 1992, 598:76-86.
127. Jansen JLR, Dragunow M, Faull RLM. Sigma receptors are highly concentrated in the rat pineal gland. Brain Res 1990, 507:158-160.
128. Jansen JLR, Dragunow M, Faull RLM, Leslie RA. Autoradiographic visualization of [^{3}H]DTG binding to σ receptors, [^{3}H]TCP binding sites, and L-[^{3}H]glutamate binding to NMDA receptors in human cerebellum. Neurosci Lett 1991a, 125:143-146.
129. Jansen KLR, Faull RLM, Storey P, Leslie RA. Loss of sigma binding sites in the CA1 area of the anterior hippocampus in Alzheimer's disease correlates with CA1 pyramidal cell loss. Brain Res 1993, 623:299-302.
130. Kawamura K, Ishiwata K, Tajima H, Ishii S, Shimada Y, Matsuno K, Homma Y, Senda M. Synthesis and in vivo evaluation of [^{11}C]SA6298 as a PET sigma$_1$ receptor ligand. Nucl Med Biol 1999, 26:915-922.
131. Maruo J, Yoshida A, Shimohira I, Matsuno K, Mita S, Ueda H. Binding of [^{35}S]GTPgS stimulated by (+)-pentazocine sigma receptor agonist, is abundant in guinea pig spleen. Life Sci 2000, 67:599-603.

132. Okuyama S, Chaki S, Yae T, Nakazato A, Muramatsu M. Autoradiographic characterization of binding sites for [^{3}H]NE-100 in guinea pig brain. Life Sci 1995, 57:PL333-337.
133. Sircar R, Nichtenhauser R, Ieni JR, Zukin SR. Characterization and autoradiographic visualization of (+)-[^{3}H]SKF10,047 binding in rat and mouse brain: Further evidence for phencyclidine/"sigma opiate" receptor commonality. J Pharmacol Exp Ther 1986, 257:681-688.
134. Soby KK, Mikkelsen JD, Meier E, Thomsen C. Lu 28-179 labels a sigma$_2$-site in rat and human brain. Neuropharmacology 2002, 43:95-100.
135. Ujike H, Akiyama K, Kuroda S. [^{3}H]YM-09151-2 (nemonapride), a potent radioligand for both sigma-1 and sigma-2 receptor subtypes. Neuroreport 1996, 7:1057-1061.
136. Walker JM, Bowen WD, Goldstein SR, Roberts AH, Patrick SL, Hohmann AG, DeCosta BR. Autoradiographic distribution of [^{3}H](+)-pentazocine and [^{3}H]1,3-di-o-tolylguanidine (DTG) binding sites in guinea pig brain: a comparative study. Brain Res 1992, 581:33-38.

Corresponding author: *Dr. Rae R. Matsumoto, Mailing address: University of Mississippi, School of Pharmacy, Department of Pharmacology, 303 Faser Hall, University, MS 38677,USA, Phone: (662) 915-1466, Fax: (662) 915-5148, Electronic mail address: rmatsumo@olemiss.edu*

Chapter 2

MEDICINAL CHEMISTRY: NEW CHEMICAL CLASSES AND SUBTYPE-SELECTIVE LIGANDS

Amy Hauck Newman[1] and Andrew Coop[2]
[1]*Section on Medicinal Chemistry, National Institute on Drug Abuse-Intramural Research Program, National Institutes of Health, Baltimore, MD 21224, USA* and [2]*Department of Pharmaceutical Sciences, University of Maryland School of Pharmacy, Baltimore, MD 21201, USA*

1. INTRODUCTION AND HISTORY

Previous reviews of the area of σ ligand structure activity relationships have covered most of the early ligands, but many of the pharmacological conclusions based on early ligands were confusing due to the ligands interacting with several other biological systems. This chapter attempts to briefly discuss the history behind the development of early σ ligands, but maintains a greater focus on the more recent σ-selective ligands, which have been developed over the past decade. For a more detailed discussion of the earlier ligands the reader is directed to these excellent reviews (1-3).

σ Receptors were initially described by Martin as a subtype of opioid receptors based on the actions of the benzomorphans, specifically racemic SKF-10,047 (**1**) (4). This was a confusing birth for the σ receptor system, as the actions attributed to the effects of SKF-10,047 at σ receptors were probably due to the interaction of the (+)-isomer of the benzomorphan with σ receptors, whereas the (-)-isomer was the agent responsible for the opioid effects (5). The situation was further confused when σ sites were believed to be part of the phencyclidine binding site (ionophore site) or polyamine site of the NMDA receptor complex.

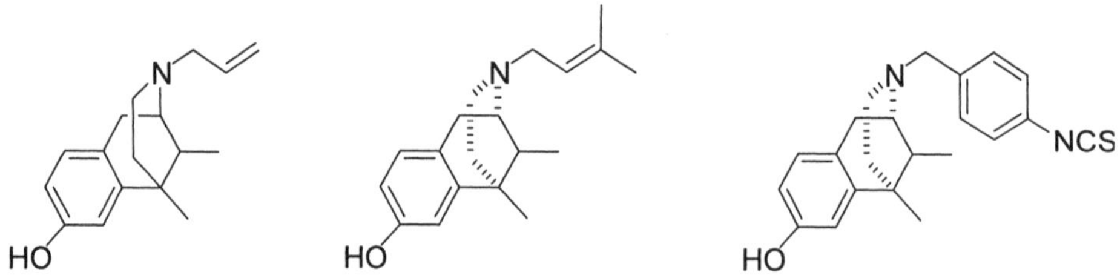

1 (+/-) SKF 10,047   **2 (+)-Pentazocine**   **3**

Figure 2-1. Benzomorphan-based σ ligands

1.1 Benzomorphans

As discussed above, the σ activity of the benzomorphan SKF-10,047 was probably due to the actions of the (+)-enantiomer, and this led to the discovery of (+)-pentazocine (**2**) as a selective σ ligand. Through the use of this ligand, and others including 3-PPP (3-(3-hydroxyphenyl)-N-(1-propyl)piperidine) and DTG (di-o-tolylguanidine), the σ receptor system was finally characterized as unique (6-10). Additional ligands of the benzomorphan class have also been found to possess affinity for the σ system, and are covered in the review by Walker et al. (2).

Since the initial cloning of the σ_1 receptor (11), studies have concentrated on the development of ligands to further characterize and purify these receptors. A recent investigation into the development of selective σ_1 receptor probes has led to a (+)-benzomorphan-based irreversible ligand. Ronsisvalle et al. (12) showed that the introduction of an isothiocyanate into the (+)-N-benzyl benzomorphan derivative gave a ligand (**3**) (Figure 2-1) which appears to show promise as such an agent (see review by Ronsisvalle on irreversible ligands in Chapter 3).

1.2 σ_1 and σ_2 receptors

It was eventually found that σ receptors consisted of a heterogeneous population of sites, now termed σ_1 and σ_2 (13-15). The discovery of the heterogeneity of σ receptors prompted concentrated efforts into the search for compounds with selectivity for each σ receptor subtype. (+)-Benzomorphans display selectivity for σ_1 receptors, and indeed tritiated (+)-benzomorphans are used in σ_1 receptor binding assays (16).

σ_2 Receptor-selective ligands have proven less common, with the currently accepted radioligand being the subtype nonselective [^{3}H]DTG in the presence of a (+)-benzomorphan to block binding to σ_1 sites. The pharmacological effects of activating both subtypes are described elsewhere in this volume. Briefly, σ_1 receptors have been associated with numerous conditions including cognitive effects, neuroprotection, and may be involved in the actions of cocaine (see Chapters 12, 15). σ_2 Receptors have been less well studied, but activation of σ_2 receptors appears to affect movement and posture and has been associated with inhibition of cell proliferation and induction of apoptotic cell death (see Chapter 11).

Many of the ligands discovered prior to about 1992 were only evaluated using binding assays against [^{3}H](+)-pentazocine, which is primarily an assay for σ_1 receptor binding affinity (17). Thus, little can be stated about the activity of these compounds for σ_2 receptors. This review will concentrate on the compounds where affinity at both receptors has been established.

2. ENDOGENOUS LIGANDS

The endogenous ligand for σ receptors remains elusive. Several laboratories have identified brain extracts, which show affinity for σ receptors (18,19). Furthermore, physiological studies have suggested depolarization- and calcium-dependent release of σ-active substances from brain slices (20,21). To date, however, none of the substances have been identified.

The search for an endogenous ligand for σ receptors did, however, lead to the discovery that certain neurosteroids possess affinity for σ_1 receptors, notably progesterone (22). From a chemical point of view, this is an interesting finding as the majority of ligands with affinity for σ receptors contain a basic nitrogen. Indeed, most models of ligand recognition include the requirement of a basic nitrogen, yet progesterone is a lipophilic steroid lacking any basic or acidic groups. This finding, along with information gleaned from cloning, lead to the hypothesis that σ_1 receptors are distantly related to enzymes of steroid biosynthesis (23). The merits of this hypothesis are discussed elsewhere in this volume.

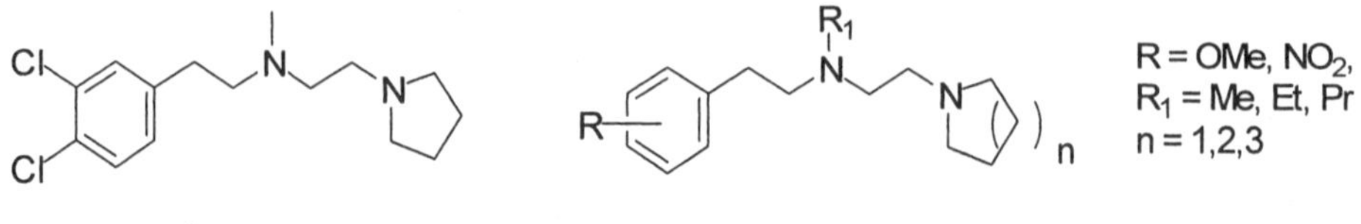

Figure 2-2. Phenylethylene diamine-based σ ligands

3. σ SELECTIVE AGENTS

Initial studies with early σ ligands were limited due to the effects on other systems influencing the pharmacology of the ligand. Obviously, what was needed were compounds that did not interact with other biological systems. One of the most widely studied class of compounds are the phenylethylene diamines: the protypical member of this class is BD1008 (**4**) (Figure 2-2) (24). BD1008 contains 3,4-dichloro substitution on the aromatic ring, a substitution pattern which leads to high affinity at both σ_1 and σ_2 receptors. Numerous other substituents have been introduced, but it appears that lipophilic substituents are preferred for high affinity agents (25). A range of substitutions that have been investigated on phenylethylene diamines (**5**) are shown in Figure 2-2.

In order to exploit the activity of the (+)-benzomorphans and phenylethylene diamines, hybrid structures were prepared where the basic amine and aromatic ring of the benzomorphan skeleton was taken as the "phenethyl" group of the phenylethylene diamines (26). Compounds such as **6** and **7** (Figure 2-3) did indeed display excellent affinity at σ_1 receptors (K_i < 10 nM), lower affinity at σ_2 receptors, and little activity at opioid receptors.

2. Medicinal chemistry

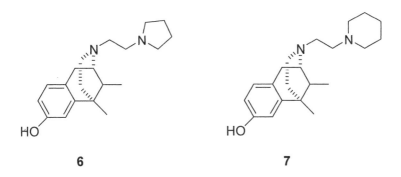

Figure 2-3. Hybrids of the benzomorphans and the phenylethylene diamines

A class of compounds that share structural similarities to the phenylethylene diamines are the phenylpentylamines (such as **8** and **9**, Figure 2-4), which show high σ receptor affinity against [^{3}H](+)-pentazocine, with a K_i of about 1 nM (27) (further discussed by Ablordeppey and Glennon elsewhere in this volume). This class can be viewed as phenylethylene diamine analogs that lack one of the basic nitrogens, and suggests that the second basic nitrogen is not essential for $σ_1$ affinity. As binding was only performed in assays to measure $σ_1$ affinity, little can be concluded about their affinity for $σ_2$ receptors. However, the recent report that AC915 (**10**) (Figure 2-4), an ester derivative of the phenylpentylamines, is a $σ_1$ ligand with excellent selectivity over $σ_2$ receptors (2000-fold), suggests that this is a class where additional $σ_1$ selective agents may be developed (28). Indeed, this compound may find use as a masking agent in $σ_2$ binding assays replacing the (+)-benzomorphans. Recently, a related class of phenoxyalkyl amines (**11**) have also been reported to possess excellent affinity for both $σ_1$ and $σ_2$ receptors, and the introduction of stereochemistry onto the alkyl chain was interestingly shown to influence affinity and selectivity (**12**) (29).

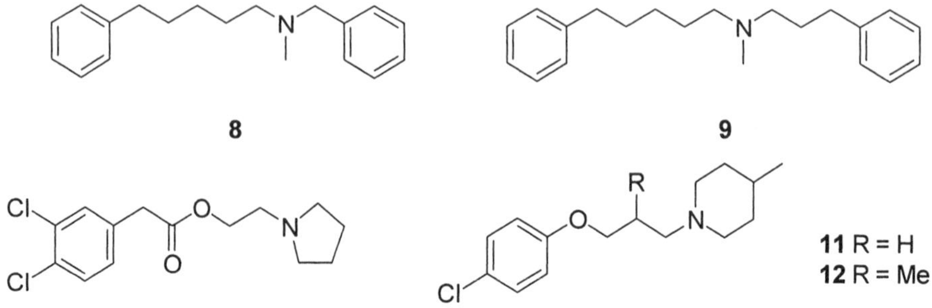

Figure 2-4. Phenyl pentyl amines, AC915, and phenoxyalkylamines

4. σ SUBTYPE SELECTIVE AGENTS

4.1 σ₁ ligands

4.1.1 Haloperidol derivatives

Compounds related to haloperidol are shown in Figure 2-5. Haloperidol (**13**) has been shown to possess high affinity for σ-receptors, with a slight preference for σ_1 over σ_2 (30). When the ketone was reduced to give reduced haloperidol (**14**), the dopamine D_2 affinity of haloperidol was greatly decreased, to give a compound relatively selective for σ receptors over other systems. These studies led to the development of the related E-5842 (**15**) as a σ_1 agent, with excellent selectivity over a range of other biological systems. E-5842 has been shown to possess promise as an antipsychotic agent (31).

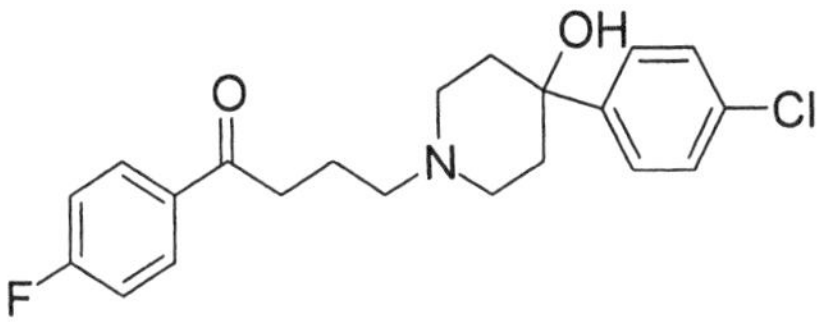

2. Medicinal chemistry

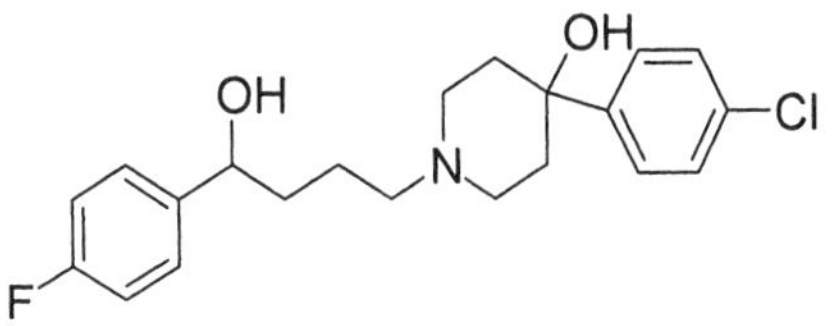

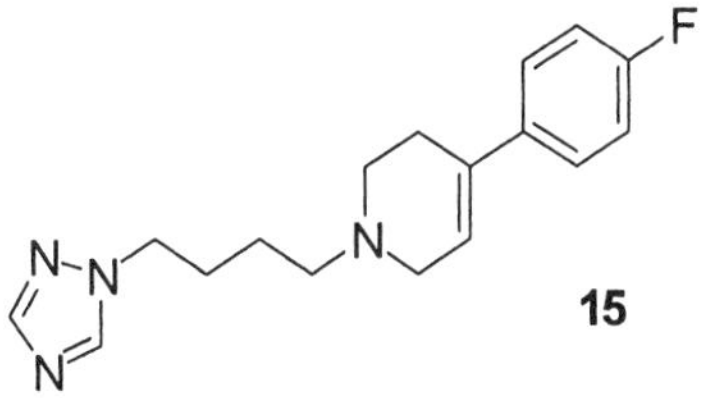

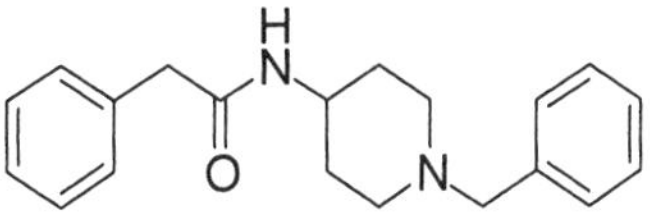

Figure 2-5. σ Ligands based on haloperidol

4.1.2 Phenylacetamides

N-(1-Benzylpiperidin-4-yl)phenylacetamides (such as **16**, Figure 2-5) share a similar skeleton to E-5842 discussed above. These compounds have been shown to possess excellent selectivity for σ_1 receptors, with affinities in the low nanomolar range, and selectivities over σ_2 up to 200-fold (32). Further studies into the structure-activity relationships of this series of compounds showed that replacing the aromatic ring with heterocyclic rings led to compounds with reduced affinity, but that the introduction of a halogen on both aromatic rings led to an increase in selectivity for σ_1 receptors over σ_2 (33).

4.1.3 NE-100

NE-100 (N,N-di-isopropyl-2-[4-methoxy-3-(2-phenylethoxy)phenyl] ethyl-amine (**17**) (Figure 2-6) is a simple amine with only two carbons between the amine and the aromatic ring. This compound shows high affinity for σ_1 receptors, and moderate selectivity over σ_2 receptors (34). Studies of this interesting class of compound have shown that both propyl groups are not necessary for affinity at σ receptors, and that the mono-propyl analog **18** possesses significant affinity (34). Further studies showed that the introduction of alkyl groups alpha to such a secondary amine (to give **19**) actually led to increases in affinity and selectivity for σ_1 receptors (Figure 2-6) (35).

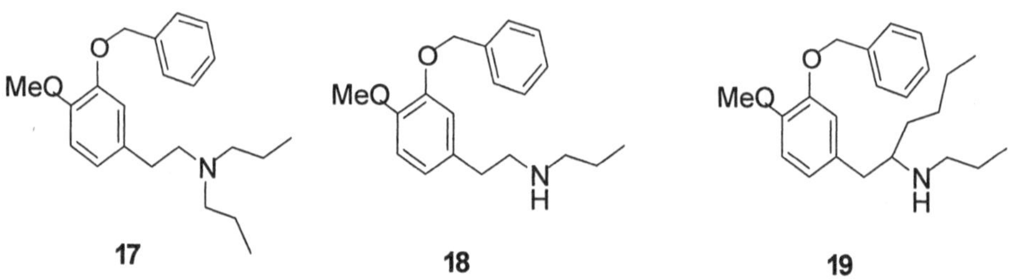

Figure 2-6. NE-100 and secondary amine analogs

4.2 σ_2 ligands

4.2.1   Benzylidine phenylmorphans

Perhaps the most widely studied σ_2 selective agents are the benzylidene phenylmorphans (Figure 2-7), typified by CB-64D (**20**) and CB-184 (**21**) (36). Both compounds show high affinity and excellent selectivity for σ_2 receptors over σ_1 receptors, with CB-184 showing the greater selectivity. Both contain the aryl morphinan skeleton present in a class of opioids, but with an additional benzylidene group. It has been suggested that this dichlorinated ring may occupy similar space on the receptor as the equivalent ring in BD1008 (37). These compounds have shown excellent activity in functional assays (38-40) and have indeed proved to be valuable tools in delineating σ_2 ligand pharmacology and the possible role of σ_2 receptors in regulation of cell growth and survival (reviewed by Bowen in Chapter 11). Even so, this class of compounds suffers from the major problem of their interaction with opioid receptors, as they display potent mu opioid agonism *in vivo*. Hence, further study of this class is required in order to develop analogs lacking the opioid component, but which maintain σ_2 receptor selectivity.

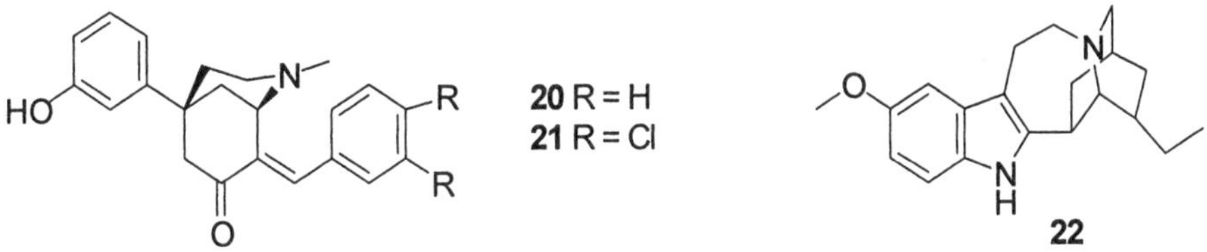

2. Medicinal chemistry

Figure 2-7. σ Ligands based on phenylmorphans and ibogaine

4.2.2 Ibogaine

Another compound that demonstrates relative selectivity for σ_2 receptors over σ_1 receptors is ibogaine (**22**) (Figure 2-7), although its affinity for σ_2 receptors is modest (41). Ibogaine gained notoriety due to its reported actions as an anti-addiction agent and has been useful as a tool to study the cytotoxicity mediated by σ_2 receptors *in vitro* (42). However, it interacts with a variety of biological systems in addition to σ_2 receptors and therefore cannot be used to study the actions of σ_2 receptors in *in vivo* assays.

4.2.3 Arylpropylamines

A recent report discussed the fact that ibogaine and CB-184 contain arylpropyl amines and display σ_2 selectivity, whereas compounds with affinity for σ_1 sites (such as NE-100) tend to possess a phenylethylamine moiety (37). Based on this observation, a simple range of phenethyl and phenylpropyl amines were studied. It was shown that phenylpropyl-piperidine (**23**) (Figure 2-8) demonstrated a preference for σ_2 sites (four-fold) and that the preference could be increased with other substituents to give **24** as a high affinity ligand for σ_2 receptors with moderate selectivity (Figure 2-8) (37). It is anticipated that this finding may lead the way to agents optimized for σ_2 receptors.

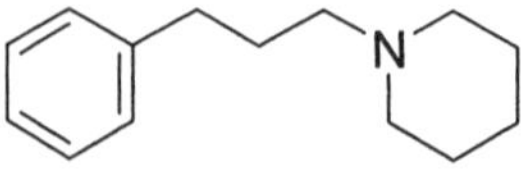

Figure 2-8. Simple phenylalkylamines

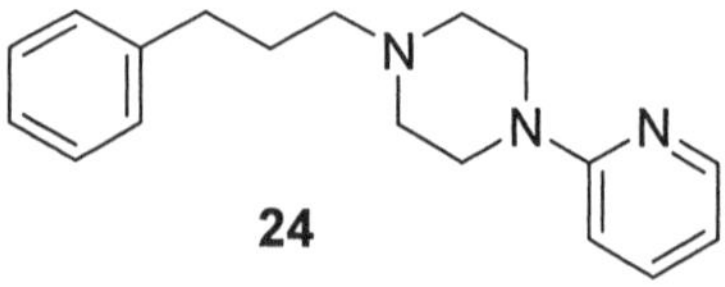

4.2.4   Tropane analogs

A recent report by Mach et al. (43) described a novel tropane-based ligand (**25**) (Figure 2-9) which is reported to possess an affinity at σ_2 receptors of 5 nM, and a selectivity over σ_1 receptors of greater than 500-fold. The para-amine substitution was shown to aid in the selectivity, as the unsubstituted phenyl analog demonstrated much reduced selectivity for σ_2 receptors.

The related tropane-containing ligand (±)-SM-21 (**26**) (Figure 2-9) has been shown to posses significant affinity for σ_2 receptors (44) and is currently used as a σ_2 preferring antagonist in behavioral assays (45).

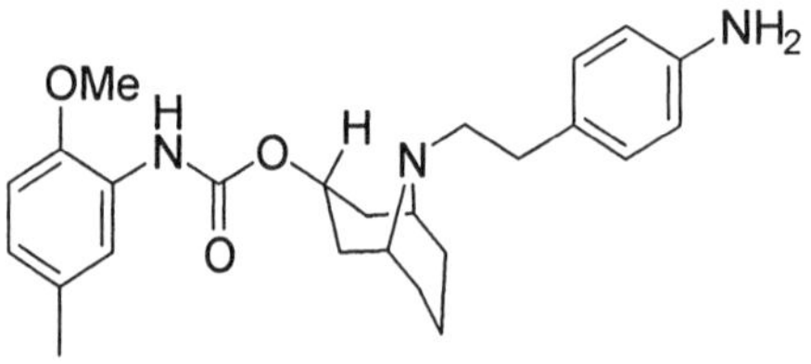

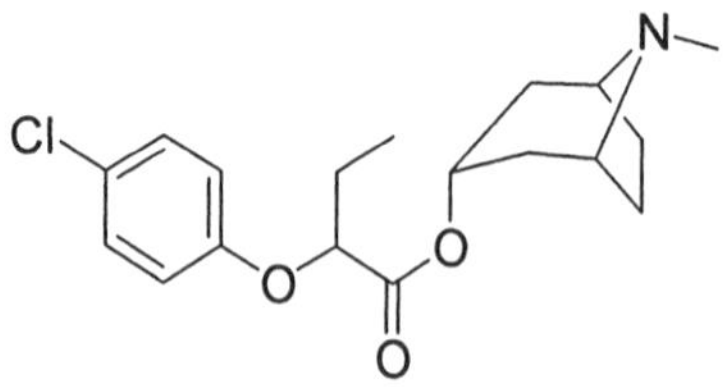

Figure 2-9. Tropane-based σ ligands

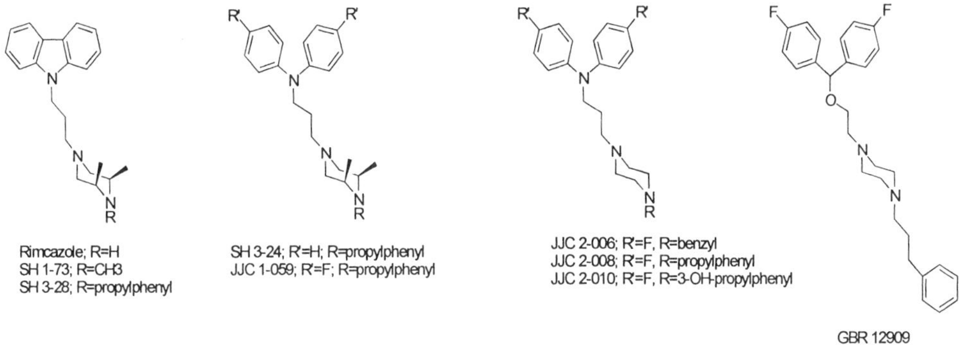

2. Medicinal chemistry

Rimcazole; R=H
SH 1-73; R=CH3
SH 3-28; R=propylphenyl

SH 3-24; R'=H; R=propylphenyl
JJC 1-059; R'=F; R=propylphenyl

JJC 2-006; R'=F, R=benzyl
JJC 2-008; R'=F, R=propylphenyl
JJC 2-010; R'=F, R=3-OH-propylphenyl

GBR 12909

Figure 2-10. Rimcazole and other piperazine analogs

5. DUAL PROBES FOR σ_1 RECEPTORS AND THE DOPAMINE TRANSPORTERS

5.1 Rimcazole analogues

Over the past decade, several lines of evidence have linked σ receptors and cocaine. For example, cocaine was reported to bind with low to moderate affinity to σ receptors and these concentrations were shown to be achievable *in vivo* (46). In addition, several σ ligands such as rimcazole and BMY 14802 (Figure 2-10) have been shown to attenuate locomotor and rewarding effects of cocaine (47,48). Recently, the σ_1 receptor antagonists NE-100 and BD1047 showed significant attenuation of cocaine-induced place preference (48). Other studies showed that σ receptor antagonists block the development of sensitization to cocaine in rats (49). Furthermore, attenuation of cocaine's convulsive and lethal effects by the selective σ antagonists BD1047, LR172 and N-alkyl substituted and conformationally restricted analogues of BD1008 has also been reported (50-53).

Curiously, there also seems to be a structural linkage to the cocaine binding site on the dopamine transporter (DAT) and the σ antagonist binding site, despite no apparent homology between the DAT and σ_1 receptor protein structures. Namely, an iodoazido-analogue of cocaine was reported to photolabel a 26 kDa polypeptide in rat brain that displayed the pharmacology of a σ receptor (54,55). Furthermore, the potent DAT inhibitor GBR 12909 was reported to potently displace [^{3}H]3-PPP from σ receptors in rat brain (IC_{50} = 48 nM) (56). More recently, an isothiocyanato analogue of the σ antagonist rimcazole has been shown to bind irreversibly to the DAT, in rat caudate-putamen (57).

These early linkages prompted an experiment evaluating nine structurally diverse σ ligands for displacement of [^{3}H]WIN 35,428 binding at DAT and inhibition of dopamine uptake, in rat caudate-putamen (58). Although most of these compounds did not bind with high affinity to DAT, rimcazole displaced [^{3}H]WIN 35,428 from DAT with an affinity of 103 nM. Rimcazole had previously been reported to attenuate the locomotor stimulant effects of cocaine at doses that were not themselves behaviorally active (47). These discoveries lead to the design and synthesis of a series of rimcazole analogues as potential dopamine uptake inhibitors and structure-activity relationships at DAT, serotonin transporter (SERT), norepinephrine transporter (NET), and σ_1 receptors were determined. It was discovered that in general, substitutions on the carbazole ring system of rimcazole served to decrease binding affinities at both σ_1 receptors and the DAT (57,59). Data for other rimcazole analogues is shown in Table 2-1. N-methylation of the terminal piperazine nitrogen (SH 1-73) resulted in a small increase in binding affinity at σ_1 receptors (K_i = 552 nM) but in a slightly less active DAT compound (K_i = 436 nM) (59). Alternatively, placing a propylphenyl group, on the terminal piperazine nitrogen (SH 3-28), as seen with GBR 12909, served to improve and restore σ_1 receptor and DAT binding affinities, respectively. Likewise, when the carbazole ring system was replaced with a diphenylamine, coupled with the N-propylphenyl substituent, a moderately potent rimcazole analogue SH 3-24 resulted (K_i = 97 nM at σ_1 and 61 nM at DAT) (59). Adding fluoro-groups to the para-positions of the diphenylamine moiety (JJC 1-059) served to significantly improve both σ_1 receptor and DAT binding (K_i =11.1 nM and 22.8 nM, respectively) (60,61). Removal of the 2,6-dimethyl groups on the piperazine ring (JJC 2-008) served to reduce lipophilicity and also reduced σ_1 receptor binding affinity (K_i = 66.2 nM) while retaining high affinity for DAT (K_i = 18 nM). Interestingly, the N-benzyl analogue (JJC 2-006) showed the highest affinity for σ_1 receptors in the demethylated series (K_i = 13.1 nM) (61).

Rice's laboratory synthesized analogues of GBR 12909 and showed that GBR 12935 and several analogues displaced [^{3}H](+)-pentazocine from σ_1 receptors with high affinity (K_i range = 8.6 - 231 nM) (62). Many of these compounds show structural similarity to the rimcazole analogues and bind with high affinity to both σ_1 receptors and DAT (62,63). The most potent σ_1 ligand in these series was the trans 2,5-dimethylpiperazinyl analogue of GBR 12909 (62). Comparing the SAR derived from the rimcazole analogues to these compounds, the presence of a dimethylated piperazine,

2. Medicinal chemistry

regardless of position and stereochemistry, appears to improve binding affinity at σ_1 receptors as compared to the unsubstituted piperazines.

Behavioral evaluation of rimcazole, SH 1-73, SH 3-24, and SH 3-28 has shown that all of these ligands produced dose-related decreases in locomotor activity and decreased cocaine-induced locomotor activity. Furthermore, rimcazole and its analogues did not generalize to the cocaine discriminative stimulus in rats trained to discriminate 10 mg/kg of cocaine from saline (64). Interestingly, SH 3-28 decreased cocaine-appropriate responding as well. Another preliminary study with JJC 1-059, in comparison to cocaine, GBR 12909 and rimcazole demonstrated that, like its parent compound, JJC 1-059 did not produce locomotor stimulation in mice (61,65). Furthermore, rimcazole and its analogues attenuated cocaine-induced convulsions in mice (66). In total, these results are curious, as all of these compounds bind to the dopamine transporter, some with higher affinity than cocaine. Hence, it has been hypothesized that despite their actions at DAT, perhaps σ_1 receptor antagonism is involved in the blockade of cocaine's actions demonstrated by rimcazole and its analogues. The recent proposal that DAT-mediated cocaine-like actions, including reinforcement, might be modulated by σ_1 receptors (67,68) further supports the development of dual DAT/σ_1 probes to investigate whether or not these combined actions might provide a novel approach to cocaine-abuse medication discovery.

Table 2-1. Binding Results at σ_1 Receptors and Dopamine Transporters (DAT)

Compound	[^{3}H](+)-Pentazocine (σ_1)	[^{3}H]WIN 35,428 (DAT)	σ_1/DAT
Cocaine	8830 ± 860^b	187 ± 19^a	47
GBR 12909	318 ± 18^a	11.9 ± 1.9^a	27
Rimcazole	908 ± 99^a	224 ± 16^a	4.1
SH 3-24	97.2 ± 14.0^a	61.0 ± 6.1^a	1.6
SH 1-73	552 ± 110^a	436 ± 44^a	1.3
SH 3-28	104 ± 0.4^a	263 ± 34^a	0.4
JJC 1-059	11.1 ± 0.8^b	22.8 ± 2.0^b	0.5
JJC 2-008	66.2 ± 3.6^b	18.1 ± 2.7^b	3.7
JJC 2-006	13.1 ± 1.2^b	27.6 ± 3.9^b	0.5
JJC 2-010	372 ± 21^b	8.5 ± 0.8^b	44

Ki in nM. Data from ref. (59)[a] and ref. (61)[b].

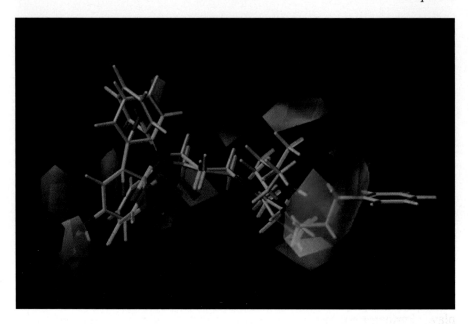

Figure 2-11. The CoMFA Contour Graphs for the Activity on the σ_1 Receptor (61). The sterically favored and unfavored (contribution at 80% and 20%, respectively) are shown as green and yellow fields and positive charges favorable and unfavorable (contribution at 80% and 20%, respectively) are shown as blue and red fields respectively.

5.2 Molecular models

Several CoMFA models were derived for σ_1 receptor binding of the rimcazole analogues and have been recently reported (61). Figure 2-11 shows the steric and electrostatic contour maps derived using σ_1 binding affinities. A sterically favored green region was observed near the terminal piperazine nitrogen substituent, supporting a strong steric interaction in this region of the molecule. Also, the scattered yellow regions around the molecule define the limits for size and shape of the substituents. Positive charge favoring regions shown as blue contours were observed in the vicinity of the para-position of the diaryl ring system. Hence small electron-withdrawing substituents, e.g. F, are predicted to improve σ_1 binding affinities. Putative binding site characteristics for the σ_1 receptor have been proposed (33,69,70) and are reviewed elsewhere in this volume by Ablordeppey and Glennon (Chapter 4). The CoMFA results describing optimal binding features of the rimcazole analogues were interpreted to be comparable to those previously described (61). As such, the substituent on the terminal piperazine nitrogen of the rimcazole analogues could be binding in the described primary hydrophobic site and the diaryl amine could be accessing the secondary binding site, which seems to tolerate bulk in this region. The region between the terminal piperazine nitrogen and the terminal phenyl ring is less tolerant to electron releasing or hydrophilic substituents as the 3-OH group of JJC 2-010 overlaps in the blue contour, which is unfavorable for activity. Likewise, comparison with the previously proposed σ model (70) would suggest that hydrophilic interactions in this region would reduce affinity towards the σ_1 receptor.

6. SUMMARY

Over the past decade, advances have been made in discovering novel σ receptor probes and developing structure-activity relationships for σ_1 and σ_2 receptor selectivity. These compounds have provided useful tools to further investigate the physiological role that central and peripheral σ receptors play. Furthermore, many of these compounds have been investigated for their *in vivo* actions, and particularly promising is their ability to attenuate cocaine-induced behaviors such as locomotor stimulation and conditioned place preference, as well as cocaine-induced toxicities. These *in vivo* studies are described in other chapters in this book and the interested reader is referred to these. Compounds that have dual actions at both σ_1 receptors and the dopamine transporter may prove to be a novel strategy for the development of a cocaine-abuse medication and is being investigated toward this goal. Compounds selective at σ_2 receptors may be useful as antineoplastic agents or for control of cell survival in neurodegenerative disease. The design and synthesis of novel and selective σ_1 and σ_2 receptor selective agonists and antagonists will undoubtedly provide the required molecular tools to elucidate both structure and function of these receptors.

REFERENCES

1. Izhak Y. Sigma Receptors. Academic Press, San Diego 1994.
2. Walker JM, Bowen WD, Walker FO, Matsumoto RR, de Costa BR, Rice KC. Sigma receptors: biology and function. Pharmacol Rev 1990, 42:355-402.
3. de Costa BR, Rothman RB, Bowen WD, Radesca L, Band L, Reid A, Paolo LD, Walker JM, Jacobson AE, Rice KC. Novel kappa opioid receptor and sigma ligands. NIDA Research Monograph 1992b, 119:76-80.
4. Martin WR, Eades CE, Thompson JA, Huppler RE. The effects of morphine- and nalorphine-like drugs in the nondependent and morphine-dependent chronic spinal dog. J Pharmacol Exp Ther 1976, 197:517-524.

5. May EL, Jaconson AE, Mattson MV, Traynor JR, Woods JH, Harris LS, Bowman ER, Aceto MD. Synthesis and in vitro and in vivo activity of (-)-(1R,5R,9R)- and (+)-(1S,5S,9S)-N-alkenyl-, -N-alkynyl-, and -N-cyanoalkly-5,9-dimethyl-2'-hydroxy-6,7-benzomorphan homologues. J Med Chem 2000, 43:5030-5036.
6. Su TP. Evidence for sigma opioid receptor: binding of [^{3}H]SKF 10,047 to etorphine-inaccessible sites in guinea-pig brain. J Pharmacol Exp Ther 1982, 223:284-290.
7. Tam SW and Cook L. Sigma opiates and certain antipsychotic drugs mutually inhibit (+)-[^{3}H]SKF 10,047 and [^{3}H]haloperidol binding in guinea pig brain membranes. Proc Natl Acad Sci USA 1984, 81:5618-5621.
8. Largent BL, Gundlach AL, Snyder SH. Pharmacological and autoradiographic discrimination of sigma and phencyclidine receptor binding sites in brain with (+)-[^{3}H]SKF 10,047, (+)-[^{3}H]-3-[3-hydroxy-phenyl]-N-(1-propyl)piperidine and [^{3}H]-1-[1-(2-thienyl)cyclohexyl]-piperidine. J Pharmacol Exp Ther 1986, 238:739-748.
9. Weber E, Sonders M, Quarum M, McLean S, Pou S, Keana JF. 1,3-Di(2-[5-^{3}H]tolyl)guanidine: a selective ligand that labels sigma-type receptors for psychotomimetic opiates and antipsychotic drugs. Proc Natl Acad Sci USA 1986, 83:8784-8788.
10. de Costa BR, Bowen WD, Hellewell SB, Walker JM, Thurkauf A, Jacobson AE, Rice KC. Synthesis and evaluation of optically pure [^{3}H](+)-pentazocine, a highly potent and selective radioligand for sigma receptors. FEBS Lett 1989, 251:53-58.
11. Hanner M, Moebius FF, Flandorfer A, Knaus HG, Striessnig J, Kempner E, Glossmann H. Purification, molecular cloning, and expression of the mammalian sigma$_1$-binding site. Proc Natl Acad Sci USA1996, 93:8072-8077.
12. Ronsisvalle G, Prezzavento O, Marrazzo A, Vittorio F, Massimino M, Murari G, Spampinato S. Synthesis of (+)-cis-N-(4-isothiocyanatobenzyl)-N-normetazocine, an isothiocyanate derivative of N-benzylnormetazocine as acylant agent for the sigma-1 receptor. J Med Chem 2002, 45:2662-2665.
13. Hellewell SB, Bowen WD. A sigma-like binding site in rat pheochromocytoma (PC12) cells: decreased affinity for (+)-benzomorphans and lower molecular weight suggest a different sigma receptor form from that of guinea pig brain. Brain Res 1990, 527:244-253.
14. Quirion R, Bowen WD, Itzhak Y, Junien JL, Musacchio JM, Rothman RB, Su TP, Tam SW, Taylor DP. A proposal for the classification of sigma binding sites. Trends Pharmacol Sci 1992, 13:85-86.
15. Hellewell SB, Bruce A, Feinstein G, Orringer J, Williams W, Bowen WD. Rat liver and kidney contain high densities of sigma-1 and sigma-2 receptors: characterization by ligand binding and photoaffinity labeling. Eur J Pharmacol Mol Pharmacol Sect 1994, 268:9-18.
16. Bowen WD, de Costa BR, Hellewell SB, Walker JM, Rice KC. [^{3}H](+)-Pentazocine: A potent and highly selective benzomorphan-based probe for sigma-1 receptors. Mol Neuropharmacol 1993, 3:117-126.
17. Ablordeppey SY, Fischer JB, Burke Howie KJ, Glennon RA. Design, synthesis and binding of sigma receptor ligands derived from butaclamol. Med Chem Res 1992, 2:368-375.
18. Contreras PC, DiMaggio DA, O'Donohue TL. An endogenous ligand for the sigma opioid binding site. Synapse 1987, 1:57-61.
19. Su TP, Weissman AD, Yeh SY. Endogenous Ligands for sigma opioid receptors in the brain ("sigmaphin"): evidence from binding assays. Life Sci 1986, 38:2199-2210.

20. Neumaier JF, Chavkin C. Calcium-dependent displacement of haloperidol-sensitive sigma receptor binding in rat hippocampal slices following tissue depolarization. Brain Res 1989, 500:215-222.
21. Connor MA, Chavkin C. Focal stimulation of specific pathways in the rat hippocampus causes a reduction in radioligand binding to the haloperidol-sensitive sigma receptor. Exp Brain Res 1991, 85:528-536.
22. Su TP, London ED, Jaffe JH. Steroid binding at sigma receptors suggests a link between endocrine, nervous, and immune systems. Science 1988, 240:219-221.
23. Moebius FF, Reiter RJ, Hanner M, Glossmann H. High affinity of sigma-1 binding sites for sterol isomerization inhibitors: Evidence for a pharmacological relationship with the yeast sterol C8-C7 isomerase. Br J Pharmacol 1997, 121:1-6.
24. de Costa BR, Radesca L, Paolo LD, Bowen WD. Synthesis, characterization, and biological evaluation of a novel class of N-(arylethyl)-N-alkyl-2-(1-pyrrolidinyl)ethylamines. J Med Chem 1992a, 35:38-47.
25. Zhang Y, Williams W, Bowen WD, Rice KC. Synthesis and evaluation of aryl-substituted N-(arylethyl)-N-methyl-2-(1-pyrrolidinyl)ethylamines and corresponding arylacetamides for sigma receptor affinity. J Med Chem 1996, 39:3564-3568.
26. Ronsisvalle G, Marrazzo A, Prezzavento O, Pasquinucci L, Vittorio F, Pittala V, Pappalardo MS, Cacciaguerra S, Spampinato S. (+)-cis-N-Ethyleneamino-N-normetazocine derivatives. Novel and selective sigma ligands with antagonist properties. J Med Chem 1998, 41:1574-1580.
27. El-Ashmawy M, Ablordeppey SY, Issa H, Gad L, Fischer JB, Burke Howie KJ, Glennon RA. Further investigation of 5-phenylpentylamine derivatives as novel sigma receptor ligands. Med Chem Res 1992:2119-2126.
28. Maeda DY, Williams W, Bowen WD, Coop A. A Sigma-1 receptor selective analogue of BD1008. A potential substitute for (+)-opioids in sigma receptor binding assays. Bioorg Med Chem Lett 2000, 10:17-18.
29. Berardi F, Loiodice F, Fracchiolla G, Colabufo NA, Perrone R, Tortorella V. Synthesis of chiral 1-[ω-(4-chlorophenoxy)alkyl]-4-methylpiperindines and their biological evaluation at sigma-1, sigma-2, and sterol Δ_8-Δ_7 isomerase sites. J Med Chem 2003, 46:2117-2124.
30. Bowen WD, Moses EL, Tolentino PJ, Walker JM. Metabolites of haloperidol display preferential activity at sigma receptors compared to dopamine D-2 receptors. Eur J Pharmacol 1990, 177:111-118.
31. Guitart X, Ballarin M, Codony X, Dordal A, Farre AJ, Frigola J, Merce R. E-5842. Drugs Fut 1999, 24:386-392.
32. Huang YS, Hammond PS, Whirret PR, Kuhner RJ, Wu L, Childers SR, Mach RH. Synthesis and quantitative structure-activity relationships of N-(1-benzylpiperidin-4-yl) phenylacetamides and related analogues as potent and selective sigma-1 receptor ligands. J Med Chem 1998, 41:2361-2370.
33. Huang Y, Hammond PS, Wu L, Mach RH. Synthesis and structure-activity relationships of N-(1-benzylpiperidin-4-yl)arylacetamide analogues as potent sigma-1 receptor ligands. J Med Chem 2001, 44:4404-4415.
34. Nakazato A, Ohta K, Sekiguchi Y, Okuyama S, Chaki S, Kawashima Y, Hatayame K. Design, synthesis, structure-activity relationships, and biological characterization of novel arylalkoxyphenylalkylamine sigma ligands as potential antipsychotic drugs. J Med Chem 1999a, 42:1076-1087.
35. Nakazato A, Kumagai T, Ohta K, Chaki S, Okuyama S, Tomisawa K. Synthesis and SAR of 1-Alkyl-2-phenethylamine derivatives designed from N,N-dipropyl-4-methoxy-

3-(2-phenylethoxy)phenethylamine to discover sigma-1 ligands. J Med Chem 1999b, 42:3965-3970.
36. Bowen WD, Bertha, CM, Vilner BJ, Rice KC. CB-64D and CB-184: Ligands with high sigma-2 receptor affinity and subtype selectivity. Eur J Pharmacol 1995a, 278:257-260.
37. Maeda DY, Williams W, Kim WE, Thatcher LN, Bowen WD, Coop A. N-Arylalkylpiperidines as high affinity sigma-1 and sigma-2 receptor ligands: phenylpropylamines as potential leads for selective sigma-2 agents. Bioorg Med Chem Lett 2002, 12:497-500.
38. Vilner BJ, Bowen WD. Modulation of cellular calcium by sigma-2 receptors: release from intercelluar stores in human SK-N-SH neuroblastoma cells. J Pharmacol Exp Ther 2000, 292: 900-911.
39. Crawford KW, Bowen WD. Sigma-2 receptor agonists activate a novel apoptotic pathway and potentiate antineoplastic drugs in breast tumor cell lines. Cancer Res 2002, 62:313-322.
40. Crawford KW, Coop A, Bowen WD. Sigma-2 receptors regulate changes in sphingolipid levels in breast tumor cells. Eur J Pharmacol 2002, 443:207-209.
41. Bowen WD, Vilner BJ, Williams W, Bertha CM, Kuehne ME, Jaconson AE. Ibogaine and its congeners are sigma-2 receptor-selective ligands with moderate affinity. Eur J Pharmacol 1995b, 279:R1-R3.
42. Bowen WD. Sigma receptors and iboga alkaloids. Alkaloids Chem Biol 2001, 56:173-191.
43. Mach RH, Vangveravong S, Huang Y, Yang B, Blair JB, Wu L. Synthesis of N-substituted 9-azabicyclo[3.3.1]nonan-3α-yl phenylcarbamate analogs as sigma-2 receptor ligands. Med Chem Res 2003, 11:380-398.
44. Mach RH, Wu L, West T, Whirrett BR, Childers SR. The analgesic tropane analogue (+/-)-SM 21 has a high affinity for sigma-2 receptors. Life Sci 1999, 64:PL131-PL137.
45. Matsumoto RR, Mack AL. (+/-)-SM 21 attenuates the convulsive and locomotor stimulatory effects of cocaine. Eur J Pharmacol 2001, 417:R1-R2.
46. Sharkey J, Glen KA, Wolfe S, Kuhar MJ. Cocaine binding at sigma receptors. Eur J Pharmacol 1988, 149:171-174.
47. Menkel M, Terry P, Pontecorvo M, Katz JL, Witkin JM. Selective sigma ligands block stimulant effects of cocaine. Eur J Pharmacol 1991, 201:251-252.
48. Romieu P, Martin-Fandon R, Maurice T. Involvement of the sigma-1 receptor in the cocaine-induced conditioned place preference. Neuroreport 2000, 11;2885-2888.
49. Ujike H, Kuroda S, Otsuki S. Sigma receptor antagonist block the development of sensitization to cocaine. Eur J Pharmacol 1996, 296:123-128.
50. McCracken K, Bowen WD, De Costa BR, Matsumoto RR. Two novel sigma receptor ligands, BD1047 and LR172, attenuate cocaine-induced toxicity and locomotor activity. Eur J Pharmacol 1999a, 370:225-232.
51. McCracken K, Bowen W, Matsumoto RR. Novel sigma receptor ligands attenuate the locomotor stimulatory effects of cocaine. Eur J Phamacol 1999b, 365:35-38.
52. Matsumoto RR, McCracken K, Friedman M, Pouw B, De Costa BR, Bowen WD. Conformationally restricted analogs of BD1008 and antisense oligodeoxynucleotide targeting sigma-1 receptors produce anti-cocaine effects in mice. Eur J Pharmacol 2001b, 419:163-174.
53. Matsumoto RR, McCracken K, Pouw B, Miller J, Bowen WD, Williams W, de Costa BR. N-Alkyl substituted analogs of the sigma receptor ligand BD1008 and traditional sigma receptor ligands affect cocaine-induced convulsions and lethality in mice. Eur J Pharmacol 2001c, 411:261-273.

54. Kahoun JR, Ruoho AE. (^{125}I) Iodoazidococaine, a photoaffinity label for the haloperidol-sensitive sigma receptor. Proc Natl Aca Sci USA 1992, 89:1393-1397.
55. Wilke RA, Mehta RP, Lupardus J, Chen Y, Ruoho A, Jackson MB. Sigma receptor photolabeling and sigma-receptor-mediated modulation of potassium channels in tumor cells. J Biol Chem 1999, 274:18387-18392.
56. Contreras PC, Bremer ME, Rao TS. GBR-12909 and fluspirilene potently inhibited binding of [^{3}H] (+) 3-PPP to sigma receptors in rat brain. Life Sci 1990, 47:133-137.
57. Husbands SM, Izenwasser S, Loeloff RJ, Katz JL, Bowen WD, Vilner BJ, Newman AH. Isothiocyanate derivatives of 9-[3-(cis-3,5-dimethyl-1-piperazinyl)propyl]-carbazole (rimcazole): irreversible ligands for the dopamine transporter. J Med Chem 1997, 40:4340-4346.
58. Izenwasser S, Newman AH, Katz JL. Cocaine and several sigma receptor ligands inhibit dopamine uptake in rat caudate-putamen. Eur J Pharmacol 1993, 243:201-205.
59. Husbands SM, Isenwasser S, Kopajtic T, Bowen WD, Vilner BJ, Katz JL, Newman AH. Structure-activity relationships at the monoamine transporters and sigma receptors for a novel series of 9-[3-cis-3,5-dimethyl-1-piperazinyl)-propyl]carbazole (rimcazole) analogues. J Med Chem 1999, 42:4446-4455.
60. Cao JJ, Husbands SM, Kopajtic T, Katz JL, Newman AH. [3-cis-3,5-Dimethyl-(1-piperazinyl)alkyl]-bis-(4'-fluorophenyl)amine analogues as novel probes for the dopamine transporter. Bioorg Med Chem Lett 2001, 11:3169-3173.
61. Cao JJ, Kulkarni SS, Husbands SM, Bowen WD, Williams W, Kopajtic T, Katz JL, George C, Newman AH. Dual probes for the dopamine transporter and sigma-1 receptors: novel piperazinyl alkyl-bis-(4'-fluorophenyl)amine analogues as potential cocaine-abuse therapeutic agents. J Med Chem 2003, 46:2589-2598.
62. Matecka D, Rothman RB, Radesca L, De Costa BR, Dersch CM, Partilla JS, Pert A, Glowa JR, Wojnick FHE, Rice KC. Development of novel, potent, and selective dopamine reuptake inhibitors through alteration of piperazine ring of 1-[2(diphenylmethoxy)ethyl]- and 1-[2-[bis(4-fluorophenyl)methoxy]ethyl]-4-(3-phenylpropyl)piperazines (GBR 12936 and GBR 12909). J Med Chem 1996, 39:4704-4716.
63. Matecka D, Lewis D, Rothman RB, Dersch CM, Wojnicki FHE, Glowa JR, DeVries AC, Pert A, Rice KC. Heteroaromatic analogs of 1-[2-(diphenylmethoxy)ethyl]- and 1-[2-[bis(4-fluorophenyl)methoxy]ethyl]-4-(3-phenylpropyl)piperazines (GBR 12935 and GBR 12909) as high-affinity dopamine reuptake inhibitors. J Med Chem 1997, 40:705-716.
64. Katz JL, Libby T, Kopajtic T, Husbands SM, Newman AH. Behavioral effects of rimcazole analogues alone and in combination with cocaine. Eur J Pharmacol 2003, 468:109-119.
65. Newman AH, Kulkarni S. Probes for the dopamine transporter: new leads towards a cocaine-abuse therapeutic - A focus on analogues of benztropine and rimcazole. Med Res Rev 2002, 5:429-464.
66. Matsumoto RR, Hewett KL, Pouw B, Bowen WD, Husbands SM, Cao JJ, Newman AH. Rimcazole analogs attenuate the convulsive effects of cocaine: correlation with binding to sigma receptors rather than dopamine transporters. Neuropharmacology 2001a, 41:878-886.
67. Romieu P, Phan V, Martin-Fardon R, Maurice T. Involvement of the sigma-1 receptor in cocaine-induced conditioned place preference: possible dependence on dopamine uptake blockade. Neuropsychopharmacology 2002, 26:444-455.

68. Maurice T, Martin-Fardon R, Romieu P, Matsumoto RR. Sigma-1 receptor antagonists represent a new strategy against cocaine addiction and toxicity. Neurosci Biobehav Rev 2002, 26:499-527.
69. Glennon RA, Ablordeppey SY, Ismaiel AM, El-Ashmawy MB, Fischer JB, Howie KB. Structural features important for sigma-1 receptor binding. J Med Chem 1994, 37:1214-1219.
70. Ablordeppey SY, Fischer JB, Glennon RA. Is a nitrogen atom an important pharmacophoric element in sigma ligand binding? Bioorg Med Chem 2000, 8:2105-2111.

Corresponding authors: *Dr. Amy Hauck Newman, Mailing address: National Institutes of Health, National Institute on Drug Abuse-Intramural Research Program, Section on Medicinal Chemistry, 5500 Nathan Shock Drive, Baltimore, MD 21224,USA, Phone: (410) 550-1455, Fax: (410) 550-1648, Electronic mail address:anewman@irp.nida.nih.gov and Dr. Andrew Coop, Mailing address: University of Maryland School of Pharmacy, Department of Pharmaceutical Sciences, 20 Penn Street, 637 HSFII, Baltimore, MD 21201, USA, Phone: (401) 706-2029, Fax: (401) 706-5017, Electronic mail address: acoop@rx.umaryland.edu*

Chapter 3

IRREVERSIBLE σ COMPOUNDS

Giuseppe Ronsisvalle and Orazio Prezzavento
Department of Pharmaceutical Sciences, University of Catania, 95125 Catania, Italy

1. AFFINITY LABELS

Irreversible ligands or affinity labels are tools that are particularly useful for characterizing the functions of receptor systems. Generally, these are compounds derived from ligands with a high affinity and selectivity. Chemical modifications are made that, on the one hand, do not obstruct binding with the same receptor site and, at the same time, allow the formation of an additional covalent bond with the receptor. These agents have in their structure particularly reactive groups (electrophilic moieties), such as the nitrogen mustards, haloacetamides, aldol esters, Michael acceptors, or the isothiocyanates (1). It is logical to suppose that a well-designed affinity label can bind to a receptor in a reversible manner only with the help of pharmacophore groups (first recognition step). Then, only after the molecule has aligned with the receptor site, does it form a covalent bond with an accessory nucleophile group that is present on the receptor site or near it (second recognition step).

The localization of the electrophile is extremely important in order to obtain irreversible compounds with high affinity and, above all, with high selectivity for a precise receptor sub-population (2). It is therefore necessary to start with compounds that are highly selective (specific would be better) and to have a good knowledge of the structure/activity correlation of the chosen compound classes. It is also clear that if one has success in the choice of the optimum electrophile group (fine tuning) and its alignment with respect to the receptor target, there is a good possibility of obtaining an increase in selectivity (recognition amplification). This is based on the high probability that this new pharmacophore will not adapt itself so well at the other sub-populations of the same receptor.

If the irreversible ligand has an electrophile group that is particularly reactive, the increase in selectivity is limited. A classic example of this is that of the aziridinium ion, its elevated reactivity, in fact, allows it to acylate any nucleophile present on the receptor site that has an ideal distance for the formation of a covalent bond, thus conferring to it a high level of aspecificity. The selectivity of a ligand that has an active electrophile group should, instead, be principally caused by the relative affinity for the different receptor populations so as to make it as specific as possible. The presence of numerous nucleophilic groups in biological tissue notably increases the probability of a non-specific covalent bond. Thus it will be necessary to re-evaluate the selectivity/specificity for the binding of the affinity label to the target sub-population, excluding the cross reaction with enzymes or other molecules present in the preparation.

2. AFFINITY LABELS BASED ON VARIOUS σ LIGAND STRUCTURAL CLASSES

2.1 Irreversible ligands derived from analogues of phencyclidine

Phencyclidine, or 1-(1-phenylcyclohexyl)piperidine (PCP), was introduced towards the end of the 1950s as a "dissociative" anesthetic. However, the side effects reported in about 33% of the patients treated led to its clinical suspension. The hallucinogenic effects produced by PCP during the 1970s led to its abuse, resulting in psychotic subjects and an increasing number of deaths due to overdose. PCP, as do its congeners, interacts centrally with several binding sites. For some of these, the importance of the interaction has not yet been fully defined. One of the mechanisms of PCP action is associated with high affinity binding to and resultant blockade of the Ca^{2+} ionophore channel present in the N-methyl-D-aspartate (NMDA)-type glutamate receptor (3,4). This high affinity site on the NMDA receptor is often referred to as the PCP receptor.

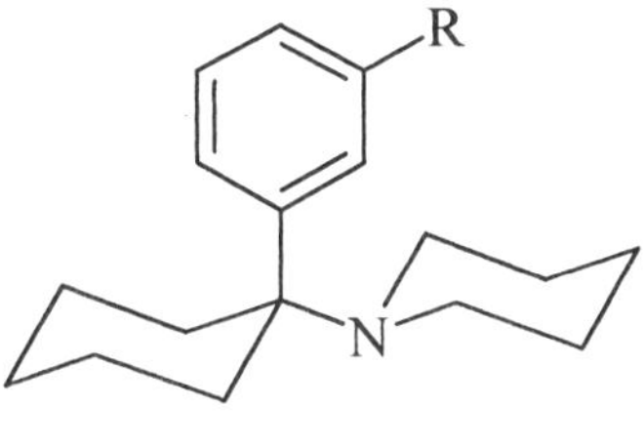

Phencyclidine; R = H
Metaphit; R = NCS

Figure 3-1. Structure of phencyclidine and metaphit

The meta-isothiocyanate derivative of phencyclidine, 1-[1-(3-isothiocyanatophenyl)cyclohexyl]piperidine (metaphit) (Figure 3-1) was synthesized in 1985 in the attempt to better characterize the PCP receptors with the help of an irreversible ligand (5-7). Metaphit significantly modifies the binding of sites labeled with [^{3}H]PCP in the rat brain and is able to antagonize numerous behavioral actions induced by phencyclidine after having been preventatively administered intracerebroventricularly (8,9). It is not clear, however, if metaphit is an agonist or an antagonist for the PCP receptors. In fact, discordant data have been published depending on the animal species and experimental model used (10-15). Metaphit reversibly binds to μ opioid receptors labeled with [^{3}H]dihydromorphine and to muscarinic receptors labeled with [^{3}H]QNB (quinuclidinyl benzilate), but only at particularly high concentrations, while it has no affinity for benzodiazepine receptors labeled with [^{3}H]diazepam (5). *In vitro* it irreversibly inhibits the binding of [^{3}H]methylphenidate (16,17) and [^{3}H]cocaine (18), two ligands representative of the dopamine transport system. Metaphit antagonizes the motor stimulation induced by cocaine, which at doses of 25 mg/kg is ineffective if the rats are treated 24 h before the behavioral test with metaphit (18). This compound also diminishes the capacity of 5-HT$_2$ receptors labeled with [^{3}H]ketanserin. If rat brain membranes are pretreated with metaphit at a concentration of 1 μM, there is a 60% decrease in the B$_{max}$ of the high affinity sites and a 23% decrease of the low affinity sites (Table 3-1) (19).

Table 3-1. Effects of metaphit on [^{3}H]ketanserin binding in rat synaptic membranes

Treatment	[^{3}H]Ketanserin			
	High		Low	
	K_d	B_{max}	K_d	B_{max}
Saline	0.207 ± 1.86 x 10^{-3}	71.56 ± 3.36	7.746 ± 0.390	895.21 ± 40.45
Metaphit	0.187 ± 9.74 x 10^{-3}	28.25 ± 1.84	5.425 ± 0.269	692.79 ± 30.62

K_i in nM and B_{max} in fmol/mg protein. Reprinted from ref. (19) with permission.

The psychotomimetic actions of phencyclidine and of some σ ligands with a benzomorphan structure show a correlation (20-22). (+)-N-Allylnormetazocine [(+)-SKF-10,047] and cyclazocine seem to share, in fact, many pharmacologic properties of phencyclidine (23,24). It should be pointed out that (–)-SKF-10,047, has an affinity for μ opioid receptors (K_i = 3.6 ± 0.05 nM), while its (+)-isomer has an affinity for both the σ and phencyclidine receptors (Table 3-2). The latter sites have, however, a preference for phencyclidine and its derivatives, rather than for the benzomorphan derivatives. On the other hand, the σ receptors show a marked preference for the (+)-cis-isomer of the N-substituted N-normetazocines, especially if they present bulky nitrogen substituents, rather than for the derivatives of phencyclidine (25-27) (Tables 3-2 and 3-3).

This cross reactivity between the σ and PCP recognition sites encouraged Bluth et al. to investigate if metaphit is able to also acylate the σ sites (28). The pretreatment of guinea pig brain membranes with metaphit at a concentration of 1 μM decreased the σ sites labeled with [^{3}H]di-o-tolylguanidine (DTG) by 40%. No significant inhibition was observed for the selective PCP receptor ligand [^{3}H]TCP at this metaphit concentration. Fifty percent inhibition of the binding of [^{3}H]TCP was found at a concentration of metaphit 10 times higher (10 μM).

Table 3-2. Binding affinities of benzomorphans at PCP and $σ_1$ receptors

Compound	[^{3}H]TCP	[^{3}H](+)-Pentazocine
(+)-SKF 10,047	225 ± 11	59.27 ± 2.5
(–)-SKF 10,047	504 ± 22	> 3000
(+)-Pentazocine	4040 ± 616	3.1 ± 0.3
(–)-Pentazocine	842 ± 124	83.1 ± 6.2
(+)-Cyclazocine	153 ± 22	17.2 ± 0.03
(–)-Cyclazocine	107 ± 34	2650 ± 385

K_i in nM. Reprinted from ref. (27) with permission.

3. Irreversible compounds

Table 3-3. Binding of benzomorphan and phencyclidine derivatives to PCP and σ receptors

Compound	[^{3}H]TCP	[^{3}H](+)-SKF-10,047 in the presence of TCP (2 µM)
(+)-SKF-10,047	405	55
(–)-SKF-10,047	820	690
Pentazocine	3310	55
PCP	66	1450
TCP	10	8330

IC$_{50}$ in nM. Reprinted from ref. (26) with permission.

It is noteworthy that metaphit causes an irreversible inhibition of the σ sites that is different according to the ligand used. The order of sensitivity is [^{3}H]DTG > [^{3}H](+)-3PPP >> [^{3}H](+)-SKF-10,047, the values of IC$_{50}$ correspond respectively to 2, 10, and 50 µM (28). [^{3}H](+)-SKF-10,047 is the most resistant to the effects of metaphit. This different order of sensitivity seems to reflect a different manner of interaction between the derivatives with a benzomorphan and non-benzomorphan structure with the σ sites. The authors put forward two hypotheses. The first is linked to the theory proposed by Bowen et al. (29), which presupposes the existence of a multi- site model for the σ receptors (allosteric model or multiple σ receptor types). Metaphit, according to this theory, could interact differently with these sites and thus have a different ability in inhibiting the binding of these ligands. In the second hypothesis, metaphit is considered as a competitive inhibitor of [^{3}H]DTG and [^{3}H](+)-3-PPP, even though the acylation of σ sites by metaphit should not be overcome by high concentrations of competing ligands as is the case for PCP receptors. Scatchard analysis of [^{3}H]TCP binding shows, in fact, a notable difference of B$_{max}$ between the membranes pre-treated with metaphit and the control membranes (182 ± 38 vs. 342 ± 16 fmol/mg) and no significant difference for binding affinity (K$_d$ 56 ± 10 vs. 56 ± 3 nM). These data suggest that metaphit is a non-competitive inhibitor of PCP receptors, as would be expected for an irreversible alkylating agent. In contrast, after treatment with metaphit, [^{3}H]DTG showed twice the value of K$_d$ (106 ± 20 nM vs. 50 ± 5 nM), and little change of B$_{max}$ (450 ± 59 vs. 523 ± 13 fmol/mg protein), relative to control membranes. Similar results were observed for [^{3}H](+)-3-PPP (K$_d$ = 80 ± 4 vs. 36 ± 2 nM and B$_{max}$ = 207 ± 33 vs. 280 ± 40 fmol/mg protein for metaphit-treated vs. control, respectively). Metaphit, therefore, as indicated by these data behaves as a competitive inhibitor of the σ ligands [^{3}H]DTG and [^{3}H](+)-3PPP. Therefore, the acylating property of metaphit is not targeted directly on the binding site of the σ receptors, but rather on a nearby nucleophilic group. This might allow competing ligands to displace metaphit from the ligand binding site while it remains covalently attached to the receptor, resulting in apparent competitive inhibition. Interestingly, a small increase in the K_d value (30-50%) is also seen in membranes not treated with metaphit, but only having undergone washing (30). The cause of this is not clear.

Reid et al. in 1990 (31) found that metaphit and other phencyclidine-based affinity ligands (Figure 3-2) produce a wash-resistant inhibition both of the σ and PCP sites, thus fortifying the idea that metaphit can interact with both sites. The qualitative differences found are probably due to variations in the experimental procedure (tissue preparation, metaphit incubation time, temperature, concentration of [^{3}H]DTG and [^{3}H]TCP, etc).

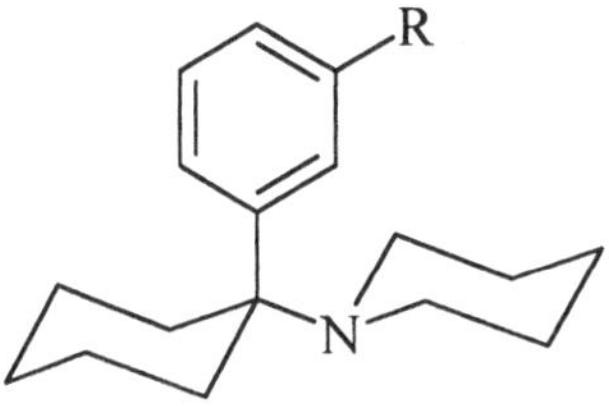

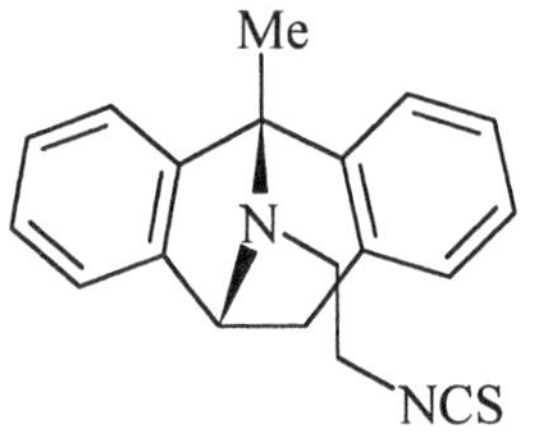

Fumaryl-PCP; R = NHCOCH=CHCOOCH$_3$
Acroyl-PCP; R = OCOCH=CH$_2$
Cinnammoyl-PCP; R = NHCOCH=CH-pBr-C$_6$H$_4$

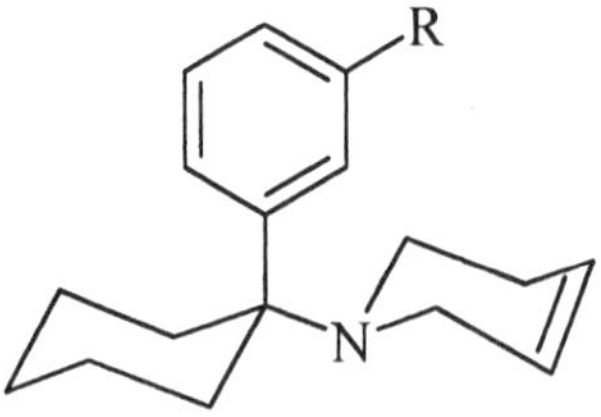

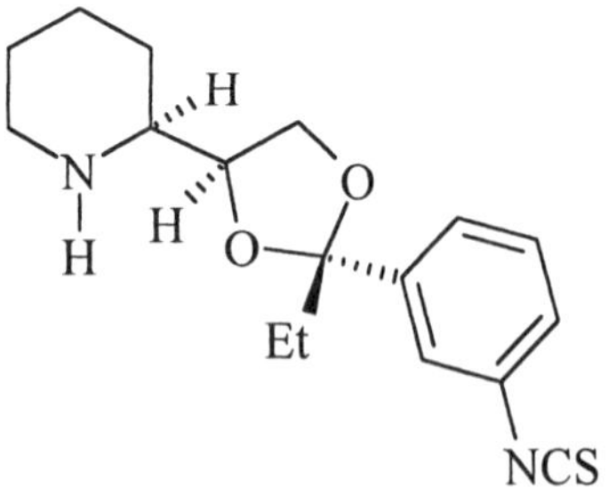

(±)-**MK801-NCS**

Bromoacetyl-DehydroPCP; R = OCOCH$_2$Br

ETOX-NCS

Figure 3-2. Phencyclidine-based ligands

3. Irreversible compounds

Table 3-4. Wash-resistant inhibition of phencyclidine and σ receptor sites by analogs of phencyclidine, MK-801 and etoxadrol

Compound	[^{3}H]DTG			[^{3}H]TCP		
	1 μM	10 μM	100 μM	1 μM	10 μM	100 μM
Phencyclidine	92.9 ± 10.5	99.4 ± 7.4	118.9 ± 11.1	91.3 ± 2.2	93.7 ± 7.7	88.7 ± 6.0
Metaphit	90.3 ± 12.6	61.0 ± 13.7	35.9 ± 1.3	90.1 ± 2.9	78.5 ± 8.3	26.2 ± 4.0
Cinnamoyl-PCP	73.6 ± 6.4	43.5 ± 2.7	13.5 ± 2.7	89.2 ± 6.7	62.9 ± 3.8	20.9 ± 4.4
ETOX-NCS	81.5 ± 6.7	46.4 ± 1.8	6.2 ± 2.6	38.1 ± 4.2	20.2 ± 2.3	9.5 ± 1.0
MK801-NCS	98.9 ± 14.5	65.3 ± 1.1	40.0 ± 2.8	102.8 ± 7.4	71.4 ± 4.1	46.2 ± 0.8
Acroyl-PCP	99.8 ± 2.4	73.8 ± 9.2	48.3 ± 4.6	93.7 ± 3.8	37.0 ± 3.1	13.7 ± 2.9
Fumaryl-PCP	104.0 ± 3.3	80.2 ± 6.3	68.9 ± 7.6	106.1 ± 1.0	114.7 ± 11.0	82.8 ± 5.2
Bromoacetyl-PCP	100.0 ± 4.8	88.9 ± 4.0	64.6 ± 1.1	66.0 ± 3.1	36.8 ± 3.5	24.1 ± 2.3

[^{3}H]DTG labels σ receptors. [^{3}H]TCP labels PCP receptors. Values represent the percent binding compared to membranes not treated with the indicated compound.

The most potent inhibitor of σ receptors in this study was cinnamoyl-PCP (Table 3-4), followed closely by the isothiocyanate derivative of the phencyclidine agonist, etoxadrol (ETOX-NCS). Both compounds at a concentration of 1 μM inhibited the binding of [^{3}H]DTG. All the other compounds examined inhibited the σ receptors only at a concentration of 10 μM, with a greater effect at 100 μM. (±)-MK-801-NCS has an equivalent effect on both the receptors at all the concentrations tested.

Recently, Zhang and Cuevas (32) have reported that the pre-incubation of neurons with metaphit (50 μM for 10 min) blocks the DTG-induced inhibition of calcium channels by ≥ 95%, showing the principle activity of metaphit as an irreversible σ antagonist.

2.2 Irreversible ligands derived from analogues of guanidine

Among the most selective σ ligands, di-o-tolyl-guanidine (DTG) must surely be included. DTG was first reported as ligand for σ receptors by Weber and colleagues (33). Though it does not discriminate between the two σ subclasses (K_i = 12 nM and 38 nM at $σ_1$ and $σ_2$, respectively), DTG shows no significant affinity for many other receptor systems (33,34). SAR studies (35,36) have revealed the importance of the presence of at least one aromatic ring on the guanidine group, since the removal of both aromatic rings of DTG leads to the complete loss of σ affinity. For example, N,N'-di(methyl)guanidine (DMG) has an IC_{50} > 100,000. The substitution of an aromatic ring with a cyclohexyl ring to give N-(o-tolyl)-N'-(cyclohexyl)guanidine (ChTG) or with an adamantyl nucleus to give N-(o-tolyl)-N'-(adamant-1-yl)guanidine (AdTG), leads to an improvement of σ affinity, with respect to DTG (IC_{50} = 13 nM and 7.6 nM, vs. 28 nM against [^{3}H]DTG). The two methyl groups present in the ortho position efficaciously contribute to the interaction process, since their removal causes a relative loss of affinity. N,N'-Di(phenyl)guanidine (DPG), exhibited an IC_{50} = 397 nM vs. 28 nM for DTG. The presence of added substituents is also well tolerated in the para position on one of the two toluene rings. For example, N-(2-methylphenyl)-N'-(2-methyl-4-nitrophenyl)guanidine and N-(2-methyl phenyl)-N'-(2-methyl-4-aminophenyl)guanidine exhibit a high to moderate affinity for the σ sites labeled with [^{3}H]DTG (IC_{50} = 37 ± 5 nM and 220 ± 14 nM, respectively).

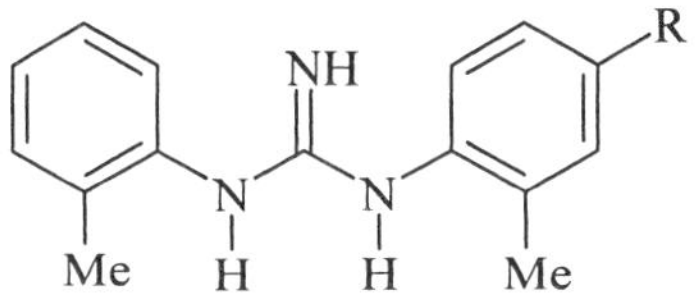

3. Irreversible compounds

DTG; R = H
DIGIT; R = NCS

Figure 3-3. Structure of DTG and DIGIT

The knowledge that some structural modifications can be made on one of the two toluene rings without altering the affinity has led to the design and synthesis of some DTG analogues as possible irreversible ligands for σ receptors. The isothiocyanate derived from DTG, 1-(2-methyl-4-isothiocyanatophenyl)-3-(2-methylphenyl)guanidine (DIGIT) (Figure 3-3) (30) has a notable power to inhibit σ sites labeled with [^{3}H]DTG. In treated membranes, 50% binding inhibition occurs at a DIGIT concentration of 28 ± 1.4 nM. Unlike metaphit, it does not have any significant activity at PCP receptors labeled with [^{3}H]TCP. In order to produce an inhibition of about 33%, a concentration of 50 µM of DIGIT is necessary. DIGIT also specifically inhibits the binding of [^{3}H](+)-3-PPP, while it has no effect on the binding of dopaminergic receptors labeled with [^{3}H]spiperone, benzodiazepine receptors labeled with [^{3}H]flunitrazepam, or opioid receptors labeled with [^{3}H]dihydromorphine.

The pretreatment of membranes with 50 nM DIGIT inhibits binding of [^{3}H]DTG by about 50%. As the recovery of [^{3}H]DTG binding in membranes pre-treated with DIGIT does not take place even after numerous washes, it is probable that DIGIT binds irreversibly. At neutral pH, the positive charge present on the guanidino group of DIGIT could result in a strong ionic interaction with the membrane that could be stable even after numerous washes. However, the inhibition of [^{3}H]DTG binding was not reversed by treatment with 2M NaCl, making an ionic interaction unlikely. Furthermore, as the binding of [^{3}H]DTG is not inhibited by the presence of KSCN at different concentrations (1-10,000 nM), it is also unlikely that the acylation of the σ receptors by DIGIT is due simply to a non-specific interaction of the isothiocyanate group with the σ receptors.

As shown in Table 3-5, DIGIT causes no variation in the number of binding sites, but rather causes a decrease in binding affinity (increase of K_d), relative to membranes treated with the reversibly binding DTG. It should be noted that the washing procedure alone caused a small decrease of affinity, since the K_d of [^{3}H]DTG in the membranes that had not undergone washing was about 50% lower compared to those treated with DTG and then washed. As these results are similar to those observed with metaphit and pose the same questions, the authors propose a series of interesting hypotheses. For example, they consider the possibility that the isothiocyanate group present in DIGIT could induce a stable partition into the lipid phase of the membrane in such a way as to make membrane association stable even after numerous washes. The ligand without the isothiocyanate group (DTG) is easily removed, clearly showing that wash-resistant binding is imparted by the isothiocyanate group. However, as this effect is specific for the σ ligands, while the binding of the other ligands at other receptors is not at all compromised, it is improbable that this is only a lipid effect.

A further hypothesis is supported by the SAR correlations carried out on derivatives of DTG. The introduction of two isothiocyanate groups, one on each of the toluene rings, leads to a loss of affinity 20 times greater with respect to DIGIT (IC_{50} = 600 nM vs. 30 nM) (30). The same binding profile was also observed with some substituted compounds on both the toluene rings of DTG. If there are two amino groups present in the para position, the potency decreases by 230 times (IC_{50} = 7150 ± 106 nM) (36). As has been stated above, the loss of affinity is less pronounced if only one of the toluene rings is substituted.

Taken together, the findings suggest that the unsubstituted toluene ring binds in the receptor pocket, whereas the ring bearing the isothiocyanate group is bound irreversibly in proximity to the pocket. Thus, the bond in the σ receptor pocket is of the reversible type, allowing an adequate quantity of a competing ligand to move DIGIT from this binding site. The irreversible bond formed by the isothiocyanate group maintains the DIGIT molecule anchored, such that in the absence of a competitive ligand, DIGIT can reoccupy the receptor pocket. The logical consequence of this would be apparent competitive inhibition, a reduction of affinity without any variation in the number of sites, as is observed in the Scatchard analysis. Finally, it is

Table 3-5. Effect of DIGIT (100 nM) on σ receptor binding parameters (K_d and B_{max}) in guinea pig brain membranes

Pretreatment	[³H]DTG	
	K_d (nM)	B_{max} (fmol/mg protein)
DIGIT	196 ± 10	3,600 ± 400
DTG	59 ± 5	3,600 ± 300
Untreated and unwashed	30	

Data extracted from ref. (30).

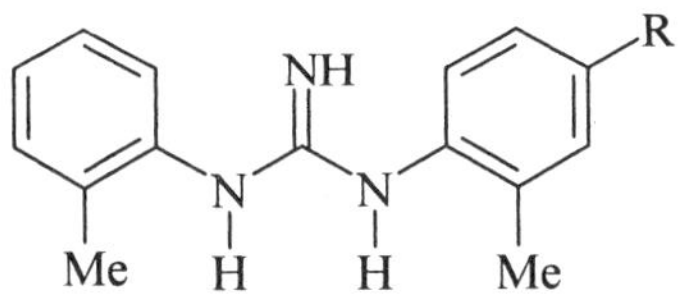

3. Irreversible compounds

DIGBA; R = NHCOCH$_2$Br
DIGIE; R = OCH$_2$CH$_2$OCH$_2$CH$_2$NCS
DIGMF; R = NHCOCH=CHCOOCH$_3$

Figure 3-4. Irreversible ligands derived from DTG

also possible the DIGIT forming a bond with a nucleophile site at the limit of the receptor implies a conformational change such as to make it unfavorable for binding. Unfortunately, in this study data are not available regarding ligands with a benzomorphan structure. It would be useful to see if the loss of σ binding is the same or different according to the structure of the σ ligands used.

DIGIT and other irreversible ligands structurally derived from DTG (Figure 3-4) have been studied by Reid et al. (31). The most interesting finding was the confirmation of the selectivity of DIGIT for the σ receptors compared to the PCP receptors. A concentration of 1 μM DIGIT is sufficient for an inhibition of σ binding, with a more marked effect at 10 μM. The same concentrations had no effect on the binding of [^{3}H]TCP (Table 3-6). Only at concentrations particularly elevated (100 μM) was a disappearance of σ/PCP selectivity seen. In this case, the loss of binding was about 90% on both the receptors. Also, 1-{4-[2-(2-isothiocyanato-ethoxy)ethoxy]-2-methyl-phenyl}-3-(2-methylphenyl) guanidine (DIGIE) caused a decrease in the binding of [^{3}H]DTG. At 10 and 100 μM, inhibition was even greater than that for DIGIT. However, DIGIE was also able to significantly inhibit the binding of [^{3}H]TCP at a concentrations of 1 μM and above. Since in DIGIE the electrophilic group occurs in a spacing chain and not directly on the toluene ring as is DIGIT, the authors believe that the nucleophile could be found in different positions on the σ site. 1-(4-Methylfumaryl-amido-2-methylphenyl)-3-(2-methyl-phenyl)guanidine (DIGMF) and 1-(4-bromo-acetylamido-2-methylphenyl)-3-(2-methylphenyl) guanidine (DIGBA), which are not isothiocyanates, produce a non-selective inhibition of σ and PCP receptor binding which occurs only at high concentrations (100 μM).

Table 3-6. Wash resistant inhibition of PCP and σ receptor sites by analogs of DTG

Compound	[^{3}H]DTG			[^{3}H]TCP		
	1 μM	10 μM	100 μM	1 μM	10 μM	100 μM
DTG	104 ± 2	99 ± 0.4	72 ± 6	97 ± 3	116 ± 8	112 ± 6
DIGIT	62 ± 0.2	31 ± 4	14 ± 1	120 ± 4	101 ± 7	12 ± 7
DIGIE	69 ± 6	21 ± 1	6 ± 1	83 ± 2	76 ± 3	51 ± 6
DIGBA	126 ± 3	92 ± 1	42 ± 3	99 ± 12	104 ± 4	68 ± 5
DIGMF	109 ± 13	118 ± 3	66 ± 9	120± 1	104 ± 14	90 ± 13

[^{3}H]DTG labels σ receptors. [^{3}H]TCP labels PCP receptors. Values represent percent of binding compared to membranes not treated with the indicated compound. Adapted from ref. (31).

2.3 Irreversible ligands derived from analogues of 3-phenylpiperidine

Irreversible σ ligands have also been synthesized starting from 3-hydroxyphenyl-N-propyl piperidine (3-PPP) (37). Of the two isomers, R-(+)-3-PPP is the one that has a preference for the σ_1 receptor (σ_1 K_i = 5 nM, against [^{3}H](+)-pentazocine; σ_2 K_i = 442 nM, against [^{3}H]DTG in the presence of (+)-pentazocine). The σ_1/σ_2 selectivity of the (-)-isomer is less pronounced (34). SAR correlations (38) have indicated that the hydroxyl group present on the aromatic ring of 3-PPP is important only for dopamine agonist activity. Substituting the 3-hydroxyl group with a OCH$_3$ group in the R-isomer of 3-PPP produces a doubling of the σ potency. The presence of a hydrogen on the basic nitrogen is detrimental for σ affinity (IC$_{50}$ > 4,500), while substituents with a greater steric bulk improve the affinity associated with the R-(+)-enantiomer until an n-propyl group is present. The bulkier homologue, N-butyl-3-(3-hydroxyphenyl)piperidine, shows no significant difference in IC$_{50}$ between the two enantiomers (R-enantiomer, IC$_{50}$ = 8 ± 1.8 nM; S-enantiomer, IC$_{50}$ = 16 ± 3 nM).

These observations have guided the synthesis of the derivatives, R-(+)-3-(4-isothiocyanato-3-methoxyphenyl)-1-propyl piperidine (**5a**) and R-(+)-1-(2-isothiocyanatoethyl)-3-(3-methoxyphenyl)piperidine (**5b**) as possible irreversible ligands for the σ sites (Figure 3-5).

Both are able to inhibit the binding of [^{3}H](+)-3-PPP, however, notable differences of IC$_{50}$ have been found between the two compounds (40 nM vs. 12,000 nM, respectively). That an irreversible bond is formed has been demonstrated by the fact that, radioligand binding is not recoverable even after numerous washings. It should be noted that none of the compounds tested was found to be able to inhibit the binding of [^{3}H]TCP.

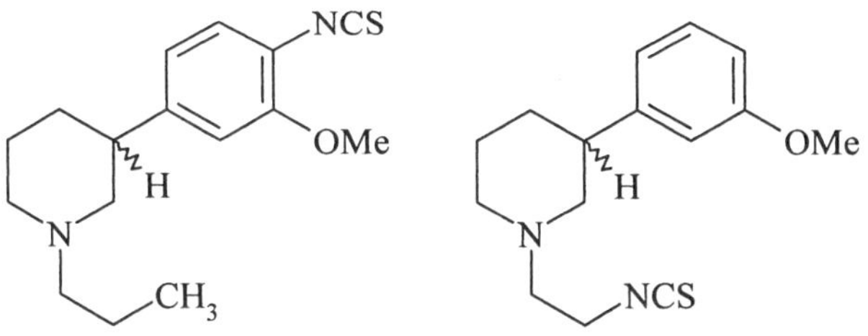

Figure 3-5. Structures of potential σ irreversible ligands

2.4 Irreversible ligands derived from benzomorphans

Though they have been useful in clarifying some aspects of the nature of σ receptors (e.g. a further distinction between σ receptors and PCP receptors, existence of more σ subtypes, etc.), the irreversible ligands described above have the disadvantage of either being designed from non-selective σ ligands or not being able to discriminate between σ_1 and σ_2 subtypes. The different order of sensitivity of the σ ligands in the studies with metaphit have shown the possibility that the derivatives with benzomorphan and non-benzomorphan structures can interact in a different way with the σ sites (28). As reported above, this different sensitivity is probably due to various factors. It is, however, important to show how many ligands with a benzomorphan structure display a high affinity for the σ_1 sites. Compounds such as (+)-pentazocine or (+)-SKF-10,047 are used in a tritiated form as selective σ_1 receptor probes. At the beginning of the 1990s, Carroll et al. synthesized a series of *N*-substituted derivatives of (+)- and (-)-cis-N-normetazocine with the aim of evaluating the affinity on three different receptor systems (σ_1, PCP, and μ opioid) (27). This study led to the synthesis of (+)-*cis*-N-benzyl-N-normetazocine, a compound with a particularly high affinity for σ_1 receptors (K_i = 0.67 nM against [^{3}H](+)-pentazocine) and with the peculiarity of having a σ_2/σ_1 ratio that is very high (σ_2 K_i = 1710 nM against [^{3}H]DTG) (39). Its dextro configuration makes it σ selective with respect to the μ opioid receptors (K_i > 1,500 nM), while the presence of a large group such as the benzyl moiety on the basic nitrogen atom eliminates the PCP receptor component (K_i > 10,000).

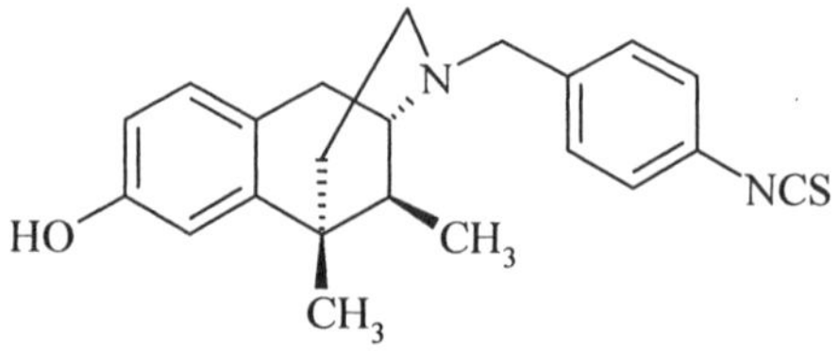

Figure 3-6. Structure of BNIT

SAR studies have further shown how it is possible to make some substitutions on the aromatic ring of the benzyl group with only a slight loss of σ_1 affinity. Relatively small substituents are able to maintain a nanomolar affinity for these receptors (40). These observations stimulated the synthesis of (+)-cis-N-(4-isothiocyanatobenzyl)-N-normetazocine (BNIT) as a covalent ligand for the σ_1 receptors (Figure 3-6) (41). Table 3-7 shows guinea pig brain membranes pre-treated with BNIT at a concentration of 0.1 µM causing a decrease in the B_{max} of [^{3}H](+)-pentazocine binding by about 40% with respect to the control. The effect is greater when the membranes are pre-incubated with BNIT at a concentration of 1 µM, where the number of sites labeled by [^{3}H](+)-pentazocine is reduced by about 96%. When membranes were pre-treated with 0.1 µM BNIT, the [^{3}H](+)-pentazocine binding affinity is decreased with respect to the control (K_d 5.2 ± 0.3 vs 34.5 ± 18.6 nM), while at a concentration of 1 µM BNIT the low number of remaining receptors did not allow the evaluation of K_d. A further increase of the concentration of BNIT (5 µM) did not allow the determination of either the B_{max} or K_d.

Table 3-7. Effect of BNIT on binding parameters of [^{3}H](+)-pentazocine to σ_1 sites in guinea pig brain membranes

Treatment	K_d (nM)	B_{max} (fmol/mg protein)
Control	5.2 ± 0.3	987 ± 78.5
BNIT 0.1 µM	34.5 ± 18.6	601 ± 139.8
BNIT 1 µM	n.d.	41.3 ± 11.3
BNIT 5 µM	n.d.	n.d.

n.d. = could not be determined. Reprinted from ref. (41) with permission.

3. Irreversible compounds

To further clarify the σ_1 selectivity of BNIT, Scatchard experiments were carried out with [^{3}H]DTG, a selective σ ligand that is not able to discriminate between σ_1 and σ_2 sites. The data listed in Table 3-8 show that BNIT at a concentration of 0.1 µM causes only an increase of K_{d1} and a reduction of the B_{max} (σ_1) of [^{3}H]DTG. In fact, even if an increase of K_{d2} was seen, the B_{max} (σ_2) remained almost the same. At higher concentrations, BNIT abolished the σ_1 binding component of [^{3}H]DTG, its capacity reduced by about 45%. It should be noted that this effect is not dependent on the BNIT concentration, the binding affinity (K_d values) do not vary significantly with respect to the control. Saturation experiments carried out with [^{3}H]DTG in the presence of (+)-pentazocine as a selective σ_1 blocker, have shown how the values of B_{max} are the same as those found when the membranes were pre-treated with BNIT (1 and 5 µM), as were the K_d values (41). BNIT, therefore, does not cause a loss of σ_2 sites labeled by [^{3}H]DTG. The binding of [^{3}H](+)-pentazocine is not recoverable, even after numerous washes of the membranes pre-treated with BNIT. It is thus possible to conclude that BNIT is able to selectively and irreversibly block only σ_1 sites. It should be pointed out, however, that both in the case of [^{3}H](+)-pentazocine and [^{3}H]DTG there was a decrease of the B_{max} and at the same time an increase in the K_d. This effect is not observed with the other irreversible isothiocyanate ligands discussed thus far. As with the other isothiocyanate ligands, it is necessary to carry out further studies in order to clarify whether this compound is able to acylate a nucleophile present on the σ_1 binding site or in an adjacent position. New studies are indispensable to clarify the complexity of these receptors and to find support for the various hypotheses formulated on the basis of the data obtained.

Table 3-8. K_i (nM) and B_{max} (fmol/mg protein) of [^{3}H]DTG binding to guinea pig brain membranes exposed or not exposed (control) to BNIT

Treatment	K_d	K_{d1} (σ_1)	K_{d2} (σ_2)	B_{max}	B_{max1} (σ_1)	B_{max2} (σ_2)
Control	13 ± 2	7 ± 1	16 ± 3	1804 ± 57	987 ± 78	1280 ± 189
BNIT (0.1 µM)		13 ± 2	21 ± 8		674 ± 36	1330 ± 210
BNIT (1 µM)	14 ± 0.3			1013 ± 148		
BNIT (5 µM)	12 ± 0.5			957 ± 59		
(+)-PENT (200 nM)	12 ± 3			1120 ± 98		

PENT = pentazocine. Reprinted from ref. (41) with permission.

3. PHOTOAFFINITY LABELS

Photoaffinity labels are a particular category of irreversible ligands. Unlike affinity labels, they have the characteristic of having a reactive group in a latent form that becomes an electrophile only when it is irradiated. It is able to form a covalent bond with a nucleophile present on the binding site or in a proximal site. Unlike affinity labels, the position of the photosensitive group and its minor reactivity only slightly condition the formation of the irreversible bond (42). Various examples document the positive applications of this technique, for example identification and isolation of the binding protein(s) of receptors. However, there are also some limitations: 1) photolabeling is a low efficiency process, 2) photolabeling is not stable, 3) UV-irradiation damages the receptor, and 4) above all, *in vivo* work cannot be performed.

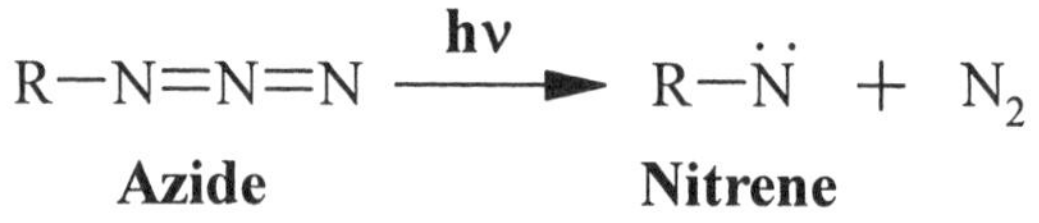

$$R-N=N=N \xrightarrow{h\nu} R-\ddot{N} + N_2$$

Azide　　　　　　**Nitrene**

Figure 3-7. Formation of reactive species from azide

In numerous photoaffinity ligands, an azide group ($-N_3$) is present as a photoactivable moiety. A property of this group is the formation, following bland doses of UV, of a nitrogen molecule and an extremely reactive species (Figure 3-7). The particular instability of the nitrenes causes a multitude of reactions. This can be an advantage in those cases where there is a shortage of functionally reactive groups at the binding site and, at the same time a disadvantage as only a few of these reactions lead to the formation of a covalent bond with the binding site. The nature of the R-substituent can condition some reactions and determine the efficiency of the photolabeling process, that is, the probability to convert a reversible ligand into an irreversible one. The aryl nitrenes ($Ar-N_3$) have the characteristic of being good absorbers of UV light and of not being very susceptible to rearrangement reactions, a particular reaction that does not result in photolabeling. The limited steric bulk of the $-N_3$ group also allows an easy positioning in the binding site without great changes to the affinity of the ligand. These qualities have allowed the aryl nitrenes to be widely used as photolabile groups. The low labeling efficacy is, however, a negative aspect if the receptors are present in a reduced concentration. In this case, the quantity of radioactivity incorporated into the receptor could be difficult to determine, especially in the case of compounds labeled with ^3H and ^{14}C. This problem is eliminated when ^{125}I is present (43).

4. PHOTOAFFINITY LABELS BASED ON VARIOUS σ LIGANDS

4.1 Azido-DTG

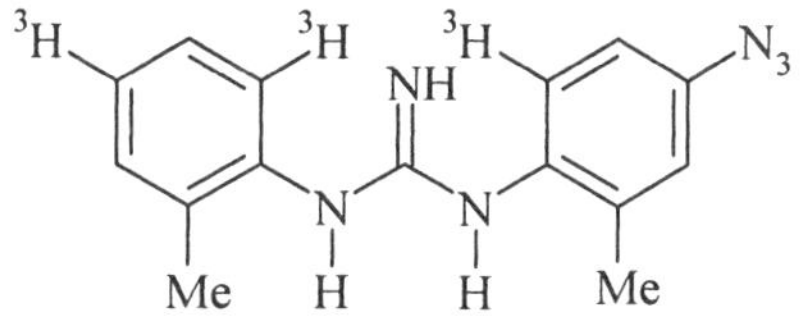

[³H]azido-DTG

Figure 3-8. Structure of [^{3}H]N_3DTG

Numerous studies have been carried out to characterize the molecular properties of the σ receptors by means of radiolabeled photoaffinity ligands containing an azide group. In 1988 Kavanaugh et al. reported the synthesis of 1-(4-azido-2-methyl[6-^{3}H]phenyl)-3-(2-methyl[4,6-^{3}H]phenyl)guanidine ([^{3}H]N_3DTG) (Figure 3-8) and the characterization of the binding subunit of the σ receptor (44). Binding studies show that [^{3}H]N_3DTG and [^{3}H]DTG bind to the same receptors. In competition studies, the rank order of potency (K_i values) for numerous σ ligands against the two radioligands exhibited a correlation factor of $r = 0.97$. Consequently, in the dark (that is, under reversible conditions), [^{3}H]N_3DTG is a highly specific radioligand that irreversibly labels σ receptor binding protein(s). When guinea pig brain membranes were irradiated in the presence of [^{3}H]N_3DTG, a protein with a molecular mass of 29 kDa was covalently labeled. Labeling of this protein was completely inhibited by several putative σ ligands at a concentration of 10 μM, but was unaffected by morphine, dopamine, serotonin, scopolamine, and γ-aminobutyric acid.

When the photoaffinity labeled guinea pig brain membrane proteins were solubilized with 20 mM sodium cholate/Tris and chromatographed on a

Sepharose CL-6B column, specific labeling of a Mr 150,000 complex was observed (44). As treatment with sodium dodecyl sulfate (NaDodSO$_4$) causes the separation of a 29 kDa σ subunit(s), it is possible to conclude that this polypeptide is not covalently associated with the Mr 150,000 protein complex. Subsequent studies confirmed that [^{3}H]DTG labeled the 29 kDa binding subunit and not the complex. The cause of this could be the dissociation of some components induced by the photoaffinity labeling technique or to the protocol used for solubilization of the native complex (45).

In studies carried out by Hellewell and Bowen (46), [^{3}H]N$_3$DTG photolabeled a polypeptide of Mr 25,000 in guinea pig brain membranes. The differences in Mr that were found between this study and that of Kavanaugh et al. (44) in the two polypeptides was justified by the different experimental conditions. That this Mr 25,000 protein is a σ polypeptide is shown by the fact that at a concentration of 1 µM, haloperidol completely blocked the labeling of this band. In PC12 cells, in contrast to guinea pig brain, two polypeptides of Mr 18,000 and Mr 21,000 were photolabeled. Both were blocked by the presence of haloperidol. Studies in the presence of enzymatic inhibitors demonstrate that the two peptides are not derived from one large protein complex.

Based on these photolabeling studies in conjunction with ligand binding data, Bowen and co-workers proposed σ$_1$ and σ$_2$ receptor subtypes, which differed in molecular size (46). However, the differences found between the guinea pig brain membranes and PC12 cells derived from rats could also, in the opinion of the authors, represent two molecular forms of σ receptors originating from species differences. Subsequent studies carried out in rat hepatic membranes showed that [^{3}H]N$_3$DTG labeled two polypeptides (Mr 25,000 and 21,500) (47). The labeling of both polypeptides was completely blocked in the presence of haloperidol, while dextrallorphan (100 nM or 500 nM) only blocked the Mr 25,000 polypeptide but not the Mr 21,500 polypeptide. Given that dextrallorphan has a clear preference for the σ$_1$ receptor with respect to σ$_2$ (K$_i$ = 16.1 nM and > 5,000 nM respectively), Mr 25,000 could be the σ$_1$ site while Mr 21,500 could be the σ$_2$ site. That the Mr 21,500 polypeptide corresponds to the σ$_2$ site is deducible from the binding data of various σ ligands.

In experiments carried out by Schuster et al. in rat and bovine cerebellar homogenates, [^{3}H]N$_3$DTG labeled two proteins, with Mr 63,000 and 65,000 (48). These probably represent two forms of the same receptor which are different only in the extents of glycosylation. Radioligand binding assays performed on the purified Mr 63,000 and 65,000 protein(s) after reconstitution into lipid vesicles demonstrated that this material exhibits characteristic σ receptor pharmacology. The photolabeling with [^{3}H]N$_3$DTG

3. Irreversible compounds

was blocked by haloperidol, DTG, and SKF 10,047 (10 µM). It is therefore possible that the protein complex of Mr 150,000 gives rise to the protein(s) of Mr 63,000 and 65,000 and that this can lead to the formation of the Mr 29,000 subunit. The preference of the Mr 63,000 and 65,000 proteins for the (+)-enantiomers of benzomorphan derivatives leads the authors to conclude that in these tissues, there could be a preponderant presence of the σ_1 receptor subtype.

4.2 Azido-Phenazocine

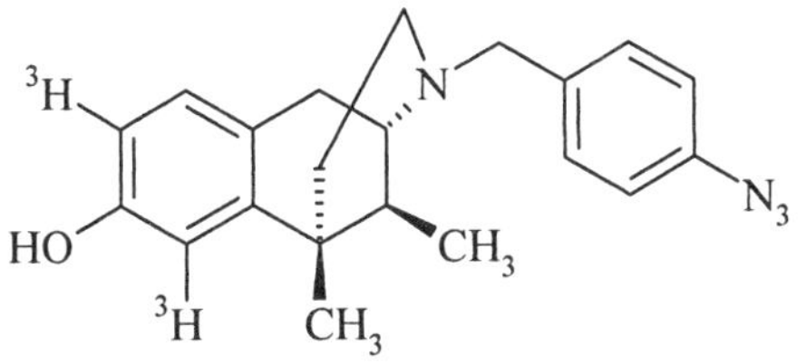

[^{3}H](+)-Azidophenazocine

Figure 3-9. Structure of [^{3}H](+)-azidophenazocine

Numerous studies have demonstrated the enantiospecificity of the (+)-benzomorphans for the σ_1 site, whereas the (-)-enantiomers sometimes show some preferential affinity for the σ_2 sites, but do not strongly discriminate the two subtypes (49). The photoactive benzomorphan derivative, [^{3}H](+)-azidophenazocine (Figure 3-9), was synthesized in an attempt to obtain a σ_1-selective probe (50). (+)-Phenazocine, structurally related to (+)-pentazocine, is a potent σ_1 selective ligand, its affinity for the σ_2 receptor being notably lower (σ_1 K_i = 3.9 nM; σ_2 K_i = 1269 nM) (51). The addition of an azide group in the para-position on the aromatic ring to give (+)-azidophenazocine, does not substantially change the affinity for the σ sites. In the dark, (+)-azidophenazocine labeled a σ_1 site in guinea pig brain membranes with high affinity (σ_1 K_i = 1.34 ± 0.21 nM, vs. [^{3}H](+)-pentazocine) (51).

(+)-Azidophenazocine is able to compete for the same receptors on splenic lymphocytes labeled with [^{3}H](+)-pentazocine (K_i = 37 ± 17 nM) (52). When [^{3}H](+)-azidophenazocine was incubated in the presence of competitive ligands (5 µM) under reversible conditions, only the σ ligands are able to block the binding to lymphocyte binding sites (52). If irradiated,

(+)-azidophenazocine labeled four proteins in lymphocyte membranes: Mr 57,000, Mr 33,000, Mr 27,000, and Mr 22,000. Experiments carried out in the presence of an excess (10 µM) of (+)- or (–)-pentazocine showed that only Mr 57,000 and Mr 22,000 were found to be stereoselective for the (+)-benzomorphas. In order to examine whether Mr 22,000 originates from the proteolysis of Mr 57,000, the experiments were repeated with 0.1 mM phenylmethanesulfonyl-fluoride (PMSF) in the lysis buffer before electrophoresis. Under both reducing and non-reducing electrophoresis conditions, proteins of Mr 200,000 and Mr 57,000 were photoaffinity labeled. Thus, it appears that the Mr 22,000 polypeptide derives from proteolysis of a larger protein. Interestingly, whereas labeling of both the Mr 200,00 and Mr 57,000 proteins is blocked by σ ligands, the Mr 200,000 polypeptide is also blocked by naloxone. Therefore it is possible to state that Mr 57,000 represents a σ receptor protein, whereas Mr 200,000 represents a site with novel pharmacology. The naloxone-sensitivity could be due to the high affinity of (+)-phenazocine for the µ opioid receptors (K_i = 55 ± 8 nM against [^{3}H]DAMGO) (27).

4.3 Iodo-Azidococaine

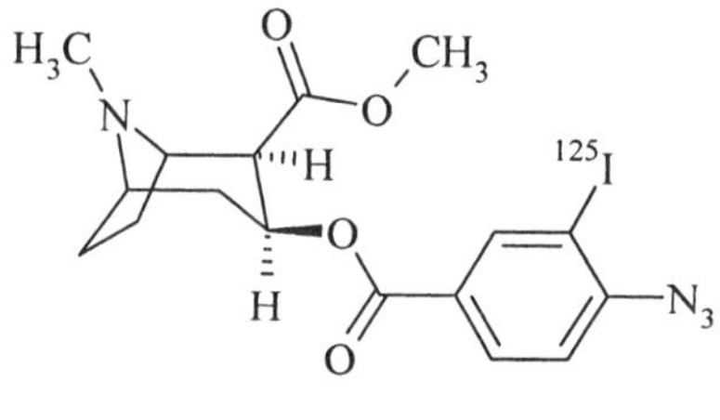

[^{125}I]Azidococaine

Figure 3-10. Structure of (-)-3-iodo-4-azidococaine

The modest affinity of (–)-cocaine for the σ receptors (K_i = 6.7 ± 0.3 µM vs. [^{3}H]haloperidol), along with a poor knowledge of the effects produced by this interaction (53) led to the synthesis of (–)-3-iodo-4-azidococaine (54) (Figure 3-10). The addition of an iodine atom and an azide group to the aromatic ring of cocaine results in a significant increase in the affinity for σ receptors (54). This compound was radiolabeled with ^{125}I to produce the photoaffinity probe [^{125}I]N$_3$-cocaine. This probe photolabeled a Mr 26,000 protein in different tissues (rat brain and liver and in human placenta), which does not seem to derive from other proteins with a higher molecular weight (54). Several σ ligands, including haloperidol, (+)- and (-)-3-PPP, DTG, dextromethorphan, and carbetapentane, inhibit its labeling, a profile consistent with labeling of σ_1 sites.

Investigations of DMS-114 cell lines confirm that $[^{125}I]N_3$-cocaine labels a protein with an apparent molecular weight of Mr 26,000 (55). A selective, concentration-dependent inhibition was observed by SKF 10,047, with a half-maximal effect at 7 μM. The observation that this photoprobe also produces an inhibition of potassium current (I_k) by greater than 90 % at 1 μM indicates that $[^{125}I]N_3$-cocaine is a potent σ receptor agonist (55).

5. CONCLUSION

Purification and cloning of the σ_1 receptor (initially from guinea pig liver) showed that the receptor is a polypeptide of 25.3 kDa (56). Thus, proteins photolabeled by $[^3H]N_3DTG$, $[^3H](+)$-azidophenazocine, and $[^{125}I]N_3$-cocaine that are in the 25–29 kDa range, as described in the various studies above, comprise the σ_1 receptor. The relationship to σ_1 receptors of observed photolabeled polypeptides in other size ranges is not yet completely clear. σ_2 Receptors have not yet been cloned. However, using photoaffinity labeling and ligand binding techniques, σ_2 receptors are proposed to comprise a polypeptide(s) of 18–21.5 kDa, depending on the tissue. Photoaffinity ligands will be useful tools in further molecular studies of σ receptor subtypes and related proteins.

Not having the limitation of needing to be exposed to light for activation, affinity agents which irreversibly bind to σ receptors via alkylation or acylation will be important tools for delineating the functions of σ receptors *in vivo*. These affinity agents could perform as long-acting agonists, long-acting antagonists, or as agents to "knock down" receptor levels in cells, tissues, and organisms. The further development of high affinity, σ subtype-specific affinity agents is therefore warranted.

REFERENCES

1. Baker BR. Design of Active-Site-Directed Irreversible Enzyme Inhibitors. New York: John Wiley and Sons, 1976.
2. Portoghese PS, Takemori AE. Affinity labels as probes for opioid receptor types and subtypes. NIDA Res Monogr 1986, 69:157-68.

3. Monaghan DT, Bridges RJ, Cotman CW. The excitatory amino acid receptors: their classes, pharmacology, and distinct properties in the function of the central nervous system. Ann Rev Pharmacol Toxicol 1989, 29:365-402.
4. Wroblewski JT, Danysz W. Modulation of glutamate receptors: molecular mechanisms and functional implications. Ann Rev Pharmacol Toxicol 1989, 29:441-474.
5. Rafferty MF, Mattson M, Jacobson AE, Rice KC. A specific acylating agent for the [^{3}H]phencyclidine receptors in rat brain. FEBS Lett 1985, 181:318-322.
6. Rice KC, Rafferty MF, Jacobson AE, Contreras P, O'Donohue TL, Lessor RA, Mattson MV. Metaphit, a specific acylating agent for the [^{3}H]-phencyclidine receptors. 1986, U.S. Patent 4,598,153, 8 pp.
7. Rice KC, Rafferty MF, Jacobson AE, Contreras PC, O'Donohue TL. Metaphit and related compounds as acylating agents for the [^{3}H]phencyclidine receptors. 1988, U.S. Patent 4,762,846, 29 pp.
8. Contreras PC, Rafferty MF, Lessor RA, Rice KC, Jacobson AE, O'Donohue TL. A specific alkylating ligand for phencyclidine (PCP) receptors antagonizes PCP behavioral effects. Eur J Pharmacol 1985, 111:405-406.
9. French ED, Jacobson AE, Rice KC. Metaphit, a proposed phencyclidine (PCP) antagonist, prevents PCP-induced locomotor behavior through mechanisms unrelated to specific blockade of PCP receptors. Eur J Pharmacol 1987, 140:267-274.
10. Koek W, Woods JH, Jacobson AE, Rice KC, Lessor RA. Metaphit, a proposed phencyclidine receptor acylator: phencyclidine-like behavioral effects and evidence of absence of antagonist activity in pigeons and in rhesus monkeys. J Pharmacol Exp Ther 1986, 237:386-392.
11. Koek W, Head R, Holsztynska EJ, Woods JH, Domino EF, Jacobson AE, Rafferty MF, Rice KC, Lessor RA. Effects of metaphit, a proposed phencyclidine receptor acylator, on catalepsy in pigeons. J Pharmacol Exp Ther 1985, 234:648-653.
12. Koek W, Woods JH, Rice KC, Jacobson AE, Huguenin PN, Burke TR Jr. Phencyclidine-induced catalepsy in pigeons: specificity and stereoselectivity. Eur J Pharmacol 1984, 106:635-638.
13. Domino EF, Gole D, Koek W. Metaphit, a proposed phencyclidine receptor acylator: disruption of mouse motor behavior and absence of PCP antagonist activity. J Pharmacol Exp Ther 1987, 243:95-100.
14. Contreras PC, Johnson S, Freedman R, Hoffer B, Olsen K, Rafferty MF, Lessor RA, Rice KC, Jacobson AE, O'Donohue TL. Metaphit, an acylating ligand for phencyclidine receptors: characterization of in vivo actions in the rat. J Pharmacol Exp Ther 1986, 238:1101-1107.
15. Davies SN, Church J, Blake J, Lodge D, Lessor RA, Rice KC, Jacobson AE. Is Metaphit a phencyclidine antagonist? Studies with ketamine, phencyclidine and N-methylaspartate. Life Sci 1986, 38:2441-2445.
16. Schweri MM, Jacobson AE, Lessor RA, Rice KC. Metaphit irreversibly inhibits [^{3}H]threo-(+/−)-methylphenidate binding to rat striatal tissue. J Neurochem 1987, 48:102-105.
17. Schweri MM, Jacobson AE, Lessor RA, Rice KC. Metaphit inhibits dopamine transport and binding of [^{3}H]methylphenidate, a proposed marker for the dopamine transport complex. Life Sci 1989, 45:1689-1698.
18. Berger P, Jacobsen AE, Rice KC, Lessor RA, Reith MA. Metaphit, a receptor acylator, inactivates cocaine binding sites in striatum and antagonizes cocaine-induced locomotor stimulation in rodents. Neuropharmacology 1986, 25:931-933.

19. Nabeshima T, Tohyama K, Noda A, Maeda Y, Hiramatsu M, Harrer SM, Kameyama T, Furukawa H, Jacobson AE, Rice KC. Effects of metaphit on phencyclidine and serotonin$_2$ receptors. Neurosci Lett 1989, 102:303-308.
20. Haertzen CA. Subjective effects of narcotic antagonists cyclazocine and nalorphine on the Addiction Research Center Inventory (ARCI). Psychopharmacologia 1970, 18:366-377.
21. Allen RM, Young SJ. Phencyclidine-induced psychosis. Am J Psychiatry 1978, 135:1081-1084.
22. Sonders MS, Keana JF, Weber E. Phencyclidine and psychotomimetic sigma opiates: recent insights into their biochemical and physiological sites of action. Trends Neurosci 1988, 11:37-40.
23. Teal JJ, Holtzman SG. Stereoselectivity of the stimulus effects of morphine and cyclazocine in the squirrel monkey. J Pharmacol Exp Ther 1980, 215:369-376.
24. Holtzman SG. Phencyclidine-like discriminative effects of opioids in the rat. J Pharmacol Exp Ther 1980, 214:614-619.
25. Sircar R, Nichtenhauser R, Ieni JR, Zukin SR. Characterization and autoradiographic visualization of (+)-[^{3}H]SKF10,047 binding in rat and mouse brain: further evidence for phencyclidine/"sigma opiate" receptor commonality. J Pharmacol Exp Ther 1986, 237:681-688.
26. Gundlach AL, Largent BL, Snyder SH. Phencyclidine and sigma opiate receptors in brain: biochemical and autoradiographical differentiation. Eur J Pharmacol 1985, 113:465-466.
27. Carroll FI, Abraham P, Parham K, Bai X, Zhang X, Brine GA, Mascarella S.W, Martin BR, May EL, Sauss C, Di Paolo L, Wallace P, Walker JM, Bowen WD. Enantiomeric N-substituted N-normetazocines: a comparative study of affinities at Sigma, PCP, and mu opioid receptors. J Med Chem 1992, 35:2812-2818.
28. Bluth LS, Rice KC, Jacobson AE, Bowen WD. Acylation of sigma receptors by Metaphit, an isothiocyanate derivative of phencyclidine. Eur J Pharmacol 1989, 161:273-277.
29. Bowen WD, Hellewell SB, McGarry KA. Evidence for a multi-site model of the rat brain sigma receptor. Eur J Pharmacol 1989, 163:309-318.
30. Adams JT, Teal PM, Sonders MS, Tester B, Esherick JS, Scherz MW, Keana JF,Weber E. Synthesis and characterization of an affinity label for brain receptors to psychotomimetic benzomorphans: differentiation of sigma-type and phencyclidine receptors. Eur J Pharmacol 1987, 142:61-71.
31. Reid AA, Kim CH, Thurkauf A, Monn JA, de Costa B, Jacobson AE, Rice KC, Bowen WD, Rothman RB. Wash-resistant inhibition of phencyclidine- and haloperidol-sensitive sigma receptor sites in guinea pig brain by putative affinity ligands: determination of selectivity. Neuropharmacology 1990, 29:1047-1053.
32. Zhang H, Cuevas J. Sigma receptors inhibit high-voltage-activated calcium channels in rat sympathetic and parasympathetic neurons. J Neurophysiol 2002, 87:2867-2879.
33. Weber E, Sonders M, Quarum M, McLean S, Pou S, Keana JF. 1,3-Di(2-[5-^{3}H]tolyl)guanidine: a selective ligand that labels sigma-type receptors for psychotomimetic opiates and antipsychotic drugs. Proc Natl Acad Sci USA. 1986, 83:8784-8788.
34. Rothman RB, Reid A, Mahboubi A, Kim CH, De Costa BR, Jacobson AE, Rice KC. Labeling by [^{3}H]1,3-di(2-tolyl)guanidine of two high affinity binding sites in guinea pig brain:evidence for allosteric regulation by calcium channel antagonists and pseudoallosteric modulation by sigma ligands. Mol Pharmacol 1991, 39:222-232

35. Campbell BG, Scherz MW, Keana JFW, Weber E. Sigma receptors regulate contractions of the guinea pig ileum longitudinal muscle/myenteric plexus preparation elicited by both electrical stimulation and exogenous serotonin. J Neurosc 1989, 9:3380-3391.
36. Scherz MW, Fialeix M, Fischer JB, Reddy NL, Server AC, Sonders MS, Tester BC, Weber E, Wong ST, Keana JFW. Synthesis and structure-activity relationships of N,N'-di-o-tolylguanidine analogues, high-affinity ligands for the haloperidol-sensitive sigma receptor. J Med Chem 1990, 33:2421-2429.
37. Grayson NA, Bowen WD, Rice KC. Chiral potentially irreversible ligands for the sigma receptor based on the structure of 3-(3-hydroxyphenyl)-N-propylpiperidine (3-PPP). Heterocycles 1992, 34:2281-2292.
38. Largent BL, Wikstrom H, Gundlach AL, Snyder S. Structural determinants of sigma receptor affinity. Mol Pharmacol 1987, 32:772-841.
39. Danso-Danquah R, Bai X, Zhang X, Mascarella SW, Williams W, Sine B, Bowen WD, Carroll FI. Synthesis and sigma binding properties of 1'- and 3'-halo- and 1',3'-dihalo-N-normetazocine analogues. J Med Chem 1995, 38:2986-2989.
40. Mascarella SW, Bai X, Williams W, Sine B, Bowen WD, Carroll FI. (+)-cis-N-(para-, meta-, and ortho-substituted benzyl)-N-normetazocines: synthesis and binding affinity at the [^{3}H]-(+)-pentazocine-labeled (sigma-1) site and quantitative structure-affinity relationship studies. J Med Chem 1995, 38:565-569.
41. Ronsisvalle G, Prezzavento O, Marrazzo A, Vittorio F, Massimino M, Murari G, Spampinato S. Synthesis of (+)-cis-N-(4-isothiocyanatobenzyl)-N-normetazocine, an isothiocyanate derivative of N-benzylnormetazocine as acylant agent for the sigma$_1$ receptor. J Med Chem 2002, 45:2662-2665.
42. Fedan JS, Hogaboom GK, O'Donnell JP. Photoaffinity labels as pharmacological tools. Biochem Pharmacol 1984, 33:1167-1180.
43. Cavalla D, Neff NH. Chemical mechanisms for photoaffinity labeling of receptors. Biochem Pharmacol 1985, 34:2821-2826.
44. Kavanaugh MP, Tester BC, Scherz MW, Keana JFW, Weber E. Identification of the binding subunit of the sigma-type opiate receptor by photoaffinity labeling with 1-(4-azido-2-methyl[6-^{3}H]phenyl)-3-(2-methyl[4,6-^{3}H]phenyl)guanidine. Proc Natl Acad Sci USA 1988, 85:2844-2848.
45. Kavanaugh MP, Parker J, Bobker DH, Keana JFW, Weber E. Solubilization and characterization of sigma-receptors from guinea pig brain membranes. J Neurochem 1989, 53:1575-1580.
46. Hellewell SB, Bowen WD. A sigma-like binding site in rat pheochromocytoma (PC12) cells: decreased affinity for (+)-benzomorphans and lower molecular weight suggest a different sigma receptor form from that of guinea pig brain. Brain Res 1990, 527:244-253.
47. Hellewell SB, Bruce A, Feinstein G, Orringer J, Williams W, Bowen WD. Rat liver and kidney contain high densities of sigma-1 and sigma-2 receptors: characterization by ligand binding and photoaffinity labeling. Eur J Pharmacol 1994, 268:9-18.
48. Schuster DI, Arnold FJ, Murphy RB. Purification, pharmacological characterization and photoaffinity labeling of sigma receptors from rat and bovine brain. Brain Res 1995, 670:14-28.
49. Di Paolo L, Carroll FI, Abraham P, Bai X, Parham K, Mascarella SW, Zhang X, Wallace P, Walker JM, Bowen WD. N-substituted derivatives of normetazocine: differentiation of sigma-1 and sigma-2 receptors. Soc Neurosci Abstr 1991, 17:814, 322.15.
50. De Costa BR, Bowen WD. Synthesis and characterization of optically pure [^{3}H](+)-azidophenazocine ([^{3}H](+)-AZPH), a novel photoaffinity label for sigma receptors. J Labelled Comp and Radiopharm 1991, 29:443-453.

51. Williams WE, Wu R, De Costa BR, Bowen WD. [^{3}H](+)-Azidophenazocine: characterization as a selective photoaffinity probe for sigma-1 receptors. Soc Neurosci Abstr 1993, 19:1553, 638.17.
52. Garza HH Jr, Mayo S, Bowen WD, DeCosta BR, Carr DJJ. Characterization of a (+)-azidophenazocine- sensitive sigma receptor on splenic lymphocytes. J Immunol 1993, 151:4672-4680.
53. Sharkey J, Glen KA, Wolfe S, Kuhar MJ. Cocaine binding at sigma receptors. Eur J Pharmacol 1988, 149:171-174.
54. Kahoun JR, Ruoho AE.(^{125}I)iodoazidococaine, a photoaffinity label for the haloperidol-sensitive sigma receptor. Proc Natl Acad Sci USA 1992, 89:1393-1397.
55. Wilke RA, Mehta RP, Lupardus PJ, Chen Y, Ruoho AE, Jackson MB. Sigma receptor photolabeling and sigma receptor-mediated modulation of potassium channels in tumor cells. J Biol Chem 1999, 274:18387-18392.
56. Hanner M, Moebius FF, Flandorfer A, Knaus H-G, Striessnig J, Kempner E, Glossmann H. Purification, molecular cloning, and expression of the mammalian sigma$_1$-binding site. Proc Natl Acad Sci USA 1996, 93:8072-8077.

Corresponding author: *Dr. Giuseppe Ronsisvalle, Mailing address: University of Catania, Department of Pharmaceutical Sciences, Viale A. Doria 6, 95125 Catania, Italy, Phone/Fax: 39-095-336722, Electronic mail address: ggrmedch@unict.it*

Chapter 4

PHARMACOPHORE MODELS FOR σ_1 RECEPTOR BINDING

Seth Y. Ablordeppey[1] and Richard A. Glennon[2]
[1]*College of Pharmacy & Pharmaceutical Sciences, Florida A&M University, Tallahassee, Florida, FL 32307, USA* and [2]*Department of Medicinal Chemistry, School of Pharmacy, Virginia Commonwealth University, Richmond, VA 23298-0540, USA*

1. INTRODUCTION

For a drug to bind to its receptor in a stereospecific manner, functional groups on the ligand must interact with complementary functional groups at the binding site of the receptor. The set of functional groups common to active molecules that bind to the same receptor binding site are referred to as the *pharmacophoric groups* or *pharmacophore elements*. Pharmacophoric groups need not be identical from one ligand to another but they must have similar physicochemical properties in order to interact appropriately with the same functional groups at the receptor site. A 3-dimensional geometric representation of the pharmacophoric groups of the ligand in space constitutes the pharmacophore model. As a result of the complementary nature of functional groups involved in receptor-ligand interactions, a pharmacophore model defines both ligands and the receptor binding site or domain.

The σ receptors represent an interesting group of receptors because of their extensive accommodation of various structural groups (1-3). Thus, they have been shown to bind a variety of drug classes including, for example, antipsychotic agents (e.g. butyrophenones, phenothiazines, thioxanthenes) anxiolytics, tricyclic antidepressants, monoamine oxidase inhibitors, antineoplastic agents, anticholinergics, inhibitors of cytochrome oxidase, and steroids. Because of the extensive structural diversity among σ ligands (see Chart 1 for some examples), one assumption is that there is some level of fluidity at the active sites of σ receptors (i.e. the receptors are dynamic structures and may be sufficiently flexible to accommodate a variety of seemingly diverse structure types). Alternatively, the diverse agents might share common pharmacophore elements that are simply not obvious upon cursory inspection. In addition, we have suggested that σ receptor ligands with symmetry elements may bind in reverse modes at the receptor sites (4). This being the case, a single, traditional, quantitative geometric pharmacophore model that precisely defines a specific 3-dimensional space for pharmacophoric groups may be difficult to identify or, in fact, may be non-existent; that is, multiple pharmacophores might be possible. Nevertheless, a number of 2-dimensional pictorial pharmacophore models have been proposed for σ receptor sites.

The medicinal chemistry of σ ligands began in the late 1970s. A search of the literature reveals that σ receptor research peaked in about 1993 and has held fairly constant since then (Figure 4-1). Of course the data are somewhat biased in that they do not reflect abstracts from scientific meetings nor articles published in journals not indexed by MEDLINE. Nevertheless, the trend shows a fairly level degree of activity during the 1990s and only in the past few years has there been a resurgence of interest in σ ligands. Numerous σ receptor ligands were reported during the 1980s to the mid 1990s and their appearance in the literature very likely parallels the overall number of papers that were being published during that time. Some of the first agents displayed affinities that now are realized to be quite modest. However, most of the effort was directed toward identification of novel ligands and structure-types rather that on formulation of pharmacophore models. This chapter is not intended to be a review of all σ ligands that have been reported; for a review of σ ligands and general structure-affinity relationships, see de Costa et al. (3) and Abou-Gharbia et al. (1). Rather, we take this opportunity to review pharmacophore development for σ binding. Emphasis is placed on work from our laboratories because our efforts have been devoted, from the very outset, to the identification of pharmacophore models (e.g. 5). Furthermore, it was felt that it might be instructive to view the logical progression of studies that began with σ receptor ligands possessing micromolar affinity and led to ligands with subnanomolar affinity. The studies also attempt to show how the binding of members of various structural classes are related to one another.

4. Pharmacophore models

Chart 1. A cross-section of potent σ_1 ligands representing the vast array of structural diversity accommodated by σ receptors. All binding constants were obtained using [^{3}H](+)-pentazocine except for BD1008 for which [^{3}H](+)-3-PPP was used as radioligand.

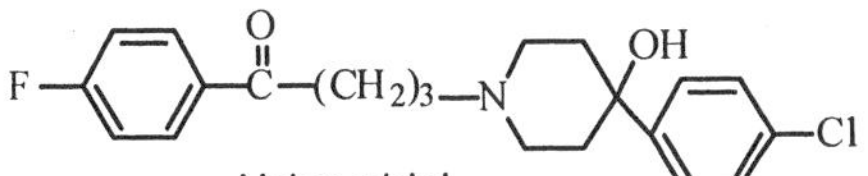

Haloperidol
Ki σ-1 = 0.65 nM

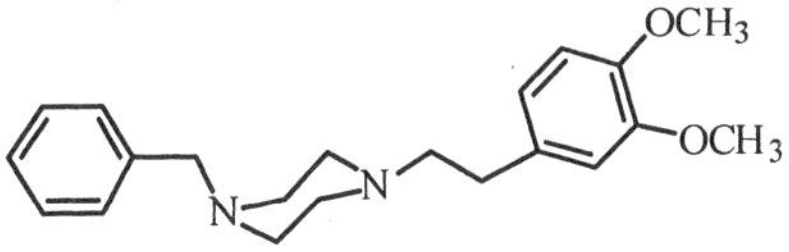

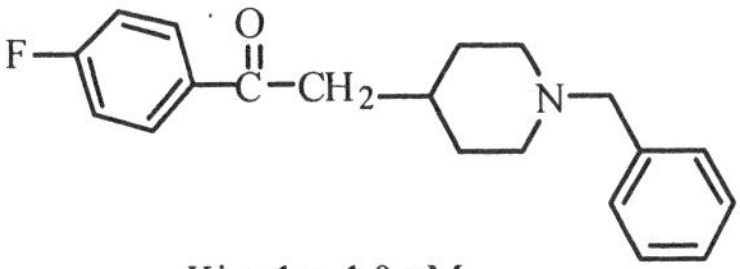

Ki σ-1 = 1.0 nM

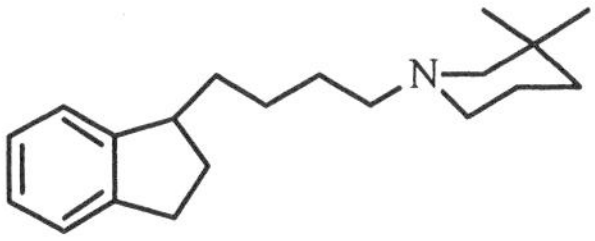

Phenylalkyl piperidine
Ki σ-1 – 0.86 μM

Phenylalkyl piperidine
Ki σ-1 = 0.07 nM

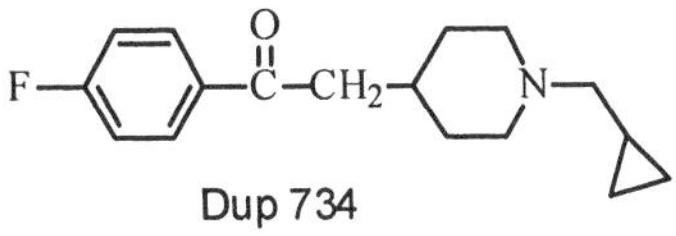

(+)-Pentazocine
Ki σ-1 = 2.0 nM

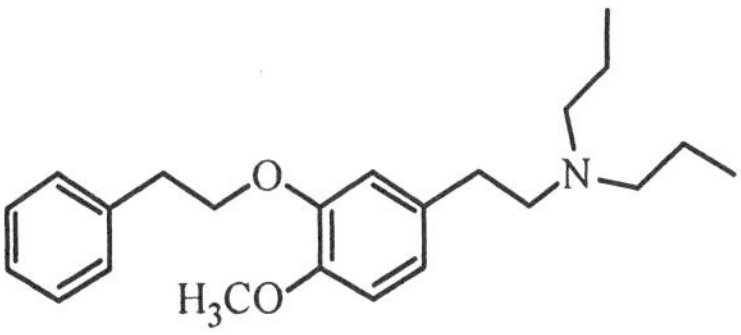

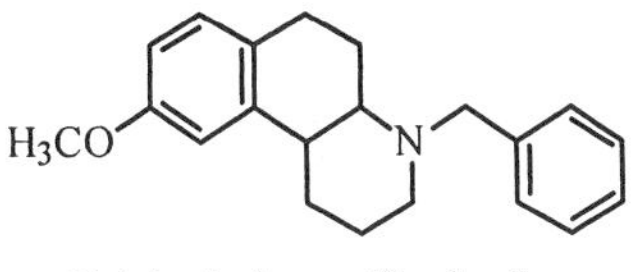

Octahydrobenzo[f]quinoline
Ki σ-1 = 5.0 nM

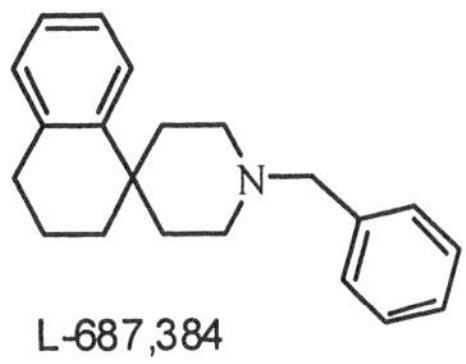

Phenylalkyl piperazine
SA4503   Ki σ-1 = 0.012 nM

Phenylalkyl piperidine
Ki σ-1 = 1.75 nM

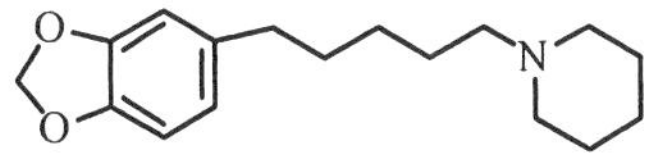

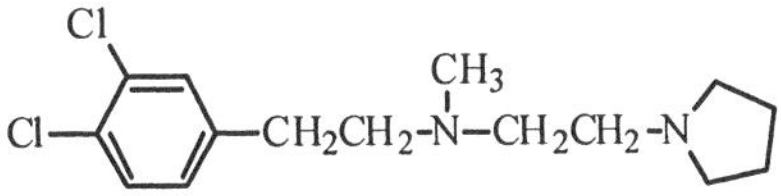

BD 1008
Ki σ-1 = 0.34 nM/@3-PPP

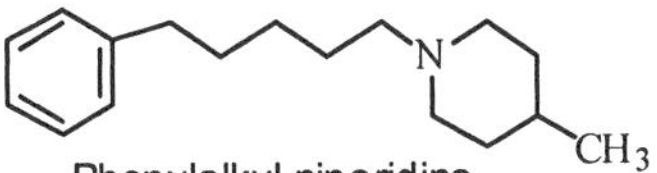

Dup 734
Ki σ-1 = 2.6 nM

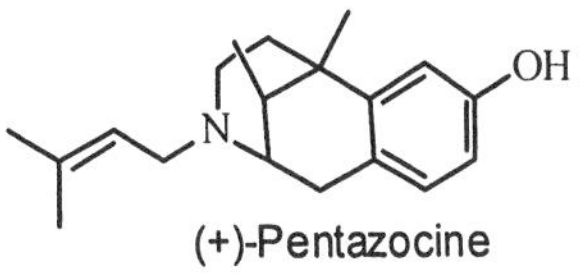

NE 100
Ki σ-1 = 1.03 nM

L-687,384
Ki σ-1 = 0.26 nM

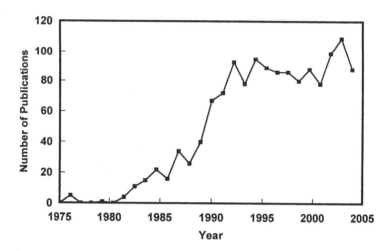

Figure 4-1. Number of publications per year on σ receptor research up to the year 2003. A MEDLINE search was conducted using the term "σ receptors."

Until relatively recently, there were at least two problems that historically hampered the development of pharmacophore models for σ receptors. First, σ ligands were assumed to bind to a single homogeneous receptor population until about 1992 when Bowen and coworkers (6) and others (7) reported the identification of an additional binding site for the σ receptors (i.e. σ_1 and σ_2 sites). Thus, ligand binding constants obtained prior to this time generally consisted of composite binding results from at least two different sites. Indeed, the term "overall binding affinity constant" was coined to distinguish such binding constants from the new binding constants at the σ_1 and σ_2 receptor binding sites. The second problem was related to the structural diversity issue. Since a large number of structural classes of compounds were found to bind at σ receptors with high affinity, any attempt to develop a model for such a diverse group of active ligands led to models that could explain one class but perhaps not the others. In addition, the fact that many of the early σ ligands displayed low (i.e. micromolar) affinity for σ receptors confounded the timely development of pharmacophore models. Thus, in the last decade several laboratories have sought to find a universal pharmacophore model to explain the diversity of binding ligands. Although it is still not possible to describe all known σ ligands by a common pharmacophore model, the present report shows how many, particularly the high-affinity, σ ligands might bind relative to one another.

2. MODELS PRIOR TO THE IDENTIFICATION OF σ_1 AND σ_2 RECEPTORS

2.1 Largent model

In their work on identifying structural determinants of σ receptor affinity, Largent and colleagues (8) found that there was a phenyl ring and an amine nitrogen atom present in most high affinity σ ligands. They subsequently examined the distances separating the phenyl rings and amine nitrogen atoms and concluded that there was a wide range of tolerance for the distances and that this could explain the considerable diversity among σ ligands with high affinity. In addition, they observed that the common pharmacophore associated with high affinity receptor binding is a phenylpiperidine moiety. By substituting the piperidine nitrogen atom with lipophilic substituents, the authors noted that affinity for σ receptors increased substantially, indicating that there is a region of high affinity binding in this area.

2.2 Manallack model

Another early model was described by Manallack et al. (9) and depicts two hypothetical receptor points (R1 and R2) 3.5Å above and below the centroid of two aromatic rings representing hydrophobic binding to the receptor. A third (N-atom position) and fourth (R3) pharmacophore element were placed at a tetrahedral distance of 2.8Å from the N atom in the molecule and represent sites that can hydrogen bond to appropriate functional groups on the receptor. The coordinates of each of the pharmacophore elements was given as R1 (0.00, 3.50, 0.00); R2 (0.00, -3.50, 0.00); R3 (6.09, 2.09, 0.00) and N (4.9, -0.12, 1.25). The distance from the centroid of the phenyl ring (origin) to the N atom was estimated to be 5.06Å and angles R1-O-N and O-N-R3 were 91.3° and 106.4° respectively. A dihedral angle of R1-O-N-R3 was found to be 34.1°. These calculations were based on the superimposition of predominantly σ_1 ligands including (+)-SKF-10,047 (N-allylnometazocine), haloperidol, DTG (di-o-tolyl-guanidine), and R(+)-3-PPP onto *trans*-4aR,10bR-9-hydroxy-N-n-propyl-octahydrobenzo[f]quinoline (*trans*-4aR,10bR-9-OH-OHBQ) as templates.

2.3 CoMFA Model

An early form of a pharmacophore model was proposed to aid the development of a CoMFA model needed to predict the activity of designed compounds for synthesis (10). *Trans*-(4aR,10bR)-7-hydroxy-N-n-propyloctahydrobenzo[f]quinoline (**1**) was selected as the template and was used to define the pharmacophore elements shown in Figure 4-2.

Although overall σ binding constants were used in developing the model, the successful predictive capacity of the derived CoMFA model [r^2(cv) = 0.843] suggested that related σ ligands may use a similar mode of binding at the σ receptor sites.

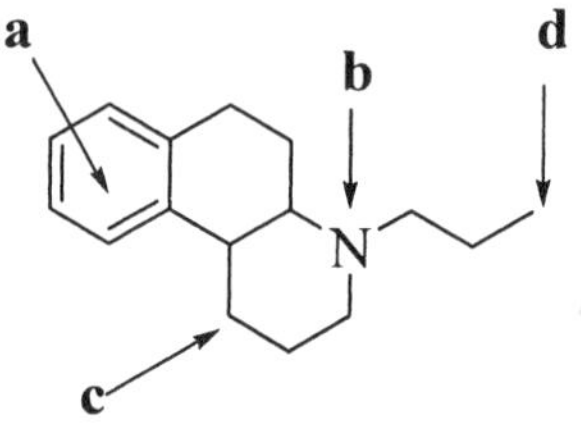

Figure 4-2. Template structure used to align σ ligands in the first 3-D QSAR model. Superimposition points included a) the centroid of an aromatic ring, b) basic nitrogen atom, c) a carbon atom separated by an ethylene space attached to the nitrogen atom and, where available, d) a small alkyl substituent on the N atom.

3. MODELS FOLLOWING THE DISCOVERY OF σ_1 AND σ_2 SITES

3.1 Gilligan model

In 1992, Gilligan and coworkers (11) identified a lead compound with selectivity for σ receptors (K_i = 6 nM). The lead compound (compound **2**) was conceptually divided into four regions, corresponding to four pharmacophore elements, as shown in Figure 4-3.

The results of testing several structural variants of compound **2** showed that there was a need to modify the initial model. Using 15 σ-selective ligands, the authors came to the conclusion that four specific pharmacophore elements contribute to optimal σ binding and oral activity in antimescaline or anti-aggression tests (Figure 4-4).

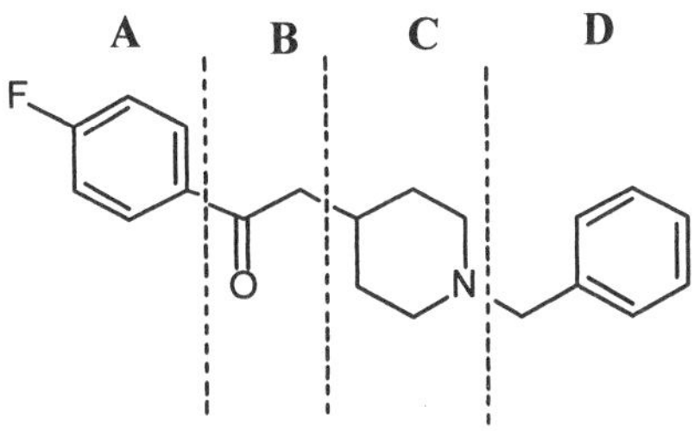

4. Pharmacophore models

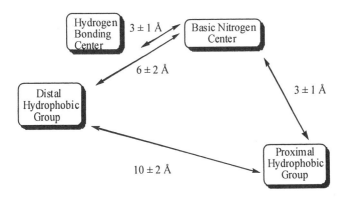

Figure 4-3. The four regions proposed by Gilligan et al. (11) for the binding of **2**: a) a distal aromatic ring (Region A), b) a nitrogen heterocycle (Region C), c) a space between the heterocycle and the distal aromatic ring (Region B), and d) a substituent on the nitrogen heterocycle (Region D).

The authors also noted that aromatic groups are preferred at the distal hydrophobic locus to achieve optimal *in vivo* activity. While there are similarities between this model and the previous ones, the hydrogen bonding center in this model replaces the electrostatic group in the Manallack proposal. Because the ligand model reflects structural features that are related to oral activity, it is unknown what role the hydrogen bonding center in the model might play in binding to σ receptors. Finally, the authors indicated that selectivity between σ binding and D_2 dopamine, or $5\text{-}HT_2$ serotonin receptor binding, was critically dependent on the chemical nature of the N-substituent and the distance between the basic nitrogen and the proximal hydrophobic moiety.

Figure 4-4. The proposed Gilligan model (11) for σ_1 receptor binding possesses: a) a basic nitrogen atom, b) two hydrophobic groups with different distances from the basic N atom, and c) a H-bonding center midway between the basic N and the distal hydrophobic site.

3.2 Glennon/Ablordeppey model

This model is the culmination of studies, originally begun in the late 1980s, aimed at identifying a pharmacophore for the binding of benzomorphan analogs at σ receptors. Benzomorphans, such as **3**, where R is a small alkyl, cycloalkyl, or alkenyl group, were among the first agents shown to bind at σ receptors. In fact, the term σ receptor was coined (derived from "σ-opioid receptors") on the basis that such compounds possess opioid-like actions that could not be explained by their binding at other known (i.e. μ-, δ-, or κ-) opioid receptors (reviewed in 7).

It was found that the phenylethylamine substructure **4** retains affinity for σ receptors and that affinity varied, depending upon stereochemistry and the nature of R and X, over a 20,000-fold range (5). One of the highest affinity agents was **5** (σ K_i = 2.6 nM), which displayed 165-fold higher affinity than N-allylnormetazocine (SKF-10,047; **3**, R = -CH_2-C_3H_5). From these investigations, it was clear that an intact benzomorphan structure was not required for high-affinity binding. Furthermore, with appropriate aryl substituents in the phenylethylamine portion of the molecule (including fused-ring structures), the length of the five-membered chain could be shortened by one or two methylene groups with retention of high (σ K_i <10 nM) affinity (12). Both secondary and tertiary amines were found to bind but, with tertiary amines, one of the amine substituents could not be much larger than a methyl group. From here, studies diverged. One direction was to further examine phenylpentylamine derivatives such as **5**, and another was to examine haloperidol hybrids of **4**.

It was quickly discovered that the phenylpentyl moiety, not a phenylethyl moiety, of **5**-type compounds was a key structure for binding at σ receptors. For example, comparison of several phenylpentylamines **6** where n was varied from n=1 to n=4 (σ K_i = 2.0-2.7 nM) showed that variation of Phenyl-**A** to N chainlength did not significantly influence affinity (13,14).

4. Pharmacophore models

[Structure: Ph-(CH₂)ₙ-HN-CH₂CH₂CH₂CH₂-Ph, labeled **6** and **1**]

In addition, either or both of the phenyl rings could be replaced by a cyclohexyl ring indicating that the interaction with σ receptors involves a hydrophobic rather than an aromatic-type interaction. Furthermore, Phenyl-A could be deleted without detriment to affinity. For example, **7** (σ K_i = 2.6 nM) and **8** (σ K_i = 2.4 nM) retained the affinity of **6** (13,14).

[Structures **7** and **8**: N,N-dimethylamino alkyl chains to cyclohexyl (7) and to phenyl (8)]

A phenylpiperidine or phenylpiperazine ring possesses dimensions approximating a phenylethylamine; thus, a second direction was to examine such derivatives (15). Subsequently, hybrid molecules were examined. Haloperidol (see Chart 1 for structure) is a piperidine-containing butyrophenone. It was reasoned, if the phenylpentylamine moiety is a significant pharmacophoric contributor, that it should be possible to extend the butyrophenone chain of haloperidol to a valerophenone. Indeed, valerophenone **9** (σ K_i = 2.3 nM) was found to bind with several-fold higher affinity than haloperidol (σ K_i = 10 nM). Furthermore, removal of polar substituents, to afford phenylpentylamine **10**, resulted in a further enhancement of affinity (**10**; σ K_i = 0.9 nM) (16). At the time, compound **10** was one of the highest affinity σ receptor ligands available.

If compound **10** binds in a manner similar to that of the original phenylpentylamines (i.e. **6**), removal of the Phenyl-A ring should not significantly decrease affinity. This was found to be the case; replacement of the piperidinyl phenyl ring of **10** by hydrogen resulted in retention of affinity (**11**; σ K_i = 1.9 nM) (16). Studies such as these established the phenylpentylamine moiety as being important for the binding of high-affinity σ ligands.

Shortly after the above had been established, Bowen et al. (6) confirmed previous speculation that σ receptors represented a heterogeneous population of sites and consisted of what are now referred to as σ_1 and σ_2 receptors. It therefore became necessary to re-examine the above pharmacophoric concepts. We found that our overall σ receptor binding data were very consistent with σ_1 receptor affinities. This was probably because of the radioligand employed and possibly because most of the available agents bind predominantly to σ_1 receptors and displayed much lower affinity for σ_2 receptors. For example, compound **8** and **11** displayed high affinity for σ_1 receptors (σ_1 K_i = 0.25 and 0.48 nM, respectively) (17). Studies with various terminal amine substituents indicated the existence on the receptor of a hydrophobic acceptor region and a nearby region of bulk tolerance; some representative results are provided in Table 4-1. N,N-Dimethyl compound **16** displayed a K_i = 14 nM; the corresponding N,N-diethyl homolog displayed higher affinity (**14**; σ_1 K_i = 6 nM). As long as one of the N-alkyl substituents was a methyl group, the nature of the second substituent had limited impact on affinity if it was at least three carbon atoms in length (e.g. compare **8** and **15**; σ_1 K_i = 0.25 nM) (Table 4-1). This was taken as evidence for the presence of a hydrophobic binding pocket of limited size, and that as long as this hydrophobic binding requirement was met, compounds possessed high affinity. Further bulk was seemingly accommodated in an associated region of bulk tolerance, and did not usually enhance affinity. Comparing the N,N-dimethyl analog **16** σ_1 K_i = 14 nM) with its monomethyl counterpart **17** (σ_1 K_i = 418 nM) might suggest that tertiary amines are favored over secondary amines. But, this was not found to be the case. Secondary amines were found to bind as well as their tertiary amine counterparts. The conclusion was reached that the N-methyl substituents of **16** and **17** were simply too short to avail themselves of an optimal interaction with the hydrophobic binding site.

4. Pharmacophore models

Table 4-1. σ Receptor affinities of representative phenylpentylamines

	R_1	R_2	σ_1 K_i (nM)
12		(8-azabicyclic) N—	1.0
13		(pyrrolidinyl) N—	0.76
14	Et	Et	6.0
8	Me	-(CH$_2$)$_2$CH$_3$	0.25
15	Me	-CH$_2$CH$_2$Ph	0.25
16	Me	Me	14
17	Me	H	418

From ref. (17).

Racemic compound **18** (σ_1 K_i = 0.6 nM), a more conformationally-constrained analog of **15**, retained high affinity. What makes this compound interesting is that it is beginning to approach the structure of the benzomorphans (e.g. **3**).

On the basis of the available data, we proposed an initial pharmacophore model for high-affinity σ_1 binding (Figure 4-5).

18

We subsequently probed the region of bulk tolerance associated with the Phenyl-**A** ring. A series of phenylpentylamines, specifically, N-[(5-phenyl)pentyl]piperidines, was prepared bearing either a phenyl or benzyl group at each of the piperidine carbon positions (Table 4-2). All of the derivatives displayed subnanomolar affinity, and an affinity comparable to that of **11**, indicating the existence of a significant multi-directional region of bulk tolerance. The 3-benzyl-substituted compound **22** (σ_1 K$_i$ = 0.08 nM; Table 4-2) actually showed about 6-fold enhanced affinity relative to **11** suggesting, perhaps, a more efficient binding of 3-substituted piperidines.

Next, we used compound **23** to re-investigate the influence of chain length on affinity. A series of N-phenylalkyl-4-benzylpiperidines **24** was examined where the length of the alkyl spacer was varied from n=5 (i.e. **23**) to n=4 and n=3; the latter two compounds displayed affinities (σ_1 K$_i$ = 0.8 and 0.4 nM, respectively) comparable to that of **23** (σ_1 K$_i$ = 0.58 nM). It would appear that there is some "slack" with respect to the length of the alkyl chain. The effect of chain-shortening might be compensated for by a more effective hydrophobic interaction on the opposite end of the molecule. Another possible explanation is a "reversed" mode of binding. Compound **24**, like **6**, can achieve a certain degree of symmetry depending upon n (i.e. chain length). Here, the possibility exists, that depending upon chain length, certain compounds might bind such that Phenyl-**A** now binds at the Phenyl-**B** site. This will be further addressed below.

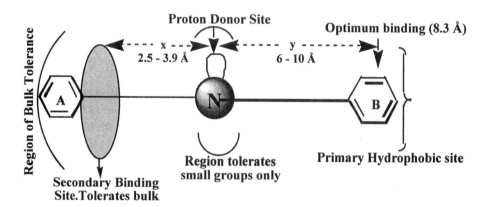

Figure 4-5. Initial Glennon/Ablordeppey pharmacophore model for high affinity σ_1 binding (18).

4. Pharmacophore models

Table 4-2. σ Receptor affinities for cyclic phenylphentylamine derivatives

	R_2	R_3	R_4	σ_1 K_i (nM)
11	H	H	H	0.48
19	Ph	H	H	0.17
20	H	Ph	H	0.14
10	H	H	Ph	0.16
21	Bn	H	H	0.40
22	H	Bn	H	0.08
23	H	H	Bn	0.58

24

There also might be some bulk tolerance associated with the Phenyl-**B** region of **6**-type compounds. This became evident upon addressing the observation that certain antipsychotics bind at σ receptors (7). One of the earliest therapeutic possibilities thought to be associated with σ receptors was antipsychotic activity due to the binding of various types of antipsychotic agents. Certain antipsychotics, such as the butyrophenone haloperidol, were discussed above. But, other classes of antipsychotic agents, such as the phenothiazines and thioxanthenes (**25**), also displayed affinity.

Thioxanthenes possess embedded in their structure a phenylalkenylamine, specifically, a phenylbuten-2-ylpiperazine moiety. We had already demonstrated that phenylalkylamines bind at σ receptors, and that the amine moiety could be replaced with a piperazine. Hence, we investigated the influence of unsaturation in the phenylalkylamine series by examining compounds **26** and **27**. Both displayed identical affinity for σ_1 receptors (K_i = 29 nM), and an affinity comparable to their phenylbutylamine counterpart **28** (K_i = 20 nM). The results suggested that unsaturation (and its attendant stereochemistry) in the side chain was not playing a major affinity-determining role.

Because one of the piperazine nitrogen atoms might be thrust into a hydrophobic region of the receptor, it was reasoned that its replacement with

4. Pharmacophore models

a methylene group should have an affinity-enhancing effect. The >10-fold enhanced affinity of **29** (σ_1 K_i = 0.7 nM) and **30** (σ_1 K_i = 1.9 nM) over that of **28** supported this concept. Reduction of the double bond in **30** (i.e. **31**; σ_1 K_i = 1.0 nM) had little additional effect. It was also reasoned that if a pentyl moiety is optimal, chain extension should further enhance affinity, and that deletion of one of the phenyl rings would be of no consequence if there is a region of bulk tolerance. Indeed, extension of **30** to **32** (σ_1 K_i = 0.13 nM) and **31** to **33** (σ_1 K_i = 0.09 nM) resulted in high-affinity compounds.

Deletion of one of the phenyl rings of **33** had no effect on affinity (**34**; σ_1 K_i = 0.07 nM). The results of this study showed how thioxanthene antipsychotics might be structurally related to the phenylpentylamines, identified a possible region of bulk tolerance associated with the aryl portion of the phenylpentylamines, and additionally revealed several very high affinity σ_1 ligands.

Another issue that required examination was the role of the basic nitrogen atom. Several studies had shown that σ ligands did not necessarily require a basic amine. Steroids, for example, had been demonstrated to bind at σ receptors (albeit with relatively low affinity) (7). Removal of the basic nitrogen atom of a phenylpentylamine would most likely result in a compound that lacks sufficient solubility for testing. Thus, it was necessary to find a location on the molecule where a solubilizing group might be incorporated so that the phenylpentylamine nitrogen atom could be removed. For this, we took advantage of the proposed region of bulk tolerance. Once again, compound **11** (σ_1 K_i = 0.48 nM) served as the lead structure. It was found that an amino group was reasonably well tolerated in the aryl ring of **11** (i.e, **35**, σ_1 K_i = 38 nM). Subsequently, the piperidine amino group was deleted to give **36**. Compound **36** (σ_1 K_i >36,000 nM) displayed >50,000-fold reduced affinity relative to **11** indicating that the presence (and location) of the basic amine is crucial to binding (19).

35

36

37 **38**

As an interesting parenthetical note, a number of aryl ring-substituted compounds were examined during the course of this study and compounds **37** and **38** were found not only to possess high affinity for σ_1 receptors (σ_1 K_i = 0.86 and 0.82 nM, respectively), they were also found to be quite selective (> 600-fold and 200-fold, respectively) for σ_1 versus σ_2 receptors (σ_2 K_i = 554 and 63 nM, respectively) (19).

Having established that the phenylpentylamine nitrogen atom is required for binding, and knowing that secondary amines and tertiary amines bind, it was important to determine if quaternary amines also bind at σ_1 receptors. Compound **23** (σ_1 K_i = 0.58 nM) was methylated to quaternary amine **39** (σ_1 K_i = 2.7 nM) and resulted only in several-fold decreased affinity, suggesting that quaternary amines are accommodated. It might be noted that σ_2 affinity was also relatively unchanged by quaternization (σ_2 K_i = 2.7 and 4.0 nM for **23** and **39**, respectively). However, when R(-)**40** was quaternized to R(-)**41**, σ_1 affinity decreased by about 10-fold (σ_1 K_i = 11 nM and 130 nM, respectively). Quaternization of S(+)**40** to S(+)**41** decreased affinity to a similar extent (σ_1 K_i = 40 and 335 nM, respectively). The effect of quaternization has not been extensively investigated, however, it is interesting to note that quaternization of R(-)**40** seemed to have a slight selectivity-reversing effect. That is R(-)**40** (σ_2 K_i = 60 nM) binds at σ_2 receptors with lower affinity than it displays for σ_1 receptors whereas its quaternary amine analog R(-)**41** (σ_2 K_i = 87 nM) possesses similar affinity for both populations. Similar results were obtained with the S(+)isomer. That is, S(+)**40** binds at σ_1 and σ_2 receptors with comparable affinity (K_i = 40 nM both for σ_1 and σ_2); quaternization to S(+)**41** (σ_2 K_i = 54 nM) had little effect on σ_2 affinity such that S(+)**41** is now about 6-fold selective for σ_2 receptors.

4. Pharmacophore models

These were not the first quaternary amines ever to be examined for σ binding. For example, PRE-079 (**42**; σ Ki = 5.1 nM) binds at σ receptors with high affinity, and quaternization to **43** (σ Ki =242 nM) resulted in 50-fold decreased affinity (20). Likewise, quaternization of haloperidol to **44** (σ_1 K_i = 274 nM, σ_2 Ki = 23 nM) resulted in a >50-fold reduction in σ_1 affinity, but in little change in σ_2 affinity. Thus, the issue of quaternization appears complex and can influence selectivity as well as affinity. These issues require further examination. For example, quaternization might have different effects on phenylpentylamines (e.g. **39**) relative to phenylpropylamines (e.g. **40**) due to differences in chain length. Compounds **42/43** might also be regarded as amines separated from an aryl group by a 5-membered chain, but here the chain contains an ester moiety. Compound **44** possesses only a 4-membered chain between the amine and aryl ring, and also might be sensitive to conformational changes upon introduction of the additional methyl group.

45   **46**

Piperazine derivatives are a special case of amines; that is, they are diamines. Piperazine **45** binds at σ_1 sites with reasonably high affinity (**45**; σ_1 K_i = 5.8 nM). Removal of the phenylalkyl phenyl group, resulting in **46** (σ_1 K_i = 82 nM), reduces affinity by about 15-fold. It would appear that a phenyl group is required for enhanced binding.

Neither **47** nor **49** (σ_1 K_i >10,000 nM) bind at σ_1 receptors. Interestingly, incorporation of a phenylbutyl chain increases the affinity of **47** (**48**; K_i = 0.2 nM) by >50,000-fold but increases the affinity of **49** (**50**; K_i = 125 nM) much less dramatically. One explanation is that the benzylic nitrogen atom of **48** is required for high-affinity binding and that reducing its basicity, as in **50**, causes **50** to bind differently than **48** (i.e. **50** must now bind using the more basic phenylalkylamine nitrogen atom). Of course, another explanation is that the carbonyl group of **50** is not tolerated by the receptor. Nevertheless, the question arises as to which of the basic nitrogens of piperazine analogs is most important for binding; that is, how do piperazines bind relative to their piperidine counterparts? Examination of **51** could provide some answers. Compound **51** lacks the "benzylic" nitrogen atom but possesses a carbonyl group. It would be required to bind in a manner similar to that of **48**. If the affinity of **51** was similar to that of **48**, this would imply that the non-benzylic nitrogen of **48** is the more important; but, if it binds with an affinity similar to that of **50**, this would suggest that the carbonyl group is not tolerated. Interestingly, the affinity of **51** (σ_1 K_i = 4 nM) is 30-fold higher than that of **50**. However, it is also 20-fold lower than that of **48**. Hence, it would appear that either nitrogen atom of piperazines might be involved in binding but that the presence of a benzylic carbonyl group can detract from affinity.

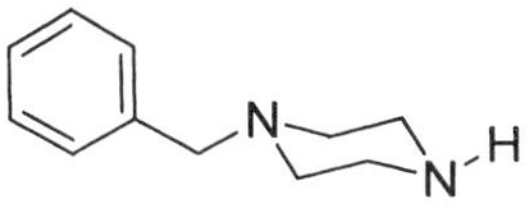

4. Pharmacophore models

47

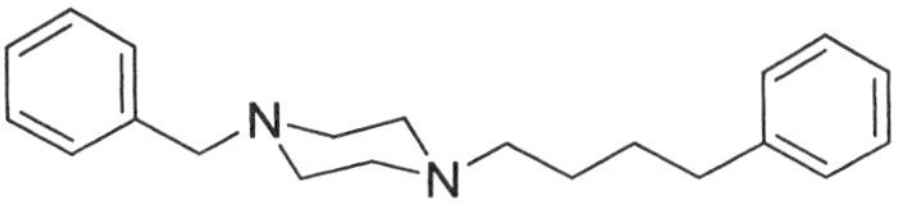

48

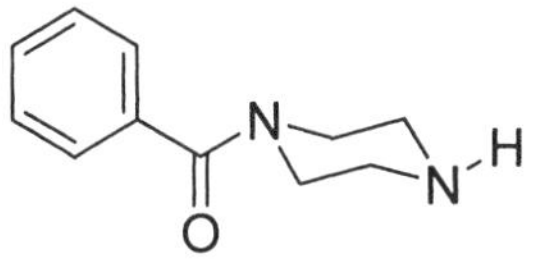

49

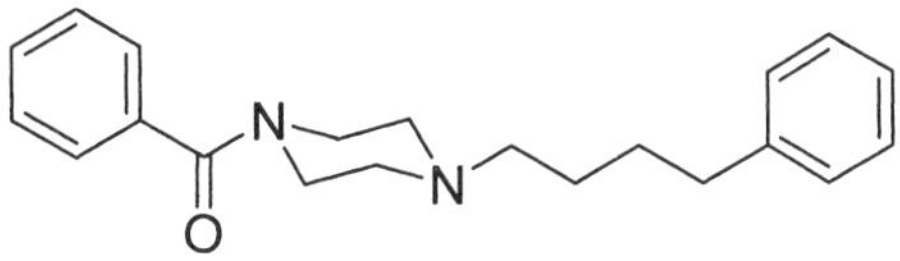

50

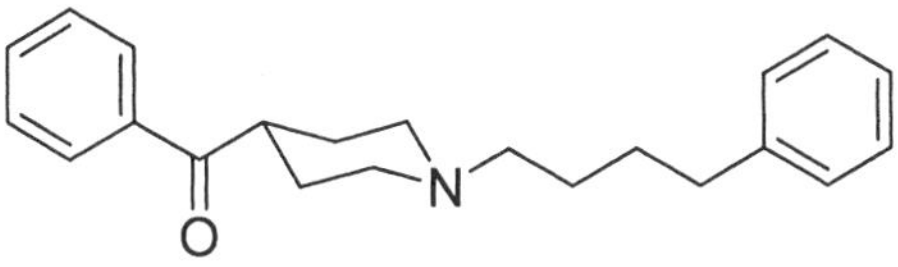

51

This was further examined by comparing the affinities of piperazine **52** with piperidines **53** and **34**. Piperazine **52** (σ_1 K_i = 1.4 nM) binds with high affinity and with an affinity comparable to piperidine **53** (σ_1 K_i = 1.3 nM), suggesting that they might be binding in a similar manner. Piperidine **34** (σ_1 K_i = 0.07 nM), however, binds with 20-fold higher affinity. On the basis of these results, it is difficult to reconcile that **52**, **53**, and **34** are binding in the same fashion.

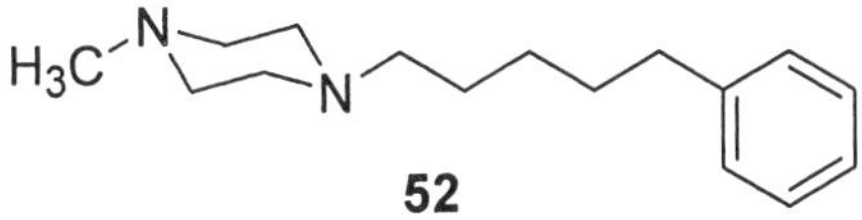

52

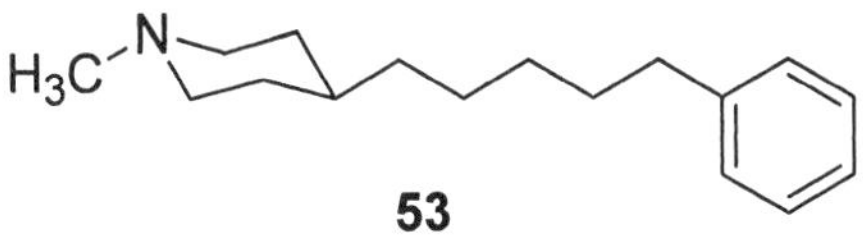

53

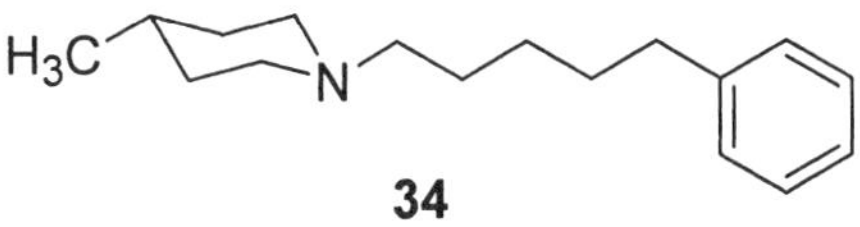

34

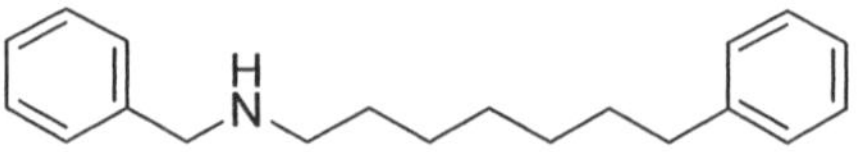

54

Numerous structurally (and therapeutically) diverse compounds have been demonstrated to bind at σ receptors. However, a cursory inspection of these compounds reveals two salient features: 1) many bind with affinity only in the μM or very high nanomolar range, and 2) most (particularly those displaying high affinity) possess an aryl or hydrophobic ring separated from a basic tertiary amine by four to seven atoms. Already mentioned above were the benzomorphans, phenylalkylamines, butyrophenones, valerophenones, thioxanthenes, and others. As evidence that long-chain compounds are accommodated, **54** (σ_1 K_i = 2.3 nM) binds in the low nanomolar range. Although a five-atom separator seems optimal, compounds with longer alkyl chains might simply interact with a hydrophobic binding site on the receptor in a less efficient manner than phenyl or cyclohexyl groups (i.e. a portion of the longer chain occupies the hydrophobic site). The possibility cannot be discounted that compounds with longer chains also might fold back somewhat to be accommodated by the receptor. In any event, long chains are accommodated.

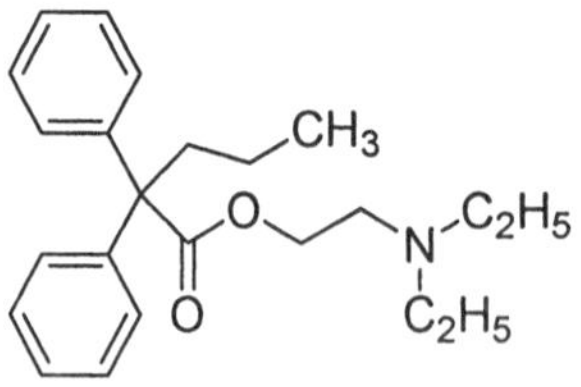

55 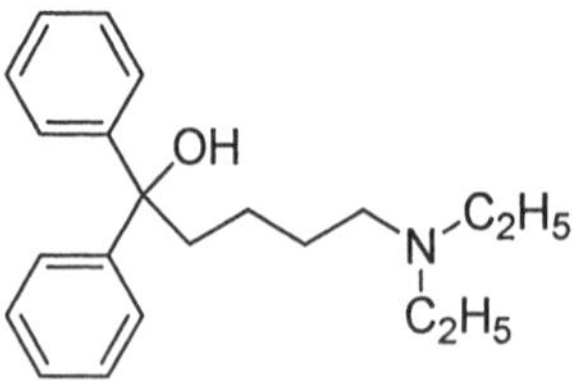   **56**

The chains, as already seen, need not be a pentyl moiety. Chain length, rather than the specific nature of the chain, is what seems important for binding; that is, an "Ar-X_5-N" moiety can substitute for an "Ar-C_5-N" moiety. A good example is SKF-525A (**55**). SKF-525A is an inhibitor of liver cytochrome monooxygenases that was demonstrated to bind with high

4. Pharmacophore models

affinity at σ_1 receptors (σ_1 K_i <10 nM; σ_2 K_i = 200 nM) (21). SKF-525A (**55**) possesses an aryl moiety separated from a basic amine by a five-atom ester-containing chain. We found SKF-525A (σ_1 K_i = 17 nM; σ_2 K_i = 1,100 nM) to bind with high affinity, and with an affinity comparable to the antidepressant benactyzine (**56**; σ_1 K_i = 12 nM; σ_2 K_i = 790 nM). From these two compounds, it is clear that the chain is not required to be a simple pentyl chain. Structurally-related compounds also bind, including **57** (σ_1 K_i = 24 nM) and **58** (σ_1 K_i = 1.4 nM), but none bind with significantly higher affinity than the simpler phenylpentylamine **59** (σ_1 K_i = 6 nM).

In 1989 de Costa and colleagues (3) discovered that the κ-opioid agonist U50,488 (**60**) binds at σ receptors (σ K_i = 81 nM) (3). Higher affinity analogs such as BD1008 (**62**; σ_1 K_i = 2 nM, σ_2 K_i = 8 nM) and BD1060 (**61**; σ_1 K_i = 3 nM, σ_2 K_i = 156 nM) have since been reported (22). Interestingly, these agents are composed of an aryl ring separated from a basic tertiary amine by a five-atom chain (i.e., the Ar-X$_5$-N moiety). In fact, we have found that **63** (σ_1 K_i = 83 nM, σ_2 K_i = 1,500 nM) binds at σ receptors and that removal of the chain NH group enhances σ_1 affinity by 100-fold (i.e., **13**; σ_1 K_i = 0.76, σ_2 K_i = 70 nM). Thus, the σ_1 affinity of U50,488 (**60**), **61**, and **62** is likely a consequence of their possessing an Ar-X$_5$-N moiety.

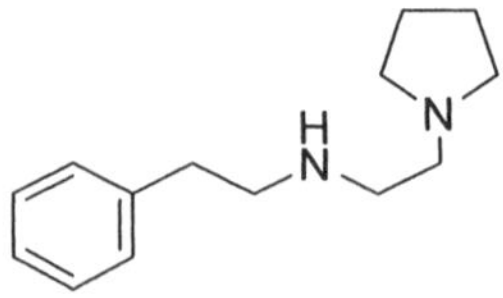

63

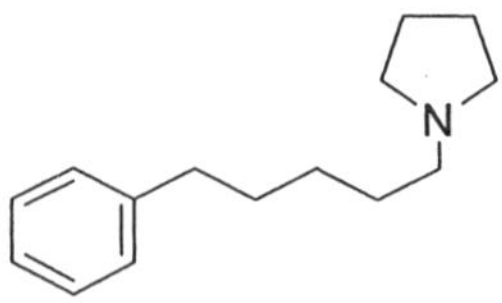

13

Various steroids bind at σ receptors, but they usually do so with μM affinity (21). Inspection of the steroids reveals they possess embedded within their structure a hydrophobic or aryl ring and a carbon chain of at least five atoms. It seemed reasonable that if an amine were to be incorporated into a cyclopentanoperhydrophenanthrene nucleus, the resulting compounds might bind at σ receptors with enhanced affinity. Consequently, we prepared and examined **64** (σ_1 K_i = 66 nM, σ_2 K_i = 24 nM). Compound **64** does not bind as well as the simple phenylpentylamines (e.g. **53**), but despite the presence of considerable bulk, stereochemical considerations, and conformational restriction (in what may or may not be a preferred conformation), **64** binds quite well. Amine-substituted steroids might offer a rich source of template molecules for further exploration of novel σ_1 and (because **64** binds with even higher affinity at σ_2 than σ_1 receptors) for σ_2 ligands.

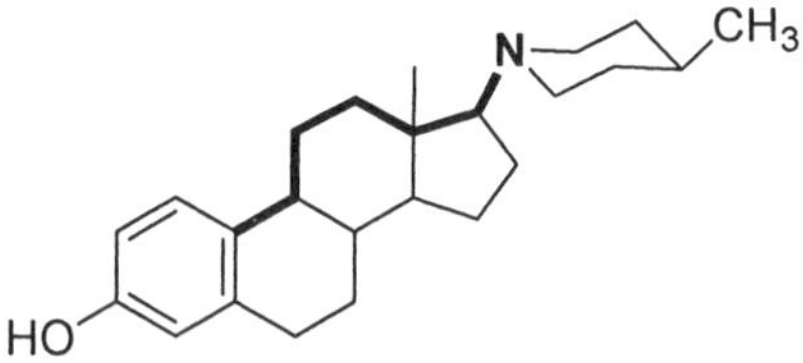

64

Evidence has been provided that an Ar-X_5-N moiety (where Ar is either an aromatic moiety or hydrophobic group such as cyclohexyl) is a common pharmacophoric feature of many high-affinity σ_1 ligands. This feature is seen again and again. However, as our work progressed, compound **65** (SC 50651; K_i = 0.075 nM) was reported to be a very high affinity σ receptor ligand. Lacking an extended phenylpentyl moiety, **65** offered a major challenge to the developing pharmacophore model. The possibility exists that **65** might represent a different type of pharmacophore, or that it was a related Ar-X_x-N pharmacophore tightly held in a preferred bioactive

4. Pharmacophore models

conformation. To examine this more closely, we synthesized **65** and both its *endo* (σ_1 K$_i$ = 66 nM) and *exo* (K$_i$ = 24 nM) isomers. The *exo* isomer of **65** was found to be the higher affinity isomer of the two, and it displayed >300-fold lower affinity than earlier reported (18). It seems possible that the original report (reported as a symposium abstract) might have contained a typographical error and that the observed affinity for **65** was actually 0.075 µM rather than 0.075 nM.

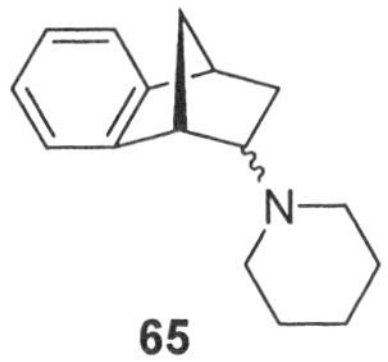

65

One additional study was focused on the hydrophobic site and resulted in novel agents with modest σ_2 selectivity. At one time, **45** (σ_1 K$_i$ = 5.8 nM) would have been considered a high-affinity ligand. Today, structurally-related compounds with 100-fold higher affinity have been identified. Nevertheless, removal of the alkyl phenyl group decreases affinity (**46**; σ_1 K$_i$ = 82 nM, σ_2 K$_i$ = 540 nM). If the phenylalkyl chain of **45** (or a portion of the alkyl chain of **46**) is interacting with a hydrophobic site, incorporation of a hydrophilic substituent should result in decreased affinity. Indeed, compound **66** (σ_1 K$_i$ = 3,300 nM; σ_2 K$_i$ = 13,500 nM) binds with very low affinity. The structurally-related 5-HT$_{1A}$ serotonin antagonist NAN-190 (**67**; σ_1 K$_i$ = 510 nM, σ_2 K$_i$ = 77 nM), on-hand as a result of other on-going studies, was found to bind with slightly enhanced σ_1 affinity, but with about 7-fold selectivity for σ_2 receptors. Replacement of the methoxy group of NAN-190 by H doubled affinity (i.e. σ_1 K$_i$ = 200 nM, σ_2 K$_i$ = 20 nM) but had little effect on selectivity.

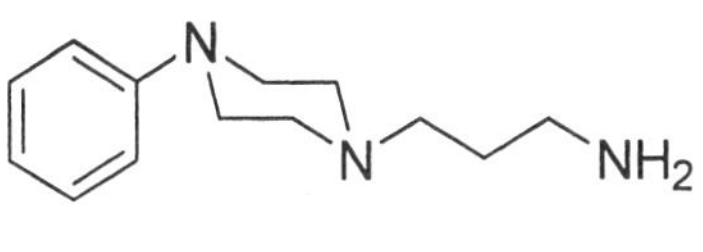

66 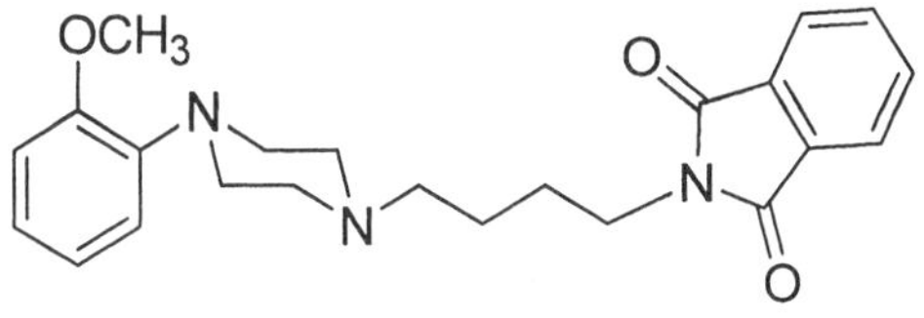   **67**

Continued investigation of these types of compounds led to **68** (σ_1 K_i = 13,100 nM, σ_2 K_i = 149 nM) and **69** (σ_1 K_i = 6,500 nM, σ_2 K_i = 2,200 nM). Compound **68** (NAN-70) displayed about 90-fold selectivity for σ_2 receptors. Apparently, chain shortening and the fused aryl-imide are beneficial to selectivity. Related compounds lacking the phthalimide moiety failed to display σ_2 selective binding. For example, **70** (σ_1 K_i = 94 nM, σ_2 K_i = 490 nM) and **71** (σ_1 K_i = 36 nM, σ_2 K_i = 90 nM) showed modest affinity and no selectivity for σ_2 receptors. These results provided useful information about σ_1 receptors, but also indicate that although similarities must exist between σ_1 and σ_2 receptors, differences also exist. Continued investigation with compounds such as NAN-70 (**68**) are likely to lead to ligands with even greater selectivity for σ_2 versus σ_1 receptors.

4. SUMMARY

Since originally proposing the Ar-C$_5$-N pharmacophore (Figure 4-5), we have attempted to confirm the various features of the model by continued investigation of different structure types. The model was challenged by investigating other ligands that might or might not be accommodated. In so doing, it was shown how members of different structural and pharmacological classes might bind at σ_1 receptors. What has been learned is that although many structure types bind at σ_1 receptors, and that from this perspective, σ receptors might be termed promiscuous, far fewer ligands bind with high affinity. Early studies were limited, for the most part, to what are now realized to be fairly low affinity (i.e. μM affinity) agents. Over the years, newer agents with higher affinity were described. With the demonstration that there now exists at least two populations of σ receptors, and many σ ligands with subnanomolar affinity, pharmacological studies and structure-affinity studies employing ligands with μM affinities might be open to reinterpretation. Some agents bind at σ receptors with mid-nanomolar affinity, but might bind at other populations of neurotransmitter receptors with subnanomolar affinity. Hence, the involvement of σ receptors in their actions requires reexamination. Indeed, the first σ ligand to enter clinical trials (i.e. rimcazole) (23) only displayed affinity in the high nanomolar to low micromolar range. In any event, numerous agents that bind at σ receptors share the Ar-X_5-N pharmacophore: an aryl (or some other hydrophobic) group separated from an amine by a five-membered chain. The spacer group "X" can be linear or branched (including unsaturation and cyclic structures), and can contain functionalities such as a ketone, amino, or ester group. Highest affinity is typically associated with an alkyl chain. Chain length seems to contribute to σ_1 affinity and a length of five atoms might be optimal. The terminal amine "N" can be secondary, tertiary, or quaternary. However, there might be only limited bulk tolerated with tertiary and quaternary amines. Typically, one hydrophobic substituent is required to be attached to the amine. Because there is evidence for regions of bulk tolerance associated with the hydrophobic sites, in some instances fairly bulky secondary amine groups are tolerated. These nearby regions of bulk tolerance might also allow the "X" group to be somewhat longer or shorter than five atoms; that is, a slight shifting in binding mode might be possible to accommodate longer and shorter chains provided that there is a more effective interaction with the hydrophobic sites.

A slight modification of the original Ar-C_5-N pharmacophore is shown in Figure 4-6. This modified pharmacophore, the Ar-X_5-N pharmacophore, attempts to accommodate the most recent findings (e.g. 24). It should be emphasized that "Ar" is not required to be an aromatic function and that its interaction with the receptor is likely of a hydrophobic nature. Likewise, "X_5" is not required to be exactly five atoms in length, with the effect of other chain lengths being offset by the specific size of the hydrophobic sites and the existence of nearby regions of bulk tolerance.

It is quite likely that a pharmacophore model for σ_2 binding will soon be described. Hitherto, too few ligands have been available to propose such a model. It is also very likely that the σ_2 pharmacophore(s) will share features in common with the σ_1 model(s), because many σ_1 ligands also bind at σ_2 sites. However, subtle differences (and in some cases, not so subtle differences) already have been shown to exist. For example, BD1008 (**62**) displays little selectivity for σ_1 versus σ_2 receptors whereas its N-desmethyl homolog **61** binds with comparable affinity at σ_1 receptors but shows 50-fold

σ_1 selectivity. Compound **37** displays >600-fold σ_1 selectivity whereas its corresponding unsaturated and unsubstituted analog **11** (σ_2 K_i = 50 nM) binds at σ_1 sites with comparable affinity but shows only 100-fold σ_1 selectivity. Cyclohexyl compound **7** (σ_1 K_i = 0.3 nM, σ_2 K_i = 195 nM) shows >600-fold σ_1 selectivity, whereas other phenylalkylamines such as **64**, **71**, or **24** where n = 4 (σ_1 K_i = 0.8 nM, σ_2 K_i = 3.1 nM) bind nearly equally well at both populations of receptors, and NAN-70 (**68**) shows 90-fold σ_2 selectivity.

With the pharmacophore data presented here, it is hopefully possible to more fully appreciate the σ_1 binding of high-affinity ligands, and the seemingly "promiscuous" nature of σ_1 receptors with regard to lower affinity agents. What remains to be accomplished is the development of models that can differentiate σ_1 from σ_2 binding.

With the pharmacophore data presented here, it is hopefully possible to more fully appreciate the σ_1 binding of high-affinity ligands, and the seemingly "promiscuous" nature of σ_1 receptors with regard to lower affinity agents. What remains to be accomplished is the development of models that can differentiate σ_1 from σ_2 binding.

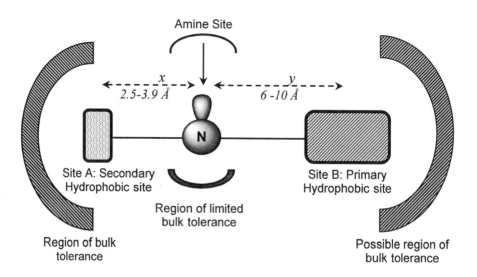

Figure 4-6. The revised Glennon/Ablordeppey "Ar-X5-N" pharmacophore model for σ_1 binding of high affinity agents. The model shown in Figure 4-5 was modified to account for the most recent findings.

ACKNOWLEDGMENTS

Much of the work from our laboratory was conducted in conjunction with Dr. James Fischer. We also wish to acknowledge Dr. Ho Law for the synthesis of **69** and **71**, and Dr. M. El-Ashwamy for the synthesis of **65**. This work was supported in part by funds from MH45225 (RAG) and MBRS GM08111 (SYA).

REFERENCES

1. Abou-Gharbia M, Ablordeppey SY, Glennon RA. Ann Rep Med Chem 1993, 28:1.
2. Walker JM, Bowen WD, Walker FO, Matsumoto RR, de Costa B, Rice KC. Sigma receptors: biology and function. Pharmacol Rev 1990, 42:355-402.
3. de Costa BR, He X. Structure-activity relationships and evolution of σ receptor ligands. In: Itzhak, Y., ed. Sigma Receptors, Academic Press, London, 1994. pp. 45-111.
4. Glennon RA, Ablordeppey SY, Ismaiel AM, El-Ashwamy MB, Fischer JB, Burke Howie KR. Structural features important for sigma-1 receptor binding. J Med Chem 1994, 37:1214-1219.
5. Glennon RA, Smith JD, Ismaiel AM, El-Ashwamy MB, Battaglia G, Fischer JB. Identification and exploitation of the sigma-opioid pharmacophore. J Med Chem 1991, 34:1094-1098.
6. Quirion R, Bowen WD, Itzhak Y Junien JL, Musacchio JM, Rothman RB, Su T, Tam SW, Taylor D. A proposal for the classification of sigma binding sites. Trends Pharmacol Sci, 1992, 13:85-86.
7. Itzhak, Y. ed. Sigma Receptors, Academic Press, London, 1994.
8. Largent BL, Wikstrom H, Gundlach AL, Snyder S.H. Structural determinants of sigma receptor affinity. Mol Pharmacol 1987, 32:772-784.
9. Manallack DT, Wong MG, Costa M, Andrews PR, Beart PM. Receptor site topographies for phencyclidine-like and sigma drugs: predictions from quantitative, conformational, electrostatic potential, and radioreceptor analyses. Mol Pharmacol 1989, 34:863-879.
10. Ablordeppey SY, El-Ashwamy MB, Glennon RA. Design, synthesis, and binding of sigma receptor ligands derived from butaclamol. Med Chem Res 1991, 1:425-438.
11. Gilligan PJ, Cain GA, Christos TE, Cook L, Drummond S, Johnson AL, Kergaye AA, McElroy JF, Rohrbach KW, Schmidt WK, Tam S W. Novel piperidine sigma receptor ligands as potential antipsychotic drugs. J Med Chem 1992, 35:4344-4361.
12. Glennon RA, Ismaiel AM, Yousif M, El-Ashwamy M, Herndon JL, Fischer JB, Howie KJ, Server AC. Binding of substituted and conformationally restricted derivatives of N-(3-phenyl-n-propyl)-1-phenyl-2-aminopropanes at sigma receptors. J Med Chem 1991, 34, 1855-1859.

13. Glennon RA, El-Ashmawy MB, Fischer JB, Burke-Howie K B, Ismaiel AM. N-Substituted 5-phenylpentylamines: a new class of sigma ligands. Med Chem Res 1991, 1:207-212.
14. El-Ashmawy MB, Ablordeppey SY, Hassan I, Gad L, Fischer JB, Burke-Howie KB, Glennon RA. Further investigation of 5-phenylethylamines derivatives as novel sigma receptor ligands. Med Chem Res 1992, 2: 119-126.
15. Glennon RA, Yousif MY, Ismaiel AM, El-Ashmawy MB, Herndon JL Fischer JB, Server AC, Burke-Howie KJ. Novel 1-phenylpiperazine and 4-phenylpiperidine derivatives as high affinity sigma ligands. J. Med. Chem 1991, 34:3360-3365.
16. Ablordeppey SY, Issa H, Fischer JB, Burke-Howie KB, Glennon RA. Synthesis and structure-affinity relationship studies of sigma ligands related to haloperidol. Med Chem Res 1993, 3:131-138.
17. Ablordeppey SY, Fischer JB, Law H, Glennon RA. Probing the proposed phenyl-A region of the sigma-1 receptor. Bioorg Med Chem 2002, 10:2759-2765.
18. Ablordeppey SY El-Ashmawy MB Fischer JB, Glennon RA. A CoMFA investigation of sigma receptor binding affinity: reexamination of a spurious sigma ligand. Eur J Med Chem. 1998, 33:625-633.
19. Ablordeppey SY, Fischer JB, Glennon RA. Is a nitrogen atom an important pharmacophoric element in sigma ligand binding? Bioorg Med Chem 2000, 8:2105-2111.
20. Su TP, Wu XZ, Cone EJ, Shukla K, Gund TM, Dodge AL, Parish DW. Sigma compounds derived from phencyclidine: identification of PRE-084, a new, selective sigma ligand. J Pharmacol Exp Ther 1991, 259:543-550.
21. Klein M, Musacchio JM. Effect of cytochrome P-450 ligands on the binding of [^{3}H]dextromethorphan and sigma receptor ligands to guinea-pig brain. In: Itzhak, Y., ed. Sigma Receptors, Academic Press, London, 1994. pp. 243-262.
22. Matsumoto RR, McKracken KA, Pouw B, Miller J, Bowen W, Williams W, de Costa BR. N-alkyl substituted analogs of the sigma receptor ligand BD1008 and traditional sigma receptor ligands affect cocaine-induced convulsions and lethality in mice. Eur J Pharmacol 2001, 411:261-273.
23. Tam SW. Potential therapeutic application of sigma receptor antagonists. In: Itzhak, Y., ed. Sigma Receptors, Academic Press, London, 1994. pp. 191-204.
24. Glennon RA, Ismaiel AM, Ablordeppey SY, El-Asmawy M, Fisher J B. Thioxanthene-derived analogs as sigma(1) receptor ligands. Bioorg Med Chem Lett 2004, 14:2217-2220.

Corresponding author: *Dr. Richard A. Glennon, Mailing address: Virginia Commonwealth University, School of Pharmacy, Department of Medicinal Chemistry, Richmond, VA 23298-0540, USA, Phone: (804) 828-8487. Fax: (804) 828-7404, Electronic mail address:glennon@hsc.vcu.edu.*

homology to any of the mammalian proteins thus far cloned. However, the receptor showed significant sequence homology with ERG2, a fungal gene product possessing sterol C8-C7 isomerase activity. This led to the suggestion that the σ_1 receptor may represent the mammalian counterpart of fungal sterol isomerase. This enzyme catalyzes a critical step in sterol synthesis by shifting the C8 double bond in zymosterol to position C7. However, subsequent studies showed that the cloned receptor lacks sterol isomerase activity as evidenced by its inability to reverse the phenotype of an ERG2 mutant strain *S. cerevisiae* which lacks the endogenous sterol isomerase activity. In contrast, overexpression of ERG2 in this mutant yeast strain is able to reverse the phenotype, indicating that the mammalian σ_1 receptor, despite the structural similarity, does not possess sterol isomerase activity similar to the fungal enzyme ERG2. This is supported by the recent findings that mammals express a totally different protein with C8-C7 sterol isomerase activity and that this protein, also known as emopamil-binding protein, shows no structural similarity to either the yeast C8-C7 sterol isomerase ERG2 or mammalian σ_1 receptor (20,21). That the cloned protein is indeed the σ_1 receptor is evident from the ability of the protein, heterologously expressed in yeast, to bind (+)-pentazocine with high affinity. Even though mammalian C8-C7 sterol isomerase possesses the ability to bind various σ ligands, the ligand specificity is quite different between this protein and the σ_1 receptor, especially with respect to (+)-pentazocine (22).

2.2 Cloning of human σ_1 receptor

The successful cloning of the guinea pig σ_1 receptor by Hanner et al. (19) led the way for the cloning of other mammalian σ_1 receptors. Kekuda et al. (23) were able to clone the human σ_1 receptor from a human placental trophoblast cell line (JAR) cDNA library. JAR cells, as well as normal human placenta, express σ_1 binding sites abundantly (24). In cloning the JAR cell σ_1 receptor, Kekuda et al. (23) used an RT-PCR product, which is specific for guinea pig σ_1 receptors, as a probe to screen the cDNA library. Independently, Jbilo et al. (25) cloned the human σ_1 receptor using a completely different strategy. These investigators first purified the human σ_1 receptor protein by monitoring the binding of the σ_1 receptor-specific ligand SR 31747A (N-cyclohexyl-N-ethyl-3-(3-chloro-4-cyclohexylphenyl) propen-2-ylamine) at each purification step, obtained a partial sequence of the purified protein, amplified a human σ_1 receptor-specific cDNA probe using degenerate nucleotide primers based on the amino acid sequence, and then used the cDNA as a probe for the isolation of the full-length cDNA coding for the receptor. This strategy is similar to that used by Hanner et al.

receptor may be a cyclophilin. However, cyclophilin A is a cytosolic protein whereas the σ-binding site is membrane-associated. This suggested that cyclophilin A must have been a contaminant in the purified preparation of the σ receptor. Alternatively, cyclophilin A might represent a σ receptor-interacting protein that co-eluted with the σ receptor from the affinity matrix. In any case, it appeared very unlikely that the σ receptor was identical to cyclophilin A at the molecular level.

The first successful attempt in the purification of the σ_1 receptor was reported by Hanner et al. in 1996 (19). These investigators used guinea pig liver microsomes as the source of σ_1 binding sites as detected by the binding of (+)-pentazocine in their purification procedure. Following solubilization of the binding sites with the detergent digitonin, the solubilized proteins were subjected to a number of protein purification procedures, and the activity of the σ_1 receptor, as detected by the binding of (+)-pentazocine, was monitored at each purification step. This strategy yielded a protein with a molecular mass of ~30 kDa. This protein was partially sequenced using one of the tryptic peptide fragments and the sequence (SEVFYPGETVVHGPGEATAVEWG) was used to raise antibodies against the σ_1 receptor. The authenticity of the antibodies was confirmed by its ability to immunoprecipitate σ_1 binding sites from detergent-solubilized guinea pig liver membranes. Hanner et al. (19) then used this partial sequence information to design degenerate nucleotide primers and, using these primers, were successful in amplifying a small RT-PCR product with guinea pig liver RNA as the template. A nucleotide primer specific for this RT-PCR product and a poly (T) 3'-oligonucleotide were then used to amplify a ~1 kbp RT-PCR product. This cDNA fragment was used to screen a cDNA library derived from guinea pig liver mRNA. This screening led to the successful cloning of the full-length σ_1 receptor cDNA. The molecular identity of this cDNA was confirmed by heterologous expression in *S. cerevisiae* and demonstration of specific high-affinity binding of (+)-pentazocine to the expressed protein.

2.1 Structural and functional features of guinea pig σ_1 receptor

The σ_1 receptor cDNA cloned from guinea pig liver is 1857 bp long and codes for a protein of 223 amino acids. The estimated molecular mass of this protein is 25.3 kDa which is similar to the values obtained by photoaffinity labeling and radiation inactivation. The sequence contains an endoplasmic reticulum retention signal (MQWAVGRR) at its N-terminus. Interestingly, the cloned σ_1 receptor exhibits no significant sequence in the presence of dextrallorphan which selectively blocks the binding of DTG to σ_1 receptors. Another distinguishing feature between the two classes of σ receptors is their interaction with ropizine and phenytoin (7,8). σ_1 Receptors are allosterically modulated by these compounds whereas σ_2 receptors are not affected. Furthermore, progesterone can differentiate between σ_1 and σ_2 receptors by its ability to interact only with the former (9, 10). These two classes of receptors are also distinguishable on the basis of their biochemical nature. The molecular mass of σ_1 receptors is 25 kDa, as detected by photoaffinity labeling (11). Similar studies have shown that σ_2 receptors have an apparent molecular mass of 18-21 kDa (12). There is evidence for the existence of yet another distinct class of σ receptors in immune cells. Splenic lymphocytes express σ receptors which interact with (+)-pentazocine with high affinity but exhibit relatively low affinity towards haloperidol and DTG (13). Photoaffinity labeling studies have shown that this third receptor has an apparent molecular mass of 57 kDa (14). There is yet another type of σ receptor, the low-affinity receptor that is often misread as σ_2 receptors (15). This fourth receptor shares the same stereoselectivity with σ_2 receptors. However, affinities of nanomolar concentrations of σ ligands are observed with σ_2 receptors in contrast to affinities of micromolar concentrations of σ ligands observed with the low-affinity receptor.

2. CLONING OF σ_1 RECEPTORS

The molecular identity of σ receptors remained a mystery for a long time. Since σ receptors are expressed at high levels in the liver and since these receptors are able to interact with a variety of xenobiotics, the idea that σ receptors may be structurally and/or functionally related to cytochrome P450s, the drug metabolizing enzymes in the liver, was entertained for some time (16). However, subsequent studies, which showed notable differences in drug selectivity between σ receptors and cytochrome P450s, ruled out this idea (4,17). A couple of years later, Schuster et al. (18) reported on their efforts to purify the σ receptor from rat liver using a haloperidol-affinity matrix. These investigators succeeded in obtaining ~500-fold enrichment of the σ receptor based on ligand binding, and the purified fraction contained two components, 28 kDa and 40 kDa in size. Since photolabeling studies already showed that the molecular mass of the σ_1 binding site is about 25 kDa, Schuster et al. focused on the 28 kDa protein. Amino acid sequencing of the N-terminus of the protein demonstrated that the sequence is identical to the N-terminal sequence of the 17 kDa rat cyclophilin A. Based upon these results, it was concluded that a critical component of the rat liver σ

Chapter 5

CLONING OF σ_1 RECEPTOR AND STRUCTURAL ANALYSIS OF ITS GENE AND PROMOTER REGION

Vadivel Ganapathy[1,2], Malliga E. Ganapathy[3] and Katsuhisa Inoue[1]
Departments of [1]Biochemistry & Molecular Biology, [2]Obstetrics & Gynecology and [3]Medicine, Medical College of Georgia, Augusta, Georgia 30912, USA

1. INTRODUCTION

Sigma (σ) receptors are defined as non-opiate, non-dopaminergic, and non-phencyclidine binding sites which interact with several psychoactive agents including benzomorphans, haloperidol, and phencyclidine (1-4). These receptors are expressed in various tissues associated with endocrine, immune, and nervous systems. Multiple functions in multiple organs suggest that the σ receptors may play a fundamental role in modulating a wide variety of responses in different tissues (1-4). Biochemical and pharmacological studies have indicated that σ receptors are heterogeneous. There are at least two classes of σ receptors that are distinguishable by pharmacological and biochemical means (5,6). Even though the ligands haloperidol, (+)-PPP [3-(3-hydroxyphenyl)-N-(1-propyl)piperidine], and DTG [1,3-di-o-tolylguanidine] are unable to differentiate between these two classes, the benzomorphan opiates such as pentazocine and SKF-10,047 (N-allylmetazocine) are able to differentiate between the two classes. σ_1 Receptors exhibit high affinity for (+)-benzomorphans and low affinity for the corresponding (-)-enantiomers. The enantioselectivity of σ_2 receptors is opposite to that of σ_1 receptors in that σ_2 receptors preferentially interact with (-)-benzomorphans. (+)-Pentazocine is a selective ligand for σ_1 receptors. There is no detectable saturable binding of this (+)-benzomorphan to σ_2 receptors. No selective ligand is available at present for σ_2 receptors. However, this class of σ receptors can be monitored by the binding of DTG

(19) in cloning the guinea pig σ_1 receptor. The functional identity of the cloned human σ_1 receptor was established by heterologous expression. Kekuda et al. (23) used a mammalian cell expression system for this purpose. The cloned receptor was expressed in HeLa cells and the binding of haloperidol was used to detect the heterologously expressed receptor. In contrast, Jbilo et al. (25) used a strain of *S. cerevisiae* (EMY47) with a disrupted ERG2 gene for this purpose. The use of this particular mutant strain was necessitated because these investigators were monitoring the activity of the cloned receptor by the binding of SR 31747A, a ligand that binds to σ_1 receptors as well as the structurally related ERG2 with almost equal affinity. The selection of the mutant strain for heterologous expression of the cloned human σ_1 receptor abolished the endogenous ligand binding activity which made it ideal for the characterization of the heterologously expressed human σ_1 receptor. It is of interest to note that Hanner et al. (19), who also used *S. cerevisiae* for heterologous expression of the guinea pig σ_1 receptor, did not employ any ERG2 mutant strain. However, these investigators used (+)-pentazocine as the ligand for the characterization of the heterologously expressed guinea pig σ_1 receptor and this ligand binds to ERG2 with a ~500-fold less affinity than to the mammalian σ_1 receptor. Consequently, the endogenous binding activity with this σ_1 receptor ligand is undetectable in control *S. cerevisiae* even with normal expression of ERG2. HeLa cells that were used by Kekuda et al. (23) express σ_1 binding sites constitutively and therefore characterization of the heterologously expressed σ_1 receptor required ligand binding studies in vector-transfected cells and in cDNA-transfected cells done in parallel to correct for the endogenous ligand binding activity. In subsequent studies however, the σ_1 receptor-defective human mammary cancer cell line (MCF-7) (26) was used in the characterization of the cloned human σ_1 receptor (27). These heterologous expression studies, either in appropriate yeast cells or in mammalian cells, confirmed the functional identity of the cloned human σ_1 receptor as evidenced by its ligand binding characteristics.

The human σ_1 receptor cDNA is ~1.7 kb long (23,25) and codes for a protein of 223 amino acids. The predicted protein is 93% identical in amino acid sequence to the guinea pig σ_1 receptor. It also exhibits significant sequence homology to the protein product of the yeast gene ERG2 as does the guinea pig σ_1 receptor. The putative amino acid sequence motif (MQWAVGRR), which is thought to be an endoplasmic reticulum retention signal, is found at the N-terminus of the human σ_1 receptor as it is in the guinea pig σ_1 receptor. The cloned human σ_1 receptor not only binds to various pharmacological agents that are considered as σ_1 selective ligands but also to progesterone, a putative endogenous σ_1 receptor-specific ligand (27).

2.3 Cloning of rodent σ_1 receptors

The σ_1 receptor from rat and mouse was cloned subsequently by homology screening (28-31). The rodent σ_1 receptor also consists of 223 amino acids. The ligand binding characteristics of these receptors have been established in heterologous expression systems in different mammalian cells.

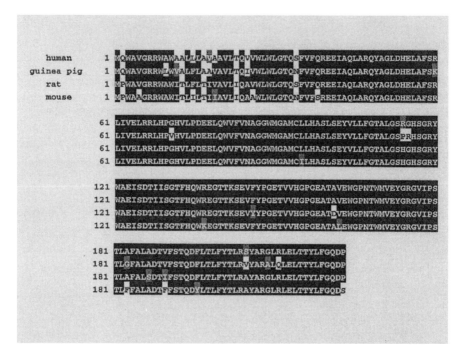

Figure 5-1. Comparison of the amino acid sequences of σ_1 receptors from different mammalian species. Regions with identical amino acids are indicated by dark shaded boxes and regions with conservative substitutions are indicated by lightly shaded boxes.

3. MEMBRANE TOPOLOGY OF σ_1 RECEPTOR

The four different mammalian σ_1 receptor proteins (guinea pig, human, mouse, and rat) thus far cloned all consist of 223 amino acids. The amino acid sequence is highly similar across different species with a sequence identity of >90% (Figure 5-1). There is no other mammalian protein thus far known that shows significant structural homology to the σ_1 receptor. Among non-mammalian proteins, the fungal ERG2 protein from different organisms (*S. cerevisiae, N. crassa, M. grisea, S. pombe,* and *U. maydis*) exhibits significant structural homology with the mammalian σ_1 receptor (Figure 5-2). The sequence identity between the human σ_1 receptor and the yeast ERG2 proteins is in the range of 34-41%.

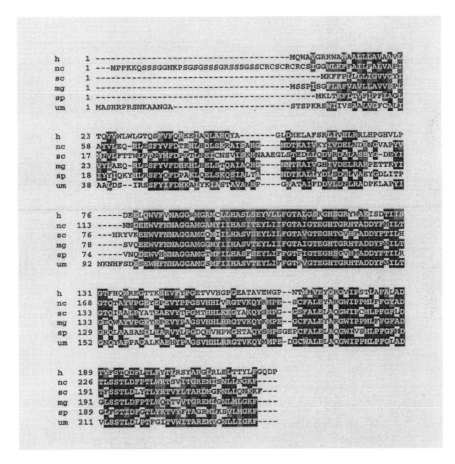

Figure 5-2. Comparison of the amino acid sequence of human σ_1 receptors and with those of sterol C8-C7 isomerases (ERG2) from different fungi. Regions with identical amino acids are indicated by dark shaded boxes and regions with conservative substitutions are indicated by lightly shaded boxes. H, human; nc, *N. crassa*; mg, *M. Grisea*; um, *U. maydis*; sc, *S. cerevisiae*; sp, *S. pombe*.

The mammalian σ_1 receptor is undoubtedly an integral membrane protein. The cloned human σ_1 receptor has been shown to localize to nuclear and endoplasmic reticulum membranes (25). Hydropathy analyses have led to the prediction of two different membrane topology models consisting of one or two transmembrane domains. Interestingly, the model with a single transmembrane domain proposed by Hanner et al. (19) identified a N-terminal region as the putative membrane-spanning domain, whereas the model with a single transmembrane domain proposed by Kekuda et al. (23) and Seth et al. (28,29) identified a region in the middle of the protein as the putative membrane-spanning domain. The alternative model with two transmembrane domains, predicted by Jbilo et al. (25) and by Pasternak and his coworkers (30,31), predict the N-terminal hydrophobic region as well as the internal hydrophobic region as potential membrane-spanning domains.

Recent studies by Aydar et al. (32) have provided evidence in support of the two transmembrane topology model. These investigators expressed the rat σ_1 receptor, which was originally cloned by Seth et al. (29) from rat brain, in *X. laevis* oocytes in the form of green fluorescent protein (GFP)-σ_1 receptor fusion proteins and assessed its membrane topology by using GFP-specific antibodies. They used two different fusion proteins, one containing GFP at the N-terminus of the σ_1 receptor and the other containing GFP at the C-terminus. Without permeabilizing the oocytes, GFP could not be detected with anti-GFP antibodies irrespective of whether GFP was present at the N-terminus or C-terminus. However, if the oocytes were permeabilized, anti-GFP antibodies could detect the protein with either of the fusion constructs. These data demonstrate that the N-terminus as well as the C-terminus of the σ_1 receptor are located on the cytoplasmic surface of the plasma membrane of the oocytes. This corroborates the two transmembrane topology model. Aydar et al. (32) used the program TMbase (www.isrec.isb-sib.ch) to predict the putative transmembrane domains of the σ_1 receptor. According to this program, the mammalian σ_1 receptor contains two transmembrane domains, one at the N-terminus (amino acid position 10-30) and the other in the middle of the protein (amino acid position 80-100). Recently, we used the program Sosui (http://sosui.proteome.bio.tuat.ac.jp/sosuiframe0.html) to predict the transmembrane domains in the σ_1 receptor. This program also predicted two putative transmembrane domains (amino acid positions 13-34 and 86-108). This model, depicted in Figure 5-3 for the human σ_1 receptor, is similar to that proposed by Aydar et al. (32).

Cloning and structural analysis of gene and promoter

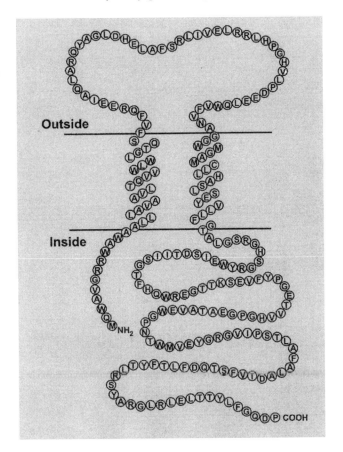

Figure 5-3. Membrane topology of human σ_1 receptor. The two putative transmembrane domains (amino acid regions 13-34 and 86-108) were identified based on the program Sosui (http://sosui.proteome.bio.tuat.ac.ip/sosuiframe().html). The individual amino acids are identified by their one letter codes.

4. EVIDENCE FOR AN ALTERNATIVE SPLICE VARIANT OF HUMAN σ_1 RECEPTOR

During the analysis of the σ_1 binding sites expressed in the human T lymphocyte cell line Jurkat, Ganapathy et al. (27) obtained evidence for the expression of an alternatively spliced variant of human σ_1 receptors in this cell line. This splice variant lacks 31 amino acids that correspond to the amino acid position 119-149 in the wild type σ_1 receptor. This region is located after the second putative transmembrane domain and therefore the membrane topology of the splice variant is likely to be similar to that of the wild type receptor. However, the splice variant does not have the ability to bind σ_1 ligands as assessed in heterologous expression systems (27). It is not known at present whether this splice variant is expressed in normal human tissues and, if it is expressed, whether it has any physiological or pathological significance.

5. CHROMOSOMAL LOCATION AND EXON-INTRON ORGANIZATION OF σ_1 RECEPTOR GENE

Prasad et al. (33) employed two different approaches to determine the chromosomal location of the gene coding for the human σ_1 receptor. Southern blot analysis of restriction fragments of genomic DNA from hybrid cell lines, each cell line harboring a single human chromosome, indicated that the gene is located on chromosome 9. Fluorescent *in situ* hybridization analysis confirmed the location of the gene on chromosome 9 and further indicated that the gene is located on the short arm of the chromosome at p13. The murine gene coding for the σ_1 receptor is located on chromosome 4 at the A5-B2 region (30).

The human gene encoding the σ_1 receptor is ~7 kp long and consists of four exons, 225, 154, 93, and 1,136 bp in size, interrupted by three introns (126, 177, and 1,138 bp in size) (33). Exon 1 contains the 5' untranslated region and part of the protein coding sequence. Exon 1 also codes for the protein region that constitutes the first putative transmembrane domain. Exon 2 codes for the region that contains the second putative transmembrane domain. Exon 3 is the shortest and codes for 31 amino acids. The alternative splice variant detected in the Jurkat cell line lacks the region coded by this particular exon. Thus, the splice variant is coded by exons 1, 2, and 4. Exon 4 contains the 3' untranslated region and also codes for part of the protein at the C-terminus. The murine gene has an exon-intron organization similar to that of the human gene (28).

6. ANALYSIS OF PROMOTER REGION OF HUMAN σ_1 RECEPTOR GENE

In our original paper describing the promoter region of the human σ_1 receptor gene, we predicted, based on the data from 5' RACE (5' rapid amplification of cDNA end) that the putative transcription start site lies 56 bp upstream of the translation start site (33). However, a recent search of the EST (expressed sequence tag) database has identified a human cDNA clone that contains a longer 5' untranslated region, indicating that the originally predicted transcription start site may not be correct. According to the EST clone, the transcription start site may actually lie at least 129 bp upstream of the translation start site. Analysis of the nucleotide sequence upstream of this putative transcription start site using the TESS-String-Based Search of the Transcription Factors Database (http://agave.humgen.upenn.edu/utess/tess) and allowing for zero mismatches indicate that the classic promoter element TATA box is absent in the expected position of the gene (i.e. within ~50 nucleotides upstream of the transcription start site). However, there are multiple GC boxes, three of them within ~100 nucleotides upstream of the transcription start site. These GC boxes are binding sites for the transcription factor SP1. This transcription factor plays an essential role in the expression of TATA-less genes. A CCAATC box in the reverse complement (GATTGG) is also present immediately upstream of the transcription start site. There are also consensus binding sites for the transcription factors nuclear factor (NF)-1/L, activator protein (AP)-1, AP-2, IL-6RE, NF-GMa, NF-GMb, steroid-responsive element, GATA-1, and Zeste, present within ~1 kb upstream of the transcription start site. Binding sites for the arylhydrocarbon receptor and the transcription factor NF-κB, both within 1 bp consensus, are also present in the putative promoter region. NF-GMa, NF-GMb, NF-κB, and IL-6RE are all cytokine responsive elements and the presence of these elements in the promoter region of the human σ_1 receptor may be relevant to the observations that the σ_1 receptor is expressed in a variety of immune cells and that the receptor plays a potential role in the modulation of immune function (4). Since steroids such as progesterone are believed to be endogenous ligands for the σ_1 receptor, the presence of a steroid-responsive element in the promoter region is of potential significance. The arylhydrocarbon receptor is a transcription factor that is involved in the regulation of gene expression in response to exposure of cells to xenobiotics. Since the σ_1 receptor binds a variety of structurally diverse xenobiotics, the presence of an arylhydrocarbon receptor binding site in the promoter region of the σ_1 receptor gene may also be potentially important.

AKNOWLEDGEMENTS

This work was supported by the National Institutes of Health Grant AI49849.

REFERENCES

1. Martin WR, Eades CG, Thompson JA, Huppler RE, Gilbert PE. The effects of morphine- and nalorphine-like drugs in the nondependent and morphine-dependent chronic spinal dog. J Pharmacol Exp Ther 1976, 197:517-532.
2. Ferris CD, Hirsch DJ, Brooks BP, Snyder SH. Sigma receptors: from molecule to man. J Neurochem 1991, 57:729-737.
3. Itzhak Y, Stein I. Sigma binding sites in the brain: an emerging concept of multiple sites and their relevance for psychiatric disorder. Life Sci 1990, 47:1073-1081.
4. Su TP. Sigma receptors. Putative link between nervous, endocrine and immune systems. Eur J Biochem 1991, 200:633-642.
5. Quirion R, Bowen WD, Itzhak Y, Junien JL, Musacchio JM, Rothman RB, Su TP, Tam SW, Taylor DP. A proposal for the classification of σ binding sites. Trends Pharmacol Sci 1992, 13:85-86.
6. Hellewell SB, Bowen WD. A σ-like binding site in the rat pheochromocytoma (PC12) cells: decreased affinity for (+)-benzomorphans and lower molecular weight suggest a different form from that of guinea pig brain. Brain Res 1990, 527:244-253.
7. Musacchio JM, Klein M, Canoll PD. Dextromethorphan and σ ligands: common sites but diverse effects. Life Sci 1989, 45:1721-1732.
8. Daven-Hudkins DL, Ford-Rice FY, Allen JT, Hudkins RL. Allosteric modulation of ligand binding to [^{3}H]-(+)-pentazocine-defined σ recognition sites by phenytoin. Life Sci 1993, 53:41-48.
9. Su TP, London ED, Jaffe JH. Steroid binding at σ receptor suggests a link between endocrine, nervous, and immune systems. Science 1988, 240:219-221.
10. McCann DJ, Su TP. Solubilization and characterization of haloperidol sensitive (+)-[^{3}H]-SKF-10,047 binding sites (σ sites) from rat liver membranes. J Pharmacol Exp Ther 1991, 257:547-554.
11. Kahoun JR, Ruoho AE. [^{125}I]Iodoazidococaine, a photoaffinity label for the haloperidol-sensitive σ receptor. Proc Natl Acad Sci USA 1992, 89:1393-1397.
12. Hellewell SB, Bruce A, Feinstein G, Orringer J, Williams W, Bowen WD. Rat and liver and kidney contain high densities of σ_1 and σ_2 receptors: characterization by ligand binding and photoaffinity labeling. Eur J Pharmacol 1994, 268:9-18.
13. Carr DJ, Mayo S, Wooley TW, De Costa BR. Immunoregulatory properties of (+)-pentazocine and σ ligands. Immunology 1992, 77:527-531.
14. Garza HH, Mayo S, Bowen WD, De Costa, BR, Carr DJ. Characterization of a (+)-pentazocine-sensitive σ receptor on splenic lymphocytes. J Immunol 1992, 151:4672-4680.
15. Wu XZ, Bell JA, Spivak CE, London ED, Su TP. Electrophysiological and binding studies on intact NCB-20 cells suggest presence of a low affinity σ receptor. J Pharmacol Exp Ther 1991, 257:351-359.

16. Klein M, Canoll PD, Musacchio JM. SKF 525A and cytochrome P450 ligands inhibit with high affinity the binding of [^{3}H]-dextromethorphan and σ ligands to guinea pig brain. Life Sci 1991, 48:543-550.
17. Basile AS, Paul IA, De Costa B. Differential effects of cytochrome P450 induction on ligand binding to σ receptors. Eur J Pharmacol 1992, 227:95-98.
18. Schuster DI, Ehrlich GK, Murphy RB. Purification and partial amino acid sequence of a 28 kDa cyclophilin-like component of the rat liver σ receptor. Life Sci 1994, 55:PL151-156.
19. Hanner M, Moebius FF, Flandorfer A, Knaus H-G, Striessnig J, Kempner E, Glossmann H. Purification, molecular cloning, and expression of the mammalian $σ_1$ binding site. Proc Natl Acad Sci USA 1996, 93:8072-8077.
20. Hanner M, Moebius FF, Weber F, Grabner M, Streissnig J, Glossmann H. Phenylalkylamine Ca^{2+} antagonist binding protein. Molecular cloning, tissue distribution, and heterologous expression. J Biol Chem 1995, 270:7551-7557.
21. Silve S, Dupuy PH, Labit-Lebouteiller C, Kaghad M, Chalon P, Rahier A, Taton M, Lupker J, Shire D, Loison G. Emopamil-binding protein, a mammalian protein that binds a series of structurally diverse neuroprotective agents, exhibits Δ8-Δ7 sterol isomerase activity in yeast. J Biol Chem 1996, 271:22434-22440.
22. Mobius FF, Streissnig J, Glossmann H. The mysteries of σ receptors: new family members reveal a role in cholesterol synthesis. Trends Pharmacol Sci 1997, 18:67-70.
23. Kekuda R, Prasad PD, Fei YJ, Leibach FH, Ganapathy V. and functional expression of the human type 1 σ receptor (hSigmaR1). Biochem Biophys Res Commun 1996, 229:553-558.
24. Ramamoorthy JD, Ramamoorthy S, Mahesh VB, Leibach FH, Ganapathy V. Cocaine-sensitive σ-receptor and its interaction with steroid hormones in the human placental syncytiotrophoblast and in choriocarcinoma cells. Endocrinology 1995, 136:924-932.
25. Jbilo O, Vidal H, Paul R, De Nys N, Bensaid M, Silve S, Carayon P, Davi D, Galiegue S, Bourrie B, Guillemot J-C, Ferrara P, Loison G, Maffrand J-P, Le Fur G, Casellas P. Purification and characterization of the human SR 31747A-binding protein. A nuclear membrane protein related to yeast sterol isomerase. J Biol Chem 1997, 272:27107-27115.
26. Vilner BJ, John CS, Bowen WD. Sigma-1 and $σ_2$ receptors are expressed in a wide variety of human and rodent tumor cell lines. Cancer Res 1995, 55:408-413.
27. Ganapathy ME, Prasad PD, Huang W, Seth P, Leibach FH, Ganapathy V. Molecular and ligand-binding characterization of the σ receptor in the Jurkat human lymphocyte cell line. J Pharmacol Exp Ther 1999, 289:251-260.
28. Seth P, Leibach FH, Ganapathy V. Cloning and structural analysis of the cDNA and the gene encoding the murine type 1 σ receptor. Biochem Biophys Res Commun 1997, 41:535-540.
29. Seth P, Fei YJ, Li HW, Huang W, Leibach FH, Ganapathy V. Cloning and functional characterization of a σ receptor from rat brain. J Neurochem 1998, 70:922-931.
30. Pan Y-X, Mei J, Xu J, Wan B-L, Zuckerman A, Pasternak GW. Cloning and characterization of a mouse $σ_1$ receptor. J Neurochem 1998, 70:2279-2285.
31. Mei J and Pasternak GW. Molecular cloning and pharmacological characterization of the rat $σ_1$ receptor. Biochem Pharmacol 2001, 62:349-355.
32. Aydar E, Palmer CP, Klyachko VA, Jackson MB. The σ receptor as a ligand-regulated auxiliary potassium channel subunit. Neuron 2002, 34:399-410.
33. Prasad PD, Li HW, Fei YJ, Ganapathy ME, Fujita T, Plumley LH, Yang-Feng TL, Leibach FH, Ganapathy V. Exon-intron structure, analysis of promoter region, and

chromosomal localization of the human type 1 σ receptor gene. J Neurochem 1998, 70:443-451.

Corresponding author: *Dr. Vadivel Ganapathy, Mailing address: Medical College of Georgia, Department of Biochemistry and Molecular Biology, Room #2208, Research and Education Building, 1459 Laney Walker Boulevard, Augusta, GA 30912 USA, Phone: (706) 721-7652, Fax: (706) 721-9947, Electronic mail address: vganapat@mail.mcg.edu*

Chapter 6

SITE-DIRECTED MUTAGENESIS

Hideko Yamamoto[1], Toshifumi Yamamoto[1,2], Keiko Shinohara Tanaka[1], Mitsunobu Yoshii[1,3], Shigeru Okuyama[4], Toshihide Nukada[5]
[1]*Department of Molecular Psychiatry, Tokyo Institute of Psychiatry, Tokyo 156-8585, Japan,*
[2]*Molecular Recognition, Yokohama City University, Yokohama 236-0027, Japan,*
[3]*Department of Neural Plasticity, Tokyo Institute of Psychiatry, Tokyo 156-8585, Japan,* [4]*The 1st Laboratory, Medicinal Research Laboratory, Taisho Pharmaceutical, Saitama 330-8530, Japan,* [5]*Department of Neuronal Signaling, Tokyo Institute of Psychiatry, Tokyo 156-8585, Japan*

1. INTRODUCTION

Sigma receptors were initially postulated to account for the psychotomimetic actions of (±)-SKF-10,047 (N-allylnormetazocine), a congener of morphine (1). Subsequent biochemical and pharmacological studies using radioligands have demonstrated that the σ receptors are distinct from opioid or phencyclidine binding sites, and now they are classified into at least two subtypes designated 'σ_1' and 'σ_2' (2). The σ_1 sites have high affinities and stereoselectivities for the (+)-isomers of SKF-10,047, pentazocine and cyclazocine, whereas the σ_2 sites have lower affinities and the opposite stereoselectivities for these agents (3). (+)-Pentazocine is a well-known ligand used as a σ_1 agonist. N,N'-Di(2-o-tolyl)-guanidine (DTG), (+)-3-(3-hydroxyphenyl)-N-(1-propyl)-piperidine [(+)-3-PPP] and haloperidol are non-discriminating ligands with high affinity for the two subtypes. There are many reports that haloperidol and N,N-dipropyl-2-(4-methoxy-3-(2-phenylethoxy)phenyl) ethylamine monohydrochloride (NE-100) act as antagonists in physiological and behavioral tests relevant to 'σ_1' pharmacology.

Several lines of evidence indicate that σ receptors in the central nervous system (CNS) are involved in drug addiction and affective disorders (4,5) and that those in the immune system are involved in immunoregulation (6). Their physiological roles, however, remain to be clarified, and their endogenous ligands have also yet to be determined.

Recently, cDNAs of the type 1 σ receptor (σ_1 receptor) were cloned, and subsequently, their amino acid sequences were deduced (7-12). The primary structure of the σ_1 receptor is highly conserved among a variety of mammalian species and tissues, indicating its importance in cellular functions. Interestingly, a protein targeted by SR 31747A, a σ ligand and novel immunosuppressant, which is also called the SR 31747A binding protein, has an identical structure to that of the σ_1 receptor and a pharmacological profile very similar to that of the σ_1 receptor (13). In addition, many antipsychotics, antidepressants and immunosuppressants have some affinity for σ receptors (2). Thus, the σ_1 receptor appears to be very important as a target for CNS drug development.

This chapter is focused on the structure-affinity relationship of ligands at the σ_1 receptor based on amino acid mutation studies. Arguably, the greatest insights into the structural basis underlying the biological function of σ_1 receptors have come from the analysis of mutant receptors: either site-directed mutagenesis or deletion. Mutant receptors have been used to examine, mainly by ligand binding assay, their entire function. The key to understanding receptor-ligand interactions is dependent on the site of point mutation. Deletion mutants can indicate the role of regions, or discrete domains within a receptor. For example, an alternatively spliced σ_1 receptor variant which lacks exon 3 is non-functional in σ_1 receptor ligand binding assays (14). More useful, at least in terms of support for molecular drug design, are single or multiple point mutants. Such mutagenesis is generally characterized as either silent, loss-of-function or gain-of-function mutants.

Mutagenesis studies implicating the involvement or non-involvement of individual amino acid residues in ligand binding to G protein coupled receptors (GPCRs) have accumulated. From these data, general rules are introduced in the control of binding of ligands to GPCRs. Before starting mutagenesis, several factors or conditions have to be considered and determined.

The first condition is which region should be chosen for mutation. Following the knowledge of general principles for GPCRs, that is small molecules, including pharmacological agents, bind within the transmembrane (TM) region of the receptor. Although this general picture is derived from accumulated mutagenesis data of GPCRs, one can use for the identification and experimental verification of the most likely site of ligand interaction in a model of the σ_1 receptor. Fortunately, as mentioned above, there is other information available to point out the coded region by exon 3. That is, the deletion of this region results in marked reduction of

[^{3}H]haloperidol binding (14). Therefore, a transmembrane region and the following C-terminal region coded by exon 3 would be chosen as a candidate region.

The next condition is which residues should be mutated. Usually, residues to be mutated are those predicted to affect or remove a molecular interaction – a hydrogen bond, charge pair or pi-stacking interaction.

The third condition is what kind of residues should substitute for. There are two directions; one is a conservative change with retaining its size and shape (i.e. serine to alanine, or tyrosine to phenylalanine); the other is a non-conservative change with abolished polarity or to a small residue (i.e. aspartate to glycine, or leusine to alanine).

At present, only two studies on mutant σ_1 receptor are available (15,16). In this article, we review the findings related to the structure-affinity relationship of σ_1 receptor based on these reports.

2. AMINO ACID RESIDUES IN THE TRANSMEMBRANE DOMAIN OF σ_1 RECEPTOR CRITICAL FOR LIGAND BINDING

In our previous study (15), guinea pig σ_1 receptor cDNA was cloned, and expressed in *Xenopus* oocytes by injection of cRNA derived from the cDNA. Using the highly selective σ_1 receptor agonist, (+)-pentazocine and the antagonist, NE-100 as well (17), the effects of site-directed mutagenesis of σ_1 receptor were investigated.

Aiming at identifying the regions on σ_1 receptor that interact with σ_1 ligands, four kinds of mutant σ_1 receptor were generated by site-directed mutagenesis and expressed in *Xenopus* oocytes (Figure 6-1A). Amino acid substitutions were performed by the analogy of GPCRs, in which ligand binding sites are assigned to the transmembrane domain (18-26). In mutant 'SA' and 'YF', Ser-99 and Tyr-103 were substituted by Ala and Phe, respectively. Mutant 'LA' had a substitution of di-Ala for di-Leu 105 and 106, and mutant 'SLA' had an additional substitution of Ala for Ser-99 on the mutant LA.

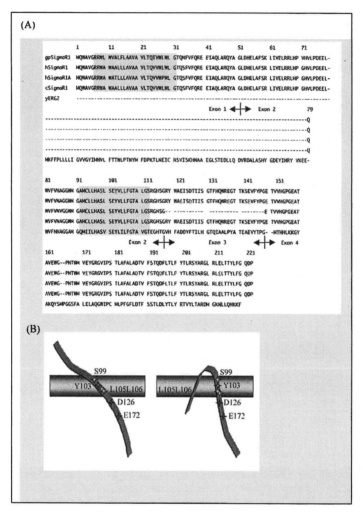

Figure 6-1. Deduced amino acid sequences of σ_1 receptor, splice variant of σ_1 receptor (σ_{1A} receptor), and ERG2 gene product. The amino acid sequences of σ_1 receptor and σ_{1A} receptor are highly conserved. Although σ_1 receptors share no homology with known mammalian proteins, the amino acid sequence is found to be related to fungal C8-C7 sterol isomerase, encoded by the ERG2 gene (35). However, $[^3H](+)$-pentazocine does not bind to Erg2 gene products (13). The putative transmembrane domains are located in the box (16). Underlined amino acids are substituted with alanine, phenylalanine or glycine in mutagenesis experiments. Also, the amino acid sequence of σ_1 receptors derived from COS-7 cells is shown for comparison. CSigmaR1, COS-7 σ_1 receptor; yERG2, yeast sterol isomerase (*Saccharomyces cerevisiae* C-8 sterol isomerase (ERG2) gene).

6. Site-directed mutagenesis

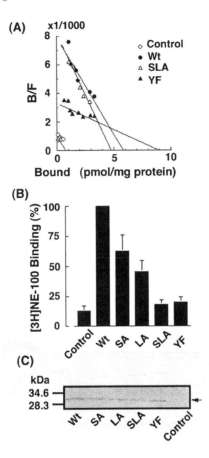

Figure 6-2. (A) Representative Scatchard plots of [^{3}H](+)-pentazocine binding using membranes (2.0-2.7 µg) prepared from *Xenopus* oocytes without cRNA injection (Control) or after injection of wild-type (Wt) or mutant (SLA or YF) σ_1 receptor cRNA. (B) Specific [^{3}H]NE-100 binding in membranes (0.7-14.1 µg) from *Xenopus* oocytes without cRNA injection (Control) or after injection of wild-type (Wt) or mutant (SA, LA, SLA or YF) σ_1 receptor cRNA. The data are represented as the ratio of [^{3}H]NE-100 binding to that obtained from oocytes injected with wild-type σ_1 receptor cRNA (Wt). The original binding before normalization was 2.43 ± 0.59, 26.8 ± 10.7, 19.0 ± 9.2, 10.6 ± 4.5, 4.0 ± 1.4 and 4.3 ± 1.9 pmol/mg protein for the Control, Wt, SA, LA, SLA and YF, respectively (n=5). To lessen seasonal or individual variations, a set of oocytes injected with cRNAs for wild-type and mutant σ_1 receptor was prepared at the same time for binding assays. (C) Immunoreactivities with the anti-σ_1 receptor antiserum to the membrane fractions prepared from *Xenopus* oocytes without cRNA injection (Control) or after injection of wild-type (Wt) or mutant (SA, LA, SLA and YF) σ_1 receptor cRNA. Note that immunoreactive 29 kDa polypeptides (arrow) were not appreciably detectable unless cRNA for wild-type or mutant σ_1 receptor was injected. SA, S99A; LA, LL105106AA; SLA, SA + LA; YF, Y103F. Reprinted from ref. (15).

2.1 Functional expression of wild-type or mutant σ_1 receptor in *Xenopus* oocyte

Based on the cDNA sequence for the guinea pig σ_1 receptor (7), wild-type σ_1 receptor cDNA was cloned from guinea pig liver. The genetic mutations of the transmembrane domain of the wild-type σ_1 receptor cDNA were performed with PCR primers containing point mutations. cRNA specific for the wild-type σ_1 receptor and mutant receptor cRNAs were synthesized *in vitro* by a MEGAscript kit (Ambion).

After removal of the follicular cell layer by treatment with 1 mg/ml collagenase for 1.5 h at 20°C (27), *Xenopus* oocytes were injected either with 0.5 µg/µl of the wild-type or a mutant σ_1 receptor cRNA: the average volume of injection was ~ 50 nl per oocyte. The injected oocytes were incubated for 3 days, and then, subjected to immunoblot analysis and binding assay.

Membrane fractions were obtained from oocytes (28). These fractions (0.5 µg) were separated by 12.5% SDS-PAGE and transferred to a FluoroTrans membrane (Pall BioSupport) (29). For immunostaining of σ_1 receptor, a polyclonal antiserum selectively staining σ_1 receptor was used (15). Labeled proteins were determined by diaminobenzidine-based HRP products with heavy metal intensification (30). In oocytes injected with the wild-type σ_1 receptor cRNA, the antiserum reacted with a polypeptide of 29 kDa (Figure 6-2C, Wt: arrow). Levels of the immunoreactivities in various mutant σ_1 receptor were comparable with those in the wild-type σ_1 receptor (Figure 6-2C).

2.2 Ligand binding activities in wild-type and mutant σ_1 receptor

Radioligand binding assays with [^{3}H](+)-pentazocine and [^{3}H]NE-100 were performed using oocyte membrane fractions according to the methods as described previously (31,32). Saturation experiments were conducted over a concentration range of 2.0 - 15 nM [^{3}H](+)-pentazocine, and NE-100 binding was carried out using 2.0 nM [^{3}H]NE-100. Nonspecific binding was determined in the presence of 1 µM of haloperidol.

To test whether mutations in the transmembrane domain affect binding activities for σ_1 ligands, saturation binding assays with [^{3}H](+)-pentazocine on *Xenopus* oocyte membranes were carried out after injection of the wild-type or a mutant σ_1 receptor cRNA. Binding of [^{3}H](+)-pentazocine to membranes prepared from oocytes injected with the wild-type σ_1 receptor cRNA was markedly increased (Figure 6-2A, Wt) as compared with those from oocytes without injection of σ_1 receptor cRNA (Figure 6-2A, control). When [^{3}H]NE-100 was used as a radioligand instead of [^{3}H](+)-pentazocine, an increase in the amount of binding was also produced by injection of the wild-type σ_1 receptor cRNA (Figure 6-2B, control and Wt).

Scatchard analysis of the [^{3}H](+)-pentazocine binding to the membranes prepared from oocytes expressing wild-type σ_1 receptor resulted in a linear plot (Figure 6-2A, Wt) suggesting a single binding site for the exogenously-expressed σ_1 receptor in *Xenopus* oocytes as reported in native tissue membranes (33). The dissociation constant (K_d) was 11.7 ± 1.2 nM (n=6) (Table 6-1). This Kd value was also comparable with that (14.2 ± 2.3 nM, n=3) obtained from binding of [^{3}H](+)-pentazocine to rat crude synaptosomal membranes prepared from rat cerebral cortex. Thus, all these findings indicate that exogenous wild-type σ_1 receptor was functionally expressed in *Xenopus* oocytes. As in the case of wild-type σ_1 receptor, binding analysis with [^{3}H](+)-pentazocine was performed using membranes from oocytes expressing each mutant receptor. A decrease in the affinity of [^{3}H](+)-pentazocine for σ_1 receptor was produced in these three kinds of mutant, SA, LA and YF (Table 6-1, Figure 6-2A: SA, LA and YF), as compared with wild-type σ_1 receptor. The rank order of binding affinity for [^{3}H](+)-pentazocine among these mutant σ_1 receptor was SA>LA>YF (Table 6-1). By contrast, the mutant SLA, having combined mutations in SA and LA, did not show a marked change in the affinity (Table 6-1, Figure 6-2A). Since both wild-type and four kinds of mutant σ_1 receptor were similarly expressed as immunoreactive polypeptides (Figure 6-2C), it seems likely that these changes in the ligand binding affinity by the mutations are due to direct perturbation of the σ ligand-σ_1 receptor interactions.

Table 6-1. Effects of amino acid substitutions in the transmembrane domain of σ_1 receptor on binding affinity for [^{3}H](+)-pentazocine

σ_1 Receptor	K_d (nM)	
None (Control)	30.4 ± 12.6	(3)
Wt	11.7 ± 1.2	(6)
SA	36.6 ± 3.9	(5)*
LA	42.0 ± 4.2	(5)*
SLA	9.9 ± 1.6	(5)
YF	49.0 ± 8.0	(6)*

The dissociation constants (K_d) were determined by saturation binding of [^{3}H](+)-pentazocine using membrane fractions prepared from *Xenopus* oocytes without cRNA injection [None (Control)] or after injection of wild-type (Wt) or mutant (SA, LA, SLA or YF) σ_1 receptor cRNA (see Fig. 6-1A and 6-2A). SA, S99A; LA, LL105106AA; SLA, SA + LA; YF, Y103F. The number of experiments is indicated in parentheses. *P < 0.01 relative to wild-type σ_1 receptor. The dissociation constant (K_d) was determined using the computer program, Ligand (36). Statistical data are represented by the mean ± S.E.M. Reprinted from ref. (15).

These mutations in the transmembrane domain of σ_1 receptor also diminished specific [^{3}H]NE-100 binding (Figure 6-2B: SA, LA and YF) as compared with wild-type σ_1 receptor (Figure 6-2B, Wt). The rank order of binding ability for [^{3}H]NE-100 was SA>LA>YF, which was identical to that for [^{3}H](+)-pentazocine, indicating that these three kinds of mutations exert similar effects on both agonist and antagonist binding to σ_1 receptor. Among these mutations, the single YF mutation yielded a most appreciable decrease in binding activities for both [^{3}H](+)-pentazocine and [^{3}H]NE-100. Therefore, it is indicated that the amino acid residue, Tyr-103, in the transmembrane domain is critical for ligand binding to σ_1 receptor. Unlike [^{3}H](+)-pentazocine binding, [^{3}H]NE-100 binding was almost completely abolished in the mutant SLA (Figure 6-2B, SLA). This discrepancy between (+)-pentazocine and NE-100 binding in the mutant SLA implies the presence of different recognition site(s) in the transmembrane domain of σ_1 receptor for different ligands such as agonist and antagonist. The amino acid residues, Ser-99, Leu-105 and Leu-106 in the domain might play an important role in differentiating ligands through interaction with the corresponding parts of these ligands. Further studies using additional σ_1 ligands and site-directed mutagenesis will be necessary to determine the direct interaction between specific amino acid residues on σ_1 receptor and specific σ_1 ligands.

3. AMINO ACID RESIDUES IN THE C-TERMINAL HALF OF σ_1 RECEPTOR CRITICAL FOR HALOPERIDOL BINDING

To identify the critical amino acid residues, Seth et al. (16) have performed two different approaches: one is chemical modification, and the other is site-directed mutagenesis. Chemical modification of anionic amino acids in σ_1 receptor with 1-ethyl-3-(3-dimethylaminopropyl) carbodiimide (EDC) markedly reduces σ ligand activities. Treatment of the JAR cell (human placental choriocarcinoma cell) membranes and MCF cell (human breast tumor cell) membrane with EDC (1 mM) causes a more than 60% inhibition of the bindings of [^{3}H]haloperidol and [^{3}H](+)-pentazocine. It is known that a splice variant of σ_1 receptor which lacks exon 3 does not have the ability to bind σ ligands (14). It suggests that existence of the C-terminal region coded by exon 3 is critical for σ_1 receptor ligand binding and is important for constructing a functional binding pocket.

Most of the σ ligands are positively charged. For this reason, Seth et al. (16) mutated each of twelve anionic amino acid residues located in the region coded by exons 3 and 4 with neutral amino acid, glycine (Figure 6-1A). Site-directed mutagenesis has been performed using the QuickChange™ site-directed mutagenesis kit (Stratagene, CA). They employ [^{3}H]haloperidol binding to assess the influence of each mutation on binding activity to the prepared membranes derived from MCF-7 cells expressing mutated σ_1 receptor.

The prepared membranes are incubated with [^{3}H]haloperidol in 5 mM potassium-phosphate buffer (pH 7.5) for 3 h. Nonspecific binding is determined in the presence of 10 µM haloperidol (14,16). Most of these mutations (Glu123Gly, Glu138Gly, Glu144Gly, Glu150Gly, Glu158Gly, Glu163Gly, Asp188Gly, Asp195Gly, Glu213Gly, Glu222Gly) have not affected the binding activities (75-120% compared to the wild type σ_1 receptor) (16). However, of interest, they have determined that two anionic amino acids, Asp126 and Glu172, are obligatory for [^{3}H]haloperidol binding to σ_1 receptor. These two mutants show a potently reduced [^{3}H]haloperidol binding (less than 10% of the wild type σ_1 receptor activity). Asp126 is coded by exon 3, and Glu172 is coded by exon 4. To exclude the possibility that these mutations might induce an enhanced degradation of protein due to decreases in the stability, they confirmed the intensity of expression of these two mutant σ_1 receptor in MCF-7 cells. These findings provide additional information on the chemical nature of the ligand-binding site of σ_1 receptor. More recently, Ablordeppy et al. (34) have represented interesting features of σ_1 receptor ligand and proposed a model to account for the binding of ligands at σ_1 receptor, in which a nitrogen atom on the longer alkyl chain appears to be an important pharmacophoric element. A nitrogen atom at a proton donor site does exist in haloperidol molecule as well as in (+)-pentazocine and NE-100 molecules (Figure 6-3). The nitrogen atom might interact with these two anionic amino acid of σ_1 receptor, Asp126 and Glu172.

4. CONCLUSIONS

Up to data, few studies on site-mutagenesis of σ_1 receptor are available for the topological assignments of the receptor (15,16). Hydropathy analysis of the primary sequences of σ_1 receptor predicts a topology with one putative transmembrane segment or more than two putative transmembrane domains (7-12) (Figure 6-1B). According to the mutagenesis studies, there appears to be a binding pocket forming proximal to Tyr-103 in the primary sequence of σ_1 receptor. The agonist and antagonist binding sites overlap in this point, although mutagenesis around Tyr-103 may lead to unpredictable consequences of receptor-ligand interactions. On the other hand, another probable binding sites are proposed by Seth et al. (16); that is, two acidic amino acid residues in the C-terminal of σ_1 receptor: Asp-126 and Glu-172. As shown in Figure 6-3, a chemical bond by static force between a protonable nitrogen of ligand and aspartate or glutamate residues of the wild-type σ_1 receptor may occur. Based on the pharmacophore model by Glennon's group (34), it can be predicted that the primary hydrophobic site or a secondary binding site (containing a phenyl group) in σ_1 receptor ligands might interact with hydrophobic transmembrane binding pocket of σ_1 receptor (Figure 6-3). Taken together, we propose a hypothesis that a typical σ_1 receptor ligand displays at least three stereotypical binding sites and interact with multiple corresponding recognition sites of σ_1 receptor molecule. To define this hypothesis, further topological assignments of σ_1 receptor are necessary using different approaches such as x-ray crystallography.

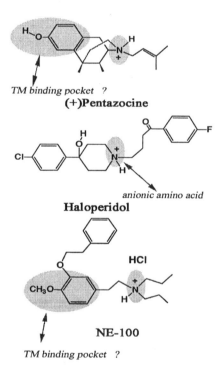

Figure 6-3. Chemical structures of (+)-pentazocine, haloperidol and NE-100

REFERENCES

1. Martin WR, Eades CG, Thomson JA, Hoppler RE, Gilbert PE. The effects of morphine and nalorphine like drugs in the nondependent and morphine dependent chronic spinal dog. J Pharmacol Exp Ther 1976, 197:517-532.
2. Walker JM, Bowen WD, Walker FO, Matsumoto RR, De Costa B, Rice KC. Sigma receptors: Biology and function. Pharmacol Rev 1990, 42:355-402.
3. Hellewell SB, Bruce A, Feinstein G, Orringer J, Williams W, Bowen WD. Rat liver and kidney contain high densities of σ_1 and σ_2 receptors: characterization by ligand binding and photoaffinity labeling. Eur J Pharmacol 1994, 268:9-18.
4. Weissman AD, Casanova MF, Kleinman JE, London ED, De Souza EB. Selective loss of cerebral cortical sigma, but not PCP binding sites in schizophrenia. Biol Psychiatry 1991, 29:41-54.
5. Witkin JM. Pharmacotherapy of cocaine abuse: Preclinical development. Neurosci Biobehav Rev 1994, 18:121-142.
6. Liu Y, Whitlock BB, Pultz JA, Wolfe Jr SA. Sigma-1 receptors modulate functional activity of rat splenocytes. J Neuroimmunol 1995, 59:143-154.
7. Hanner M, Moebius FF, Flandorfer A, Knaus H-G, Striessnig J, Kempner E, Glossmann H. Purification, molecular cloning, and expression of the mammalian σ_1-binding site. Proc Natl Acad Sci USA 1996, 93:8072-8077.
8. Kekuda R, Prasad PD, Fei Y-J, Leibach FH, Ganapathy V. Cloning and functional expression of the human type 1 σ receptor (σ_1 receptor). Biochem Biophys Res Commun 1996, 229:553-558.
9. Seth P, Leibach FH, Ganapathy V. Cloning and structural analysis of the cDNA and the gene encoding the murine type 1 σ receptor. Biochem Biophys Res Commun 1997, 241:535-540.
10. Prasad PD, Li HW, Fei Y-J, Ganapathy ME, Fujita T, Plumley LH, Yang-Feng TL, Leibach FH, Ganapathy V. Exon-intron structure, analysis of promoter region, and chromosomal localization of the human type 1 receptor gene. J Neurochem 1998, 70:443-451.
11. Seth P, Fei Y-J, Li HW, Huang W, Leibach FH, Ganapathy V. Cloning and functional characterization of a σ receptor from rat brain. J Neurochem 1998, 70:922-931.
12. Pan Y-X, Mei J, Xu J, Wan B-L, Zuckerman A, Pasternak GW. Cloning and characterization of a mouse σ_1 receptor. J Neurochem 1998, 70:2279-2285.
13. Jbilo O, Vidal H, Paul R, De Nys N, Bensaid M, Silve S, Carayon P, Davi D, Galiegue S, Bourrie B, Guillemot J-C, Ferrara P, Loison G, Maffrand J- P, Le Fur G, Casellas P. Purification and characterization of the human SR 31747A-binding protein: A nuclear membrane protein related to yeast sterol isomerase. J Biol Chem 1997, 272:27107-27115.
14. Ganapathy ME, Prasad PD, Huang W, Seth P, Leibach FH, Ganapathy V. Molecular and ligand-binding characterization of the σ-receptor in the Jurkat human T lymphocyte cell line. J Pharm Exp Ther 1999, 289:251–260.
15. Yamamoto H, Miura R, Yamamoto T, Shinohara K, Watanabe M, Okuyama S, Nakazato A, Nukada T. Amino acid residues in the transmembrane domain of the type 1 σ receptor critical for ligand binding. FEBS Lett 1999, 445:19-22.
16. Seth P, Ganapathy ME, Conway SJ, Bridges CD, Smith SB, Casellas P, Ganapathy V. Expression pattern of the type 1 σ receptor in the brain and identity of critical anionic amino acid residues in the ligand-binding domain of the receptor. Biochim Biophys Acta 2001, 1540:59-67.

17. Okuyama S, Imagawa Y, Ogawa S, Araki H, Ajima A, Tanaka M, Muramatsu M, Nakazato A, Yamaguchi K, Yoshida M, Otomo S. NE-100, a novel σ receptor ligand: in vivo tests. Life Sci 1993, 53:PL285-290.
18. Javitch JA, Li X, Kaback J, Karlin A. A cysteine residue in the third membrane-spanning segment of the human D_2 dopamine receptor is exposed in the binding-site crevice. Proc Natl Acad Sci USA 1994, 91:10355-10359.
19. Page KM, Curtis CAM, Jones PG, Hulme EC. The functional role of the binding site aspartate in muscarinic acetylcholine receptors, probed by site-directed mutagenesis. Eur J Pharmacol 1995, 289:429-437.
20. Dixon RA, Sigal IS, Rands E, Register RB, Candelore MR, Blake AD, Strader CD. Ligand binding to the beta-adrenergic receptor involves its rhodopsin-like core. Nature 1987, 326:73-77.
21. Strader CD, Sigal IS, Register RB, Candelore MR, Rands E, Dixon RA. Identification of residues required for ligand binding to the beta-adrenergic receptor. Proc Natl Acad Sci USA 1987, 84:4384-4388.
22. Kikkawa H, Isogaya M, Nagao T, Kurose H. The role of the seventh transmembrane region in high affinity binding of a beta 2-selective agonist TA-2005. Mol Pharmacol 1998, 53:128-134.
23. Strader CD, Sigal IS, Candelore MR, Rands E, Hill WS, Dixon RA. Conserved aspartic acid residues 79 and 113 of the beta-adrenergic receptor have different roles in receptor function. J Biol Chem 1988, 263:10267-10271.
24. Fraser CM, Wang C-D, Robinson DA, Gocayne JD, Graig Venter J. Site-directed mutagenesis of m_1 muscarinic acetylcholine receptors: Conserved aspartic acids play important roles in receptor function. Mol Pharmacol 1989, 36:840-847.
25. Strader CD, Candelore MR, Hill WS, Sigal IS, Dixon RA. Identification of two serine residues involved in agonist activation of the beta-adrenergic receptor. J Biol Chem 1989, 264:13572-13578.
26. Strader CD, Sigal IS, Dixon RA. Structural basis of beta-adrenergic receptor function. FASEB J 1989, 3:1825-1832.
27. Takao K, Yoshii M, Kanda A, Kokubun S, Nukada T. A region of the muscarinic-gated atrial K^+ channel critical for activation by G protein subunits. Neuron 1994, 13:747-755.
28. Goodhardt M, Ferry N, Buscaglia M, Baulieu EE, Hanoune J. Does the guanine nucleotide regulatory protein Ni mediate progesterone inhibition of Xenopus oocyte adenylate cyclase? EMBO J 1984, 3:2653-2657.
29. Towbin H, Staehelin T, Gordon J. Electrophoretic transfer of proteins from polyacrylamide gels to nitrocellulose sheets: Procedure and some applications. Proc Natl Acad Sci USA 1979, 76:4350-4354.
30. Adams JC. Heavy metal intensification of DAB-based HRP reaction product. J Histochem Cytochem 1981, 29:775.
31. Chaki S, Tanaka M, Muramatsu M, Otomo S. NE-100, a novel potent σ ligand, preferentially binds to $σ_1$ binding sites in guinea pig brain. Eur J Pharmacol 1994, 251:R1-R2.
32. Sagi N, Yamamoto H, Yamamoto T, Okuyama S, Moroji T. Possible expression of a $σ_1$ site in rat pheochromocytoma (PC12) cells. Eur J Pharmacol 1996, 304:185-190.
33. DeHaven-Hudkins DL, Fleissner LC, Ford-Rice FY. Characterization of the binding of [^{3}H](+)-pentazocine to σ recognition sites in guinea pig brain. J Pharmacol 1992, 227:371-378.
34. Ablordeppey SY, Fischer JB, Law H, Glennon RA. Probing the proposed phenyl-A region of the sigma-1 receptor. Bioorg Med Chem 2002, 10:2759-2765.

6. Site-directed mutagenesis

35. Ashman WH, Barbuch RJ, Ulbright CE, Jarrett HW, Bard M. Cloning and disruption of the yeast C-8 sterol isomerase gene. Lipids 1991, 26:628-632.
36. Munson PJ, Rodbard D. Ligand: a versatile computerized approach for characterization of ligand-binding systems. Anal Biochem 1980, 107:220-239.

Corresponding author: *Dr. Hideko Yamamoto, Mailing address: Tokyo Institute of Psychiatry, Department of Molecular Psychiatry, 2-1-8 Kamikitazawa, Setagaya-ku, Tokyo 156-8585, Japan, Phone: 81-3-3304-5701 ext. 568, Fax: 81-3-3329-8035, Electronic mail address: yamahide@prit.go.jp*

Chapter 7

σ RECEPTOR MODULATION OF ION CHANNELS

Chris P. Palmer, Ebru Aydar and Meyer B. Jackson
Department of Physiology, University of Wisconsin, Madison, WI 53706, USA

1. INTRODUCTION

Many of the proposed functions of σ receptors relate to electrical excitability. At the cellular level, σ receptors modulate action potential firing in neurons (1,2), and contraction of various kinds of muscle fibers (3,4). σ Receptors have been implicated in the modulation of the release of dopamine (5,6), acetylcholine (7), and glutamate (8), as well as muscarinic receptor-stimulated phosphoinositide turnover (9). σ Receptors have a variety of effects at the system level, including antagonism of meth-amphetamine and cocaine sensitization, and antagonism of locomotor stimulation by cocaine (10). σ Receptor ligands have neuroprotective and antiamnesic activity (11). A link between σ receptors and schizophrenia has been implied by some studies of σ receptors, and by the antipsychotic and psychotomimetic actions of σ receptor ligands (10,12). Functions for σ receptors in endocrine and immune systems have also been suggested (10,13). σ Receptors are present in excitable cells such as neurons, heart muscle cells, and endocrine cells, as well as nonexcitable tissues such as liver and kidney (10). σ Receptors are also implicated in cancer cell biology (14,15), although the molecular mechanisms are unknown. Among the many functions attributed to σ receptors, some clearly have electrical manifestations. In others the electrical dimension is more remote, but ion channel involvement remains a possibility. Ion channels represent important functional targets of σ receptors, and this chapter will explore the nature and mechanisms of σ receptor interactions with members of this important class of membrane proteins.

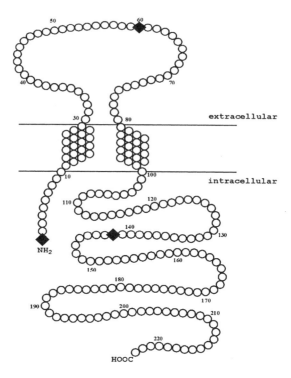

Figure 7-1. Topology of the σ receptor. The accessibility of various parts of the receptor, probed with GFP tags and surface biotinylation (21) indicated that the σ_1 receptor forms two membrane spanning helices, with cytoplasmic N- and C-termini and an extracellular loop. The diamonds represent primary amino groups targeted for biotinylation.

What is known about the structure of the σ receptor indicates that it is a novel protein for which there are no obvious effector targets or signal transduction pathways. Two subtypes of σ receptor, termed σ_1 and σ_2 are distinguishable pharmacologically, functionally, and by molecular size (16). The σ_2 receptor has been identified as an 18–21 kDa protein but has not yet been cloned. The σ_1 receptor has been cloned and its sequence sets it apart from all other known classes of receptor proteins (17-19). A σ receptor photolabel identified a protein with a molecular weight of 26 kDa (20) similar to that of 25.3 kDa predicted from the deduced amino acid sequence of the σ_1 receptor. This protein lacks significant homology with known mammalian proteins, but possesses weak homology with fungal sterol isomerase.

7. Modulation of ion channels

Hydropathy analysis of the deduced amino acid sequence identified three segments with significant hydrophobicity. It was originally suggested that only the most hydrophobic of these segments spanned a lipid bilayer. Aydar et al. (21) showed that two of these segments span the plasma membrane when the protein is expressed in *Xenopus* oocytes. Both the N- and C-termini reside at the cytoplasmic face of the membrane, as shown in a topological illustration (Figure 7-1). σ_1 Fusion proteins with GFP at either terminus maintained the capacity for ion channel modulation when coexpressed with channels in *Xenopus* oocytes. The GFP tags were inaccessible to anti-GFP antibodies until the oocytes were permeabilized. Conversely, surface biotin labeling of primary amines on the σ_1 receptor (indicated by the filled diamonds in Figure 7-1) demonstrated that the loop between the two predicted transmembrane domains was accessible at the extracellular surface, but the N- and C-termini were not (21). The topological arrangement of the domains of the protein with respect to the membrane serves as a useful guide in the interpretation of physiological experiments on the regulation of ion channels by σ_1 receptors.

2. MODULATION OF POTASSIUM CHANNELS

σ Receptors modulate a variety of potassium channels in many different types of cells. The σ receptor ligand (+)-3-(3-hydroxyphenyl)-N-(1-propyl) piperidine (3-PPP) depolarizes sympathetic neurons of the mouse hypogastric ganglion (22). Three distinct potassium channels, the channel underlying the M-current, the channel underlying a calcium-activated potassium current, and the channel underlying the A-current were all inhibited by 3-PPP. These channels exhibited different sensitivities to 3-PPP in the order M-current > calcium-activated potassium current > A-current. The modulation of multiple channel types by σ receptors is a recurring theme in σ receptor physiology.

Whole cell voltage clamp studies by Wu et al. (23) in NCB-20 cells showed that several σ receptor ligands, haloperidol, BMY 14802 (α-(4-fluorophenyl)-4-(5-fluoro-2-pyrimidinyl)-1-piperazine-butanol), pentazocine, SKF-10,047 (N-allylnormetazocine), 3-PPP, phencyclidine, TCP, and MK-801 (5-methyl-10,11-dihydro-5H-dibenzo[a,d]cyclohepten-5,10-iminie hydrogen maleate), activated an apparent inward current, which was actually due to blockade of a tonic outward potassium current. [^{3}H](+)-SKF-10,047 labeled two sites in intact NCB-20 cells (K_d = 49 nM, B_{max} = 1.0 pmol/mg protein and K_d = 9.6 microM, B_{max} = 69 pmol/mg protein). The high affinity site was similar pharmacologically to the σ receptor assayed in membrane fragments from these cells. However, the low affinity site showed a slightly different profile, highlighted by a reverse stereoselectivity. Further work in NCB-20 cells using the whole-cell patch-clamp technique tested antipsychotics and naloxone, and found a rank order of potency of bromperidol > haloperidol > mosapramine = clocapramine > carpipramine > chlorpromazine > remoxipride > naloxone (24). Sulpiride, which does not bind σ receptors, induced only small inward currents. Additionally the authors investigated the effects of various inhibitors on the modulation of currents. Haloperidol responses were not affected by pretreatments with the non-σ receptor ligands sulpiride, dopamine, atropine, N-methyl-D-aspartate, 2-amino-7-phospho-noheptanoic acid, morphine, or ICS 205-930 (endo-8-methyl-8-azzbicyclo[3.2.1]oct-3-olindol-3-yl-carboxylate hydrobromide hemicarbonate).

The peptidergic nerve terminals of the rat neurohypophysis contain three distinct types of potassium channel with roles in the regulation of neuropeptide release (25). Two of these, a large conductance calcium-activated potassium channel and a potassium channel underlying a transient A-current, were reversibly inhibited by the σ receptor specific agonists SKF-10,047, pentazocine, and DTG (1,3-di-o-tolylguanidine) as well as the nonspecific σ receptor ligands haloperidol, (±)-2-(N-phenylethyl-N-propyl)-amino-5-hydroxy-tetralin (PPHT), and apomorphine (26) (Figure 7-2). Haloperidol reduced potassium current by the same fraction at all voltages without shifting the voltage dependence of activation and inactivation. In the rat neurohypophysis, all of the σ receptor ligands tested inhibited the A-current and the calcium-activated potassium current equally, suggesting that both channels are modulated by a single population of σ receptors (Figure 7-2C and 7-2D). By contrast, in mouse peptidergic nerve terminals one drug, U101958 (1-benzyl-4-aminomethyl-N-[(3'-isopropoxy)-2'-pyridyl] piperidine maleate) inhibited the calcium-activated potassium current but spared the A-current. This is an intriguing observation. It may mean that the mouse neurohypophysis, but not the rat neurohypophysis, has two receptors. However, if it is a single receptor that differentially couples to two of its ion channel targets in a ligand dependent fashion, this may hold a clue for the mechanism of signal transduction. This could reflect an additional allosteric transition of the receptor, or, alternatively, a contribution of domains on the channel protein to the binding of ligand (see Figure 7-9).

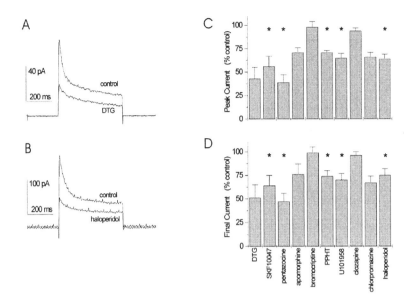

Figure 7-2. Inhibition of potassium current in neurohypophysial nerve terminals. Potassium current was evoked by 500 ms voltage steps from -80 mV to 10 mV. (A) Di-o-tolylguanidine (DTG, 100 μM) reduced both peak and sustained current. (B) Haloperidol (100 μM) had a similar effect. Peak current is predominantly the transient, A-current, with some contribution from calcium-actviated potassium current. Sustained current at the end of a pulse is predominantly the calcium-activated potassium current. Peak (C) and sustained (D) potassium currents were measured in the presence of various σ receptor ligands (all at 100 μM), using the pulse protocol of (A) and (B). Peak and sustained current were normalized to the pre-drug control for each nerve terminal, averaged, and plotted as mean + S.E.M.; * indicates P < 0.05. No statistics are given for DTG and chlorpromazine because only three measurements were made (26).

To link the pharmacological action of σ ligands to the σ receptor protein, Wilke et al. (27) combined patch clamp techniques with photoaffinity labeling in DMS-114 cells (a tumor cell line known to express σ receptors). SKF-10,047 (Figure 7-3A), ditolylguanidine, and PPHT all inhibited voltage-activated potassium current in these cells. Iodoazidococaine (IAC), a high affinity σ receptor photoprobe, produced a similar inhibition of potassium current (Figure 7-3B), and when cell homogenates were illuminated with high intensity ultra-violet light in the presence of IAC, a protein with a molecular mass of 26 kDa was covalently labeled (Figure 7-3C). Photolabeling of this protein by IAC was inhibited by SKF-10,047 with half-maximal effect at 7 μM (Figure 7-3D). The effect of SKF-10,047 on potassium current was also concentration dependent, with an EC_{50} of 14 μM. This study showed that physiological responses to σ receptor ligands are mediated by a protein with the same molecular size as the cloned σ_1 receptor. This provided an important link between the physiological response to σ receptor ligands, and the molecule identified as the σ_1 receptor. IAC also labeled a 26 kDa protein in the rat neurohypophysis, as well as exhibiting irreversible agonist activity (28).

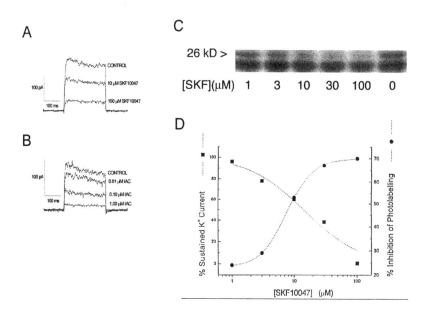

Figure 7-3. Inhibition of potassium current in cancer cells. (A) Whole-cell potassium current was measured in voltage-clamped DMS-114 cells. The membrane potential was stepped from -80 to 10 mV for 250 ms, and then returned to -80 mV. A concentration dependent reduction in potassium current was produced by the σ receptor ligand, SKF-10,047. (B) Iodoazidococaine (IAC) produced a similar reduction in potassium current but was significantly more potent (concentrations indicated to the right of the traces). (C) [^{125}I]IAC photolabeled a DMS-114 cell homogenate. SDS-PAGE and autoradiography revealed the σ receptor at ~26 kDa. The labeling of this band was selectively inhibited by SKF-10,047 in a concentration dependent fashion (concentrations indicated below each lane). (D) The SKF-10,047 concentration dependence of potassium current inhibition (left axis) and block of photolabeling (right axis) are plotted and compared. The curves are fit to single site binding expressions with EC_{50} = 7 μM for current inhibition and K_b = 14 μM for blocking of photolabeling.

3. MODULATION OF CALCIUM CHANNELS

A number of studies have implicated σ receptors in various forms of calcium signaling in both neuronal and nonneuronal cells (29-33). A series of σ receptor ligands was tested for modulation of voltage-gated Ca^{2+} channels in cultured hippocampal pyramidal neurons. Although many ligands strongly inhibited multiple subtypes of Ca^{2+} channels, the rank order of potency showed a marked departure from the rank order of σ receptor binding, indicating that σ receptors could not account for all of these actions (34).

A recent study in intracardiac and superior cervical ganglia indicated that σ receptors modulate multiple types of voltage-gated Ca^{2+} channels in both sympathetic and parasympathetic neurons (35). The high potency of ibogaine suggested that these responses were mediated by $σ_2$ receptors. However, PCR suggested the presence of $σ_1$ receptor mRNA in these cells. The modulation was voltage dependent, with drugs shifting the voltage dependence of channel activation. Pilot studies of various channel types in posterior pituitary nerve terminals indicated that the σ receptor ligand PPHT can modulate voltage-gated Ca^{2+} channels (Wilke, Lupardus, and Jackson, unpublished observations) as well as voltage-gated Na^+ channels (Hsu and Jackson, unpublished observations), but the pharmacological characterization of these responses is incomplete.

4. SIGNAL TRANSDUCTION MECHANISMS

Ion channels can be modulated by a vast number of receptor types, and the transduction mechanisms are the subject of an enormous body of research. The overwhelming majority of these cases follow a few basic mechanistic patterns involving activation of G proteins, or protein kinases, or both in combination (36-38). More recently, signal transduction mechanisms have been identified in which ion channels can be modulated by phospholipids such as phosphatidyl inositides (39) and reactive gases such as nitric oxide (40). The unique molecular structure of the σ receptor defies classification along the lines of established transduction mechanisms, and raises the question of whether σ receptors employ novel mechanisms to modulate ion channels.

Although some evidence suggests that σ receptors can modulate responses mediated by second messenger systems, the direct transduction pathway employed by σ receptors remains unclear (9). Some studies suggest that, despite their lack of homology with G protein coupled receptors, σ receptors utilize G proteins. In a patch clamp study in cultured frog pituitary melanotrophs, the σ receptor agonists DTG and (+)-pentazocine inhibited both a tonic potassium current and a delayed rectifier potassium current, and treatment with cholera toxin abolished this response (41). Further experiments determined that the inhibition of A-current by (+)-pentazocine was irreversibly prolonged when GTPγS was dialyzed into a cell (42). The authors hypothesized that the σ_1 receptor functionally interacts with G proteins through a mechanism that differs from that of classical G protein coupled receptors.

The studies in frog melanotrophs are exceptional; most physiological experiments suggest the opposite - that σ receptor signal transduction does not depend on G proteins. Morio et al. (24) showed that the inhibition of potassium channels in NCB-20 cells was not affected by pretreatments with A23187, 100 microM forskolin, 1 microM phorbol-12,13-dibutyrate, 100 ng/ml cholera toxin, or 100 ng/ml pertussis toxin. These results thus suggest that second messenger systems within a cell are not essential for the modulation of potassium channels by σ receptors.

Experiments in rat neurohypophysis also produced negative results for second messenger and G protein mediation (28). Pentazocine and SKF-10,047 modulation of potassium channels persisted while nerve terminals were internally perfused with GTP-free solutions, the G protein inhibitor GDPβS (Figure 7-4A and 7-4B), and the G protein activator GTPγS (Figure 7-4C and 7-4D). GDPβS and GTPγS modulated ion channels effectively on their own. The potassium current increased as GDPβS entered a nerve terminal (Figure 7-4A and 7-4B) and fell as GTPγS entered a nerve terminal (Figure 7-4C and 7-4D), providing a positive control for G protein manipulation. The inhibition of both peak and sustained potassium current by σ receptor ligands was quantitatively comparable to that seen in control recordings (Figure 7-4E and 7-5F). In DMS-114 cells perfusion with GDPβS also failed to alter responses to SKF-10,047 (27).

Along similar lines, the σ receptor-mediated modulation of voltage-gated Ca^{2+} channels in parasympathetic neurons remained strong in cells after internal dialysis with GTP free solutions, or solutions containing GDPβS (35). In the same study, the efficacy of these manipulations in blocking G proteins was validated by testing the modulation of Ca^{2+} channels by muscarinic receptors.

7. Modulation of ion channels

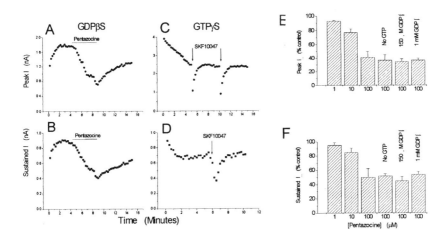

Figure 7-4. G protein independent modulation of potassium current. Potassium current was evoked in rat neurohypophysial nerve terminals by 250 ms voltage steps from -100 to 50 mV at 15 s intervals. As 1 mM GDPβS perfused a nerve terminal, both peak (A) and sustained (B) current increased. Application of 100 μM pentazocine (during the bar) reversibly inhibited both components. As 100 μM GTPγS perfused a nerve terminal, peak (C) and sustained (D) current decreased. Application of 100 μM SKF-10,047 (at the arrows) reversibly inhibited both components in two successive applications. (E) Peak potassium current normalized to control is shown for inhibition by 1, 10 and 100 μM pentazocine for control recording solutions with 100 mM GTP. The bars labeled as no GTP and GDPβS are for all responses to 100 μM pentazocine. (F) Same as (E) for sustained potassium current (28).

Similar negative results were obtained in tests for protein kinase mediation. Internal perfusion of nerve terminals with the non-hydrolysable ATP analogue AMPPcP had no effect on potassium current inhibition by σ receptor ligands (Figure 7-5A). Of particular significance was the observation that channels in excised outside-out patches were modulated by SKF-10,047 (Figure 7-5B and 7-5C). This excludes a role for any soluble cytoplasmic factor. In contrast, channels within cell-attached patches were not modulated by ligand application outside a patch, indicating that receptors and channels must be in close proximity for functional interactions (Figure 7-5D and 7-5E). These experiments indicate that σ receptor-mediated signal transduction is membrane delimited, and requires neither G protein activation nor protein phosphorylation.

These results point toward a transduction mechanism mediated by membrane proteins in close proximity, and possibly in direct contact. Signal transduction by σ receptors can proceed without employing any of the well-established molecular systems known to operate in the receptor-mediated modulation of membrane excitability. The elimination of G proteins and protein kinases in the modulation of potassium channels therefore implies a novel mechanism of signal transduction, in keeping with the novel structure of the σ receptor protein.

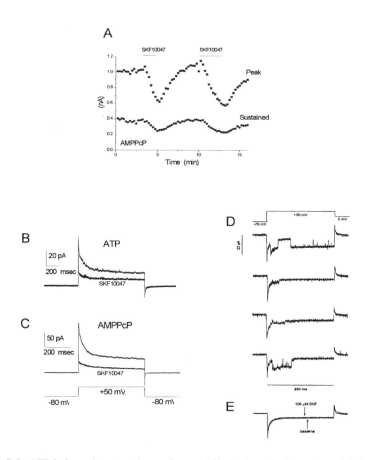

Figure 7-5. ATP-independence and membrane delimitation of channel modulation. Rat neurohypophysial potassium current was evoked by pulses in Figure 7-4. (A) Peak (upper) and sustained (lower) potassium current were reduced by 100 µM SKF-10,047 applied at the bars. (B) In outside-out patches, voltage pulses of 300 ms from -80 to 50 mV activated potassium current. With pipette solutions containing 2 mM ATP, 100 µM SKF-10,047 reduced potassium current. (C) Potassium current reduction by SKF-10,047 in outside-out patches with 50 nM okadaic acid and 2 mM AMPPcP in place of ATP. (D) Cell-attached patches were hyperpolarized to -110 mV for 250 ms, and then depolarized to 50 mV for 300 ms to activate potassium channels. Voltage was estimated by assuming a resting membrane potential of -70 mV. Four traces show single channel activity through A-current and calcium-activated potassium channels under control conditions. (E) Averages of 25 sweeps were recorded before and after application of 100 µM SKF-10,047 to the bathing solution. SKF-10,047 reduced current only slightly in these experiments because the drug could not reach the membrane under the pipette tip in which channel activity was recorded (28).

5. RECONSTITUTION OF ION CHANNEL MODULATION IN XENOPUS OOCYTES

To clarify the molecular basis of signal transduction, Aydar et al. (21) employed the *Xenopus laevis* oocyte system to express σ receptors together with voltage-gated potassium channels. In oocytes injected with mRNA encoding the voltage-gated potassium channel Kv1.4 (a channel that is thought to be responsible for the A-current in nerve terminals) or mRNA encoding the closely related potassium channel Kv1.5, depolarizing voltage steps evoked large outward potassium currents (Figure 7-6A and 7-6B). Potassium currents could also be evoked in cells co-expressing potassium channels and σ receptors (Figure 7-6C and 7-6D). In these oocytes, the potassium currents were highly sensitive to σ receptor ligands (SKF-10,047 and DTG). By contrast, the modulation of potassium current by these ligands was very weak in cells in the absence of exogenous σ receptor (Figure 7-6A and 7-6B). Thus, potassium channels show consistent and robust responses to σ receptor ligands only when the σ receptor is coexpressed.

In oocytes injected only with mRNA encoding a potassium channel, σ receptor ligands weakly inhibited the potassium current (Figure 7-6A and 7-6B). This effect was found to be due to endogenous *Xenopus* σ receptors that are expressed in the oocytes. When the biosynthesis of these receptors was suppressed by injecting σ receptor antisense mRNA, the action of σ receptor ligands on expressed potassium channels was almost completely abolished (Figure 7-6E and 7-6F). No modulation of potassium channels was seen in oocytes expressing Kv1.4 potassium channels, and modulation was seen in only 10% of the oocytes expressing Kv1.5 potassium channels. Suppression of the endogenous *Xenopus* σ receptor reduced the responses to σ receptor ligands to levels not significantly different from zero. Thus, all of the modulation of expressed potassium channels could be accounted for by σ receptors. These experiments ruled out the possibility that σ receptor ligands interact directly with the potassium channels. The σ receptor and voltage-gated potassium channel are the only exogenous proteins needed for the formation of a fully functional signal transduction apparatus in the *Xenopus* oocyte.

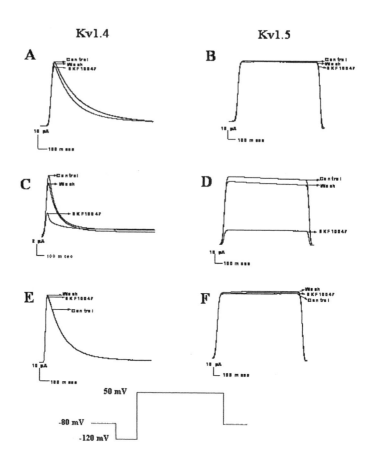

Figure 7-6. Channel modulation in oocytes. Currents through potassium channels formed by the Kv1.4 or Kv1.5 gene products were evoked by 900 ms voltage pulses to 50 mV, following a 200 ms conditioning pulse from -80 to -120 mV (pulse sequence shown schematically below). SKF-10,047 (100 µM) slightly reduced outward current in both Kv1.4 (A) and Kv1.5 (B) expressing oocytes. Coexpression of the σ receptor together with a channel enabled SKF-10,047 to reduce current by more than half (C and D). Injection of σ receptor antisense mRNA together with channel mRNA eliminated the response to SKF-10.047 with both Kv1.4 (E) and Kv1.5 (F). Drug tests were recorded 2 min after drug addition. Reversal was recorded 30 min after returning to control solution (21).

7. Modulation of ion channels

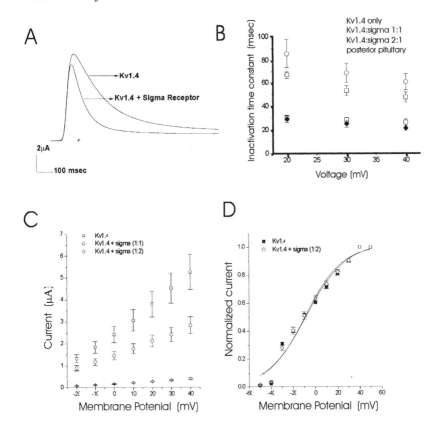

Figure 7-7. Ligand independent channel modulation. (A) Current traces are compared from oocytes expressing Kv1.4 potassium channels with and without σ receptor. Currents in oocytes coexpressing σ receptor inactivate more rapidly than currents in oocytes expressing only Kv1.4. Currents were elicited by voltage pulses as in Figure 7-6. (B) Time constants of inactivation obtained from single exponential fits were plotted versus voltage. The slowest inactivation was seen in oocytes expressing Kv1.4 alone. In oocytes injected with equal quantities of Kv1.4 and σ receptor mRNA, inactivation was faster. Increasing the ratio of σ receptor RNA to Kv1.4 RNA (2:1) accelerated inactivation further. Inactivation time constants were measured for the A-current in neurohypophysial nerve terminals from the fast component of a double exponential fit. These values are very close to those obtained from oocytes injected with Kv1.4 and excess σ receptor mRNA. (C) Effect of the σ receptor on Kv1.4 activation. Current-voltage plots were constructed for peak currents as in Figure 7-6. For Kv1.4 alone, n=7; for Kv1.4 to σ receptor mRNA = 1:1, n=8; for Kv1.4 to σ receptor mRNA = 1:2, n=7. Increasing amounts of σ receptor depressed current at all voltages tested. (D) Normalized conductance-voltage plots. Peak current for Kv1.4 alone and Kv1.4 + σ receptor mRNA at a ratio of 1:2 were divided by driving force (V - E_K; with E_K = -80) to obtain conductance, and normalized to the maximum conductance. The plots are essentially superimposable. Fits to the Boltzmann equation gave parameters that were not significantly different (21).

6. LIGAND INDEPENDENT CHANNEL MODULATION

The oocyte expression system made it possible to distinguish between ligand-dependent and ligand-independent effects of σ receptors on channel function. Expressing Kv1.4 potassium channels in the presence and absence of σ receptors showed that the channel had different biophysical properties in the presence of the σ receptor. Thus, even in the absence of ligand, σ receptors can modulate channels (Figure 7-7) (21). Kv1.4 potassium channels exhibit strong inactivation in response to sustained depolarization, but when σ receptors are present these channels inactivate about three times more rapidly (Figure 7-7A). For voltage pulses to 40 mV, fitting the inactivating current to a single exponential gave a time constant of 61 ± 7 ms (n = 8) in oocytes expressing Kv1.4 alone, and 47 ± 4 ms (n = 6) in oocytes injected with equal amounts of Kv1.4 and σ receptor mRNA. Increasing the ratio of σ receptor mRNA to Kv1.4 mRNA to 2:1 further reduced the time constant for inactivation to 26 ± 3 ms (n = 4). The same trend was seen at other voltages. Increasing amounts of σ receptor accelerated Kv1.4 channel inactivation at all voltages tested (Figure 7-7B). These results indicate that σ receptors can modulate potassium channels even in the absence of σ receptor agonists.

Kv1.4 potassium channels have been detected in nerve terminals (43), and the inactivation of this channel in oocytes resembles that of the A-current of neurohypophysial nerve terminals (25). Furthermore, as noted above, the neurohypophysial A-current can be modulated by σ receptor ligands (26). Thus, at a superficial level, it would appear that the potassium current in these nerve terminals can be accounted for by Kv1.4. However, quantitative comparison revealed that the pituitary nerve terminal A-current more closely resembled Kv1.4 potassium channels expressed in oocytes when the σ receptor was coexpressed. A-current in nerve terminals inactivated with a time constant of 21.2 ± 2.1 ms (n = 9) at 40 mV (Figure 7-7A and 7-7B), nearly 3 times faster than expressed Kv1.4 channels at the same voltage. However, this value was very similar to the value of 26 msec, noted above for Kv1.4 potassium channels in the presence of excess σ receptor. This trend was seen for other voltages as well (Figure 7-7B). Thus, coexpression of Kv1.4 with σ receptors produces channels with inactivation behavior in oocytes that is essentially the same as that seen *in vivo*. This suggests that the channel responsible for the nerve terminal A-current is a complex of Kv1.4 and the σ receptor.

7. Modulation of ion channels 141

Ligand-independent effects of σ receptors were seen on other biophysical properties of Kv1.4 potassium channels as well. Plots of current versus voltage showed that co-expression of the σ receptor reduced the current at all voltages between -20 and 40 mV (Figure 7-7C) (21). Biotinylation experiments showed that the surface expression of Kv1.4 was unaffected by σ receptors. Thus, the reduction in current reflects a reduction in activity of expressed channels. Normalizing to the maximum current and dividing by the driving force (V - E_K, assuming E_K = -80 mV) yielded plots of conductance versus voltage, allowing a comparison of the voltage dependence of channel activation (Figure 7-7D). These plots showed that the σ receptor had no effect on the channel activation curve. $V_{1/2}$ = -9.0 ± 3.0 mV for Kv1.4 alone and -7.9 ± 4.0 mV for Kv1.4 + σ receptor (mRNA in a ratio of 1:2). The steepness factors were also unchanged. Thus, Kv1.4 channels interact with σ receptors in the absence of ligand to accelerate channel inactivation and reduce overall channel activity.

7. MOLECULAR INTERACTIONS BETWEEN σ RECEPTORS AND ION CHANNELS

Physiological studies reviewed above indicated that σ receptors can modulate ion channels without assistance from other proteins commonly involved in signal transduction, eg. G proteins and protein kinases. These results could be explained by a direct physical interaction between σ receptors and channels. To investigate this possibility, immunoprecipitation experiments (21) were performed to detect the association between Kv1.4 and σ receptor (Figure 7-8). Membrane lysates prepared from the posterior pituitary of rats were immunoprecipitated using an anti-Kv1.4 antibody. Immunoprecipitated complexes were resolved with SDS-PAGE and probed with rabbit antibodies against Kv1.4 or with antibodies against the σ receptor (44). Antibodies against Kv1.4 revealed this channel at the appropriate position of ~87 kDa (Figure 7-8A – right). This same sample revealed a band at ~ 25 kDa when probed with antibodies against the σ receptor (Figure 7-8A – left, lane 2). No σ receptor bands were observed in the control sample prepared without the anti-Kv1.4 immunoprecipitating antibody (Figure 7-8A - left, lane 4). Thus, immunoprecipitation of Kv1.4 from rat posterior pituitary membrane lysate pulls down the σ receptor.

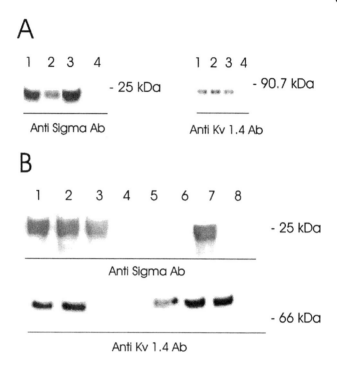

Figure 7-8. Coimmunoprecipitation of σ receptors and Kv1.4 channels. (A) Membrane lysates were prepared from rat posterior pituitary, immunoprecipitated with anti-Kv1.4 antibody, and resolved with SDS-PAGE. Immunoprecipitated samples were run on duplicate gels and probed with either anti-σ receptor or anti-Kv1.4 antibodies (as labeled). Total membrane preparations were run in lanes 1 and 3, and the eluates from the immunoprecipitations were run in lanes 2 and 4. (B) Membrane lysates were prepared from *Xenopus* oocytes expressing Kv1.4 (lanes 5 and 6), σ receptors (lanes 3 and 4), and both Kv1.4 and σ receptors (lanes 1, 2, 7 and 8). The lysates were immunoprecipitated with anti-Kv1.4 antibody and resolved with SDS-PAGE. Immunoprecipitated samples were run on duplicate gels, and blots were probed with either anti-σ receptor or anti-Kv1.4 antibodies (as labeled). Total membrane lysates were run in lanes 1, 3, 5 and 7. Immunoprecipiated eluates were run in lanes 2, 4, 6, and 8. The eluates were concentrated 4-fold, compared to the total membrane lysates. Control samples without an immunoprecipitation antibody were run in lane 8 (21).

A similar experiment was carried out in oocytes expressing Kv1.4 and σ receptor (Figure 7-8B). Western blots revealed the σ receptor at ~25 kDa in the sample immunoprecipitated with anti-Kv1.4 antibody, but not in the control sample. This interaction was specific for Kv1.4. Another potassium channel, a G protein coupled inward rectifier, was expressed with a myc tag in oocytes. Anti-myc antibodies precipitated the channel but σ receptor was not detected in the resolved immunoprecipitate (21). The interaction is thus specific for a channel that can be modulated, and can be detected both in posterior pituitary nerve terminals and in the oocyte heterologous expression system.

A lower percentage of σ receptor was coimmunoprecipitated with Kv1.4 potassium channels from the neurohypophysis than from oocytes (8% and 26%, respectively). This can be explained by the presence of additional ion channel binding partners for the σ receptors in the neurohypophysis (21). Other channels in the neurohypophysis can be modulated by σ receptors, including calcium-activated potassium channels (26), and possibly sodium and calcium channels as well. σ Receptors associated with these other channel proteins should not be coprecipitated by antibodies against Kv1.4.

8. CONCLUSIONS

A number of results converge to support the hypothesis that σ receptor signal transduction is mediated by protein-protein interactions between the receptor and target channels. Lupardus et al. (28), Wilke et al. (27), Morio et al. (24), and Zhang and Cuevas (35) were all unable to alter the transduction process with reagents and manipulations that eliminate or alter G protein function. Similar experiments showed that ATP hydrolysis and protein phosphorylation play no role (28). Thus, these processes do not participate in signal transduction during σ receptor-mediated responses (9). Aydar et al. (21) reconstituted σ receptor-mediated responses in oocytes with only the σ receptor and Kv1.4 or Kv1.5 channels, and without any other heterologously expressed proteins. Immunoprecipitation of the Kv1.4 K^+ channel coprecipitated the σ receptor in native tissue (the neurohypophysis) and in a heterologous expression system (*Xenopus* oocytes). Finally, the σ receptor can alter channel function independently of σ receptor ligands. Although these experiments do not exclude roles for additional proteins present in both native tissue and *Xenopus* oocytes, they do indicate the existence of a protein complex that comprises both the σ receptor and one of its channel targets. Within this complex, the state of the receptor is transmitted to the channel. If the receptor binding sites are vacant, then the σ receptor accelerates voltage-dependent channel inactivation and reduces channel current. If the receptor binding sites are occupied, then the σ receptor reduces channel current further to ~25% of that seen when the receptor is present but unliganded.

The investigations of σ receptor signal transduction indicate a unique mechanism of modulation of ion channel function, in keeping with the novel structure of the σ receptor protein. This mechanism depends on protein-

protein interactions within a multi-protein complex, and does not engage cytoplasmic second messengers, G proteins, or ATP hydrolysis. An important unresolved question is whether additional molecular components are required, and if so, what are their identities. The fact that the response is membrane delimited suggests that any additional proteins required for signal transduction are membrane bound (28). The fact that the response can be reconstituted in oocytes suggests that additional proteins can replace one another easily between species (21). If there are no other proteins, then the minimal functional unit for σ receptor action is the receptor, an ion channel, and a lipid bilayer.

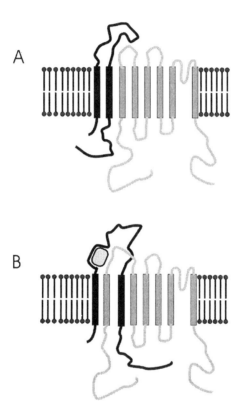

Figure 7-9. Signal transduction. Schematic structural representations of the σ receptor (black) and a generic Kv1 potassium channel (gray). (A) In the absence of ligand, the two proteins are associated and the channel behavior is altered. (B) The presence of ligand (light gray) alters the interaction and strengthens an inhibitory modulation of the channel.

Figure 7-9 presents a drawing of this minimal system for the σ receptor and a Kv1-type potassium channel. The interaction in the absence of ligand is indicated by the contact between the two proteins within the lipid bilayer. The binding of ligand alters this interaction, and in the fanciful drawing offered here, a swapping of the locations of two membrane spanning segments is depicted to indicate a functional consequence. A portion of the channel is shown interacting with the ligand. There is no direct evidence for this at present, but the target-specific actions of some ligands (26) could be accounted for by channel participation in drug binding. The many controversies regarding different activities and efficacies of various σ receptor ligands in different systems may reflect contributions of channel targets to ligand binding. A test of this simple hypothesis may lead the field in a very fruitful direction.

In general, models for ion channel modulation by σ receptors will have to account for the following findings and properties of σ receptor signal transduction: (i)G protein independent, (ii) ATP hydrolysis independent, (iii) soluble cytoplasmic factor independent, (iv) membrane delimited, (v) dependent on close proximity between receptor and channel, (vi) reconstitution in *Xenopus* oocytes requires only receptors and channels, and (vii) receptors and channels form a stable complex.

σ Receptors have been implicated in a wide range of functions. The challenge before us is to formulate a hypothesis for the molecular mechanism of signal transduction by σ receptors which satisfies the conditions enumerated above and also explains the varied functions of σ receptors in the many tissues and cell types where it is found. Given the wide distribution of σ receptors, it will be interesting to see if ion channel modulation is a general physiological role for σ receptors or if other classes of target proteins give rise to distinct forms of response.

The functional interaction between σ receptors and channels is altered by drugs, but σ receptors modulate ion channels in the absence of ligands. It is quite possible that many of the biological functions of σ receptors are ligand independent, with the σ receptor acting like an auxiliary ion channel subunit (21). The extent to which the biological role of σ receptors is dependent on activation by ligand will be aided by progress in the identification of endogenous σ receptor ligands.

Perhaps the greatest challenge for research on the mechanisms of ion channel modulation by σ receptors is understanding the extraordinary promiscuity of the σ receptor in its interactions with membrane targets. Channels as structurally divergent as Kv1.4, calcium-activated potassium channels (presumably products of the *Slo* gene), and calcium channels, are all modulated by σ receptors. It is not clear how one molecule can recognize so many targets by direct protein-protein interactions. In other instances, this kind of divergence in signal transduction was accounted for by G proteins and protein kinases. In the case of the σ receptor, a direct interaction with targeted channels implies that these channels possess a σ receptor recognition motif. Such a motif would have to be shared by many structurally distant channel proteins. An alternative possibility is that a scaffolding protein holds the receptor and channel together. Ankyrin is a good candidate for such a scaffolding protein as it forms a complex with σ receptors and IP_3 receptors (45). Another possibility is that the action of σ receptors on ion channels is mediated by lipids. The homology of the σ receptor with sterol isomerase raises the possibility of cholesterol binding activity that could have relevance to channel function. The idea has been proposed that a lipid binding activity could enable σ receptors to alter the organization of lipid domains or rafts (46). Channel modulation would then depend on the channels residing within these domains and being sensitive to the state of the lipids. We can hope that further study of the structural basis for σ receptor modulation of ion channels will shed light on these questions.

REFERENCES

1. Monnet FP, Debonnel G, de Montigny C. In vivo electrophysiological evidence for a selective modulation of N-methyl-D-aspartate-induced neuronal activation in rat CA3 dorsal hippocampus by σ ligands. J Pharmacol Exp Ther 1992, 261:123-130.
2. Morin-Surun MP, Collin T, Denavit-Saubie M, Baulieu E-E, Monnett FP. Intracellular $σ_1$ receptor modulates phospholipase C and protein kinase C activities in the brainstem. Proc Natl Acad Sci USA 1999, 96:8196-8199.
3. Campbell BG, Scherz MW, Keana JF, Weber E. Sigma receptors regulate contractions of the guinea pig ileum longitudinal muscle/myenteric plexus preparation elicited by both electrical stimulation and exogenous serotonin. J Neurosci 1989, 9:3380-3391.
4. Ela C, Barg J, Vogel Z, Hasin Y, Eilam Y. Sigma receptor ligands modulate contractility, Ca^{2+} influx and beating rate in cultured cardiac myocytes. J Pharmacol Exp Ther 1994, 269:1300-1309.
5. Gonzalez-Alvear GM, Werling LL. Regulation of [$_3$H]DA release from rat striatal slices by σ receptor ligands. J Pharmacol Exp Therapeut 1994, 271:212-219.
6. Weatherspoon JP, Gonzalez-Alvear GM, Frank AR, Werling LL. Regulation of [^{3}H] dopamine release from mesolimbic and mesocortical areas of guinea pig brain by σ receptors. Schizophren Res 1996, 21:51-62.
7. Horan B, Gifford AN, Matsuno K, Mita S, Ashby CR. Effect of SA4503 on the electrically evoked release of ^{3}H-acetylcholine from striatal and hippocampal rat brain slices. Synapse 2002, 46:1-3.
8. Lobner D, Lipton P. σ-Ligands and non-competitive NMDA antagonists inhibit glutamate release during cerebral ischemia. Neurosci Lett 1990, 117:169-174.

9. Bowen WD. (1994) Interaction of sigma receptors with signal transduction pathways and effects of second messengers. In: Sigma Receptors (Y. Itzhak, ed), pp 139-170. San Diego: Academic Press.
10. Su T-P. Delineating biochemical and functional properties of σ receptors: Emerging concepts. Crit Rev Neurobiol 1993, 7:187-203.
11. Maurice T, Lockhart BP. Neuroprotective and anti-amnesic potentials of σ receptor ligands. Prog Neuro-Psychopharm Biolog Psych 1997, 21:69-102.
12. Ferris CD, Hirsch DJ, Brooks BP, Snyder SH. Sigma receptors: from molecule to man. J Neurochem 1991, 57:729-737.
13. Wolfe SA, Culp SG, De Souza EB. σ-receptors in endocrine organs: identification, characterization, and autoradiographic localization in rat pituitary, adrenal, testis, and ovary. Endocrinology 1989, 124:1160-1172.
14. Crawford KW, Bowen WD. Sigma-2 receptor agonists activate a novel apoptotic pathway and potentiate antineoplastic drugs in breast tumor cell lines. Cancer Res 1992, 62:313-322.
15. Vilner BJ, John CS, Bowen WD. Sigma-1 and σ_2 receptors are expressed in a wide variety of human and rodent tumor cell lines. Cancer Res 1995, 55:408-413.
16. Quirion R, Bowen WD, Itzhak Y, Junien JL, Musacchio LM, Rothman RB, Su T-P, Tam SW, Taylor DP. A proposal for the classification of σ binding sites. Trends Pharmacol Sci 1992, 13:85-86.
17. Hanner M, Moebius FF, Flandorfer A, Knaus H-G, Striessnig J, Kempner E, Glossman H. Purification, molecular cloning, and the expression of the mammalian σ_1 binding site. Proc Natl Acad Sci USA 1996, 93:8072-8077.
18. Kekuda R, Prasad PD, Fei Y-J, Leibach FH, Ganapathy V. Cloning and functional expression of the human type 1 σ receptor. Biochem Biophys Res Comm 1996, 229:553-558.
19. Seth P, Leibach FH, Ganapathy V. Cloning and structural analysis of the cDNA and the gene encoding the murine type 1 σ receptor. Biochem Biophys Res Comm 1997, 241: 535-540.
20. Kahoun JR, Ruoho AE. (^{125}I) Iodoazidococaine, a photoaffinity label for the haloperidol-sensitive σ receptor. Proc Natl Acad Sci USA 1992, 89:1393-1397.
21. Aydar E, Palmer CP, Klyachko VA, Jackson MB. The σ receptor as a ligand-regulated auxiliary potassium channel subunit. Neuron 2002, 34:339-410.
22. Kennedy C, Henderson G. Inhibition of potassium currents by the σ receptor ligand (+)-3-(3-hydroxyphenyl)-N-(1-propyl)piperidine in sympathetic neurons of the mouse isolated hypogastric ganglion. Neuroscience 1990, 35:725-733.
23. Wu X-Z, Bell JA, Spivak CE, London ED, Su T-P. Electrophysiological and binding studies on intact NCB-20 cells suggest the presence of a low affinity σ receptor. J Pharmacol Exp Ther 1991, 257:351-359.
24. Morio Y, Tanimoto H, Yakushiji T, Morimoto Y. Characterization of the currents induced by σ ligands in NCB20 neuroblastoma cells. Brain Res 1994, 637:190-196.
25. Bielefeldt K, Rotter JL, Jackson MB. Three potassium channels in rat posterior pituitary nerve endings. J Physiol 1992, 458:41-67.
26. Wilke RA, Lupardus PJ, Grandy DK, Rubinstein M, Low MJ, Jackson MB. K^+ channel modulation in rodent neurohypophysial nerve terminals by σ receptors and not by dopamine receptors. J Physiol 1999, 517:391-406.
27. Wilke RA, Mehta RP, Lupardus PJ, Chen Y, Ruoho AE, Jackson MB. Sigma receptor photolabeling and σ receptor-mediated modulation of potassium channels in tumor cells. J Biol Chem 1999, 274:18387-18392.

28. Lupardus PJ, Wilke RA, Aydar E, Palmer CP, Chen Y, Ruoho AE, Jackson M.B. Membrane-delimited coupling between σ receptors and K^+ channels in rat neurohypophysial terminals requires neither G-proteins nor ATP. J Physiol 2000, 526.3:527-539.
29. Hayashi T, Kagaya A, Takebayashi M, Shimizu M, Uchitomi Y, Motohashi N, Yamawaki S. Modulation by σ ligands of intracellular free Ca^{2+} mobilization by N-methyl-D-aspartate in primary culture of rat frontal cortical neurons. J Pharmacol Exp Ther 1995, 275:207-214.
30. Brent PJ, Pang G, Little G, Dosen PJ, VanHelden DF. The σ receptor ligand, reduced haloperidol, induces apoptosis and increases intracellular-free calcium levels $[Ca^{2+}]_i$ in colon and mammary adenocarcinoma cells. Biochem Biophys Res Comm 1996, 219:219-226.
31. Brent PJ, Herd L, Saunders H, Sim ATR, Dunkley PR. Protein phosphorylation and calcium uptake into rat forebrain synaptosomes: modulation by the σ ligand, 1,3-ditolylguanidine. J Neurochem 1997, 68:2201-2211.
32. Vilner BJ, Bowen WD. Modulation of cellular calcium by sigma-2 receptors: release from intracellular stores in human SK-N-SH neuroblastoma cells. J Pharmacol Exp Ther 2000, 292:900-911.
33. Hayashi T, Maurice T, Su T-P. Ca^{2+} signaling via sigma 1 receptors: Novel regulatory mechanism affecting intracellular Ca^{2+} concentration. J Pharmacol Exp Ther 2000, 293:788-798.
34. Church J, Fletcher EJ. Blockade by σ site ligands of high voltage-activated Ca^{2+} channels in rat and mouse cultured hippocampal pyramidal neurones. Br J Pharmacol 1995, 116:2801-2810.
35. Zhang H, Cuevas J. Sigma receptors inhibit high-voltage-activated calcium channels in rat sympathetic and parasympathetic neurons. J Neurophysiol 2002, 87:2867-2879.
36. Neher E. The use of the patch clamp technique to study second messenger-mediated cellular events. Neuroscience 1988, 26:727-734.
37. Hille B. G Protein-coupled mechanisms and nervous signaling. Neuron 1992; 9:187-195.
38. Levitan IB. Modulation of ion channels by protein phosphorylation and dephosphorylation. Ann Rev Physiol 1994, 56:193-212.
39. Hilgemann DW, Feng S, Nasuhoglu C. The complex and intriguing lives of PIP2 with ion channels and transporters. Science's STKE [Electronic Resource]: Signal Transduction Knowledge Environ 2001:RE19.
40. Ahern GP, Klyachko VA, Jackson MB. cGMP and S-nitrosylation: two routes for modulation of neuronal excitability by NO. Trends Neurosci 2002, 25:510-517.
41. Soriani O, Vaudry H, Mei YA, Roman F, Cazin L. Sigma ligands stimulate the electrical activity of frog pituitary melanotrophs through G-protein dependent inhibition of potassium conductances. J Pharmacol Exp Ther 1998, 286:163-171.
42. Soriani A, Le Foll F, Roman F, Monnet FP, Vaudry H, Cazin L. A-current downmodulated by σ receptor in frog pituitary melanotrope cells through a G protein-dependent pathway. J Pharmacol Exp Ther 1999, 289:321-328.
43. Sheng M, Liao YJ, Jan YN, Jan LY. Presynaptic A-current based on heteromultimeric K^+ channels detected in vivo. Nature 1994, 365:72-75.
44. Yamamoto H, Miura R, Yamamoto T, Shinohara K, Watanabe M, Okuyama S, Nakazato A, Nukada T. Amino acid residues in the transmembrane domain of the type 1 sigma receptor critical for ligand binding. FEBS Lett 1999, 445:19-22.
45. Hayashi T, Su T-P. Regulating ankyrin dynamics: Roles of sigma-1 receptors. Proc Natl Acad Sci USA 2001, 98:491-496.
46. Kaczmarek LK, McKay SE. Act locally: New ways of regulating voltage-gated ion channels. Mol Interventions 2002, 2:215-218.

Corresponding author: *Dr. Meyer Jackson, Mailing address: University of Wisconsin Medical School, Department of Physiology, SMI 127, 1300 University Avenue, Madison, WI 53706, USA, Phone: (608) 262-9111, Fax: (608) 262-9072, Electronic mail address: Mjackson@physiology.wisc.edu*

Chapter 8

SUBCELLULAR LOCALIZATION AND INTRACELLULAR DYNAMICS OF σ_1 RECEPTORS

Teruo Hayashi and Tsung-Ping Su
National Institutes of Health, National Institute on Drug Abuse, Intramural Research Program, Baltimore, MD 21224, USA

1. INTRODUCTION

σ_1 Receptors are present in the central nervous system (CNS) as well as in peripheral organs. Although early binding assay studies demonstrated the enrichment of σ_1 receptors in microsomes of the brain, the exact cellular or subcellular localization of σ_1 receptors was unclear. In a highly polarized neuron, whether σ_1 receptors localize on pre- or post-synaptic regions has been an important issue for understanding the functions of this receptor. Until the 1990s, because only ligand binding assays were available for identifying σ_1 receptors, it was not easy to clarify subcellular distributions and specific membrane domains where σ_1 receptors may target. However, by combining membrane fractionation and radioligand binding assays, some groups investigated the subcellular distribution of σ_1 receptors and have obtained a fairly consistent data. Most studies showed the striking enrichment of σ_1 receptors in microsomes from either the brain or peripheral organs, suggesting that σ_1 receptors are present in the endoplasmic reticulum (ER) (4,5,14,17-19,21,34).

Owing to the success of the cloning of σ_1 receptors, it becomes possible to explore the cellular localization of σ_1 receptors by immunocytochemical or molecular biological techniques. Of important information obtained from cloning studies is that the deduced amino acid sequence of σ_1 receptors has at least one transmembrane domain and a double-arginine ER retention signal at the N-terminus (8,16,26,27). The retention signal is known to direct the retrieval of membrane proteins from Golgi apparatus to the ER via a retrograde transport pathway (25). This aspect supports that the ER membrane is the main loci where σ_1 receptors exist. Recently some studies used specific antibodies against σ_1 receptors and explored the cellular and subcellular distribution of σ_1 receptors. Even though these studies are at the beginning phase, they demonstrated a quite unique subcellular distribution pattern of σ_1 receptors in a variety of cells (1,7,10,11,15,20,22,28,30,35). Furthermore, some studies indicate that σ_1 receptors have altered subcellular distribution pattern under certain conditions or with applications of receptor ligands (9,10,12,20). In this chapter, we first discuss the early fractionation studies using radiolabeled ligands. Next, the recent evidences on subcellular localization of σ_1 receptors from immunocytohistochemical studies are presented. Studies using expression of green fluorescence protein (GFP)-tagged σ_1 receptors indicate that σ_1 receptors target the unique lipid-enriched loci on the ER. We also discuss the intracellular dynamics of σ_1 receptors and its potential roles in regulation of signal transduction and lipid transport by σ_1 receptors.

2. MEMBRANE FRACTIONATION STUDIES FOR SUBCELLULAR DISTRIBUTION OF σ_1 RECEPTORS

2.1 Approaches using radioligand binding assays

Several membrane fractionation studies have shown that σ_1 binding sites (e.g. (+)-SKF-10,047 or (+)-pentazocine-binding sites) are enriched in microsomal fractions in both brain and liver, but they are also present in the P_1 and P_2 fractions. In 1989, McCann et al. (17) first reported that $[^3H](+)$-SKF-10,047 binding sites are present in a high level at the microsomal fraction. McCann and Su (18) further examined subcellular distribution of $[^3H](+)$-SKF-10,047 binding sites. They showed that $[^3H](+)$-SKF-10,047 binding sites are three to four times more concentrated in microsomes than in synaptosomes in the rat brain. Although the P_2 fraction contains less $[^3H](+)$-SKF-10,047 binding sites than P_1 and P_3 fractions in the differential centrifugation, they found that the binding sites are also enriched in the myelin fraction (P_{2A}) obtained by P_2 subfractionation. $[^3H](+)$-SKF-10,047 binding sites in synaptosome (P_{2B}) and mitochondria (P_{2C}) fractions are less than one fourth of those in the myelin fraction. These results suggest that σ_1 receptors may be predominantly localized on the ER and possibly on myelin, and not concentrated at synapses in the rat brain. Further, they found that [^{3}H](+)-SKF-10,047 binding sites exhibit significantly lighter buoyant density than plasma membrane markers such as 5'-nucleotidase and ATP-stimulated [^{3}H]ouabain binding sites in a continuous sucrose gradient. This result further supports that the majority of σ_1 receptor localization is inside cells rather than the plasma membrane. Although σ_1 receptors are shown to be enriched in the P_3 fraction in the differential centrifugation, distribution of [^{3}H](+)-SKF-10,047 binding sites did not match to those of the ER resident protein markers in a extensive fractionation study using a continuous sucrose gradient. Therefore, they raised one possibility that σ_1 receptors are localized in the as yet to be identified specialized domains on a certain organelle membrane. Itzhak et al. (14) also found the enrichment of [^{3}H](+)-SKF-10,047 binding sites in the microsome of the C57BL/6 mouse brain, but also detected a moderate level of the binding sites in the mitochondrial fraction. Cagnotto et al. (4) used the more selective σ_1 ligand (+)-pentazocine for the subcellular fractionation study. [^{3}H](+)-Pentazocine binding sites were mostly enriched in the microsomal fraction, and next in the nuclear fraction in the rat brain. Among subfractions of P_2, the myelin fraction (P_{2A}) are most enriched in [^{3}H](+)-pentazocine binding sites, consistent with results from a subcellular fractionation study using [^{3}H](+)-SKF-10,047 (17). [^{3}H](+)-Pentazocine binding sites were low in synaptosomal and mitochondrial fractions. [^{3}H](+)-Pentazocine binding sites in the microsomal fraction (P_3) were more than 10-fold enriched than that in the P_2 fraction in their study. Taken together, these subcellular fractionation experiments indicated that localization of the σ_1 receptor to synaptic regions of plasma membrane or to mitochondria is minimum, if any. A few studies tempted to compare differences in the subcellular distribution between σ_1 and σ_2 receptors (3,19). Although both subtypes are enriched in the microsomal fraction, the enrichment of the σ_1 receptor in microsomes is more prominent; σ_1 receptors in microsomes are several folds higher than those in other fractions, whereas the levels of σ_2 receptors in P_1 and P_2 fractions are close to that in the microsomal fraction (3).

It was shown that σ_1 receptors exist in liver, spleen, adrenal gland and placenta. Samovilova and Vinogradov (24) found that the distribution profile of (+)-SKF-10,047 binding sites in the liver coincided with that of NADPH-cytochrome c reductase, an ER marker. [^{3}H](+)-Pentazocine binding sites were shown to be enriched in microsomal fractions in liver, testis, and the heart (5). On the other hand, it was shown that progesterone-sensitive [^{3}H]haloperidol binding sites, which possess binding characteristics of σ_1 receptors, are present in the plasma membrane of human placenta as well as in intracellular membranes (23).

In summary, subcellular fractionation studies combining with σ_1 ligand binding assays indicate that σ_1 receptors exhibit multi-organelle distributions, but are predominantly enriched in microsomes (suggesting localization on the ER). A study using a continuous sucrose gradient suggests that σ_1 receptors may target specific regions of the membranes which show unique high buoyancy in a sucrose gradient (18).

2.2 Subcellular fractionation studies using immunoassays for σ_1 receptor

Specific antibodies against σ_1 receptors enabled us to identify more specifically the distribution and enrichment of these receptors. Western blotting using σ_1 receptor antibodies showed that σ_1 receptors are highly enriched in microsomes, but much less in the synaptosomal fraction in rat brain (35). This result is perfectly in a line with those shown in previous binding assay studies. Similar results were also obtained from a subcellular fractionation study using a neuronal cell culture. σ_1 Receptors in the microsomal fraction was shown to be 3- to 4-fold higher than those in P_1 or P_2 fractions in the NG108-15 neuroblastoma x glioma hybrid cell line (9). When the P_3 fraction was separated by a 3-layer discontinuous sucrose gradient, σ_1 receptors were separated into an NADPH-cytochrome P450 reductase-enriched fraction, suggesting that σ_1 receptors are localized on the smooth ER in NG108 cells (10). However, when NG108 cell homogenates were extensively fractionated by using a 13-layer sucrose gradient, the σ_1 receptor-enriched fractions (0.6-0.73M sucrose) were apparently different from those containing NADPH-cytochrome P450 reductase or plasma membrane proteins (>0.85 M sucrose) in NG108 cells (12). Interestingly, these σ_1 receptor-containing unique low-density membranes contained high amounts of lipids such as cholesterol, fatty acids and neutral lipids (12). These results are in agreement with previous data from a rat brain sucrose gradient study (18), and explain why σ_1 receptor-containing ER membrane is separated as a low-density membrane from bulk ER membranes in sucrose gradients.

Thus, subcellular fractionation studies using either specific antibodies or σ_1 ligand (+)-benzomorphans such as (+)-SKF-10,047 and (+)-pentazocine indicate that σ_1 receptors localize on the ER membrane, but at an unique low-density domain on the ER (12). In the next section, this aspect will be supported further by the immunocytochemistry results.

3. IMMUNOCYTOHISTOCHEMICAL STUDIES FOR SUBCELLULAR LOCALIZATION OF σ_1 RECEPTORS

3.1 Immunocytochemistry using specific σ_1 receptor antibodies

Unique intracellular distributions of σ_1 receptors have been demonstrated in some immunocytochemical studies using monoclonal or polyclonal anti-σ_1 receptor antibodies. The first immunocytochemical studies were examined in a human white blood cell line using monoclonal anti-σ_1 receptor antibodies (7,15). In agreement with results of fractionation studies, σ_1 receptors are specifically localized on the ER membrane and on nuclear envelopes of white blood cells from immunofluorescence and electron microscopic examinations. In the promonocytic cell line THP, more σ_1 receptors targeted nuclear envelopes than ER membranes, but no σ_1 receptors were observed in mitochondria, endosome and plasma membrane (7). Immunofluorescence studies have shown that polyclonal σ_1 receptor antibodies predominantly stain cell body cytoplasmic areas in neuronal and retinal cells, indicating an ER localization of σ_1 receptors (1,10,20,28). σ_1 Receptors in human breast gland cells show a cytoplasmic granular staining, very often with a perinuclear localization (30). The first electron microscopic examination of σ_1 receptors in the adult animal brain was done recently by Maurice and coworkers (1). σ_1 Receptors are highly expressed in neuron cell bodies and dendrites in the granular layer of the olfactory bulb, hypothalamic nuclei, septum, central gray, motor nuclei of the hindbrain, dorsal horn of the spinal cord and dentate gyrus of hippocampus. They also observed that σ_1 receptors are in the ER cisternae, in the limiting plasma membrane, and the membrane of mitochondria and postsynaptic thickening of neurons. σ_1 Receptors were not seen in the axon in their study. These results confirm the enrichment of σ_1 receptors on the ER of neurons in the adult animal brain and confirm their postsynaptic localization.

However, some of these results, especially localization of σ_1 receptors on mitochondria and the plasma membrane, are contradictory to the previously published results. So far, no systematic examination on the localization of σ_1 receptors at mitochondria has been done. Future examinations should clarify whether σ_1 receptors can target mitochondria. However, plasma membrane localization of σ_1 receptors, especially at the postsynaptic thickness, seems possible because the smooth ER membrane connects physically to postsynaptic membranes in the adult animal brain (31). Interestingly, Jackson and coworkers recently demonstrated that transfected σ_1 receptors target plasma membranes of *Xenopus* oocytes and are coupled to Kv1.4 potassium channels (2). The localization of σ_1 receptors on the plasma membrane might be a possibility in particular cell types and/or specialized domains on the plasma membrane.

Recently, an immunohistochemical study demonstrated that σ_1 receptors are also present in oligodendrocytes in rat brains (13,22). This result is consistent with the fractionation studies of the rat brain that showed the enrichment of $[^3H](+)$-SKF-10,047 or $[^3H](+)$-pentazocine binding sites in the myelin fraction (4,18). We also found that σ_1 receptors highly express in both neurons and oligodendrocytes, but few in type-1 astrocytes in rat hippocampal primary culture (13, Figure 8-1). Interestingly, σ_1 receptors on the sheet of differentiated oligodendrocytes were highly clustered and form cholesterol-enriched lipid microdomains (lipid rafts) with galactocylceramides (13, Figure 8-1c). Further, knockdown of σ_1 receptors by small interfering RNA is shown to inhibit myelination of oligodendrocytes (13). These results suggest that σ_1 receptor may play a role in formation/maintenance of myelin (6,13).

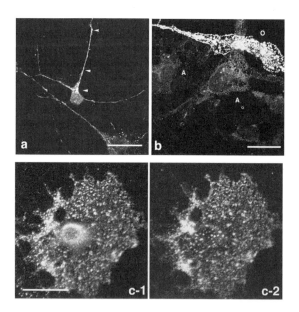

Figure 8-1. Cellular distribution of σ_1 receptors in rat hippocampus primary Culture. (a) σ_1 Receptors show dense clusters (arrowheads) in a cell body and dendrites in hippocampal neurons. (b) Cellular distribution of σ_1 receptors in glia from the rat hippocampus. σ_1 Receptors highly express and cluster in an oligodendrocyte (O), but the level is very low in type-1 astrocytes (A). (c) Clusters of σ_1 receptors (c-1) on the sheet of an oligodendrocyte colocalize with free cholesterol (filipin staining in c-2). σ_1 Receptors were detected by immunofluorescence using anti- σ_1 receptor antibodies. Scale bars = 10 μm. From ref. (11).

8. Subcellular localization and intracellular dynamics

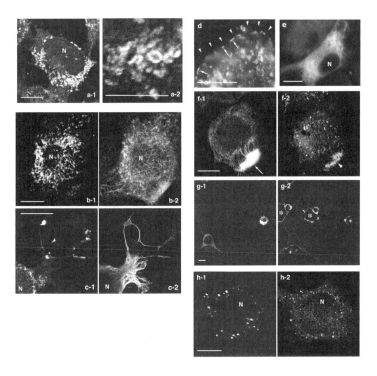

Figure 8-2. σ_1 Receptors target specialized areas of the endoplasmic reticulum (ER) in NG108-15 cells. (a) Intracellular distribution of endogenously expressed σ_1 receptors in NG108-15 cells. σ_1 Receptors were detected by immunofluorescence using anti- σ_1 receptor antibodies. σ_1 Receptors highly clustered in the perinuclear area (a-1) and target globular structures inside cells. σ_1 Receptor-enriched globules in higher magnification are shown in a-2. (b) σ_1 Receptor-EYFP-targeting globular structures are associated with NADPH-cytochrome P450 reductase-positive smooth ER network. Transfected C-terminally EYFP-tagged σ_1 receptors (σ_1 receptor-EYFP) are shown in b-1. NADPH-cytochrome P450 reductase immunostaining in the same cell shown in b-2. (c) Localization of σ_1 receptor-EYFP in varicosities and tips of neuritis (c-1). β-Tubulin immunostaining in the same cell (c-2). (d) σ_1 Receptor-EYFP clustered in the plasmalemma cortices (arrows). Arrowheads locate the position of the plasma membrane. A star indicates a σ_1 receptor-EYFP-containing ER globular structure. (e) N-terminally EYFP-tagged σ_1 receptors (EYFP- σ_1 receptors) fail to target ER globules. Note: a uniform distribution on the entire ER. (f) The bulbous ER aggregation induced by overexpression of EYFP- σ_1 receptors (arrows in f-1). Retention of neutral lipids in the aggregation (arrowheads in f-2; Nile red staining of the same cell in f-1). (g) Reduction of plasma membrane lipid rafts by overexpression of EYFP- σ_1 receptors. EYFP- σ_1 receptor (g-1); cholera toxin subunit β-labeled plasma membrane lipid rafts of the same cells in g-1 (g-2). Note: EYFP- σ_1 receptor overexpressing cells (+) contain less lipid rafts than non-transfected cells (*). (h) Cellular localization of σ_1 receptor-EYFP lacking N-terminal 7 amino acids including an ER retention signal (h-1). These σ_1 receptor mutants specifically target cytosolic lipid droplets that are stained by antibodies against adipose differention related protein, a cytosolic lipid droplet-coating protein (g-2). N, nucleus; scale bars = 10 μm.

3.2 Intracellular distribution of GFP-tagged σ_1 receptors

By using confocal fluorescence microscopy, we found recently that endogenously expressed σ_1 receptors in NG108 cells localize on both ER tubular elements and nuclear envelopes, but are highly clustered and predominantly targeting unique globular structures inside cells (Figure 8-2a). C-terminally enhanced yellow fluorescence protein (EYFP)-tagged σ_1 receptors (σ_1 receptor-EYFP) also showed the same distribution pattern to that of endogenously expressed σ_1 receptors. The confocal microscopic colocalization study confirmed that a majority of σ_1 receptors is present on the smooth ER membrane stained with NADPH-cytochrome P450 reductase antibodies and that σ_1 receptor-enriched globular structures are a part of ER membranes (Figure 8-2b; 11). These σ_1 receptor-enriched globules are demonstrated to contain high amounts of free cholesterol and neutral lipids as confirmed by filipin and Nile red stainings (11). The fatty acid treatment caused significant enlargement of these unique ER globules, suggesting that they function as lipid storage sites on the ER (eg. lipid droplets associated with the ER). Thus, σ_1 receptors target specialized subcompartments of ER membranes. A portion of σ_1 receptors also exists in peripheries of cells. Clustered σ_1 receptor-EYFPs are present on varicosities or tips of neurites (Figure 8-3c; 10). In some cells, an accumulation of σ_1 receptor-EYFP is seen on the plasmalemmal cortices, which consist of cytoskeleton lattices (Figure 8-2d). These results suggest that σ_1 receptors can distribute in close proximity to the plasma membrane, and possibly on the plasma membrane although they cannot accumulate specifically on the plasma membrane of NG108 cells. Further, we found that tagging the EYFP to the N-terminus of σ_1 receptors (EYFP-σ_1 receptors) largely disturbed the proper subcellular distribution of σ_1 receptors and their functions. EYFP-σ_1 receptors diffusely distribute on the entire ER tubules (Figure 8-2e), but cannot target lipid-rich globules. Highly expressing EYFP-σ_1 receptors in NG108 cells caused a pathological bulbous aggregation of the ER and a large retention of cholesterol and neutral lipids therein (Figure 8-2f; 11). In these cells, there exhibit no compartmentalization of neutral lipids in lipid-rich globular structures, a decrease of cytosolic lipid droplets as well as a decrease of cholesterol on the plasma membrane (11). These findings indicate that σ_1 receptors on the ER play major roles in the compartmentalization of lipids in ER lipid storage sites (ER globular structures), and export them to peripheries of cells. Cholesterol synthesized at the ER is known to be transported to the plasma membrane and form detergent-insoluble microdomain lipid rafts with glycosphingolipids (29). Lipid rafts are involved in a variety of cellular functions such as vesicle transport, receptor clustering and internalization, and signal transduction (29). In fact, in NG108 cells expressing functionally negative EYFP-σ_1 receptors, levels of lipid rafts on plasma membrane as well as cholesterol were significantly decreased (Figure 8-2g). Moreover, a recent study demonstrated that overexpression of σ_1 receptors in PC-12 cells causes an increase in lipid rafts and the reconstitution of lipids in plasma membranes (33), suggesting that σ_1 receptors regulate plasma membrane lipid raft formation by regulating lipid transport. These results also indicate that the N-terminal of σ_1 receptors contains a critical domain for their localization and biological functions. Indeed, when 7 amino acids on the N-terminus containing an ER retention signal was truncated, σ_1 receptors cannot locate on the ER any more and are exported to cytosolic lipid droplets (Figure 8-2h; 11).

σ_1 Receptors show a similar localization pattern in hippocampal rat primary neurons and in differentiated oligodendrocytes. σ_1 Receptors form clusters in cell bodies or dendrites in neurons and on the sheets of oligodendrocytes. On the oligodendrocyte sheets, σ_1 receptors form cholesterol-enriched subcompartments on the ER similar to those observed in NG108 cells (ER globules, see Figure 8-1c). Taken together, recent evidences suggest that σ_1 receptors predominantly localize on the ER in varieties of cells, but show a quite unique distribution pattern on the ER that has never been demonstrated for any other ER proteins. Studies using EYFP-tagged σ_1 receptors strongly indicate the involvement of σ_1 receptors in ER lipid transport.

4. INTRACELLULAR DYNAMICS OF σ_1 RECEPTORS

Recent studies demonstrated the dynamic changes of intracellular distribution of σ_1 receptors when stimulated by σ receptor ligands. Morin-Surun et al. (20) first demonstrated the translocation of σ_1 receptors. They found that perfusing (+)-pentazocine, a σ_1 receptor agonist, through the basilar artery can cause a shift of σ_1 receptors from cytoplasm toward the vicinity of the cytoplasmic membrane by immunofluorescence examination. Thus, they suggested that (+)-pentazocine triggers σ_1 receptor translocation from the ER to the cytoplasmic membrane to exert a subsequent regulation of neural excitability through a heterotrimeric G protein/phospholipase C (PLC)/protein kinase C (PKC) cascade (20).

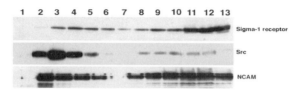

Figure 8-3. σ_1 Receptors in detergent-resistant lipid rafts. NG108 cell lysates were prepared by Tris-NaCl buffer containing 0.5% Triton X-100 at 4° C. Lipid rafts were separated by flotation sucrose gradient centrifugation in the presence of Triton X-100. Protein levels in rafts were detected by Western blotting using respective specific antibodies. Samples were fractionated from the top (fraction #1) to the bottom (fraction #13). Lipid rafts and accompanying proteins were enriched in fractions 1-7. Note the lower buoyancy of σ_1 receptor-containing lipid rafts than Src- or NCAM-containing rafts. NCAM, neuronal cell adhesion molecule.

We also found that σ_1 receptors translocate by stimulation with receptor ligands in NG108 cells. However, σ_1 receptors in our study translocate from the ER to either plasmalemma or nuclear membranes. Translocation of σ_1 receptors consequently affects the distribution of a cytoskeletal protein (10,12). We found that a portion of σ_1 receptors on the ER forms a complex with ankyrin, a cytoskeletal adaptor protein, and this complex tonically inhibits calcium release through inositol 1,4,5-trisphosphate (IP_3) receptors on the ER in NG108 cells (9,10). σ_1 Receptor agonists cause a dissociation of the σ_1 receptor and ankyrin complex from IP_3 receptors and, as a result, potentiates calcium release through IP_3 receptors (32). Thus, translocation of σ_1 receptors is a critical step to exert their modulatory action on the regulation of intracellular calcium signaling. In a sucrose fractionation study, (+)-pentazocine was shown to decrease σ_1 receptors in the smooth ER fraction (P_{3L}) and concomitantly increase them in P_1 and P_2 fractions in a period of 10 min (10,12). The (+)-pentazocine treatment decreased the number of σ_1 receptor clusters at ER globular structures and caused a uniformed distribution pattern of σ_1 receptors over the entire ER network (12).

The rapid translocation of σ_1 receptors is also examined in living cells by using real-time monitoring of EYFP-tagged σ_1 receptors. This technology clearly demonstrated that EYFP-tagged σ_1 receptors move out from lipid-enriched ER globular structures and slide on the ER reticular network toward the nuclear membrane, Golgi, or plasmalemmal areas in living NG108 cells (12). σ_1 Receptors on neurites moved anterogradely toward the tip of a neurite. Fluorescence recovery after photobleaching analysis showed that in ER globular structures only 3% of σ_1 receptors are mobile, whereas 76.3% are on ER tubular elements. On ER tubular elements, σ_1 receptors move, both in the presence and absence of σ_1 ligands, at approximately 8-10 µm/min. The reason why the movement of σ_1 receptors in ER lipid storage sites is highly restricted is unclear at present. Because σ_1 receptors reside in cholesterol-enriched lipid rafts on the membrane of ER globules (Figure 8-3; 11), the liquid-ordered lipid environment of rafts may affect mobility of σ_1 receptors. The σ_1 receptor-containing ER lipid rafts possess some distinct differences in their properties when compared to plasma membrane lipid rafts that contain Src or neuronal cell adhesion molecules (NCAM) (Figure 8-3; 11).

σ_1 Receptor-containing ER rafts show lower buoyancy in sucrose gradients containing Triton X-100, and are associated tightly with actin cytoskeletons. Importantly, (+)-pentazocine caused translocation of σ_1 receptors from raft to non-raft membranes (11). Therefore, mobility of σ_1 receptors might be regulated, at least in part, by lipid raft microdomains where tight protein-lipid interactions are present under an environment of low membrane fluidity. As mentioned in a previous section, σ_1 receptor may affect transport of lipids from the ER. Functionally negative EYFP-σ_1 receptors were shown to fail in forming lipid rafts on the ER and also fail to translocate. Therefore, it is proposed that translocation of σ_1 receptors may play a crucial role in the regulation of lipid transport by σ_1 receptors (11,12). This role of σ_1 receptor translocation is in addition to the regulation of calcium signaling at the ER (10).

5. CONCLUSIONS

Recent studies using specific σ_1 receptor antibodies, in corroboration with early subcellular fractionation studies, begin to unveil the unique cellular localization of σ_1 receptors. It is clear now that σ_1 receptors localize predominantly on the ER in varieties of mammalian cells, although σ_1 receptors can show a multi-organelle distribution pattern inside the cell. Details of σ_1 receptor distribution in highly polarized neurons in the brain will be examined in future studies. Understanding the protein topology and lipid components of specific membrane domains where σ_1 receptors target (eg. lipid-rich globules on the ER) must help to unveil the exact molecular mechanism of σ_1 receptors. σ_1 Receptors show a very unique ability to translocate that is evoked by varieties of psychotropic drugs possessing σ_1 receptor affinities. The dynamics of σ_1 receptors appears to be related intimately to diverse actions exerted by σ_1 ligands, particularly at the plasma membrane and at the ER. These actions include regulations of channel activity, exocytosis, signal transduction, plasma membrane remodeling (raft formation), and lipid transport/metabolism. All these actions may have a bearing in promoting neuronal plasticity in the brain.

REFERENCES

1. Alonso G, Phan V, Guillemain I, Saunier M, Legrand A, Anoal M, Maurice T. Immunocytochemical localization of the σ_1 receptor in the adult rat central nervous system. Neuroscience 2000, 97:155-170.
2. Aydar E, Palmer CP, Klyachko VA, Jackson MB. The σ receptor as a ligand-regulated auxiliary potassium channel subunit. Neuron 2002, 34:399-410.
3. Basile AS, Paul IA, de Costa B. Differential effects of cytochrome P-450 induction on ligand binding to σ receptors. Eur J Pharmacol 1992, 227:95-98.
4. Cagnotto A, Bastone A, Mennini T. [^{3}H](+)-pentazocine binding to rat brain σ_1 receptors. Eur J Pharmacol 1994, 266:131-138.
5. DeHaven-Hudkins DL, Lanyon LF, Ford-Rice FY, Ator MA. Sigma recognition sites in brain and peripheral tissues. Characterization and effects of cytochrome P450 inhibitors. Biochem Pharmacol 1994, 47:1231-1239.
6. Demerens C, Stankoff B, Zalc B, Lubetzki C. Eliprodil stimulates CNS myelination: new prospects for multiple sclerosis? Neurology 1999, 52:346-350.
7. Dussossoy D, Carayon P, Belugou S, Feraut D, Bord A, Goubet C, Roque C, Vidal H, Combes T, Loison G, Casellas P. Colocalization of sterol isomerase and σ_1 receptor at endoplasmic reticulum and nuclear envelope level. Eur J Biochem 1999, 263:377-386.
8. Hanner M, Moebius FF, Flandorfer A, Knaus HG, Striessnig J, Kempner E, Glossmann H. Purification, molecular cloning, and expression of the mammalian σ_1-binding site. Proc Natl Acad Sci USA 1996, 93:8072-8077.
9. Hayashi T, Maurice T, Su TP. Ca^{2+} signaling via σ_1 receptors: novel regulatory mechanism affecting intracellular Ca^{2+} concentration. J Pharmacol Exp Ther 2000, 293:788-798.
10. Hayashi T, Su TP. Regulating ankyrin dynamics: Roles of σ_1 receptors. Proc Natl Acad Sci USA 2001, 98:491-496.
11. Hayashi T, Su TP. Sigma-1 receptors form raft-like microdomains and target lipid droplets on the endoplasmic reticulum (ER): roles in ER lipid compartmentalization and export. J Pharmacol Exp Ther 2003a, 306:718-725.
12. Hayashi T, Su TP. Intracellular dynamics of σ_1 receptors in NG108-15 cells. J Pharmacol Exp Ther 2003b, 306:726-733.
13. Hayashi T, Su TP.Sigma-1 receptors at galactosylceramide-enriched lipid microdomains regulate oligodendrocyte differentiation. Proc Natl Acad Sci USA 2004, 101:14949-14954.
14. Itzhak Y, Stein I, Zhang SH, Kassim CO, Cristante D. Binding of σ-ligands to C57BL/6 mouse brain membranes: effects of monoamine oxidase inhibitors and subcellular distribution studies suggest the existence of σ-receptor subtypes. J Pharmacol Exp Ther 1991, 257:141-148.

15. Jbilo O, Vidal H, Paul R, De Nys N, Bensaid M, Silve S, Carayon P, Davi D, Galiegue S, Bourrie B, Guillemot JC, Ferrara P, Loison G, Maffrand JP, Le Fur G, Casellas P. Purification and characterization of the human SR 31747A-binding protein. A nuclear membrane protein related to yeast sterol isomerase. J Biol Chem 1997, 272:27107-27115.
16. Kekuda R, Prasad PD, Fei YJ, Leibach FH, Ganapathy V. Cloning and functional expression of the human type 1 σ receptor (hSigmaR1). Biochem Biophys Res Commun 1996, 229:553-558.
17. McCann DJ, Rabin RA, Rens-Domiano S, Winter JC. Phencyclidine/SKF-10,047 binding sites: evaluation of function. Pharmacol Biochem Behav 1989, 32:87-94.
18. McCann DJ, Su TP. Haloperidol-sensitive (+)[^{3}H]SKF-10,047 binding sites (σ sites) exhibit a unique distribution in rat brain subcellular fractions. Eur J Pharmacol 1990; 188:211-218.
19. McCann DJ, Weissman AD, Su TP. Sigma-1 and $σ_2$ sites in rat brain: comparison of regional, ontogenetic, and subcellular patterns. Synapse 1994, 17:182-189.
20. Morin-Surun MP, Collin T, Denavit-Saubie M, Baulieu EE, Monnet FP. Intracellular $σ_1$ receptor modulates phospholipase C and protein kinase C activities in the brainstem. Proc Natl Acad Sci USA 1999, 96:8196-8199.
21. Night AR, Noble A, Wong EHF, Middlemiss DN. The subcellular distribution and pharmacology of the sigma recognition site in the guinea-pig brain and liver. Mol Neuropharmacol 1991, 1:77-82.
22. Palacios G, Muro A, Vela JM, Molina-Holgado E, Guitart X, Ovalle S, Zamanillo D. Immunohistochemical localization of the $σ_1$-receptor in oligodendrocytes in the rat central nervous system. Brain Res 2003, 961:92-99.
23. Ramamoorthy JD, Ramamoorthy S, Mahesh VB, Leibach FH, Ganapathy V. Cocaine-sensitive σ-receptor and its interaction with steroid hormones in the human placental syncytiotrophoblast and in choriocarcinoma cells. Endocrinology 1995, 136:924-932.
24. Samovilova NN, Vinogradov VA. Subcellular distribution of (+)-[^{3}H]SKF 10,047 binding sites in rat liver. Eur J Pharmacol 1992, 225:69-74.
25. Schutze MP, Peterson PA, Jackson MR. An N-terminal double-arginine motif maintains type II membrane proteins in the endoplasmic reticulum. EMBO J 1994, 13:1696-1705.
26. Seth P, Leibach FH, Ganapathy V. Cloning and structural analysis of the cDNA and the gene encoding the murine type 1 σ receptor. Biochem Biophys Res Commun 1997, 241:535-540.
27. Seth P, Fei YJ, Li HW, Huang W, Leibach FH, Ganapathy V. Cloning and functional characterization of a σ receptor from rat brain. J Neurochem 1998, 70:922-931.
28. Shamsul Ola M, Moore P, El-Sherbeny A, Roon P, Agarwal N, Sarthy VP, Casellas P, Ganapathy V, Smith SB. Expression pattern of σ receptor 1 mRNA and protein in mammalian retina. Brain Res Mol Brain Res 2001, 95:86-95.
29. Simons K, Toomre D. Lipid rafts and signal transduction. Nat Rev Mol Cell Biol 2001, 1:31-39.
30. Simony-Lafontaine J, Esslimani M, Bribes E, Gourgou S, Lequeux N, Lavail R, Grenier J, Kramar A, Casellas P. Immunocytochemical assessment of $σ_1$ receptor and human sterol isomerase in breast cancer and their relationship with a series of prognostic factors. Br J Cancer 2000, 82:1958-1966.
31. Spacek J, Harris KM. Three-dimensional organization of smooth endoplasmic reticulum in hippocampal CA1 dendrites and dendritic spines of the immature and mature rat. J Neurosci 1997, 17:190-203.
32. Su TP, Hayashi T. Cocaine affects the dynamics of cytoskeletal proteins via $σ_1$ receptors. Trends Pharmacol Sci 2001, 22:456-458.

33. Takebayashi M, Hayashi T, Su TP. Sigma-1 receptors potentiate epidermal growth factor signaling towards neuritegenesis in PC12 cells: potential relation to lipid raft reconstitution. Synapse 2004, 53:90-103.
34. Tanaka M, Shirasaki T, Kaku S, Muramatsu M, Otomo S. Characteristics of binding of [^{3}H]NE-100, a novel sigma-receptor ligand, to guinea-pig brain membranes. Naunyn Schmiedeberg Arch Pharmacol 1995, 351:244-251.
35. Yamamoto H, Miura R, Yamamoto T, Shinohara K, Watanabe M, Okuyama S, Nakazato A, Nukada T. Amino acid residues in the transmembrane domain of the type 1 σ receptor critical for ligand binding. FEBS Lett 1999, 445:19-22.

Corresponding author: *Dr. Teruo Hayashi, Mailing address: National Institutes of Health, National Institute on Drug Abuse, Intramural Research Program, 333 Cassell Drive, Baltimore, MD 21224, USA, Phone: (410) 550-6568, ext. 118, Fax: (410) 550-6856, Electronic mail address: thayashi@intra.nida.nih.gov*

Chapter 9

INTRACELLULAR SIGNALING AND SYNAPTIC PLASTICITY

Francois P. Monnet
Institut National de la Sante et de la Recherche Medicale Unite 705 – CNRS UMR 7157, Hôpital Fernand Widal, 75475 Paris cedex 10, France

1. INTRODUCTION AND HISTORICAL PERSPECTIVE

The pharmacological characteristics, biochemical impacts, and physiological roles of σ receptors have been predominantly documented during the last two decades for the nervous system, although both receptor subtypes (σ_1 and σ_2) (1) are more abundant in peripheral organs and systems (e.g. liver, digestive tract, heart, kidney, steroidogenic glands) than in neural tissue (2; see Chapters 1, 12, 18). Indeed, the quantities (B_{max}) of σ receptor proteins in the periphery are at least equivalent to the highest concentrations found in the nervous system (B_{max} between 600-770 fmol/mg protein in motor function-related nuclei, hippocampal pyramidal layers, and blood cells), but reach up to B_{max} values ≈ 10 pmol/mg protein, for example, in the liver. From that point of view, the strategy which has consisted of focusing on σ receptors in the central nervous system has failed to be a determinant for understanding the implication of these proteins in cell functioning. It is useful to remember that exploring σ receptor function in the brain has not been conclusive with regard to definitively identifying their biological importance, and to note that the initial successful cloning procedures for the [^{3}H]SKF-10,047-sensitive σ_1 receptor were for liver (3), placenta (4) and lymphocytes (5).

This peculiar orientation of most groups, including mine, with respect to exploring the function of σ proteins in fact originates from the initial studies performed by the group of Martin who focused on drug discrimination studies by comparing the potency and pharmacological profiles of N-allylnormetazocine (SKF-10,047) and pentazocine with those of various analgesics and opioid antagonists, especially with regard to dependence-producing properties, from chronic spinal dogs (6). During the following decade, it was recognized that radioligand binding studies and functional approaches showed that σ receptors belong neither to opiate, cocaine-sensitive dopaminergic nor phencyclidine (PCP)-associated N-methyl-D-aspartate (NMDA) receptor subtypes although they very potently modulate each of these neurotransmitter systems as well as the other central monoaminergic ones, suggesting their involvement in most of the major neuropsychiatric disorders (see for recent review 7).

2. PERSISTENCE OF THE "σ ENIGMA"

Although these binding studies were not very informative with regard to what the real function of σ receptors was, they constituted a very useful biochemical basis for later exploration of the biological actions of σ receptors. Indeed, it was at this time that a subtype of σ receptor proteins (which was later denoted σ_1) was demonstrated to be associated with guanine nucleotide binding proteins (G proteins; 8). It was also in the late 1980s and early 1990s that neurochemical and electrophysiological studies were carried out to crucially conceptualize that σ receptors may interact with glutamatergic neurons, especially via the NMDA/PCP receptor subtype in the hippocampus (9,10), the dopaminergic neurons via the D_2 receptor subtype in the motor control-related brainstem nuclei and dopamine transporters in cortical structures (11,12), the noradrenergic neurons via both uptake sites and α_2 receptor subtypes (13,14), the cholinergic neurons predominantly via muscarinic receptor subtypes (15,16), and the opiate system via µ, κ and δ receptor subtypes (17). The role of σ drugs in the regulation of NMDA-sensitive glutamatergic, catecholaminergic, opiate and cholinergic pathways was thereafter proposed to account for their behavioral effects on thought and motor control, but beneficial ones for learning and memory performances, neuroprotection and against ischemic damage (e.g. stroke, seizures), concomitantly with their putative role in the pathophysiology of major neuropsychiatric disorders such as drug abuse, psychoses, mood and anxiety disorders (18), neurodegenerative diseases (e.g. Alzheimer's and Parkinson's) (19), brain cancer and immune deficiency (2,20).

It is interesting to note that since Chavkin (21) introduced the concept of the "σ enigma," the huge progress in characterizing the cellular and molecular actions of the σ_1 receptor has unexpectedly hampered the elucidation of this system since the multitude of biological effects ascribed to σ ligands has led to attenuating their specific interest in neurobiology and thereby clinical medicine. Indeed, this protein, present in almost all living organisms with a molecular structure very similar between yeast and mammals, interferes with biological events that lead amongst those most essential for cell survival and functions, has finally questioned the specificity of the σ_1 binding. This attitude was also and unfortunately accentuated by the cloning of the σ_1 protein, which indicated that the σ_1 protein belongs to a unique family of small (28 kDa) intracellular proteins anchored at rest to the endoplasmic reticulum membrane and binding multiple classes of pharmaceutical agents including neuroprotective drugs, analgesics, antiamnesics, psychotropics, steroids, regulators of endocrine and immune functions, muscular and gastrointestinal motility (2,3,20).

It is nevertheless from these abovementioned studies that has emerged the concept that at least some mechanisms of action of the σ_1 receptor subtype might affect intracellular signal transduction pathways and second messenger cascades. These include calcium (Ca^{2+}) influx through the plasma membrane, mobilization of intracellular Ca^{2+} ($[Ca^{2+}]_i$) stores, recruitment of heterotrimeric G proteins and activation of phosphatidylinositide metabolism and their related cellular impacts (8,15,22). The aim of this brief review is to shed light on recent data supporting the notion that σ proteins, especially of the σ_1 subtype, affects these distinct systems, which constitute to date the most efficient biochemical pathways for neuronal plasticity. This will open a new perspective for understanding their interrelationships and their cooperativity with regard to σ_1 receptor functions.

3. NEURONAL PLASTICITY AND SIGNAL TRANSDUCTION

The central nervous system was generally believed capable of little plasticity. Over the past few years, however, there have been enormous advances in identifying factors that can increase its plasticity, in particular after injury and during diseases. Neuronal damage occurring during stroke, hypoglycemia and neurodegenerative disorders (e.g. Alzheimer's disease and dementia, Parkinson's disease) is associated, among various factors, with excessive concentrations of $[Ca^{2+}]_i$ and extracellular glutamate and with reactive gliosis.

Amplification or restriction in dynamic adaptations in the structural characteristics of the postsynaptic membrane, and in particular of dendritic spine shape, is proposed to occur as the consequence of the massive entry of Ca^{2+} or its blockade, respectively. This impact of $[Ca^{2+}]_i$ is actually considered to constitute the first step of a long cascade of biochemical events that may lead to either cell proliferation, differentiation, elongation of existing dendritic spines, creation of new ones or to spine retraction, cellular process collapse, apoptosis and ultimately cell death when the Ca^{2+} entry is too excessive. The required enhancement of the cytosolic cation concentration is predominantly mediated through activated NMDA receptor channels but also involves plasma membrane bound Ca^{2+} channels and $[Ca^{2+}]_i$ mobilization from the endoplasmic reticulum and the mitochondria, all of which often occur synergistically (23,24). The already well-identified cellular adaptive mechanisms are the activation of protein kinases that are globally responsible for the neurite outgrowth or protein phosphatases which are predominantly responsible for the involution processes. It is also currently believed that this is essential for increasing the efficacy at excitatory synapses that underlie many forms of adaptive behavior, including learning and memory (25).

Experimental evidence indicates that a major role of σ ligands is to modulate glutamatergic, catecholaminergic and cholinergic receptor functions, as well as $[Ca^{2+}]_i$ mobilization. These effects of σ drugs although involving primarily neurons might also be effective on glial cells. Indeed, it has recently been shown that oligodendrocytes and astrocytes, which have σ_1 proteins, are very sensitive to glutamate, acetylcholine and $[Ca^{2+}]_i$ (26). However, the functional link between neurons and glial cells with regard to the σ_1 receptor remains elusive. Also unknown are the consequences of an overstimulation of these later neurotransmitters on the ability of glial cells to affect steroids synthesized and released, some of them being proposed to be endogenous ligands of the σ_1 receptor (27).

Neurochemical and electrophysiological studies have indicated that in the central nervous system, the σ_1 receptor regulates NMDA receptor activity, potassium (K^+) currents and $[Ca^{2+}]_i$ signaling, responsible for the facilitation of both neuronal excitation and synaptic neurotransmitter release (e.g. glutamate, acetylcholine and catecholamines). There is considerable evidence (e.g. from pharmacological challenges) that the glutamate, acetylcholine, and catecholamine pathways and $[Ca^{2+}]_i$ signaling play important roles in neuronal plasticity, learning and memory abilities, and in aging processes. The blockade of certain forms of neuronal plasticity (e.g. long term potentiation or LTP) and several learning processes (e.g. spatial learning or passive avoidance) by NMDA or acetylcholine antagonists, or Ca^{2+} depletion as well as the poor capacity of acquisition and storage of spatial memory combined with the lack of hippocampal LTP in transgenic strains lacking, for example, subtypes of NMDA, α-amino-3-hydroxy-5-methylisoxazole-4-propionate (AMPA) and ryanodine receptors, inositol-1,4,5-triphosphate (IP_3) kinase, Ca^{2+}-calmodulin kinase, protein kinase C (PKC) isoforms further supports this notion (for recent reviews see 24,25). In the hippocampal CA1 region, it is already well known that the amplitude and reliability of LTP induction parallels NMDA receptor-mediated postsynaptic Ca^{2+} accumulation (robust Ca^{2+} influx plus massive Ca^{2+} release from internal stores) and is reinforced by subsequent activation of kinases (e.g. PKC) (28).

Interestingly, the impact of σ_1 drugs on the Ca^{2+}-triggered intracellular machinery parallels their regulatory action on neurotransmitter-induced excitatory action, neuroprotective, motor and mnesic effects demonstrated at the biochemical, physiological and behavioral levels (2,3,11,20,29-32). Previous reports have shown that σ_1 agonists improve learning effectiveness in accelerated-senescence-prone mice (SAM-P/8) to a level similar to that of the senescent-resistant (SAM-R/1) controls although remaining ineffective in young adults and in rats treated with amyloid β_{25-35} protein (33). This latter protein is known to induce a long lasting gliosis and a reduction of the choline acetyltransferase activity similar to those reported in Alzheimer's disease. This unusual spectrum has been confirmed at the clinical level for igmesine (JO 1784), which appears more efficient among the elderly in cognitive defects associated with depressive mood (34). Therefore, this has suggested that σ_1 agonists are effective in normal and pathological aging associated with learning and mnesic deficits associated with alterations in glutamate and acetylcholine autoreceptor functions (35). This might also be consistent with the proposed role of this receptor in the pathophysiology of neurodegenerative diseases (e.g. Alzheimer's, Parkinson's) (36), psychotic and affective disorders (18,33).

However, the cloning of this protein, which is identical in the central nervous system and periphery (3), had failed to solve the functional dilemma of an intracellular protein with a single transmembrane domain recruiting heterotrimeric G proteins and regulating electrical or secretary activities. Furthermore, for a number of transmitter systems affected by σ_1 drugs (i.e. glutamate, acetylcholine, catecholamines), it has been suggested that distinct σ_1 receptor subtypes may be targeted to specific brain regions, possibly enabling cells to respond in different ways to the same ligand. This is clearly exemplified in both motor and limbic systems where prototypic σ drugs act either as agonists or antagonists according to the experimental model (3,11,27,29). This brings into question the functional consequences of an exclusive activation of the σ_1 receptor. It is also intriguing that the functionality of the σ_1 receptor has been revealed so far solely in biological systems already active (spontaneously, pharmacologically, or electrically). All has considerably hampered the identification of the biological mode of action of the σ_1 receptor, which has been partially dissected and understood. Several lines of evidence have suggested that the σ_1 receptor may affect Ca^{2+} influx, phosphoinositide turnover, and protein phosphorylation. Two recent reports by Morin-Surun et al. (32) and Hayashi and Su (37) have documented these biological processes demonstrating that the σ_1 receptor translocates to activate the Ca^{2+}-dependent phospholipase C (PLC)-PKC cascade and facilitates both Ca^{2+} release from endoplasmic reticulum stores and Ca^{2+} influx. This constitutes evidence for a novel mode of rapid recruitment of a membrane-bound second messenger cascade and regulation of Ca^{2+} signaling via an intracellular receptor.

Pursuing its exploration of the biological action of σ_1 receptors, the group of Su (38), has shown that the σ_1 agonist (+)-pentazocine, as well as the σ receptor-binding antidepressants imipramine and fluvoxamine potentiate nerve growth factor (NGF)-mediated but not cyclic adenosine monophosphate(cAMP)-mediated sprouting in PC-12 cells. This effect is prevented by pretreatment with NE-100 (N,N-dipropyl-2-[4-methoxy-3-(2-phenylethoxy)phenyl]ethylamine, a prototypic σ_1 antagonist) and a σ_1 receptor antisense oligonucleotide, indicating that the σ_1 receptor is involved. In addition, these authors have also shown that overexpressing σ_1 binding proteins in PC-12 cells results in a facilitation of the NGF-mediated sprouting. These data are of the greatest relevance for supporting the beneficial effects of σ_1 drugs on neuronal plasticity and have to be put in relation with those of Vilner and Bowen (39-41; see below) suggesting a role for σ proteins in cell development and trophicity.

4. G PROTEIN COUPLING

Attention toward elucidation of the coupling of the σ_1 receptor protein to G proteins, in particular in the nervous system, has been a topic of great interest over the last decade. Molecular studies on σ_1 receptor function initially focused mainly on the mechanisms that underlie protein coupling, suggesting a conversion of the σ_1 protein from a high to a low affinity state (8,42,43). This emphasis was due, in large part, to the success of simple biochemical and electrophysiological models. Although there are many unanswered questions about the role and dynamics of G proteins as cellular effectors of the σ_1 receptor, it is clear that the study of G proteins in σ_1 receptor-mediated effects has provided a way to identify and characterize molecular mechanisms that potentially underlie σ_1 receptor action. Previous authors have reviewed the biochemical evidence supporting the notion that the σ_1 receptor was associated with pertussis toxin-sensitive heterotrimeric $G_{i/o}$ proteins (see for example 44). Aside from ligand binding studies, my group was the first to provide *in vivo* and *in vitro* functional data indicating that in a brain structure, (+)-benzomorphans and igmesine potentiate either presynaptic or postsynaptic neuronal NMDA receptors via $G_{i/o}$ proteins since both *in vivo* preinjection of pertussis toxin or *in vitro* preadministration of N-ethyl-maleimide acid prevented these effects (43,45). Here, I will expose most recent advances in some of the neural systems and molecular processes involving or regulating the implication of heterotrimeric G proteins that may be responsible at least in part for the biological effects of σ drugs.

Using primary rat hippocampal culture exposed to glutamate, Lesage et al. (46) have shown that glutamate-activated cGMP formation was reduced concentration-dependently by some, but not all, σ drugs (i.e. ifenprodil > fenpropimorph > sabeluzole > opipramole > carbetapentane > PD-128298 > haloperidol) after prolonged (7 days) treatment. This provides relevant information supporting the notion that some σ drugs may exhibit neuroprotective actions via attenuating the NMDA receptor-mediated glutamate-activated nitric oxide synthase pathway. Exploring the regulation of intracellular Ca^{2+} flux by (+)-pentazocine and PRE-084 (2-(4-morpholino)ethyl-1-phenylcyclohexane-1-carboxylate) from NG-108 cells in culture, Hayashi et al. (47) observed that pertussis toxin pretreatment, via inactivating $G_{i/o}$ proteins, abolished the σ_1 agonist-induced modulation of depolarizing-induced $[Ca^{2+}]_i$ mobilization, further supporting the functional link between $G_{i/o}$ proteins and the σ_1 receptor. More recently, Ueda et al. (48) have shown that σ_1 receptor agonist-induced $[^{35}S]GTP\gamma S$ binding can be abolished by reconstitution with recombinant G_{i1} but not G_{oA} proteins. Further documenting the interrelationship between G proteins and σ_1 receptors, Ueda et al. (49) have also proposed, using a peripheral pain-producing flexor test in mice, that $G_{\alpha i}$ was involved in the σ_1 receptor agonist-induced nociceptive response via activating a likely small number of IP_3 molecules generated by PLC then gating IP_3 receptor to cause a transient $[Ca^{2+}]_i$ efflux from the endoplasmic reticulum. The group of Jackson (see Chapter 7) has refuted that inhibition of voltage-activated K^+ channels by (+)-benzomorphans (understood via σ_1 receptors) was coupled to GTP-coupled signaling systems. This recent study (50), performed on rat neurohypophysial peptidergic nerve terminals using the patch-clamp technique had the interest of examining the functionality of this interaction at the molecular level. It should however be noted that previous groups had come to a similar conclusion, most of them having performed binding studies with membrane extracts (whereas binding studies using whole cell preparations had predominantly found the linkage between σ_1 receptor and G proteins). In addition, it is most probable that the existence of the coupling between σ_1 receptors and G proteins varies from one brain region to another, as our group documented for the rat hippocampal subfields (51). Hence, the exact relationship between the σ_1 protein and either $G_{i/o}$, $G_{q/11}$ or G_s proteins remains elusive. In fact, it is still questioned how a receptor with a single (or two) transmembrane domain(s) might be coupled to heterotrimeric G proteins; the current dogma being that seven transmembrane regions are mandatory for a metabotropic receptor to be coupled with such effectors. The study of Lupardus et al. (50) suggests moreover that it is likely that the σ_1 receptor may not be directly associated with these G proteins. According to the discovery that σ_1 receptor migrates in the cell following its activation (32,37), it may be suggested that it is during its translocation that the σ_1 protein may interfere with this signaling cascade. This would be consistent with the finding that σ_1 agonists recruit plasma membrane-bound PLC although the exact nature of this pathway remains unknown.

Biochemical evidence supporting the notion that the σ_1 receptor signaling pathway interferes with the production of adenylate cyclase has been lacking. With this regard, Sommermeyer et al. (52) have shown with LZR-1 cells in culture (which express both cAMP-associated D_2 dopaminergic receptors and σ_1 sites) that neither (+)-SKF-10,047, (+)-pentazocine or DTG affected the foskolin-induced cAMP production, nor did they modulate the ability of D_2 dopaminergic receptors to affect adenylate cyclase. This conclusion supporting the absence of a relationship between the σ_1 receptor function and adenylate cyclase contrasts however with our recent observation with cultures of frog pituitary melanotrope cells using the whole-cell patch-clamp configuration that has shown that internal dialysis of guanosine-5'-O-3-thiophosphate irreversibly prolonged the transient outward K^+ current (I_A) obtained with (+)-pentazocine and igmesine (53). This latter report poses the question of the linkage between G proteins and the σ_1 receptor.

5. REGULATION OF INTRACELLULAR CALCIUM MOBILIZATION

The first assumption suggesting an interaction between σ receptors and the regulation of Ca^{2+} entry at the plasma membrane emerged from indirect data showing that several inorganic Ca^{2+} channel blockers such as Cd^{2+}, Ni^{2+} and La^{2+} as well as the nonselective Na^+/Ca^{2+} channel blockers phenylamine, cinnarizine, amidirone and amiloride affected the equilibrium binding of σ sites labeled with [^{3}H]dextromethorphan or [^{3}H]DTG (1,3-di-o-tolylguanidine) in guinea pig brain membranes (54,55). Basile et al. (56), investigating the link between Ca^{2+} influx, $[Ca^{2+}]_i$ regulation and σ binding, have shown with guinea pig cerebellum extracts that the divalent cations Zn^{2+}, Ni^{2+}, La^{2+}, Sr^{2+}, Mg^{2+} and Ca^{2+} inhibited [^{3}H](+)-pentazocine-sensitive $σ_1$ receptor binding in a biphasic manner within the millimolar concentration range (slowing solely the rate of association), but [^{3}H]DTG-sensitive $σ_2$ receptor binding in a monophasic manner within the micromolar range concentration. Subsequent dissociation experiments performed with [^{3}H]DTG have shown that verapamil and amidirone but neither nifedipine, BAY-K8644 nor amiloride, enhanced the dissociation of [^{3}H]DTG from σ binding sites while Cd^{2+}, Ni^{2+} and La^{2+} were much more active on the low affinity site, suggesting that a Ca^{2+} channel could be associated with $σ_2$ receptors. It has to be remembered that neuronal voltage sensitive Ca^{2+} channels (VSCCs) have been differentiated into L-, N- and P/Q-types. The L-type Ca^{2+} channels preferentially bind 1,4-dihydropyridine and phenylalkylamine analogues, such as nifedipine and verapamil, respectively. They would mainly participate in the generation of Ca^{2+} influx in depolarized neurons and thereafter in the regulation of intracellular Ca^{2+} concentrations. By contrast to L-type VSCC blockers, the N- and P/Q-type Ca^{2+} channel blockers, i.e. the venom ω-conotoxin fractions GVIA and MVIIC, respectively, have been helpful for documenting the molecular events related to glutamate-, NMDA- and KCl-evoked neurotransmitter release and exocytosis. In addition, the unusual structure-activity relationship of the Ca^{2+} antagonists on the dissociation of [^{3}H]DTG did not support the notion that a neuronal low-threshold inactivating Ca^{2+} channel might be involved in this effect (57), but supported the hypothesis that both σ receptor subtypes are associated in some way with Ca^{2+} channels.

Zhang and Cuevas (58) have more recently brought further support to this notion by showing that (+)-pentazocine, haloperidol and DTG inhibited peak currents and facilitated current inactivation of L-, N-, P/Q- and R-type VSCCs likely via acting at the level of $σ_2$ receptors in neurons from neonatal rat intracardiac and superior cervical ganglia. Prior to this latter study, Church and Fletcher (59), with whole-cell voltage clamp in cultured mouse hippocampal pyramidal neurons have shown the micromolar concentration efficacy of BD737 [(+)-cis-N-methyl-N-[2-(3,4-dichlorophenyl)ethyl]-2-(1-pyrrolidinyl)cyclohexylamine] as well as of ifenprodil and haloperidol to reduce (at concentrations ranging from 20 to 100 µM) both L- and N-type voltage-activated Ca^{2+} channel currents (I_{BA}) and to increase the rate of inactivation of both VSCC subtypes. Interestingly, DTG and the prototypic $σ_1$ agonists (+)-pentazocine, (+)-3-PPP (3-(3-hydroxyphenyl)-N-(1-propyl) piperidine) and dextromethorphan were ineffective in their patch-clamp paradigm.

Carpenter et al. (22) were however the first to provide dynamic information with regard to σ agonists affecting $[Ca^{2+}]_i$ equilibrium by showing that micromolar concentrations of dextromethorphan reduced KCl-induced Ca^{2+} uptake in rat brain synaptosomes, which is a more functional approach to investigating the biochemical effect of σ drugs on Ca^{2+} influx and $[Ca^{2+}]_i$.

Paul et al. (60), using adrenal chromaffin cells, have indicated that (+)-pentazocine, (+)-SKF-10,047 and haloperidol within the micromolar range of concentrations selectively inhibited the increase of $[Ca^{2+}]_i$ following nicotine administration, bringing support to the notion that $σ_1$ receptors are involved in the regulation of $[Ca^{2+}]_i$. However, since nicotine interfered with the equilibrium constants of $[^3H](+)$-pentazocine binding, these authors proposed that the $σ_1$ receptor was likely coupled to the nicotine receptor-associated Ca^{2+} ionophore.

A similar hypothesis has also been suggested to explain the potentiating action of σ drugs on NMDA receptor-mediated responses solely by means of electrophysiological recordings (45) and not by measuring $[Ca^{2+}]_i$ variations. Whether the potentiating action of σ drugs on the NMDA receptor response involves $[Ca^{2+}]_i$ mobilization has recently been studied. To answer this question, the regulation of $[Ca^{2+}]_i$ by σ drugs, using the fluorescent probe Fura-2 or Fluo-3 in primary cultures of preloaded rat frontocortical (61,62) and hippocampal neurons (59) responding to NMDA (61,62) and KCl (59) has been investigated. Hayashi et al. (61) showed that micromolar amounts of igmesine, (+)-pentazocine, DTG, and haloperidol concentration-dependently reduced the NMDA receptor-mediated $[Ca^{2+}]_i$ increase. The two former drugs, inactive on the initial $[Ca^{2+}]_i$ peak, affected the sustained phase of the $[Ca^{2+}]_i$ mobilization that is responsible for the rapid desensitization of the Ca^{2+} response to NMDA. By contrast, the two latter drugs reduced both the initial and the sustained phases of $[Ca^{2+}]_i$ increase. The rank order of potencies for these compounds was compatible with the involvement of the $σ_1$ receptor subtype although two distinct profiles for $σ_1$ receptors could be suggested (61). Klette et al. (62) reported that benzomorphans acting preferentially on the $σ_1$ site such as dextromethorphan, (+)-SKF-10,047 and (+)-pentazocine exhibited a short-lasting inhibitory effect on the NMDA-induced $[Ca^{2+}]_i$ increase whereas DTG and haloperidol remained almost inactive. However, these authors clearly stated that the neuronal response to NMDA in the presence of both (+)-SKF-10,047 (10 μM) and (+)-pentazocine (100 μM) rapidly desensitized, conversely to that evoked with DTG, hence suggesting that the $σ_1$ but not the $σ_2$ receptor signaling pathway most likely involves $[Ca^{2+}]_i$ mobilization.

In their study, Church and Fletcher (59) found that BD737, ifenprodil and haloperidol reversibly attenuated the increase of $[Ca^{2+}]_i$ in response to KCl.

The concentrations required being substantially higher than their IC_{50} for σ binding and the atypical profile of the σ drugs rather indicates that this σ receptor subtype corresponds to the σ receptor denoted non-$σ_1$, non-$σ_2$ site (63). It is noteworthy that in colon and mammary adenocarcinoma cells, Brent et al. (64,65) have shown a similar non-$σ_1$, non-$σ_2$ profile of action of σ drugs on $[Ca^{2+}]_i$ mobilization by showing that DTG decreased, while reduced haloperidol increased $[Ca^{2+}]_i$ independent of Ca^{2+} entry through the plasma membrane and independent of any protein kinases or phosphatases. This later observation has thus brought support to the notion that the atypical σ receptor subtype might interfere with $[Ca^{2+}]_i$ homeostasis. Further exploring the role of $σ_1$ receptors on $[Ca^{2+}]_i$, Hayashi et al. (47) reported that the impact of (+)-pentazocine, PRE-084, and pregnenolone sulfate (considered a powerful endogenous $σ_1$ agonist) also occurred at the level of the endoplasmic reticulum since a cocktail of bradykinin, thapsigargin (an inhibitor of Ca^{2+}-ATPase in the endoplasmic reticulum), and caffeine (an endoplasmic reticulum-bound ryanodine receptor agonist) was required for σ drugs to elicit their potentiating effect. The pharmacological profile and the mode of action of the atypical σ receptor described by Brent et al. on $[Ca^{2+}]_i$ mobilization differ in fact from those of the $σ_1$ receptor that mediates $[Ca^{2+}]_i$ mobilization in NG-108 cells and guinea pig neurons, which has been shown to implicate the endoplasmic reticulum and subsequently the recruitment of PLCβ isoforms and the IP_3 receptor (32,37,47).

Interestingly, the data of Novakova et al. (66) and those of Vilner and Bowen (41) using primary cultures of rat cardiac myocytes and human SK-N-SH neuroblastoma cells, respectively, would reconcile both notions since BD737, reduced haloperidol, and BD1047 (N-[2-(3,4-dichlorophenyl)ethyl]-N-methyl-2-(dimethylamino)ethylamine) that act on the non-$σ_1$, non-$σ_2$ receptor enhanced both electrically-evoked $[Ca^{2+}]_i$ mobilization independently of any Ca^{2+} influx through the plasma membrane but dependently on plasma membrane phospholipases and IP_3 production. In this regard, it is noteworthy that Hayashi et al. (37,47) reported that both high affinity $σ_1$ receptor ligands, (+)-pentazocine and PRE-084 at concentrations < 10 μM enhanced the bradykinin-induced $[Ca^{2+}]_i$ mobilization and IP_3 release also involving both L-type VSCCs and endoplasmic reticulum Ca^{2+} stores (see also Chapter 8).

To clarify the apparent controversy with regard to the opposite effects of σ drugs on NMDA-induced $[Ca^{2+}]_i$ mobilization, exhibiting either facilitatory (47) or inhibitory (61,62) profiles, a recent study was performed with the same approach using microspectrofluorometry of the Ca^{2+}-sensitive indicator Fura-2 in primary cultures of embryonic rat hippocampal pyramidal neurons (67). This showed that (+)-SKF-10,047, (+)-pentazocine and igmesine reversibly and time-dependently increased and then attenuated the $[Ca^{2+}]_i$ mobilization triggered by glutamate, acting on the NMDA receptor.

Interestingly, in this latter study, (+)-SKF-10,047-, (+)-pentazocine- and igmesine-induced potentiation of the glutamate-induced $[Ca^{2+}]_i$ increase faded rapidly and completely in hippocampal pyramidal neurons when drug perfusions were repeated. Moreover, the potentiations were prevented by preadministration of a Ca^{2+}-dependent conventional PKC (cPKC) isoform inhibitor, Go-6976 (12-(2-cyanoethyl)-6,7,12,13-tetrahydro-13-methyl-5-oxo-5H-indolo[2,3-a]pyrrolo [3,4-c]carbazole) (Figure 9-1), leading to the conclusion that the potentiating counterpart of the glutamate response to σ_1 agonists was cPKC dependent. As shown in Figure 9-1, (+)-benzomorphan-mediated potentiation of both spontaneous (after a 1 min pulse of glutamate) and glutamate-induced $[Ca^{2+}]_i$ mobilization occurred simultaneously with the same rapid time course, suggesting that both effects might be triggered by the same biochemical cascades. Since, preincubation with Go-6976 prior to drug exposures also resulted in complete prevention of the "spontaneous" effect of σ_1 agonists, cPKC was likely involved in the σ_1 receptor-mediated modulation, as has been reported in the motor brainstem (32).

This rapid desensitization of the potentiating action of σ_1 agonists on the NMDA receptor-mediated increase of $[Ca^{2+}]_i$ most likely explains why Hayashi et al. (61) and Klette et al. (62) have failed to show it, since the drug application procedures of both these groups stimulated the neurons with NMDA following a brief pre-perfusion of the σ_1 agonists. In addition, the facilitatory action of (+)-benzomorphans and igmesine faded completely within less than three 1 min applications as shown in Figure 9-1. This suggests that following the procedure of Hayashi et al. and Klette et al., the σ_1 agonist-mediated response was already desensitized when NMDA was perfused, which supports the lack of effectiveness of their ligands. In addition, in their study, Klette et al. (62) clearly showed that (+)-SKF-10,047 (10 µM) and (+)-pentazocine (100 µM) rapidly desensitized the response to NMDA, which was not the case with DTG, further supporting our hypothesis. Moreover, it can also be assumed that the final attenuation of the glutamate-mediated $[Ca^{2+}]_i$ mobilization, revealed by the desensitization of the potentiating effect of (+)-benzomorphans, also implicates the σ_1 receptor, since NE-100 prevented the (+)-SKF-10,047-mediated effect on the glutamate response. Another possibility would be that the resulting profile of σ agonists on glutamate-mediated $[Ca^{2+}]_i$ mobilization corresponds to the recruitment of the NE-100-sensitive non-σ_1, non-σ_2 site (63). It is noteworthy that Novakova et al. (66,68) reported a reduced effectiveness of (+)-3-PPP, BD737, and BD1047, three high affinity σ_1 receptor ligands on $[Ca^{2+}]_i$ transients in isolated cardiac myocytes, whose function is driven by K^+, loaded with Indo-1 following successive drug applications. It could thus support the notion that the σ_1 receptor protein likely acts as a co-allosteric modulatory protein with variable functionalities forming an unstable complex with the NMDA receptor-associated channel, depending on the presence or absence of bound drug and phosphorylation state, as has recently been suggested for K^+ channels (69).

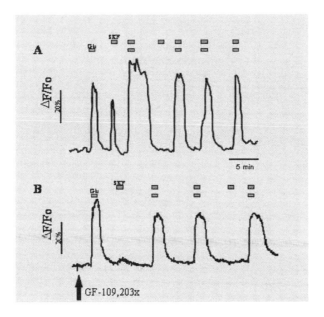

Figure 9-1. (+)-SKF-10,047 initially potentiates NMDA receptor-mediated glutamate-induced $[Ca^{2+}]_i$ mobilization in hippocampal pyramidal neurons. (A) The selective and high affinity benzomorphan (+)-SKF-10,047 (SKF, 1 μM) initially potentiated glutamate (Glu, 50 μM)-mediated $[Ca^{2+}]_i$ mobilization. After several applications of both glutamate and (+)-SKF-10,047, the $[Ca^{2+}]_i$ response returned to baseline. (B) The protein kinase C inhibitor Go-6976 prevented the (+)-benzomorphan-induced modulation of the NMDA receptor-mediated $[Ca^{2+}]_i$ mobilization in primary cultures of hippocampal pyramidal neurons. Stimulations of the Ca^{2+}-dependent Fura-2 fluorescence was induced by a brief (1 min) bath application of glutamate in the presence of (+)-SKF-10,047 (1 μM) via one inlet tube converging on the neurons. The increase in $[Ca^{2+}]_i$ was obtained by subtracting the basal $[Ca^{2+}]_i$ from the glutamate-induce $[Ca^{2+}]_i$. Adapted with permission from ref. (67).

Together, these observations hence show that σ_1 agonists exhibit two actions in modulating the plasma membrane-dependent NMDA receptor-mediated glutamate-mediated $[Ca^{2+}]_i$ increase. The first, involving activation of cPKC, is responsible for a potentiation of the glutamate response (67). The second one does not involve cPKC and is responsible for a reduction of the NMDA response, as previously illustrated with cortical neurons in culture (61,62). Ca^{2+} influx through the plasma membrane is mandatory for both actions. The present results however do not fit with those of Hayashi et al. (47) who have shown a facilitatory role of σ_1 agonists on the bradykinin-induced $[Ca^{2+}]_i$ increase, since this later effect requires at least a 10 min perfusion of the σ_1 drugs and occurs independently of the NMDA receptor and a Ca^{2+} influx through the plasma membrane.

As pointed out in Chapter 11, Vilner et al. (39,40) have established that 3-6 hr exposure to σ drugs such as BD737, BD1008, reduced haloperidol and analogs but not (+)-benzomorphans, DTG and (+)-3-PPP, at concentrations ≥ 100 μM triggered a response whereby cells became spherical, and that this effect was associated with a cessation of cell division, and a loss of cell processes in 13 tumor-derived cell lines of human and nonhuman origin. Interestingly, GTP and Gpp(NH)p had no significant impact on σ_1 binding, suggesting that the binding site was unlikely of the σ_1 subtype, which is fully consistent with the unusual structure-activity profile, which suggests more a non-σ_1, non-σ_2 site (63). These observations however support the notion that at least a subpopulation of σ receptors might be involved in cell proliferation, death, and differentiation.

In addition, it is noteworthy that both trophic and cytotoxic effects of σ_1 ligands have been reported exclusively at concentrations > 100 μM, which parallel the ability of these drugs to enhance $[Ca^{2+}]_i$ mobilization in carcinoma (41,64). It is thus tempting to suggest that when σ_1 ligands by themselves enhance $[Ca^{2+}]_i$ mobilization, they lead to cytotoxicity and likely to ontogeny, whereas when they regulate a pharmacologically-induced $[Ca^{2+}]_i$ increase, they participate in the control of cell function, e.g. neuronal firing. Further support for this hypothesis is provided by the observation that, for (+)-benzomorphans to produce the $[Ca^{2+}]_i$ mobilization once the $[Ca^{2+}]_i$ has returned to baseline, a prior neuronal activation was required within the 5-10 min preceding the administration of the σ_1 ligand. In addition, in spontaneously active cell preparations [e.g. isolated cardiac myocytes (66,68) and brainstem hypoglosse (32)], (+)-benzomorphans have demonstrated their ability to affect $[Ca^{2+}]_i$ mobilization, subsequently triggering Ca^{2+}-dependent enzymes and cell excitability. Together, this indicates that σ_1 receptor activation would have a facilitatory role in quiescent neuronal and non-neuronal cells for the recruitment of intracellular signaling cascades that lead to the facilitation of the depolarizing process and cell trophicity.

6. IMPACT ON PHOSPHOLIPASES AND PROTEIN KINASES

It is definitively accepted that activation of σ_1 receptors interferes with $[Ca^{2+}]_i$–dependent protein signaling cascades. This assertion was initially postulated given the demonstration that the σ agonists (+)-SKF-10,047 and (+)-pentazocine as well as DTG and haloperidol inhibit carbachol- and oxotremonine-stimulated phosphoinositide turnover in gut and brain synaptoneurosomes (15,70). Although the precise mechanism of this regulation has not yet been elucidated, functional bioassays as well as binding studies have favored an indirect modulation of phosphoinositide metabolism by these drugs via the σ_1 receptor subtype. It was however from 1997 that further studies showed the close interrelationship between the σ_1 receptor and phosphoinositide metabolism via the PLC pathway. Almost simultaneously, my group using the *ex vivo* guinea pig brainstem preparation (32,71), Novakova et al. using human SH-SY-5Y cells and rat cardiac myocytes (66) and Hayashi et al. using NG-108 cells (47) provided evidence that σ_1 receptor activation affects IP_3 via the recruitment of the PLCβ isoform. The notion that activation of the σ_1 receptor affects the PLCβ activity results from the brainstem preparation (32). Taking advantage of the spontaneous rhythmic activity of the neuronal networks in charge of the motor hypoglossal and C1-C4 cervical root firing, we demonstrated that prolonged perfusion (15-30 min) or successive and brief ones (3 min) with (+)-pentazocine or (+)-SKF-10,047 produced a transient σ_1 receptor-mediated decrease of the spontaneous rhythmic motor activity that was followed by a full recovery of the bursting hypoglossal firing occurring during the drug perfusion period. Focusing on the rapid attenuation of the hypoglossal response to σ_1 drugs, we administered the cell membrane bound phosphatidyl inositol-specific PLCβ inhibitor U-73,122 (1-[6-[[(17β)-3-methoxyestra-1,3,5 (10)-trien-17-yl]amino]hexy]-1H-pyrrole-2,5-dione) prior to and during a 3 min perfusion of (+)-SKF-10,047 or (+)-pentazocine. The aminosteroid U-73,122 (300 nmol for 10 min), which mildly reduced the basal burst amplitude of motor rhythmic firing, totally prevented the subsequent inhibitory effect of the (+)-benzophorphans. By contrast, the structurally analog of U-73,122, U-73,343 (1-[6-[[(17β)-3-methoxyestra-1,3,5(10)-trien-17-yl]amino]hexy]-2,5-pyrrolidine-dione), inactive on PLC, was ineffective on both basal and (+)-pentazocine-induced basal spontaneous rhythmic activity. Our paradigm has thus appeared of the greatest physiological relevance by dissecting on-line and step-by-step the successive biochemical events recruited by the activation of the neuronal σ_1 receptor.

The following step forward aiming at elucidating the signaling cascades recruited by the σ_1 receptor was provided by Brent et al. (72) who explored the modulation by (+)-pentazocine, (+)-SKF-10,047, DTG, and haloperidol of the phosphorylation rate of vesicular proteins such as synapsin 1B and dynamin, which play a prominent role in neurotransmitter release at the presynaptic level. This was the first study showing that the σ_1 receptor might affect protein phosphorylation. However, the fact that all σ drugs favored the presynaptic release by increasing the phosphorylation of both proteins reactivated the questions about the nomenclature and functional characteristics of either σ agonists or σ antagonists. It is nevertheless noteworthy that in the brainstem motor system, all σ drugs behaved similarly, which is not the case in cerebral structures such as the hippocampus or the cortex (10,32,43,45,73).

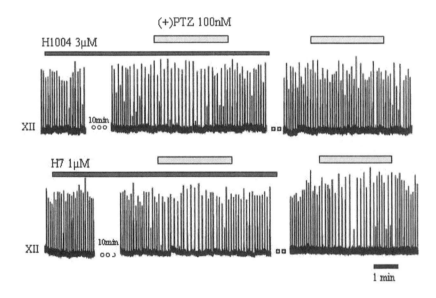

Figure 9-2. The desensitization process recruited by σ_1 agonists in the motor brainstem involves protein kinase C and not protein kinase A. The nonselective PKA inhibitor, H-1004, does not prevent (+)-pentazocine (PTZ)-induced desensitization of the neuronal response, whereas the nonselective PKA/PKC inhibitor, H-7, prevents this process.

The notion that activation of the σ_1 receptor affects precisely PKC activity also resulted from the brainstem preparation (32). Several biochemical cascades have been identified to mediate the rapid attenuation of drug-induced neuronal or neural responses - the protein kinase cascades have been extensively studied with this regard. Taking into consideration that PLC was implicated in the suddenness of the present σ_1 ligand-induced spontaneous hypoglossal rhythmic activity desensitization favors these pathways, it was thus logical to then assess the involvement of protein kinases, the following steps in the known biochemical cascade. In fact, two classical routes leading to sudden desensitization have been identified in the nervous system which are initiated by the stimulation of plasma membrane-bound seven transmembrane-bound receptors and lead to the activation of protein kinases coupled to either adenylate cyclase (i.e. protein kinase A or PKA) or to Ca^{2+}/diacylglycerol (i.e. PKC). Accordingly, to examine their putative role in the regulation of the effect of σ_1 ligands, H-7 (1-(5-isoquinolinesulfonyl)-2-methylpiperazine) and H-1004 (N-(2-guanidinoethyl)-5-isoquinoline-sulfonamide, 1 µM), a nonselective but competitive inhibitor of both protein kinase families acting by binding to their ATP-catalytic moiety were administered prior to (+)-SKF-10,047 or (+)-pentazocine. H-7 failed to reduce either the (+)-pentazocine- or (+)-SKF-10,047-induced inhibition of the response but prevented both (+)-pentazocine- and (+)-SKF-10,047-mediated motor desensitization (Figure 9-2). Unfortunately, H-7 does not discriminate between protein kinase subtypes. Therefore, H-1004 (3 µM), which inhibits the same protein kinases blocked by H-7 but is devoid of activity on PKC at the concentration used, was tested in the brainstem preparation. H-1004 also remained inactive on both (+)-benzomorphan-mediated inhibition of the hypoglossal response and the subsequent desensitization (Figure 9-2).

The potency of H-1004 for preventing PKA activation was confirmed as it prevented the enhancement of spontaneous motor firing frequency induced by isoproterenol, a β-adrenergic drug that recruits the PKA cascade. The present ineffectiveness of H-1004, which in other respects was biologically active on the isoproterenol-induced SRA response, contrasting the effectiveness of H-7 in preventing (+)-pentazocine-induced desensitization, privileged the hypothesis that the PKC cascade may be the main protein kinase system involved in the σ_1 receptor agonist-induced neural desensitization of the motor response. Further exploring this intracellular cascade, we have shown which subclass of PKCs the σ_1 receptor was coupled. Indeed, H-7 is not a selective PKC inhibitor, and does not discriminate between the three groups of PKC that are differentiated by distinct cofactor regulations (Ca^{2+} binding sites and diacylglycerol/phorbol ester binding site; 74,75). Three groups of PKC isoenzymes have been differentiated according to their cofactor regulation. The ubiquitous one discovered first, i.e. the conventional group, is sensitive to diacylglycerol/phorbol ester, a property shared with the novel PKCs, and selectively sensitive to Ca^{2+} (74,75). The already stated Ca^{2+}-dependency of most σ-mediated effects prompted my group to assess whether cPKCs were the target of $σ_1$ receptor biological action. In the central nervous system, both cPKCs and novel PKCs have been clearly identified. To determine which members of the PKC family were involved, we used two selective and high affinity PKC inhibitors that compete at the ATP-binding site. Go-6976 inhibits selectively cPKC while GF-109,203x (3-[1-[3-(dimethylamino) propyl]-1H-indol-3yl]-4-(1H-indol-3yl)-1H-pyrrole-2,5-dione) and tamoxifen (2-2-[4-(1,2-diphenyl-1-butenyl)phenoxy]-N,N-dimethyl-ethan amine) block both novel and cPKCs (76,77). Following a first 3 min perfusion of (+)-pentazocine or (+)-SKF-10,047 (as for control values), GF-109,203x (50 nM,) or Go-6976 (50 nM) were administered. The subsequent 3 min perfusion of either (+)-pentazocine or (+)-SKF-10,047 every 20 min produced a brief and reproducible inhibition of the rhythmic hypoglossal activity (Figure 9-3). The maintenance of the responsiveness to selective $σ_1$ drugs lasted for up to five hours. After this duration, the inhibition of the spontaneous hypoglossal rhythmic activity by the selective $σ_1$ ligand slowly faded. It was noteworthy that a subsequent 3 min perfusion of GF-109,203x was responsible for full recovery of the inhibitory potency of $σ_1$ receptor agonists on the motor bursting for additional hours. To complement these experiments, the PKC activator phorbol-12-myristate-13-acetate (PMA) was also tested in the presence of the $σ_1$ agonists. As expected, PMA produced a marked effect on the motor bursting activity, mimicking that of $σ_1$ drugs and also occluded their latter effects when both PMA and (+)-SKF-10,047 or (+)-pentazocine were applied simultaneously.

To provide additional support for the involvement of cPKCs in this $σ_1$ ligand-induced modulation, we studied the subcellular distribution of the cPKC isoenzymes in response to a 30 min perfusion of either (+)-pentazocine or (+)-SKF-10,047. This thus complemented the pharmacological characterization of the intracellular signaling events consecutive to $σ_1$ receptor activation. Accordingly, we performed immunohistochemistry using specific antibodies directed against the cPKC isoforms as well as the $σ_1$ receptor itself. Indeed, since an efficient marker of the activation of PKC is its translocation from the cytosol to the plasma membrane compartment (74), we used confocal immunofluorescence microscopy to characterize which cPKC isoforms were involved in the $σ_1$ ligand-induced neural desensitization process, using isoform-specific

9. Intracellular signaling and synaptic plasticity

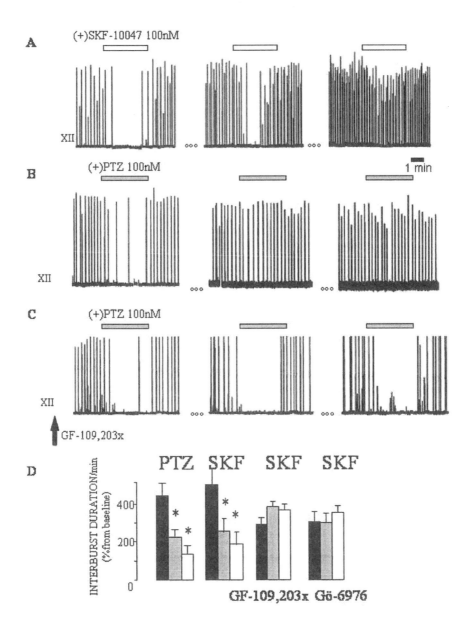

Figure 9-3. Desensitization of the σ_1 response via cPKC. Desensitization of the σ_1 effect on integrated hypoglossal (XII) activity after successive 3 min perfusions (open bars) of (+)-SKF-10,047 (100 nM; A) and (+)-pentazocine (100 nM; B). The desensitization process induced by successive perfusions of (+)-pentazocine (100 nM; B) was prevented by a pre-perfusion of GF-109,203x (50 nM, 5 min; C). (D) Mean values of the effects of successive perfusions of σ_1 (+)-benzomorphans in control conditions and after GF-109,203x and Go-6976 perfusions whcih prevented the (+)-benzomorphan-induced desensitization. Adapted in part with permission from ref. (32).

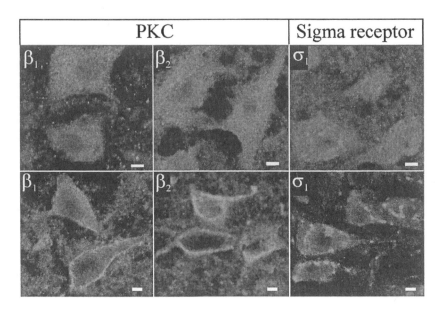

Figure 9-4. Translocation from the cytoplasmic to membrane compartment of β_1 and β_2 cPKC isoforms and of the σ_1 receptor in response to (+)-pentazocine (100 nM). In basal conditions (upper line), β_1 and β_2 cPKC isoforms, and the σ_1 receptor immunofluorescence staining was found in the cytoplasm. A long perfusion of (+)-pentazocine (lower line) produced a selective translocatin of β_1 and β_2 cPKC isoforms, and of the σ_1 receptor to the membrane vicinity. Neuronal images are single confocal sections. Scale bars = 5 µm. Adapted with permission from ref. (32).

antibodies. Figure 9-4 illustrates that the cPKC isoforms β_1 and β_2 translocated from the cytosol to the plasma membrane compartment in response to (+)-benzomorphans, whereas the cell distribution of both α and γ isoforms were not different between the control and treated conditions.

We also studied the subcellular distribution of the σ_1 receptor itself in response to (+)-benzomorphans. It was already known, from an immunohistochemical approach, that the σ_1 receptor is present intracellularly, predominantly anchored to the endoplasmic reticulum (3-5). In addition, Northern blot analysis has shown the presence of the σ_1 receptor in guinea pig and human central nervous system (3,5). It was hence all the more relevant to assess the cellular distribution of the σ_1 protein in the hypoglossal preparation that the paradigm focused on a spontaneous functional property of the σ_1 receptor. Previous binding experiments performed with either $[^3H](+)$-pentazocine or $[^3H](+)$-SKF-10,047 had been apparently controversial with regard to the cellular distribution of σ_1 binding sites as most studies concluded that the σ_1 receptor can be present simultaneously in all cell compartments, e.g. the microsomes, the mitochondria, the plasma and nuclear membrane levels or predominantly in one or two of them (see for example 3,5,78,79). It should however be noticed that depending on the preparation (intact cells or cellular membrane extracts, procedure for cell fixation, etc.), the results differed greatly. This is particularly obvious when σ_1 receptor/G protein coupling was investigated. With this respect, our electrophysiological paradigm of the guinea pig brainstem favored the exploration of the σ_1 binding protein (it is known however that in the rat, the σ_2 subtype is predominant; see 2) in a functional bioassay.

In fact, our study provided evidence that all previous binding approaches were correct when it was claimed that the σ_1 protein was predominant in such and such a cell compartment. Unexpectedly, we demonstrated that the location of the σ_1 receptor depends on its level of activation. Taking advantage of the specific antibody directed against the guinea pig liver σ_1 receptor (3), we found that the cellular distribution of the σ_1 receptor was affected by its subsequent activation. As indicated in Figure 9-4, the σ_1 receptor, anchored to the endoplasmic level at rest, translocates at the vicinity of the plasma membrane level following σ_1 receptor activation induced by (+)-benzomorphans. This prompted us to conclude that when the σ_1 receptor remains inactivated, i.e. in the absence of exposure to σ_1 drug, the labeling is located predominantly at the endoplasmic reticulum level, whereas when it is activated, i.e. in response to (+)-pentazocine or (+)-SKF-10,047 (free or labeled), the σ_1 receptor translocates in the vicinity of the plasma membrane level. Autoradiography and Western blot experiments then further identified that the (+)-pentazocine-induced activation of cPKC triggered protein phosphorylation. Figure 9-5 shows that (+)-pentazocine triggered a phosphorylation of a brainstem protein migrating at the same level as that expected for the σ_1 receptor. Preadministration of GF-109,203x totally prevented the (+)-benzomorphan-mediated protein phosphorylation and in particular erased that of the protein with the same molecular weight as the σ_1 receptor. The subsequent Western blot step allowed detection of a single band of 29 kDa as the (+)-pentazocine phosphorylated band which was suppressed in the presence of pbp45, a synthetic peptide directed against the cloned σ_1 receptor (3). These results tend to indicate that the σ_1 receptor would be phosphorylated through a process involving its own activation. Finally, this study showed that an intracellular receptor bound at rest to the endoplasmic reticulum is able to translocate to the plasma membrane. These observations are of prominent clinical relevance as they provide further insight into and better understanding of the cellular events triggered by σ_1 ligands, which are

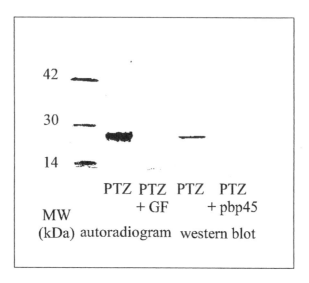

Figure 9-5. Autoradiogram reveals that a perfusion of (+)-pentazocine (PTZ) triggers the phosphorylation of the σ_1 protein (28 kDa), which is prevented following a pre-perfusion with GF-109,203x (GF). In the Western blot, this band disappeared in the presence of the pbp45 antipeptide. Adapted with permission from ref. (32).

currently under investigation for their purported neuroprotective and cognitive abilities, as well as for their beneficial action in several neurological and psychiatric disorders. Since that, a serine/threonine-rich protein belonging to the nuclear matrix-associated proteins (Ramp; 80) has been shown to translocate similarly to the σ_1 protein. This former protein is involved in patterning and neurogenesis during development and interferes with the cell differentiation of postmitotic neurons of the central nervous system. It is thus tempting to speculate whether σ_1-induced cPKC activation has a more fundamental regulatory role for known PKC functions, e.g. regulating cell growth and survival, and in learning and memory. Indeed, it is well established that PKC participates in cell proliferation, differentiation, and cellular death. It is also known that this protein kinase family is mandatory for a full processing of the hippocampal long term potentiation and long term depression, two models related to cognition and memory. The fact that the σ_1 receptor participates in the intracellular signaling cascades also brings support to the notion that the σ_1 protein might play an important role in cell proliferation for example during cancer. It remains to be demonstrated whether in such conditions, the σ_1 receptor translocation also occurs. In addition, it is tempting to speculate that such a function of the σ_1 protein might be of significance for its purported beneficial action in cardiac and immune functions. Indeed, it is most likely that this outstanding property of the σ_1 protein is not restricted to the motor brainstem system, as it has recently been demonstrated in the hippocampus (67), striatum (81), and likely occurs in the periphery, Ela et al. (82) having suggested that the myorelaxant action of (+)-pentazocine described in neonate rat cardiomyocytes desensitizes with a similar dynamics to that in brain structures.

7. CONCLUSIONS

Pharmacological, biochemical, and genetic approaches have done much to define the nature of the intracellular σ receptor and how it is recruited during the process of neuronal activation. Experiments outlined in this review indicate that most of the biochemical cascades triggered by the activation of the endoplasmic reticulum-bound σ_1 protein share many of the molecular events occurring during synaptic plasticity at both presynaptic and postsynaptic levels. This supports the notion that this receptor plays a crucial role in the adaptative changes mandatory for the memory process as is documented in Chapter 12. It remains, however, to determine whether and how these molecular mechanisms, involved in learning and recruited by σ drugs, operate independently or synergically (most of them usually proceeding in parallel). This will require the demonstration that σ drugs can also control action potential-independent mechanisms, and are involved in Ca^{2+} influx and modulation of docking-fusioning processes of the synaptic vesicles in the presynaptic terminals, respectively. Important areas for future research include further study on the interplay between presynaptic and postsynaptic structural changes occurring in response to σ drugs and the roles of the cytoskeleton and cell-adhesion molecules in mediating these changes. There is no doubt that the identification of the nature of the proteins synthesized in response to σ drugs will also need to be soon addressed.

The physiological implication of the fact that σ_1 agonists *in vitro* favor the phosphorylation of proteins implicated in presynaptic neurotransmitter release and development of the dendritic tree has to be realized from studies involving functionally active neurons to ascertain that such mechanisms are operational in the synapses *in situ*, the results of which may thus further support not only the behavioral effectiveness of the drugs but also the clinical relevance of their use as therapeutic tools. Imaging techniques have recently emerged for giving information on the identification of gene expression and neural activity and exploring the role of specific intracellular signaling pathways in the processes of neuronal plasticity by using appropriate reported genes. The combination of these techniques with the development of rapidly regulated genetic systems promises to define the molecular mechanism by which σ drugs affect neuronal function.

REFERENCES

1. Quirion R, Bowen WD, Itzhak Y, Junien JL, Musacchio JM, Rothman RB, Su TP, Tam SW, Taylor DP. Classification of σ binding sites: a proposal. Trends Pharmacol Sci 1992, 13:85-86.
2. Walker JM, Bowen WD, Walker FO, Matsumoto RR, deCosta BR, Rice KC. Sigma receptors: biology and function. Pharmacol Rev 1990, 42:355-402.
3. Hanner M, Moebius FF, Flandorfer A, Knaus H-G, Striessnig J, Kemper E, Glossmann H. Purification, molecular cloning, and expression of the mammalian $σ_1$-binding site. Proc Natl Acad Sci USA 1996, 93:8072-8077.
4. Kekuda R, Prasad PD, Fei YJ, Leibach FH, Ganapathy V. Cloning and functional expression of the human type 1 σ receptor (hSigmaR1). Biochem Biophys Res Commun 1996, 229:553-558.
5. Jbilo O, Vidal H, Paul R, de Nyst N, Bensaid M, Silve S, Carayon P, Davi D, Galiègue S, Bourrié B, Guillemot JC, Ferrara P, Loison G, Maffrand JP, LeFur G, Casellas P. Purification and characterization of the human SR-31747A-binding protein. J Biol Chem 1997, 272:27107-27115.
6. Martin WR, Eades CG, Thompson JA, Huppler RE, Gilbert PE. The effects of morphine and nalorphine like drugs in the non dependent and morphine dependent chronic spinal dog. J Pharmacol Exp Ther 1976, 197:517-532.
7. Hayashi T, Su TP. Sigma-1 receptor ligands: potential in the treatment of neuropsychiatric disorders. CNS Drugs 2004, 18:269-284.
8. Itzhak Y. Multiple affinity binding states of the σ receptor: effect of GTP-binding protein-modifying agents. Mol Pharmacol 1989, 36:512-517.
9. Monnet FP, Debonnel G, de Montigny C. Potentiation by haloperidol of the antagonism by MK 801 of the excitatory effect of dicarboxylic amino acids: an electrophysiological study in the rat dorsal hippocampus. Soc Neurosci Abst 1988, 14:381.14.
10. Monnet FP, Debonnel G, Junien JL, de Montigny C. N-methyl-D-aspartate-induced neuronal activation is selectively modulated by σ receptors. Eur J Pharmacol 1990, 179:441-445.
11. Golstein SR, Matsumoto RR, Thomson TL, Patrick RL, Bowen WD, Walker JM. Motor effects of two σ ligands mediated by nigrostriatal dopamine neurons. Synapse 1989, 4:254-258.
12. Iyengar S, Mick S, Dilworth V, Michel J, Rao TS, Farah JM, Wood PL. σ receptors modulate the hypothalamic-pituitary-adrenal (HPA) axis centrally: evidence for a functional interaction with NMDA receptors, in vivo. Neuropharmacology 1990, 29:299-303.

9. Intracellular signaling and synaptic plasticity

13. Rogers CA, Lemaire S. Role of the σ receptor in the inhibition of [^{3}H]noradrenaline uptake in brain synaptosomes and adrenal chromaffin cells. Br J Pharmacol 1991, 103:1917-1922.
14. Kim MB, Bickford PC. Electrophysiological effects of phencyclidine and the σ agonist ditolylguanidine in the cerebellum of the rat. Neuropharmacology 1992, 31:77-83.
15. Bowen WD, Kirschner BN, Newman AH, Rice KC. σ Receptor negatively modulate agonist-stimulated phosphoinositide metabolism in rat brain. Eur J Pharmacol 1988, 149:399-400.
16. deHaven-Hudkins DL, Hudkins RL. Binding of dexetimide and levetimide to [^{3}H](+)-pentazocine and [^{3}H]1,3-di(2-tolyl)guanidine-defined σ recognition sites. Life Sci 1991, 49:PL135-PL139.
17. Kobayashi T, Ikeda K, Ichikawa T, Togashi S, Kumanishi T. Effects of σ ligands on the cloned mu-, delta- and kappa-opioid receptors co-expressed with G-protein-activated K$^+$(GIRK) channel in Xenopus oocytes. Br J Pharmacol 1996, 119:73-80.
18. Olney JW, Farber NB. Glutamate receptor dysfunction and schizophrenia. Arch Gen Psychiatry 1995, 52:998-1007.
19. Tam SW. Potential therapeutic application of σ receptor antagonists. In Sigma Receptors. Y Itzhak, ed. Academic Press, San Diego, pp 191-204, 1994.
20. Su TP. Delineating biochemical and functional properties of σ receptors: emerging concepts. Crit Rev Neurobiol 1993, 7:187-203.
21. Chavkin C. The σ enigma: biochemical and functional correlates emerge for the haloperidol-sensitive σ binding site. Trends Neurosci 1990, 11:213-215.
22. Carpenter CL, Marks SS, Watson DL, Greenberg DA. Dextromethorphan and dextrorphan as calcium channel antagonists. Brain Res 1988, 449:372-375.
23. Korkotian E, Segal M. Release of calcium from stores alters the morphology of dendritic spines in cultured hippocampal neurons. Proc Natl Acad Sci USA 1999, 96:12068-12072.
24. Lüscher C, Nicoll RA, Malenka RC, Muller M. Synaptic plasticity and dynamic modulation of the postsynaptic membrane. Nature Neurosci 2000, 3:545-550.
25. Abel T, Lattal KM. Molecular mechanisms of memory acquisition, consolidation and retrieval. Curr Opinion Neurobiol 2001, 11:180-187.
26. Palacios G, Muro A, Vela JM, Molina-Holgado E, Guitart X, Ovalle S, Zamanillo D. Immunohistochemical localization of the σ$_1$-receptor in oligodendrocytes in the rat central nervous system. Brain Res 2003, 96:92-99.
27. Monnet FP, Mahé V, Robel P, Baulieu EE. Neurosteroids, via σ receptors, modulate the [^{3}H]norepinephrine release evoked by NMDA in the rat hippocampus. Proc Natl Acad Sci USA 1995, 92:3774-3778.
28. Cammarota M, Bernabeu R, Levi De Stein M, Izquierdo I, Medina JH. Learning-specific, time-dependent increases in hippocampal Ca^{2+}/calmodulin-dependent protein kinase II activity and AMPA GluR1 subunit immunoreactivity. Eur J Neurosci 1998, 10:2669-76.
29. Maurice T, Su TP, Parish DW, Nabeshima T, Privat A. PRE-084, a selective PCP derivative, attenuates MK-801-induced impairment of learning in mice. Pharmacol Biochem Behav 1994, 49:859-869.
30. Maurice T, Phan VL, Urani A, Kamei H, Noda Y, Nabeshima T. Neuroactive neurosteroids as endogenous effectors for sigma-1 (σ1) receptor: pharmacological evidence and therapeutic opportunities. Jpn J Pharmacol 1999, 81:125-155.
31. Matsumoto RR, Bowen W, Tom M, Vo V, Truong D, De Costa B. Characterization of two novel σ receptor ligands: antidystonic effects in rats suggest σ receptor antagonism. Eur J Pharmacol 1995, 280:301-310.

32. Morin-Surun MP, Collin T, Denavit-Saubié M, Baulieu EE, Monnet FP. Sigma-1 receptor modulate PLC/PKC cascade in the motor brainstem. Proc Natl Acad Sci USA 1999, 96:8196-8199.
33. Maurice T, Lockhart BP, Privat A. Amnesia induced in mice by centrally administered beta-amyloid peptides involves cholinergic dysfunction. Brain Res 1996, 706:181-193
34. Pande AC, Genève J, Scherrer B. Igmésine, a novel σ ligand, has antidepressant properties XXI CINP Abst 1998, 30S-30M 0505.
35. Matsumoto RR, Bowen WD, Walker JM. Age-related differences in the sensitivity of rats to a selective σ ligand. Brain Res 1989, 504:145-148.
36. Hemstreet MK, Matsumoto RR, Bowen WD, Walker JM. Sigma binding parameters in developing rats predict behavioral efficacy of a σ ligand. Brain Res 1993, 627:291-198.
37. Hayashi T, Su TP. Regulating ankyrin dynamics: roles of σ_1 receptors. Proc Natl Acad Sci USA 2001, 98:491-496.
38. Takebayashi M, Hayashi T, Su TP. Nerve growth factor-induced neurite sprouting in PC-12 cells involves σ_1 receptors: implications for antidepressants. J Pharmacol Exp Ther 2002, 303:1227-1237.
39. Vilner BJ, deCosta BJ, Bowen WD. Cytotoxic effects of σ ligands: σ receptor-mediated alterations in cellular morphology and variability. J Neurosci 1995, 15:117-134.
40. Vilner BJ, John CS, Bowen WD. Sigma-1 and σ_2 receptors are expressed in a wild variety of human and rodent tumor cell lines. Cancer Res 1995, 55:408-413.
41. Vilner BJ, Bowen WD. Modulation of cellular calcium by σ_2 receptors: release from intracellular stores in human SK-N-SH neuroblastoma cells. J Pharmacol Exp Ther 2000, 292:900-911.
42. Itzhak Y, Khouri M. Regulation of the binding of σ- and phencyclidine(PCP)-receptor ligands in rat brain membranes by guanine nucleotides and ions. Neurosci Lett 1988, 85:147-152.
43. Monnet FP, Blier P, Debonnel G, de Montigny C. Modulation by σ ligands of N-methyl-D-aspartate-induced [^{3}H]norepinephrine release in the rat hippocampus: G-protein dependency. Naunyn-Schmiedeberg Arch Pharmacol 1992a, 346:32-39.
44. Bowen WD. Interaction of σ receptors with signal transduction pathways and effects on second messengers, In Sigma Receptors. Y Itzhak, ed. Academic Press, San Diego, pp 139-170, 1994.
45. Monnet FP, Debonnel G, de Montigny C. In vivo electrophysiological evidence for a selective modulation of N-methyl-D-aspartate-induced neuronal activation in rat CA_3 dorsal hippocampus by σ ligands. J Pharmacol Exp Ther 1992b, 261:123-130.
46. Lesage AS, De Loore KL, Peeters L, Leysen JE. Neuroprotective σ ligands interfere with the glutamate-activated NOS pathway in hippocampal cell culture. Synapse 1995, 20:156-164.
47. Hayashi T, Maurice T, Su TP. Ca^{2+} signaling via σ_1 receptors: novel regulatory mechanism affecting intracellular Ca^{++} concentration. J Pharmacol Exp Ther 2000, 293:788-798.
48. Ueda H, Yoshida A, Tokuyama S, Mizuno K, Maruo J, Matsuno K, Mita S. Neurosteroids stimulate G protein-coupled σ receptors in mouse brain synaptic membrane. Neurosci Res 2001a, 41:33-40.
49. Ueda H, Inoue M, Yoshida A, Mizuno K, Yamamoto H, Maruo J, Matsuno K, Mita S. Metabotropic neurosteroid/sigma-receptor involved in stimulation of nociceptor endings of mice. J Pharmacol Exp Ther 2001b, 298:703-710.
50. Lupardus PJ, Wilke RA, Aydar E, Palmer CP, Chen Y, Ruoho AE, Jackson MB. Membrane-delimited coupling between sigma receptors and K^+ channels in rat

neurohypophysial terminals requires neither G-protein nor ATP. J Physiol 2000, 526:527-539.
51. Monnet FP, Debonnel G, de Montigny C. The effects of σ ligands and of neuropeptide Y on N-methyl-D-aspartate-induced neuronal activation are differentially affected by pertussis toxin in the rat CA_3 dorsal hippocampus. Br J Pharmacol 1994, 112:709-715.
52. Sommermeyer H, Dompert WU, Glaser T. Signalling via rat dopamine D_2-receptors expressed in mouse fibroblasts is not influenced by compounds binding to the σ sites of these cells. Cell Signal 1993, 5:747-752.
53. Soriani O, LeFoll F, Roman F, Vaudry H, Monnet FP, Cazin L. A-current down-regulated by σ ligands in frog pituitary melanotrope cells through a G-protein-dependent pathway. J Pharmacol Exp Ther 1999, 289:321-328.
54. Klein M, Santiago LJ, Musacchio JM. Effect of calcium and other ion channel blocking agents on the high affinity binding of dextromethorphan to guinea pig brain. Soc Neurosci Abst 1985, 14:86.
55. Rothman RB, Reid A, Mahboubi A, Kim CH, deCosta BJ, Jacobson AE, Rice KC. Labeling by [^{3}H]1,3-di(2-tolyl)guanidine of two high affinity binding sites in guinea pig brain: evidence for allosteric regulation by calcium channel antagonists and pseudoallosteric modulation by σ ligands. Mol Pharmacol 1991, 39:222-232.
56. Basile AS, Paul IA, Mirchevich A, Kuijpers G, deCosta BJ. Modulation of (+)[^{3}H]pentazocine binding to guinea pig cerebellum by divalent ions. Mol Pharmacol 1992, 42:882-889.
57. Kostyuk P, Akaike N, Osipchuk Y, Savchenko A, Shuba Y. Gating and permeation of different types of Ca channels. Ann N Y Acad Sci 1989, 560:63-79.
58. Zhang H, Cuevas J. Sigma receptors inhibit high-voltage-activated calcium channels in rat sympathetic and parasympathetic neurons. J Neurophysiol 2002, 87:2867-79.
59. Church J, Fletcher EJ. Blockade by σ site ligands of high voltage-activated Ca^{2+} channels in rat and mouse cultured hippocampal pyramidal neurons. Br J Pharmacol 1995, 116:2801-2810.
60. Paul IA, Basile AS, Rojas E, Youdim MBH, deCosta BJ, Skolnick P, Pollard HB, Kuijpers GAJ. Sigma receptors modulate nicotine receptor function in adrenal chromaffin cells. FASEB J 1993, 7:1171-1178.
61. Hayashi T, Kagaya A, Takebayashi M, Shimizu M, Uchitomi Y, Motohashi N, Yamawaki S. Modulation by sigma ligands of intracellular free Ca^{2+} mobilization by N-methyl-D-aspartate in primary culture of rat frontal cortical neurons. J Pharmacol Exp Ther 1995, 275:207-214.
62. Klette KL, Lin Y, Clapp LE, de Coster MA, Moreton JE, Tortella FC. Neuroprotective sigma ligands attenuate NMDA and trans-ACPD-induced Ca signaling in rat primary neurons. Brain Res 1997, 756:231-240.
63. Monnet FP, de Costa BR, Bowen WD. Differentiation of σ ligand-activated receptor subtypes that modulate NMDA-evoked [^{3}H]noradrenaline release in rat hippocampal slices. Br J Pharmacol 1996, 119:65-72.
64. Brent PJ, Pang G, Little G, Dosen PJ, van Helden DF. The σ receptor ligand, reduced haloperidol, induces apoptosis and increases intracellular-free Ca levels [Ca^+]$_i$ in colon and mammary adenocarcinoma cells. Biochem Biophys Res Commun 1996, 219:219-226.
65. Brent PJ, Herd L, Saunders H, Sim ATR, Dunkley PR. Protein phosphorylation and Ca uptake into rat forebrain synaptosomes: modulation by σ ligands. J Neurochem 1997, 68:2201-2211.

66. Novakova M, Ela C, Bowen WD, Hasin Y, Eilam Y. Highly selective σ receptor ligands elevate inositol 1,4,5-triphosphate production in rat cardiac myocytes. Eur J Pharmacol 1998, 353:315-327.
67. Monnet FP, Morin-Surun MP, Leger J, Combettes L. Protein kinase C-dependent potentiation of intracellular Ca mobilization by σ_1 receptor agonists in rat hippocampal neurons. J. Pharmacol Exp Ther 2003, 307:705-712.
68. Novakova M, Ela C, Barg J, Vogel Z, Hasin Y, Eilam Y. Ionotropic action of σ receptor ligands in isolated cardiac myocytes from adult rats. Eur J Pharmacol 1995, 286:19-30.
69. Aydar E, Palmer CP, Klyachko VA, Jackson MB. The σ receptor as a ligand-regulated auxiliary potassium channel subunit. Neuron 2002, 34:399-410.
70. Candura SM, Coccini T, Manzo L, Costa LG. Interaction of σ-compounds with receptor-stimulated phosphoinositide metabolism in the rat brain. J Neurochem 1990, 55:1741-1748.
71. Monnet FP, Morin-Surun MP. Electrophysiological evidence for the role of protein kinase C in the action of σ_1 receptor ligands. Soc Neurosci Abst 1997, 23:905.5.
72. Brent PJ, Haynes H, Jarvie PE, Mudge L, Sim AT, Dunkley PR. Phosphorylation of synapsin I and dynamin in rat forebrain synaptosomes: modulation by sigma (σ) ligands. Neurosci Lett 1995, 1-2:71-74.
73. Matsumoto RR, Walker JM. Inhibition of rubral neurons by a specific ligand for σ receptors. Eur J Pharmacol 1988, 158:161-165.
74. Tanaka C, Nishizuka Y. The protein kinase C family for neuronal signaling. Ann Rev Neurosci 1994, 17:551-567.
75. Newton AC. Protein kinase C: structure, function, and regulation. J Biol Chem 1995, 270:28495-28498.
76. Toullec D, Pianetti P, Coste H, Bellevergue P, Grand-Perret T, Ajakane M, Baudet V, Boissin P, Boursier E, Loriolle F, Duhamel L, Charon D, Kirilovsky J. J Biol Chem 1991, 266:15771-15781.
77. Martiny-Baron G, Kazanietz MG, Mischak H, Blumberg PM, Kochs G, Hug H, Marme D, Schächtele C. Selective inhibition of protein kinase C isozymes by the indolocarbazole Go-6976. J Biol Chem 1993, 268:9194-9197.
78. McCann DJ, Su TP. Solubilization and characterization of haloperidol-sensitive (+)-[^{3}H]-SKF-10,047 binding sites (σ sites) from rat liver membranes. J Pharmacol Exp Ther 1991, 257:547-554.
79. Klouz A, Sapena R, Liu J, Maurice T, Tillement JP, Papadopoulos V, Morin D. Evidence for σ_1-like receptors in isolated rat liver mitochondrial membranes. Br J Pharmacol 2002, 135:1607-1615.
80. Cheung WM, Chu AH, Chu PW, Ip NY. Cloning and expression of a novel nuclear matrix-associated protein that is regulated during the retinoic acid-induced neuronal differentiation. J Biol Chem 2001, 276:17083-17091.
81. Nuwayhid SJ, Werling LL. Sigma-1 receptor agonist-mediated regulation of N-methyl-D-aspartate-stimulated [^{3}H]dopamine release is dependent upon protein kinase C. J Pharmacol Exp Ther 2003, 304:364-369.
82. Ela C, Hasin Y, Eilam Y. Apparent desensitization of a σ receptor subpopulation in neonatal rat cardiac myocytes by pre-treatment with σ receptor ligands. Eur J Pharmacol 1996, 295:275-280.

Corresponding author: Dr. Francois Monnet, Mailing address: Institut National de la Sante et de la Recherche Medicale Unite 705 – CNRS UMR

9. Intracellular signaling and synaptic plasticity 193

7157, Hôpital Fernand Widal, 200 rue du Faubourg Saint-Denis, 75475 Paris cedex 10, France, Phone: 33 1 40 05 43 45, Fax: 331 40 05 43 42, Electronic mail address : francois.monnet@fwidal.inserm.fr

Chapter 10

MODULATION OF CLASSICAL NEUROTRANSMITTER SYSTEMS BY σ RECEPTORS

Linda L. Werling, Alicia E. Derbez and Samer J. Nuwayhid
Department of Pharmacology & Physiology, The George Washington University Medical Center, Washington, DC 20037, USA

1. INTRODUCTION

Although actions of σ receptors on several physiological processes (Chapters 11-18) have been described, very little is known about σ receptor-mediated neurotransmission. This is largely due to the lack of unequivocal identification of an endogenous ligand. Recent evidence has implicated neurosteroids as potential endogenous transmitters at σ receptors (reviewed in Chapter 1). Despite these recent provocative data, more information is available regarding the modulation of other, classical neurotransmitter systems via activation of σ receptors by prototypical and novel σ receptor ligands. Even when ligands that have been well characterized in radioligand binding assays are used, there is not complete agreement on which drugs act as agonists, and which as antagonists. σ Receptors are unlikely to be "classical" transmitter receptors. The data reported in the ion channel studies of Jackson and colleagues (1-3) as well as those from the ankyrin/IP$_3$ receptor dynamics studies of Su and colleagues (4,5) clearly demonstrate that protein-protein interactions are important in σ receptor signaling, and at least some σ receptor-mediated processes probably rely on the σ receptor associating with one or more additional proteins to cause a physiological effect. Such interactions might or might not follow the agonist/antagonist relationships that are the hallmark of traditional pharmacology. Another often debated aspect of σ receptor modulation of neurotransmitter function is that of G protein coupling. While the cloned σ$_1$ receptor protein is too small and does not have the appropriate site for coupling directly to a G protein, several reports of G protein- or pertussis toxin-sensitivity of σ agonist-mediated processes have been reported (6). However, most studies show no effect of guanyl nucleotides on the binding of haloperidol or (+)-pentazocine to σ receptors (7). Recent evidence associating σ_1 receptor activation with phospholipase C (PLC)/protein kinase C (PKC) pathways may partially explain these findings, as some G proteins involved in PLC function are pertussis toxin sensitive (8).

Even though there is not complete agreement in various studies about identification of σ agonists versus antagonists, the evidence shows that some drugs that bind to σ receptors antagonize the actions of others that also bind, so at some level, receptor-like properties exist. Discrepancies also likely arise from the probability that there are more σ receptor subtypes than have been unequivocally identified, as well as from the possibility that σ receptor-mediated signaling is complex and multiple pathways may be activated depending upon the neuron studied. Additionally, tonic actions of endogenous ligand probably influence the experimental results. In this chapter, modulation of several well-characterized transmitter systems via σ receptor-mediated actions will be described, and where these processes have been linked to signaling pathways, those will be mentioned in an attempt to develop a unifying picture of possible σ receptor function. We will concentrate on direct actions of σ ligands on central systems. Much work has been done on central effects of peripherally administered σ ligands, and these studies are critically important in the development of therapeutic strategies that might utilize σ ligands in the future. However, since many steps are likely to exist between peripheral administration and central neuronal activity, primarily the direct effects on central nervous system neurons or cells in culture will be addressed here, except in cases in which peripheral and local central administration of the σ drugs produced similar results.

σ Receptors are distributed throughout the central nervous system as well as in peripheral tissues. Bouchard and Quirion (9) studied autoradiographically the distribution of σ_1 and σ_2 receptors as labeled by [^{3}H](+)-pentazocine, and [^{3}H]di-o-tolylguanidine (DTG) in the presence of unlabeled (+)-pentazocine, respectively. They found enrichment of σ_1 binding in brainstem nuclei, especially the oculomotor, trigeminal and facial cranial nerve nuclei. The red nucleus and the substantia nigra were also highly labeled, as was the pyramidal layer of the hippocampus. σ_2 Sites were generally in lower density than σ_1, but several areas of brain, including substantia nigra pars reticulata, central gray, oculomotor nucleus, nucleus accumbens, cerebellum and motor cortex showed a relatively greater density of σ_2, as compared to σ_1, sites. These data vary somewhat from those of

Leitner et al. (10) who, using homogenate binding, observed a ratio of greater than 1.0 for σ_2 to σ_1 receptors in all brain areas examined. In their study, the highest densities of σ_1 receptors were found in hindbrain and midbrain, while σ_2 receptors were enriched in those areas in addition to cortex and cerebellum. Alonso et al. (11) used an antibody raised to amino acids 143-162 of the σ_1 protein to label σ_1 sites in rat brain. They detected high σ_1 binding in the granular layer of the olfactory bulb, several hypothalamic nuclei, the septum, the central gray, and motor nuclei of the hindbrain and the dorsal horn of the spinal cord. In general, expression of σ_1 receptor mRNA coincided with distributions first described by quantitative autoradiography (12). Logically, one might examine regulation of transmitter systems in areas of high σ receptor density, but several areas where σ receptor-mediated effects have been described are not especially enriched in σ receptors, a reminder that a few receptors are capable of profound effects if the amplification system is robust.

Subcellularly, σ receptors have been localized to plasma membrane, endoplasmic reticulum (ER), mitochondria and cytoplasm (11,13). Trafficking of σ receptors has been demonstrated in guinea pig hypoglossal neurons (14) and NG108 cells (15), and by extrapolation, is likely to occur in other neurons and cells as well. Trafficking allows for regulation of multiple cellular processes, and subsequently of neural systems at several levels.

An issue in the σ receptor field has been that while many studies demonstrate a profile of a variety of σ ligands with a rank order of potency identical or nearly identical to the binding affinity at σ receptors, supporting actions of these ligands via σ receptors, the potency of the compounds in physiological assays is orders of magnitude lower than that in binding assays. Ideally, one would be able to see effects of σ ligands at concentrations that are commensurate with their affinities at σ receptors in radioligand binding assays. Most σ-active compounds bind with affinities in the nanomolar to low micromolar range, whereas in many reports on function, concentrations required to observe effect are in the micromolar to high micromolar range. Explanations offered for this phenomenon have included a compelling argument for pH and cell permeability (16), intermediate steps, and the disruption of accessory protein ensembles required for function by preparation of tissue for binding assays. Yet, correlation of potency with K_i in some assays is quite direct. For instance, in regulation of catecholamine release, IC_{50} values for regulation by σ receptors are virtually identical to K_i values in binding studies (6,17). This would imply that pH and its effects on protonation status are less in important in some physiological functions of σ receptors, and perhaps that those receptors mediating functions at concentrations similar to K_i values occur via σ

receptors that are located on the plasma membrane. In contrast, those actions that require binding to an intracellular σ receptor, such as one on the ER, would require a non-protonated form of the ligand, which would be necessary for it to cross the cell membrane to gain access to intracellular receptors (e.g. 18).

One way that σ ligands could exert their effects on classical neurotransmitter systems is via direct modification of ion channels (see Chapter 7). Since theoretically all neurons bear potassium and calcium channels, both of which are modified by application of σ ligands, if a particular neuron also bears σ receptors, effects of σ ligands could be quite profound. σ Receptor activation has also been linked to intracellular calcium homeostasis, a critical regulator of cellular function (15,18). For example, we have demonstrated that (+)pentazocine and several neurosteroids enhance bradykinin-induced increases in intracellular calcium in SH-SY5Y cells, and these enhancements are blocked by haloperidol (82). Again, such effects could mediate σ regulation of multiple neurotransmitter systems.

2. REGULATION OF CLASSICAL TRANSMITTER SYSTEMS

Several neurotransmitter systems have been found to be modified by σ ligands, including catecholaminergic, glutamatergic, and opioidergic systems. Cell firing, neurotransmitter uptake and release, and signaling including intracellular calcium homeostasis have been studied. Signaling is discussed in more detail in Chapters 8, 9 and 11.

2.1 Effects of σ receptors on glutamatergic neurotransmission

Debonnel, deMontigny and coworkers have described in detail the modulation of N-methyl-D-aspartate (NMDA)-induced electrophysiological responses by σ receptors (19-23). They have shown that application of σ ligands enhances the responsiveness of pyramidal neurons in hippocampal CA1 and CA3 regions to microiontophoretic applications of NMDA (21). In general, the σ ligands produce bell-shaped dose-response curves, with enhancement at lower concentrations, and decline toward control values at higher concentrations. Assuming that the production of an enhancement can be interpreted as an agonist property, DTG, igmesine (JO 1784), dehydroepiandrosterone (DHEA), and (+)-pentazocine appear to be agonists in their assays. In contrast, haloperidol, NE-100 (N,N-dipropyl-2-[4-methoxy-3-(2-phenylethoxy)phenyl]ethylamine), and the neurosteroids progesterone and testosterone act as antagonists in their system (24). Manipulation of the hormone status of the experimental animal also affects σ receptor function in these studies. For instance, pregnancy reduced brain σ receptor function (22), causing a requirement for 10-fold higher doses on σ ligands to potentiate the NMDA-evoked responses, whereas in ovariectomized rats treated for three weeks with progesterone, the neuronal response to σ ligands was enhanced. These findings also support the hypothesis that neurosteroids are endogenous ligands at σ receptors. Finally, Monnet et al. (25) found that the enhancement of response of CA3 pyramidal neurons to some σ ligands, such as DTG, igmesine and neuropeptide Y were abolished by pertussis toxin, whereas responses to others, such as (+)-pentazocine, persisted following pertussis toxin treatment. Again, these findings support that σ receptors signal via both G protein and non-G protein-mediated pathways.

Several σ receptor ligands have also been found to protect against NMDA-induced neurotoxicity in several models, including primary cultures of rat cortical neurons (26) and organotypic dopaminergic midbrain slice cultures (27). In these studies, σ ligands classified previously as agonists (e.g. (+)-SKF-10,047, N-allylnormetazocine) and antagonists (e.g. haloperidol) both were neuroprotectant. σ Receptor ligands themselves can, however, be neurotoxic, and the toxicity appears to depend upon the degree of mobilization of intracellular calcium from ER (28).

Exactly how modification of the NMDA receptor-mediated response is achieved by σ ligands is not known. While a binding site for σ drugs on the NMDA receptor/channel complex has been postulated, the evidence from autoradiographic and membrane radioligand binding studies does not indicate that there is colocalization between NMDA and σ sites (compare 29 and 30). As discussed above, the anatomical distribution of the two is not convergent. It has been shown that chronic treatment with the experimental antipsychotic E-5842 (4-[4-fluorophenyl]-1,2,3,6-tetrahydro-1-[4-[1,-2,4-triazol-1-il]butyl]pyridine), a σ_1 receptor antagonist, upregulates glutamate receptor subunits in limbic brain areas (31). This indicates that feedback loops exist between σ and glutamatergic systems as well.

σ Ligands may also modulate glutamate release. Annels et al. (32) examined the effects of dextromethorphan, caramiphen, and carbetapentane, all ligands at σ receptors, for their effects on potassium-stimulated release of endogenous glutamate from rabbit hippocampal slices. All three reduced stimulated release, although the concentrations required were fairly high, and no reversal by σ antagonists was demonstrated. Ellis and Davies (33) reported that haloperidol, reduced haloperidol, rimcazole and ifenprodil also reduced potassium-evoked endogenous glutamate release from rat striatal slices, but again at high concentrations.

The interactions of σ ligands with the glutamate systems are intriguing, but difficult to assemble into a unified picture. Glutamatergic pathways are widespread, so while effects of σ ligands are likely to be profound, it is not obvious what the entire picture will eventually reveal. Nevertheless, σ/glutamate interactions appear to be important in several areas of mental health, including depression and psychosis.

2.2 Effects of σ receptors on cholinergic neurotransmission

Much behavioral evidence links σ receptor activity to regulation of cholinergic processes, including cognition and memory. These data are discussed in Chapter 12. Subcutaneous application of the σ receptor agonist (+)-SKF-10,047 as well as DTG has been shown to enhance the release of acetylcholine from prefrontal cortex in a haloperidol-reversible manner as measured by microdialysis (34,35). Kobayashi et al. (36) found that hippocampal acetylcholine release was similarly enhanced by both (+)-SKF-10,047 and DTG, but that striatal acetylcholine levels were unaffected. More recent studies using the selective σ_1 agonist SA4503 (1-(3,4-dimethoxyphenethyl)-4-(3-phenylpropyl)piperazine) have yielded similar results, i.e. an increase in acetylcholine release from frontal cortex and hippocampus (37). Junien et al. (38) reported a haloperidol-reversible increase in potassium-stimulated [^{3}H]acetylcholine release from hippocampal slices by igmesine. Horan et al. (39) extended these findings to show that when SA4503 was administered directly into the hippocampus, acetylcholine release was increased but to a much lesser extent than was seen with subcutaneous injection. These data indicate that the σ agonist has both direct and indirect effects on cholinergic systems. Doses of both (+)-SKF-10,047 (40) and SA4503 (41) that enhanced acetylcholine release from frontal cortex and hippocampus reduced scopolamine-induced amnesia.

All these data indicate that the behavioral effects of σ ligands in cognition are likely to be grounded to a large extent in their ability to regulate cholinergic systems. This indicates enormous therapeutic potential in the areas of Alzheimer's, possibly other forms of dementia, and cognitive enhancement in general.

2.3 Effects of σ receptors on opioidergic neurotransmission

Although σ receptors were initially thought to be members of the opioid receptor family (42), they were later recognized as naltrexone-insensitive (43). Despite some lingering confusion, σ receptors are now not generally considered opioid. The exception to this rule is exemplified by recent studies from Tsao and Su (44), who purified a naloxone- and haloperidol-sensitive binding site from rat liver and brain. Couture and Debonnel (45) found some evidence for potential involvement of this unique site in some electrophysiological responses to (+)-pentazocine, but not to other prototypical σ_1 receptor ligands. However, for the most part, the majority of σ receptors appear not to share opioid receptor properties.

While there are apparently very few, if any, studies on direct interactions between σ receptors and opioid pathways that explore the relationship at a neurochemical level, in functional tests, σ receptors have been shown to have an anti-opioidergic action (46-48). In those studies, the σ systems exert a tonic suppression of opioid activity. (+)-Pentazocine can activate the tonic system and block surpaspinal or spinal analgesia, and this action is reversed by haloperidol.

The nociceptin/orphanin FQ receptor has been compared with the σ receptor in that both are pronociceptive, although under certain circumstances, nociceptin/orphanin FQ can actually be analgesic as well (49,50). Kobayashi et al. (51) reported that the σ ligands carbetapentane and rimcazole act as antagonists at the nociceptin/orphanin FQ receptor that regulates the G protein-activated potassium channel expressed with it in *Xenopus* oocytes. Rossi et al. (50) were able to detect supraspinal nociceptin/orphanin FQ analgesia in mice only when haloperidol was present. Furthermore, in D_2 receptor knockout mice, nocicpetin/orphanin FQ analgesia was enhanced (52). It appears therefore that the σ and nociceptin/orphanin FQ systems may converge on opioid pathways mediating analgesia.

While there is relatively little biochemical data on opioid/σ interactions, the behavioral data from analgesia studies are provocative and could provide important new approaches to pain management.

2.4 Effects of σ receptors on catecholaminergic neurotransmission

2.4.1 σ receptors and dopamine

Dopamine is a major transmitter in motor and limbic pathways. The substantia nigra and the nucleus accumbens have significant numbers of σ receptors. Many antipsychotics and experimental antipsychotics have σ receptor antagonist properties. Regulation of dopaminergic activity has obvious clinical potential in the treatment of schizophrenia (see Chapter 13) and motor function.

σ Receptor ligands have been tested for their effects on dopaminergic neuronal activity as well as dopamine release. Many of these studies have employed peripheral administration of σ ligands coupled with measurement of central activity, which is critical for therapeutic development, but does not address mechanism of action. Intraperitoneal injection of the σ receptor agonists (+)-pentazocine and (+)-SKF-10,047 were shown to increase dopamine release in the striatum and prefrontal cortex, terminal fields of A9 and A10 nuclei, respectively (53). However, the σ_1 receptor antagonists DuP 734 (1-(cyclopropylmethyl)-4-(2'-oxoethyl)piperidine) and (-)-butaclamol also increased release in that study. Recent studies on chronic treatment with a σ receptor antagonist, E-5842, indicated a decreased spontaneous activity of A10, but not A9, neurons (54). This could have important implications in the antipsychotic actions of E-5842. Booth and Baldessarini (55) have reported that (+)-pentazocine and (+)-SKF-10,047 stimulate dopamine synthesis in preparations of rat striatum. The stimulation they observed was only about 20%, but was antagonized by BMY 14802 (α-(4-fluorophenyl)-4-(5-fluoro-2-pyrimidinyl)-1-piperazine-butanol). Gronier and Debonnel (56) reported that the σ_2 receptor antagonist Lu 28-179 (1'-[4-[1-(4-fluorophenyl)-1H-indol-3-yl]-1-butyl]spiro[isobenzofuran-1(3H),4'-piperidine) alone did not affect firing rate of A9 or A10 neurons, but enhanced NMDA-evoked increase in A10 neuronal activity. In the nucleus accumbens, a terminal field of A10 neurons, σ_1 agonists potentiated the NMDA-evoked response.

In our laboratory, we have studied the effects of σ agonists on dopamine release using a superfusion system. We found that the σ agonists (+)-pentazocine and BD737 [(+)-cis-N-methyl-N-[2-(3,4-dichlorophenyl)ethyl]-2-(1-pyrrolidinyl)cyclohexylamine], at concentrations consistent with their K_i values at σ_1 receptors, inhibit NMDA-stimulated [^{3}H]dopamine release from slices of rat and guinea pig striatum, nucleus accumbens and prefrontal cortex (57-59). The σ agonists must be present for several minutes prior to administration of the NMDA stimulus in order to inhibit it;

an acute application concurrent with the NMDA does not inhibit release (60). This suggests that some time-dependent process must occur in order for the σ receptor activation to affect NMDA-stimulated dopamine release.

Inhibition by σ agonists yields IC_{50} values consistent with reported K_i values from radioligand binding studies characterizing competition for $σ_1$ sites. Furthermore, the inhibition of release of [^{3}H]dopamine by σ agonists is reversed by several identified σ antagonists, including haloperidol and the $σ_1$-selective antagonist DuP 734. In rat striatum, a second component of inhibition is seen at higher concentrations of (+)-pentazocine, also with IC_{50} values comparable to K_i values from binding studies, consistent with actions via $σ_2$ receptors. This component of inhibition is reversed by the non-subtype-selective antagonist BD1008 (N-[2-(3,4-dichlorophenyl)ethyl]-N-methyl-2-(1-pyrrolidinyl)ethylamine). Interestingly, cocaine also appears to inhibit dopamine release from rat striatal slices via interaction with $σ_2$ receptors. This inhibition is blocked by the $σ_2$ antagonist Lu28-179 and is separate from cocaine's action at the dopamine transporter (83).

σ Receptors that modulate NMDA-stimulated release of dopamine from striatum appear to reside on dopaminergic nerve terminals. The experiments were performed using striatal slices, which mean that no effects of σ ligands on dopaminergic cell bodies would have been observed. Regulation of release was not abolished by tetrodotoxin, which blocks propagation of action potentials (57), presumably excluding any effects of σ ligands on interneurons that might have been activated by the NMDA stimulus. Regulation of release can also be observed in synaptosomal preparations (61), supporting σ receptor localization on dopaminergic nerve terminals.

We have also investigated the signaling mechanism through which $σ_1$ and $σ_2$ receptors inhibit NMDA-stimulated [^{3}H]dopamine release. Several observations led us to test the PLC/PKC system as a possible signaling mechanism for σ receptors. Vilner and Bowen (18) had shown that (+)-pentazocine and several other $σ_2$ receptor ligands mobilize inositol 1,4,5-triphosphate (IP_3)-sensitive calcium stores from SK-N-SH cells. Hayashi et al. (15) demonstrated that in NG108 cells, $σ_1$ ligands and pregnenolone sulfate release calcium from ER stores, and showed that the release was blocked by the $σ_1$ antagonist NE-100. As discussed above, Hayashi et al. (4) have shown that $σ_1$ receptors, when stimulated with agonist, associate with the IP_3 receptor and ankyrin on the ER to regulate the release of calcium. Novakova et al. (62) reported that treatment of rat cardiac myocytes with the selective $σ_1$ agonist BD737 increased $[IP_3]_i$, and also potentiated electrically evoked amplitudes of contraction and calcium transients as well as increase in spontaneous twitch. The latter effects were blocked by prior depletion of IP_3-sensitive calcium stores with thapsigargin, and also by the PLC inhibitor neomycin, and were reversed by a σ receptor antagonist. Collectively, these studies suggested to us that the PLC/PKC system could be involved in σ receptor actions.

PLC is activated when agonist binds to a receptor coupled to G_q, and the α, and β/γ subunits subsequently dissociate. PLC can be activated by the $G\alpha_q$ subunit, or sometimes by the $G_{i/o}$ β/γ subunit, in which case the response can be pertussis toxin sensitive (64). PLC produces the intracellular signaling molecules IP_3 and diacylglycerol (DAG) via cleavage of membrane bound phosphatidylinositol 4,5-diphosphate (PIP_2). IP_3 binds to a receptor on the ER and causes the release of ER calcium. The released calcium can activate the conventional, calcium-dependent PKCs (α, β and γ). DAG activates two classes of PKCs, the conventional PKCs (cPKCs) and the novel PKCs (nPKCs) δ, ε, η and θ. These classes of PKC isozymes, but not the atypical PKCs (aPKCs), which are calcium and DAG-independent, have been associated with neurotransmitter release in brain.

We tested several inhibitors of PKC, as well as a PLC inhibitor. A general inhibitor of PKCs, chelerythrine, an inhibitor of the calcium-dependent PKCs, GF109203x (2-[1-(3-dimethylaminopropyl)indol-3-yl]-3-(indol-3-yl)maleimide), and a selective inhibitor of PKCβ, LY379196, all blocked the inhibition of stimulated [^{3}H]dopamine release produced by (+)-pentazocine at concentrations at which it acted through the σ_1 receptor. Therefore, it appears that inhibition of release by σ_1 receptor agonists was dependent upon a functional PKC, most likely of the PKCβ subtype (60). If a functional PKC was required for σ receptor-mediated inhibition, we reasoned that a prior downregulation of PKC with phorbol-12-myristate-13-acetate (PMA) should also prevent σ agonist regulation. A 30 min pretreatment with 1 μM PMA, which binds to the DAG site, did indeed abolish the ability of the σ_1 agonist (+)-pentazocine to regulate release. The neurosteroids progesterone and pregnenolone similarly inhibit NMDA-stimulated release and their actions are reversed by σ antagonists, as well as the PKCβ inhibitor, LY379196 (64).

We also investigated the effects of neuropeptide Y (NPY), which has been proposed as an endogenous σ agonist (65). Interestingly, NPY had opposite effects to those of (+)-pentazocine on NMDA-stimulated [^{3}H]dopamine release, i.e. NPY enhanced stimulated release in a concentration-dependent manner (66-68). However, NPY enhancement as well as (+)-pentazocine-mediated inhibition of stimulated release were both blocked by the σ_1-selective antagonist DuP 734 and the NPY antagonist PYX-1 (Ac-[3-(2,6-dichlorobenzyl)Tyr^{27},D-Thr^{32}]NPY-(27-36)amide). This led us to hypothesize that NPY might be an inverse agonist at σ_1 sites. However, our subsequent autoradiographic analysis of binding data did not support the common receptor hypothesis (69). Instead, our other more recent studies using PLC and PKC inhibitors suggest that the two receptors,

σ_1 and NPY are not the same, but converge on the PLC signaling pathway (60,61). About 30% of NPY receptors are coupled via the PLCβ system (70). Our data show that PKC inhibitors block the enhancement in NMDA-stimulated dopamine release by NPY, suggesting that it is this PLC-coupled population of NPY receptors that are activated in our experiments. The inhibition in release by the σ receptor agonist (+)-pentazocine is blocked by the same PKC inhibitors. It is possible that both the NPY receptors and σ receptors work at different points along the PLC/PKC pathway, such that the PLC system must be activated to see σ agonist effects, and that interfering with the downstream PKC system also disrupts the modulation by either NPY or σ agonist.

Our recent experiments using concentrations of (+)pentazocine > 100 nM or 10 μM cocaine to study σ_2 receptor-mediated inhibition of NMDA-stimulated dopamine release indicate involvement of a calcium-dependent PKC as well, but not a β isoform (83).

Other evidence implicates the PLC/PKC pathway in σ_1 receptor signaling. Using a guinea pig hypoglossal neuron preparation, Morin-Surun et al. (14) showed that (+)-pentazocine inhibited neuronal firing rate, and that the effect rapidly desensitized. Blockade of PLC with U-73,122 (1-[6-[[(17β)-3-methoxyestra-1,3,5(10)trine-17-yl]amino]hexyl]-1H-pyrrole-2,5-dione), or blockade of PKC with GF109203x or Go-6976 (12-(2-cyanoethyl)-6,7,12,13-tetrahydro-13-methyl-5-oxo-5H-indolo[2,3-a]pyrrolo [3,4-c]carbazole) prevented the desensitization of the response to (+)-pentazocine. Also in that study, using immunofluorescent microscopy, the authors demonstrated that activation of σ_1 receptors with (+)-pentazocine initiated a translocation of both σ_1 receptor and the βI and βII isoforms of PKC from cytosolic to plasma membrane.

We propose a general model for σ_1 receptor regulation of release of dopamine in Figure 10-1. In that model, σ_1 agonist, which may be a neurosteroid, binds to the σ receptor, either at the plasma membrane location, or intracellularly upon entry into the cell or neuron. Since regulation of release is dependent upon the β isozyme of PKC in our studies, one could postulate that the PKCβ is activated by agonist binding to the σ_1 receptor. Generally, PKCs translocate upon activation, and may be desensitized or downregulated. One would further postulate a phosphorylation of a relevant vesicular or other protein involved in exocytotic dopamine release, as the NMDA-stimulated release of dopamine is inhibited by σ_1 receptor activation.

Gudelsky (71) measured extracellular dopamine concentrations in the striatum of rat after intrastriatal infusion of σ ligands. He found that (+)-pentazocine at 300 μM and 1 mM produced a biphasic response; first a brief increase of about 70% in dopamine release occurred, followed by a prolonged decrease of about 65%. Similar effects were observed with the (-)-isomer of pentazocine, as well as with DTG. The NMDA antagonist 3-(2-carboxypiperazine-4-yl)propyl-1-phosphonic acid (CPP) blocked the initial increase, but not the inhibition by σ agonists. Gudelsky's findings are thus similar to our own in that both σ_1 and σ_2 receptors appear to regulate release, and the stimulation of release is via activation of NMDA receptors.

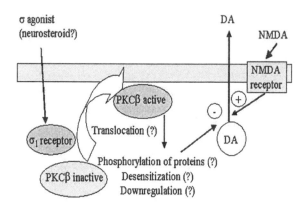

Figure 10-1. Proposed model of σ_1 receptor regulation of vesicular release of dopamine

We also have found that σ_2 receptor activation enhances [^{3}H]dopamine release via the dopamine transporter (DAT) stimulated by amphetamine (72-75). The regulation of DAT activity is dependent upon protein kinases. Protein kinases are known to mediate trafficking of the DAT to and from the plasma membrane. In PC12 cells, the protein kinase involved appears to be Ca^{2+}/calmodulin kinase II (74), as it is blocked by selective inhibitors of that enzyme, while in brain tissue, the kinase involved appears to be a member of the Ca^{2+}-dependent PKC family, but not the β isozyme (74). The enhancement by (+)-pentazocine in striatal slices is blocked by the Ca^{2+}-dependent PKC inhibitor GF109203x, but not by the β-isozyme-selective PKC inhibitor LY379196. Depletion of ER calcium stores eliminated σ_2 receptor-mediated regulation of dopamine release via the DAT, indicating that the ER is the source of calcium required for σ receptor regulation. In Izenwasser et al. (72), we showed that only non-selective and σ_2 selective antagonists blocked (+)-pentazocine-mediated enhancement of amphetamine-stimulated [^{3}H]dopamine release. In the studies described in Liu et al. (75), we tested new trishomocubane agonists (76) for their activity

10. Modulation of classical neurotransmitter systems

in our assay of amphetamine-stimulated release. Several of these drugs enhanced release similarly to (+)-pentazocine at concentrations consistent with their K_i values at σ_2 receptors. Interestingly, we found that not only a very selective σ_2 antagonist, Lu28-179, but also the highly selective σ_1 receptor antagonist DuP 734 blocked the enhancement by these compounds. This may be evidence for an interaction between σ_1 and σ_2 binding sites in regulation of some physiological functions.

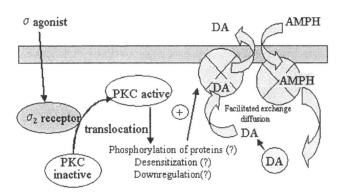

Figure 10-2. Proposed model of σ_1 receptor regulation of transporter-mediated release of dopamine

We propose a general model for σ_2 receptor function that is shown in Figure 10-2. In this model, amphetamine enters the cell or neuron via the DAT, and initiates facilitated exchange diffusion of dopamine. The activation of a σ_2 receptor enhances that release in a calcium- and PKC-dependent manner. Generally, PKCs translocate when activated, and their function is to phosphorylate the relevant protein(s). The PKC may be desensitized and/or downregulated, although we have no direct evidence for this at present. PKCs are often implicated in protein trafficking, and it is known that DAT is trafficked to and from plasma membrane locations in response to various signals (78), one of which may be σ_2 receptor activation.

2.4.2 σ receptors and norepinephrine

Noradrenergic pathways arise from locus coeruleus and terminate in hippocampus, cortex, and cerebellum. In our hands, the effects of σ receptor agonists on NMDA-stimulated release of [^{3}H]norepinephrine from terminal fields of locus coeruleus neurons was similar to that of σ agonists on dopamine release, i.e. σ agonists inhibited stimulated [^{3}H]norepinephrine release. Release of [^{3}H]norepinephrine from hippocampus or cerebellum was stimulated by 25 μM NMDA. Both (+)-pentazocine and BD737 inhibited release from rat hippocampus in a concentration-dependent manner with IC_{50} values consistent with actions via σ_1 and σ_2 receptors (78,79). Inhibition was reversed by σ receptor antagonists, supporting a role for both σ_1 and σ_2 receptors in regulation of release. Unlike σ_1 receptor-mediated regulation of [^{3}H]dopamine release, σ_1 receptor-mediated regulation of [^{3}H]norepinephrine release was tetrodotoxin-sensitive. This suggests that σ_1 receptors in the hippocampus are likely not located on noradrenergic nerve terminals. In rat cerebellum, regulation of stimulated [^{3}H]norepinephrine release was primarily via σ_1, but not σ_2, receptors (79). In guinea pig hippocampus, regulation seems to be primarily via σ_2 receptors (80).

In contrast to our findings on inhibition of release by σ agonists, Monnet et al. (81) found that the σ agonists (+)-pentazocine and BD737 enhanced NMDA-stimulated release at concentrations between 30 and 300 nM, but had no effect at higher concentrations. In their experiments, DTG inhibited NMDA-stimulated release. The enhancement of release by (+)-pentazocine as well as the inhibition by DTG, were sensitive to reversal by haloperidol and BD1063 (1-[2-(3,4-dichlorophenyl)ethyl]-4-methylpiperazine), whereas the enhancement by BD737 was not. Neither reduced haloperidol nor BD1008 antagonized the enhancement by (+)-pentazocine nor the inhibition by DTG, but were effective against the BD737-mediated enhancement. They concluded that their effects were mediated by a σ receptor other than σ_1 or σ_2. Differences in the experimental protocols used by Monnet's group and our group included the use by Monnet and coworkers of a four fold higher concentration of NMDA for twice as long as our stimulation period, which may have mobilized intracellular calcium as well as admitting calcium via the NMDA receptor-operated channel. Also, their fractional release of [^{3}H]norepinephrine produced by NMDA was about 3.5%, compared to an approximately 11% fractional release in our studies. Despite the pharmacological differences in the two sets of experiments, it does appear that σ receptors regulate norepinephrine release, and that the results obtained are very sensitive to the experimental conditions used.

Monnet et al. (6) also examined the possible regulation of norepinephrine release by neurosteroids that have been postulated to act as endogenous ligands at σ receptors. In that study, haloperidol blocked both an enhancement in NMDA-stimulated release by DHEA sulfate, as well as an inhibition by pregnenolone sulfate and DTG. BD1063 also antagonized both the DHEA sulfate and pregnenolone sulfate responses. The effects of the neurosteroids were also sensitive to pertussis toxin, suggesting that the σ receptor involved is G protein linked, and because there is evidence to suggest that σ_2 receptors are not G protein linked, their responses are more likely to be via σ_1 receptors.

While fewer studies on signal transduction mediating σ receptor regulation of norepinephrine release have been published compared with dopamine release studies, the data collected thus far suggest that such experiments are feasible and should yield interesting results. For instance, the demonstration of pertussis toxin sensitivity in the noradrenergic system, as well as the differential localization of the σ receptors regulating release from the two sets of pathways could yield important clues to potential further subdivision of σ receptor subtypes, and the generality or specificity of signaling systems utilized.

In summary, probably the greatest body of data has been collected for regulation of catecholaminergic systems by σ receptors. While the individual studies do not agree completely in detail, some general conclusions can be drawn. The relationship of σ receptors to the dopaminergic and noradrenergic systems indicates that σ receptors could have key roles in depression (see Chapter 14) and schizophrenia (Chapter 13). In the current chapter, we summarize the neurochemical mechanisms through which these roles are likely to be mediated.

3. GENERAL CONCLUSIONS

σ Receptors have been studied by multiple approaches since their discovery in 1976 by Martin et al. (42). While much progress has been made, the field is not nearly as advanced as others that emerged in a similar time frame. Recent data should contribute greatly to the advancement of our understanding. Especially important are the implications of steroids as natural transmitters. While this hypothesis is not completely proven, it changes the thinking regarding intracellular location of σ receptors, signaling pathways, and role of σ receptors in such major problems as drug abuse and mental illness. At the heart of these clinical issues is the need for a more complete description of the actions of σ receptors with other neurotransmitter systems. It is hoped that more and better work will emerge in this field as a natural progression of the studies described herein.

REFERENCES

1. Aydar E, Palmer CP, Klyachko VA, Jackson MB. The σ receptor as a ligand-regulated auxiliary potassium channel subunit. Neuron 2002, 34:399-410.
2. Wilke RA, Luparadus PJ, Grandy DK, Rubenstein M, Low MJ, Jackson MB. K^+ channel modulation in rodent neurohypophysial nerve terminals by σ receptors and not by dopamine receptors. J Physiol 1999a, 517:391-406.
3. Wilke RA, Mehta RP, Luparadus PJ, Chen Y, Ruoho AE, Jackson MB. Sigma receptor photolabeling and σ receptor-mediated modulation of potassium channels in tumor cells. J Biol Chem 1999b, 18387-18392.
4. Hayashi T, Su TP. Regulating ankyrin dynamics: Roles of σ_1 receptors. Proc Natl Acad Sci USA 2001, 98:491-496.
5. Su TP, Hayashi T. Cocaine affects the dynamics of cytoskeletal proteins via σ_1 receptors. Trends Pharmacol Sci 2001, 22:456-8.
6. Monnet FP, Mahe V, Robel P, Balieu EE. Neurosteroids via σ receptors, modulate [^{3}H]norepinephrine release evoked by N-methyl-D-aspartate in the rat hippocampus. Proc Natl Acad Sci USA 1995, 92:3774-3778.
7. Hong W, Werling LL. Evidence that σ_1 receptors may not be directly coupled to G proteins. Eur J Pharmacol 2000, 408:117-125.
8. Boyer JL, Graber SG, Waldo GL, Harden K, Garrison JC. Selective activation of phospholipase C by recombinant G-protein α- and β/-subunits. J Biol Chem 1994. 269:2814-2819.
9. Bouchard P, Quirion R. [^{3}H]1,3-ditolylguanidine and [^{3}H](+)-pentazocine binding sites in the rat brain: autoradiographic visualization of the putative σ_1 and σ_2 receptor subtypes. Neuroscience 1997, 76:467-477.
10. Leitner ML, Hohmann AG, Patrick SL, Walker JM. Regional variation in the ratio of σ_1/σ_2 binding in rat brain. Eur J Pharmacol 1994, 259:65-69.
11. Alonso G, Phan V-L, Guillemain I, Saunier M, Legrand A, Anoal M, Maurice T. Immunocytochemical localization of the σ_1 receptor in the adult rat central nervous system. Neuroscience 2000, 97:155-170.
12. Kitaichi K, Chabot JG, Moebius FF, Flandorfer A, Glossman H, Quirion R. Expression of the purported σ_1 (sigma-1) receptor in the mammalian brain and its possible relevance in definits induced by antagonism of the NMDA receptor complex as revealed using an antisense strategy. J Chem Neuroanat 2000, 20:375-387.
13. McCann DJ, Weissman AD, Su T-P. Sigma-1 and σ_2 sites in rat brain: comparison of regional, ontogenetic, and subcellular patterns. Synapse 1994, 17:182-189.
14. Morin-Surun MP, Collin T, Denavit-Saubie M, Baulieu EE, Monnet FP. Intracellular σ_1 receptor modulates phospholipase C and protein kinase C activities in the brainstem. Proc Natl Acad Sci USA 1999, 96:8196-8199.
15. Hayashi T, Maurice T, Su TP. Ca^{2+} signalling via σ_1 receptors: novel regulatory mechanism affecting intracellular Ca^{2+} concentration. J Pharmacol Exp Ther 2000, 293:788-798.
16. Bowen WD. Sigma receptors: recent advances and new clinical potentials. Pharmaceutica Acta Helvetica 2000, 74:211-218.
17. Gonzalez-Alvear GM and Werling LL. Regulation of [^{3}H]dopamine release from rat striatal slices by σ receptor ligands. J Pharm Exp Ther 1994, 271:212-219.
18. Vilner BJ, Bowen WD. Modulation of cellular calcium by σ_2 receptors: release from intracellular stores in human SK-N-SH neuroblastoma cells. J Pharmacol Exp Ther 2000, 292:900-911.

19. Bergeron R, Debonnel G, DeMontigny C. Modification of the N-methyl-D-aspartate response by antidepressant σ receptor ligands. Eur J Pharmacol 1993; 240:319-323.
20. Bergeron R, dcMontigny C, Debonnel G. Biphasic effects of σ ligands on the neuronal response to N-methyl-D-aspartate. Naunyn-Schmeid Arch Pharmacol 1995, 351:252-260.
21. Bergeron R, de Montigny C, Debonnel G. Effect of short-term and long-term treatments with σ ligands on the N-methyl-D-aspartate response in the CA3 region of the rat dorsal hippocampus. Br J Pharmacol 1997, 120:1351-1359.
22. Bergeron R, deMontigny C, Debonnel G. Pregnancy reduces brain σ receptor function. Br J Pharmacol 1999, 127:1769-1776.
23. Monnet FP, Debonnel G, Junien JL, DeMontigny C. N-Methyl-D-aspartate-induced neuronal activation is selectively modulated by σ receptors. Eur J Pharmacol 1990, 79:441-445.
24. Debonnel G, Bergeron R, deMontigny C. Potentiation by dehydroepiandosterone of the neuronal response to N-methyl-D-aspartate in the CA3 region of the rat dorsal hippocampus: an effect mediated via σ receptors. J Endocrinol 1996, 150 Suppl:S33-S42.
25. Monnet FP, Debonnel G, Bergeron R, Gronier B, deMontingny C. The effects of σ ligands and of neuropeptide Y on N-methyl-D-aspartate-induced neuronal activation of CA3 dorsal hippocampus neurons are differentially affected by pertussis toxin. Br J Pharmacol 1994, 112:709-715.
26. Klette KL, Lin Y, Clapp LE, DeCoster MA, Moreton JE, Tortella FC. Neuroprotective σ ligands attenuate NMDA and trans-ACPD-induced calcium signaling in rat primary neurons. Brain Res 1997, 756:231-240.
27. Shimazu S, Katsuki H, Takenada C, Tomita M, Kuma T, Kaneko S, Akaike A. Sigma receptor ligands attenuate N-methyl-D-aspartate cytotoxicity in dopaminergic neurons of mesencephalic slice cultures. Eur J Pharmacol 2000, 388:139-146.
28. Vilner BJ, deCosta BR, Bowen WD. Cytotoxic effects of σ ligands: σ receptor-mediated alterations in cellular morphology and viability. J Neurosci 1995, 15:643-654.
29. Largent BL, Gundlach AL, Snyder SH. Pharmacological and autoradiographic discrimination of σ and phencyclidine receptor binding sites in brain with (+)-[^{3}H]SKF 10,047, (+)-[^{3}H]-3-[3-hydroxyphenyl]-N-(1-propyl)piperidine and [^{3}H]-1-[1-(2-thienyl)cyclohexyl]piperidine. J Pharmacol Exp Ther 1986, 238:739-748.
30. Monaghan DT, Bridges RJ, Cotman CW. The excitatory amino acid receptors: their classes, pharmacology, and distinct properties in the function of the central nervous system. Ann Rev Pharmacol Toxicol 1989, 29:365-402.
31. Guitart X, Mendez R, Ovalle S, Andreu F, Carceller A, Farre AJ, Zamanillo D. Regulation of ionotropic glutamate receptor subunits in different rat brain areas by a preferential σ$_1$ receptor ligand and potential antipsychotic. Neuropsychopharmacology 2000, 23:539-546.
32. Annels SJ, Ellis Y, Davies JA. Non-opioid antitussives inhibit endogenous glutamate release from rabbit hippocampal slices. Brain Res 1991, 15:341-343.
33. Ellis Y, Davies JA. The effects of σ ligands on the release of glutamate from rat striatal slices. Naunyn-Schmeidebergs Arch Pharmacol 1994, 350:143-148.
34. Matsuno K, Matsunaga K, Mita S. Increase of extracellular acetylcholine level in rat frontal cortex induced by (+)N-allylnormetazoxine as measured by brain microdialysis. Brain Res 1992, 575:315-219.
35. Matsuno K, Matsunaga K, Senda T, Mita S. Increase in extracellular acetylcholine level by σ ligands in rat frontal cortex. J Pharm Exp Ther 1993, 26:851-859.

36. Kobayashi T, Matsuno K, Mita S. Regional differences of the effect of σ receptor ligands on the acetylcholine release in the rat brain. J Neural Transm Gen Sect 1996a, 103:661-669.
37. Kobayashi T, Matsuno K, Nakata K, Mita S. Enhancement of acetylcholine release by SA4503, a novel $σ_1$ receptor agonist, in the rat brain. J Pharm Exp Ther 1996b, 279:106-113.
38. Junien JL, Roman FJ, Brunells G, Pascaud X. JO1784, a novel $σ_1$ ligand, potentiates [^{3}H]acetylcholine release from rat hippocampal slices. Eur J Pharmacol 1991, 200:343-345.
39. Horan B, Gifford AN, Matsuno K, Mita S, Ashby CR Jr. Effect of SA4503 on the electrically evoked release of ^{3}H-acetylcholine from striatal and hippocampal rat brain slices. Synapse 2002, 46:1-3.
40. Matsuno K, Senda T, Matsunaga K, Mita S. Ameliorating effects of σ receptor ligands on the impairment of passive avoidance tasks in mice: involvement in the central acetylcholinergic system. Eur J Pharmacol 1994, 11:43-51.
41. Senda T, Matsuno K, Okamto K, Kobayashi T, Nakata K, Mita S. Ameliorating effect of SA4503, a novel $σ_1$ receptor agonist, on memory impairments induced by cholinergic dysfunction in rats. Eur J Pharmacol 1996, 315:1-10.
42. Martin WR, Eades CG, Thompson JA, Huppler RE, Gilbert PE. The effects of morphine and nalorphine-like drugs in the non-dependent and morphine-dependent chronic spinal dog. J Pharm Exp Ther 1976, 197:517-532.
43. Vaupel DB. Naltrexone fails to antagonize the σ effects of PCP and SKF10,047 in the dog. Eur J Pharmacol 1983, 92:264-269.
44. Tsao L-I, Su T-P. Naloxone-sensitive, haloperidol-sensitive, [^{3}H](+)SKF-10,047-binding protein partially purified from rat liver and rat brain membranes: An opioid/σ receptor? Synapse 1997, 25:117-124.
45. Couture S, Debonnel G. Some of the effects of the selective σ ligand (+)pentazocine are mediated via a naloxone-sensitive receptor. Synapse 2001, 39:323-331.
46. Chien C-C, Pasternak GW. Selective antagonism of opioid analgesia by a σ system. J Pharm Exp Ther 1994, 271:1583-1590.
47. Chien C-C, Pasternak GW. (-)-Pentazocine analgesia in mice: interactions with a σ receptor system. J Pharm Exp Ther 1995a, 271:1583-1590.
48. Chien C-C, Pasternak GW. σ Antagonists potentiate opioid analgesia in rats. Neuroscience Lett 1995b, 190:137-139.
49. Rossi G, Leventhal L, Bolan EA, Pasternak GW. Pharmacological characterization of orphanin FQ/nociceptin and its fragments. J Pharm Exp Ther 1997, 282:858-865.
50. Rossi G, Leventhal L, Pasternak GW. Naloxone-sensitive orphanin/FQ-induced analgesia in mice Eur J Pharmacol 1996, 311:R7-R8.
51. Kobayashi T, Ikeda K, Togashi S, Itoh N, Kumanishi T. Effects of σ ligands on the nociceptin/orphanin FQ receptor co-expressed with the G-protein-activated K+ channel in Xenopus oocytes. Br J Pharmacol 1997, 120:986-987.
52. King MA, Bradshaw S, Chang AH, Pintar JE, Pasternak GW. Potentiation of opioid analgesia in dopamine-2 receptor knock-out mice: evidence for a tonically active anti-opioid system. J Neurosci 2001, 21:7788-7792.
53. Gudelsky GA. Effects of σ receptor ligands on the extracellular concentration of dopamine in the striatum and prefrontal cortex of the rat. Eur J Pharmacol 1995, 286:223-228.
54. Sanchez-Arroyos R, Guitart X. Electrophysiological effects of E-5842, a $σ_1$ receptor ligand and potential atypical antipsychotic, on A9 and A10 dopamine neurons. Eur J Pharmacol 1999, 378:31-37.

55. Booth RG, Baldessarini RJ. (+)-6,7-benzomorphan σ ligands stimulate dopamine synthesis in rat corpus striatum tissue. Brain Res 1991, 557:349-352.
56. Gronier B, Debonnel G. Involvement of σ receptors in the modulation of glutamatergic/NMDA neurotransmission in the dopaminergic systems. Eur J Pharmacol 1999, 368(2-3): 183-196.
57. Gonzalez-Alvear GM, Werling LL. Sigma-1 receptors in rat striatum regulate NMDA-stimulated [^{3}H]dopamine release via a presynaptic mechanism. Eur J Pharmacol 1995b, 294:713-719.
58. Gonzalez-Alvear GM, Werling LL. Release of [^{3}H]dopamine from guinea pig striatal slices is modulated by σ_1 receptors. Naunyn-Schmeideberg Arch Pharmacol 1997; 356:455-461.
59. Weatherspoon JK, Gonzalez-Alvear GM, Frank AR, Werling L. Regulation of [^{3}H]dopamine release from mesolimbic and mesocortical and areas of guinea pig brain by σ receptors. Schizophrenia Res 1996, 21:51-62.
60. Nuwayhid SJ, Werling LL. Sigma-1 receptor agonist-mediated regulation of N-methyl-D-aspartate-stimulated [^{3}H]dopamine release is dependent upon PKC. J Pharm Exp Ther 2002, 364-369.
61. Sanchez C, Nuwayhid SJ, Werling LL. Signaling of NPY receptors and σ receptors via and PLC/PKC system. (In preparation) 2005.
62. Novakova M, Ela C, Bowen WD, Hasin Y, Elim Y. Highly selective σ receptor ligands elevate inositol 1,4,5-trisphosphate production in rat cardiac myocytes. Eur J Pharmacol. 1998, 353:315-327.
63. Rebecchi MJ, Pentyala SN. Structure, function, and control of phosphoinositide-specific phospholipase C. Physiol Rev 2000, 80:1291-1335.
64. Werling LL. Sigma receptor-active steroids and neurotransmitter release. Int J Neuropsychopharmacology 2002, 5:S29.2
65. Roman FJ, Pascaud X, Duffy O, Vauche D, Martin B, Junien JL. Neuropeptide Y and peptide YY interact with rat brain σ and PCP binding sites. Eur J Pharmacol 1989, 174:301-302.
66. Ault DT, Radeff JM, Werling LL. Modulation of [^{3}H]dopamine release from rat nucleus accumbens by neuropeptide Y via a σ_1-like receptor. J Pharm Exp Ther 1998, 284:553-560.
67. Ault DT, Werling LL. Differential modulation of NMDA-stimulated [^{3}H]dopamine release from rat striatum by neuropeptide Y and σ receptor ligands. Brain Res 1997, 760:210-217.
68. Ault DT, Werling LL. Neuropeptide Y-mediated enhancement of [^{3}H]dopamine release from rat prefrontal cortex is reversed by σ_1 receptor antagonists. Schizophrenia Res 1998, 31:27-36.
69. Hong W, Werling LL. Lack of effects by σ ligands on neuropeptide Y-induced G-protein activation in rat hippocampus and cerebellum. Brain Res 2001, 901:208-218.
70. Parker SL, Parker MS, Swaetman T, Cowley WR. Characterization of the G protein and phospholipase C-coupled agonist binding to the Y-1 neuropeptide Y receptor in rat brain: sensitivity to G protein activators and inhibitors and to inhibitors of phospholipase C. J Pharm Exp Ther 1998, 286:382-391.
71. Gudelsky GA. Biphasic effect of σ receptor ligands on the extracellular concentration of dopamine in the striatum of rat. J Neural Transm 1999, 106:849-856.
72. Izenwasser S, Thompson-Montgomery DT, Deben SE, Chowdhury IN, Werling LL. Modulation of amphetamine-stimulated (transporter-mediated) dopamine release by σ_2 receptor agonists and antagonists in vitro. Eur J Pharmacol 1998, 346: 189-196.

73. Weatherspoon JK, Werling LL. Modulation of amphetamine-stimulated [^{3}H]dopamine release from rat pheochromocytoma (PC12) cells by σ_2 receptors. J Pharm Exp Ther 1999, 289:278-284.
74. Derbez AE, Werling LL. Sigma-2 receptor regulation of dopamine transporter activity via protein kinase C. J Pharmacol Exp Ther 2002, 301:306-314.
75. Liu X, Nuwayhid S, Christie M, Kassiou M, Werling LL. Trishomocubanes: novel σ ligands modulating amphetamine-stimulated [^{3}H]dopamine release in vitro. Eur J Pharmacol 2001, 422:39-45.
76. Nguyen VH, Kassiou M, Johnston GA, Christie MJ. Comparison of binding parameters of σ_1 and σ_2 binding sites in rat and guinea pig brain membranes: novel subtype-selective trishomocubanes. Eur J Pharmacol 1996, 311:233-240.
77. Zahniser NR, Doolen S. Chronic and acute regulation of Na^+/Cl^--dependent neurotransmitter transporters: drugs, substrates, presynaptic receptors, and signaling systems. Pharmacol Ther 2001, 92(1): 21-55.
78. Gonzalez-Alvear GM, Werling LL. Sigma receptor regulation of norepinephrine release from rat hippocampal slices. Brain Res 1995a, 673:61-69.
79. Gonzalez-Alvear GM, Thompson-Montgomery D, Deben SE, Werling LL. Functional and binding properties of σ receptors in rat cerebellum. J Neurochem 1995, 65: 2509-2516.
80. Weatherspoon JK, Gonzalez-Alvear GM, Werling LL. Regulation of [^{3}H]norepinephrine release from guinea pig hippocampus by σ_2 receptors. Eur J Pharmacol 1997, 326:133-138.
81. Monnet FP, deCosta BR, Bowen WD. Differentiation of σ ligand-activated receptor subtypes that modulate NMDA-evoked [^{3}H]noradrenaline release in rat hippocampal slices. Br J Pharmacol 1996, 119:65-72.
82. Hong W, Nuwayhid SJ, Werling LL. Modulation of bradykinin-induced calcium changes in SH-SY5Y cells by neurosteroids and sigma receptor ligands via a shared mechanism. Synapse 2004 54:102-110.
83. Nuwayhid SJ, Werling LL. Sigma$_2$ (σ_2) receptors as a target for cocaine action in the rat striatum. Eur J Pharmacol 2006 535:98-103.

Corresponding author: *Dr. Linda L. Werling, Mailing address: George Washington University Medical Center, Department of Pharmacology & Physiology, 2300 Eye St. NW, Washington, DC 20037, USA, Phone: (202) 994-2918, Fax: (202) 994-2780, Electronic mail address: phmllw@gwumc.edu*

Chapter 11

σ_2 RECEPTORS: REGULATION OF CELL GROWTH AND IMPLICATIONS FOR CANCER DIAGNOSIS AND THERAPEUTICS

Wayne D. Bowen
Department of Molecular Pharmacology, Physiology & Biotechnology, Division of Biology and Medicine, Brown University, Providence, RI 02912, USA

1. INTRODUCTION

σ Receptors comprise a unique, pharmacologically defined family of proteins that bind psychotropic agents from a variety of structural classes (1-7). These receptors have thus far been divided into σ_1 and σ_2 subtypes. They are noted for having high affinity for haloperidol and other typical antipsychotic agents, but can be distinguished pharmacologically by affinity for dextrorotary benzomorphans, with σ_1 receptors having high affinity and σ_2 receptors having low to negligible affinity for these compounds (2,8). σ_1 Receptors are 25 kDa proteins, have been cloned, and are significantly homologous to the sterol isomerase enzyme of yeast and other fungi (9,10). Though lacking enzymatic activity and any structural relationship to the mammalian sterol isomerase, σ_1 receptors share some pharmacology with this enzyme. The structure of the σ_2 receptor is not yet known, but is a protein of 21.5 kDa by photoaffinity labeling (8,11). No endogenous ligands have been conclusively demonstrated for σ receptors. However, there is significant evidence that such substances exist, and that they may be related to sterols (7,12,13). σ_1 Receptors are selectively radiolabeled using [^{3}H](+)-pentazocine (14). There is no available selective σ_2 radioligand. σ_2 Receptors are labeled using [^{3}H]1,3-di-o-tolyguanidine (DTG), a subtype non-selective ligand, in the presence of unlabeled (+)-pentazocine or dextrallorphan to block binding to σ_1 receptors (11).

Most of what is known regarding the functions of σ receptors concerns their actions in the brain. However, $σ_1$ and $σ_2$ receptors are also widely distributed outside of the central nervous system. They are expressed in peripheral tissues including liver, kidney, heart, gut, and tissues of the immune and endocrine systems (11,15). In fact, σ receptors have been found in every organ system examined thus far. Thus, in addition to certain behavioral functions mediated through expression in the brain, σ receptors are likely to subserve other more general functions via expression in sites throughout the organism. Several lines of evidence suggest that σ receptors, particularly $σ_2$ receptors, may be involved in regulation of cell proliferation and cell survival.

2. EXPRESSION OF σ RECEPTORS IN TUMORS AND CELL LINES

σ Receptors are highly expressed in tumors and tumor cell lines of various tissue origins. σ Receptors were first reported in solid neural and non-neural tumors and were found to be more highly expressed than in surrounding normal tissue (16,17). A preliminary report indicated that σ receptors were expressed in very high levels in breast tumor biopsy samples, while surrounding normal tissue taken at the same time showed no detectable σ receptor binding activity (18).

Both $σ_1$ and $σ_2$ receptors are expressed in tumor cell lines (8,19,20). We have reported that σ receptor subtypes are present in neuroblastomas, gliomas, melanomas, and in breast, prostate, lung, and leukemia cell lines (20). In fact, every tumor cell line examined to date has been found to express σ receptors.

Although both σ receptor subtypes are present in cell lines, $σ_2$ receptors are most consistent in terms of binding affinity and number of receptors present. The levels of $σ_2$ receptors present in these cell lines were estimated to average 300,000 to 1 million receptors per cell (20). Interestingly, MCF-7 cells show either no specific $[^3H](+)$-pentazocine binding or very low binding relative to $σ_2$ receptor labeling, suggesting that this cell does not express active $σ_1$ receptors (20). However, an apparent $σ_1$ receptor splice variant is present in these cells (21).

3. REGULATION OF σ_2 RECEPTOR EXPRESSION

σ_2 Receptor levels have been shown to vary with state of cell proliferation. Wheeler and colleagues investigated σ receptors in rapidly proliferating and quiescent mouse mammary adenocarcinoma 66 breast tumor cells (22,23). σ_2 Receptor levels were 10-fold higher in the rapidly dividing cells compared to the quiescent cells. Furthermore, the high receptor levels in the rapidly dividing cells were found to decrease when the proliferating cells returned to the quiescent state. Much less change was observed with σ_1 receptors, levels being only about two-fold higher in rapidly dividing, compared to quiescent cells. Similar results for σ_2 receptor upregulation were obtained with mammary line 66 tumor xenografts in nude mice (24). Together, these findings are consistent with some relationship of σ_2 receptors with cell growth regulating signals.

4. EFFECTS OF σ RECEPTOR LIGANDS ON CELLS IN CULTURE

σ Receptor ligands have dramatic effects on the morphology and viability of cells in culture. When rat C6 glioma cells were treated with micromolar concentrations of various σ ligands for 24-36 hours, cells stopped proliferating and underwent a dramatic change in morphology which was characterized by withdrawal of cell processes, cell rounding, and detachment from the substratum (25). This was followed by cell death. This morphological effect was observed with compounds from various structural classes, including butyrophenones (e.g. haloperidol), phenothiazines (e.g. fluphenazine), and the highly σ receptor selective arylethylene diamines (e.g. BD737 [(+)-cis-N-methyl-N-[2-(3,4-dichlorophenyl)ethyl]-2-(1-pyrrolidinyl)cyclohexylamine]), a class of compounds resulting from modification of κ opiate receptor-selective arylacetamides (26,27). This effect was both dose- and time-dependent, and was extraordinarily specific for compounds that exhibit σ receptor affinity. Various ligands for other receptors, ion channels, and enzymes had no effect on these cells, even at doses up to 300 μM. Subsequent analysis of the pharmacological profile showed a correlation to σ_2 receptor binding, as opposed to σ_1. For example, σ_1 selective (+)-benzomorphans were nearly inactive, whereas σ_1/σ_2 ligands were active. Similar effects of σ ligands on cell morphology were observed in SK-N-SH neuroblastoma, SH-SY5Y neuroblastoma, NG108-15 neuroblastoma-glioma, PC12 pheochromocytoma, and COS-7 cells (25).

In addition to morphology changes consistent with apoptosis, apoptotic cell death was confirmed using three independent measures: 1) annexin V binding to determine inversion of phosphatidyl serine in the early stages of apoptosis, 2) TUNEL staining to measure DNA fragmentation, and 3) bisbenzimide nuclear staining to show chromatin condensation (28-30). The pharmacological profile of this effect was again consistent with σ_2 receptor activation, since the σ_2 selective ligands CB-64D [(+)-1R,5R-(E)-8-benzylidene-5-(3-hydroxyphenyl)-2-methyl-morphan-7-one], CB-184 [(+)-1R,5R-(E)-8-(3,4-dichlorobenzylidene)-5-(3-hydroxyphenyl)-2-methyl-morphan-7-one], and ibogaine are cytotoxic while selective σ_1 receptor agonists such as (+)-pentazocine and dextrallorphan had no effect (28-30).

CB-64D and CB-184 are σ_2 subtype-selective ligands derived from 5-phenylmorphan opiates (31,32). We have shown that these ligands induce dose dependent apoptotic cell death in the drug-sensitive breast tumor cell line MCF-7, and in drug-resistant MCF-7/Adr, SKBr3, and T47D breast tumor cells (33). Treatment of T47D and MCF-7 breast tumor cells with haloperidol, reduced haloperidol, or CB-64D induced cell death and positive TUNEL staining, indicating DNA fragmentation (33). CB-64D also caused inversion of phosphatidyl serine, as indicated by annexin V staining in MCF-7 cells. Thus, both combination σ_1/σ_2 agonists and σ_2 subtype-selective agonists are able to induce apoptotic cell death in breast tumor cell lines. This was the first demonstration that specific σ_2 receptor activation induces apoptosis in breast tumor cells.

Though this will not be addressed in the current review, there is accumulating evidence that σ_1 receptors also play a role in cell proliferation. Brent and colleagues showed that various σ ligands inhibited proliferation of MCF-7 breast tumor cells, WIDr colon adenocarcinoma cells, and melanoma and induced apoptosis as indicated by Hoechst 33258 nuclear staining (34,35). However, the subtype mediating these effects was not clearly defined. Other studies have implicated σ_1 receptor blockade or σ_1 inhibition of potassium channels as a mediator of antiproliferative activity (36-40; see Chapter 17). We have attempted to utilize σ_2 receptor-selective ligands (CB-64D, CB-184, ibogaine) in our studies of σ receptor-mediated apoptosis. However, there is a paucity of σ_2-selective compounds. Continuing efforts to develop σ_2 receptor-selective agonists and antagonists will aid in distinguishing σ_1 receptor mechanisms from σ_2 receptor mechanisms. Furthermore, the MCF-7 breast tumor cells provide a model system in which σ_1 receptors appear to be absent or in very low density and thus antiproliferative and apoptotic effects of σ ligands in these cells must be mediated via σ_2 receptors (20,33).

5. σ₂ RECEPTORS AND SIGNALING MECHANISMS RELATED TO CELL PROLIFERATION AND SURVIVAL

5.1 Effects on calcium

Calcium is known to play an important role in regulation of cell proliferation (41-44). Transient increases in cytoplasmic free calcium levels occur when cells are stimulated with various hormones and growth factors. This calcium can cause induction of genes important for proliferation. However, when calcium homeostasis in cells becomes dysregulated, causing high and sustained levels of cytoplasmic calcium, pathological effects result. High calcium levels result in activation of proteases, nucleases, and other enzymes that degrade key cellular components. High calcium levels can also cause mitochondrial dysfunction which leads to cell death.

We have shown that σ_2 agonists regulate calcium in two ways in the SK-N-SH neuroblastoma cell line (45). Direct effects on cell calcium occur when σ_2 receptors are activated by agonists. Treatment of SK-N-SH neuroblastoma cells with σ ligands resulted in a rapid and transient rise in calcium, beginning immediately upon addition of ligand, peaking at 1.5 to 3 min, and returning to near baseline within 5 min after reaching peak. This calcium derived from the thapsigargin-sensitive pool in the endoplasmic reticulum. The pharmacological profile of this effect was strongly consistent with mediation by σ_2 receptors since σ_2 subtype-selective compounds (CB-64D and ibogaine) and σ_1/σ_2 compounds (e.g. BD737, JL-II-147 (2-[N-[2-[1-pyrrolidinyl]ethyl]-N-methylamino]-6,7-dichlorotetralin), reduced haloperidol) produced effects while σ_1 selective agonists (e.g. (+)-pentazocine, dextrallorphan) were virtually inactive. Removal of σ ligand after the calcium transient produced no further effect on calcium and no changes in cellular morphology.

However, if the cells remained exposed to σ_2 agonist over a period of around 20 min, a second, sustained rise of cytosolic calcium was observed which continued to rise for up to 60 min. This calcium derived from a thapsigargin-insensitive intracellular store, most likely mitochondria. The pharmacological profile for production of this sustained phase of calcium was identical to that for the transient phase. During this time period, the cells were beginning to show morphological signs of apoptosis including initial loss of processes and cell rounding. Annexin V staining of these same cells under conditions of the calcium assay revealed no signs of apoptosis during the time period of the calcium transient (10 min), whereas over the time period of the latent, sustained calcium signal there were early signs of apoptosis (46). Thus, the transient calcium signal is not a trigger for apoptosis, whereas the latent signal is temporally associated with the onset of apoptosis.

σ Ligand-induced increases in cytosolic free calcium in a breast adenocarcinoma and colon carcinoma cell line have also been reported (35). These effects were also likely mediated by σ_2 receptors, although the subtype involved was not determined. Thus σ_2 receptors may utilize calcium signals in wide variety of cell types.

5.2 Effects on ceramide

Ceramide is a sphingolipid second messenger that has been directly linked to regulation of cell growth. Depending on cell type, ceramide can either stimulate cell proliferation or cause inhibition of cell proliferation and induction of apoptosis (47,48). Ceramide can be formed in cells through *de novo* synthesis from palmitate and serine. However, various apoptotic signals result in formation of ceramide via activation of neutral or acidic sphingomyelinases, which cleave the phosphorylcholine head group of sphingomyelin to form ceramide (49). Ceramide can stimulate cell proliferation by downstream activation of the mitogen-activated protein (MAP) kinase pathway via a ceramide-activated protein kinase (or kinase suppressor of ras) (47). However, ceramide is most often associated with the induction of apoptosis. Apoptosis can result from activation of a ceramide-activated protein phosphatase (CAPP), a member of the protein phosphatase 2A (PP2A) family (47,50-53). Ceramide can also have direct effects on caspases and on mitochondria which result in apoptosis, or it can induce apoptosis in a caspase-independent manner (54,55).

We have shown that σ_2 receptor agonists cause an increase in ceramide in breast tumor cell lines. CB-184 caused a dose dependent increase in ceramide levels in MCF-7/Adr and T47D cells, where the sphingolipid pools were metabolically prelabeled with [^{3}H]palmitic acid and [^{14}C]serine (56). This was accompanied by a concomitant decrease in sphingomyelin levels. At the highest dose of CB-184 (100 µM), [^{3}H]ceramide levels increased by 342% and [^{3}H]sphingomyelin decreased by 590%, compared to baseline levels in MCF-7/Adr cells. The σ_2 antagonist, N-phenethylamine (AC927), attenuated both effects. In T47D breast tumor cells, 100 µM CB-184 induced a 265% increase in [^{3}H]ceramide and an 87% decrease in [^{3}H]sphingomyelin, compared to baseline levels. Similar effects were observed in SK-N-SH neuroblastoma cells (57).

The mechanism of ceramide formation is currently under investigation. Although a decrease in sphingomyelin with an increase in ceramide is consistent with activation of sphingomyelinase, this appears not to be the case. Sphingomyelinase activity was not increased upon treatment of cells with σ_2 agonists (unpublished result). Ceramide appears to form via acylation of sphingosine with a fatty acid (58; Crawford and Bowen, submitted). When detergent extracts of SKBr3 or MCF-7/Adr breast tumor cells were incubated with D-*erythro*-sphingosine and [^{3}H]palmitic acid, in presence of various concentrations of CB-184, a dose dependent formation of [^{3}H]ceramide was observed (58; Crawford and Bowen, submitted). Furthermore, the decrease in sphingomyelin appears to be due to hydrolysis to produce sphingosylphosphoryl-choline (see below). Therefore, σ_2 receptors may activate an enzyme similar to sphingolipid-ceramide N-deacylase, a bacterial enzyme known to acylate sphingosine while also having the ability to hydrolyze sphingomyelin (59,60). Alternatively, ceramide could be formed via other mechanisms such as reversal of ceramidase activity or activation of *de novo* synthesis. However, these mechanisms appear less likely since ceramide can be formed in a cell-free system.

As mentioned above, ceramide has several targets in cells. Perhaps the best characterized is the type 2A phosphatase, CAPP (52,53,61). Certain substrates for CAPP have been identified via experiments showing that ceramide promotes the dephosphosphorylation of these proteins in various cells. These include, Akt, Bad, and Bcl-2 (52,53,61). Dephosphorylation of any of these substrates would promote loss of cell viability and lead to apoptosis. More work will be required to identify the targets of ceramide in cells treated with σ_2 agonists, but ceramide formation is likely to play a key role in σ_2 receptor-mediated apoptosis.

5.3   Effects on sphingosylphosphorylcholine

Another class of sphingolipid-derived messengers that regulate cell proliferation and survival are the lysosphingolipids (62-66). These compounds are related to ceramide, but have a free amino group moiety and derivatized head group. These include sphingosine-1-phosphate and sphingosylphosphoryl choline. The most well studied is sphingosine-1-phosphate (S-1-P), which is formed by phosphorylation of sphingosine by sphingosine kinases. Sphingosine kinases are activated by various stimuli to produce S-1-P, which in turn usually has proliferative and anti-apoptotic effects on cells. Less is known regarding the related compound, sphingosylphosphorylcholine (SPC). Studies have shown SPC to have either proliferative effects or apoptotic effects, depending on the system (66). S-1-P and SPC have multiple modes of action in cells. S-1-P can act through a family of cell surface G protein coupled receptors (Edg or SP receptors) to regulate both cAMP and calcium levels in cells (62,63). SPC has been recently shown to also interact with G protein coupled receptors, termed GPR4, OGR1, and G2A, to regulate calcium levels via inositol 1,4,5-triphosphate (IP_3) formation (65,66). In addition, both S-1-P and SPC have intracellular targets, both producing calcium release from stores in the endoplasmic reticulum or sarcoplasmic reticulum in an IP_3-independent manner (64,66,67).

Treatment of SKBr3 breast tumor cells with σ_2 receptor agonists resulted in the formation of SPC. CB-184 caused dose dependent [^{3}H]SPC formation with an EC_{50} of 1.7 µM, in SKBr3 cells where the sphingolipid pool had been prelabeled with [^{3}H]palmitic acid (58; Crawford and Bowen, submitted). [^{3}H]SPC formation was blocked by the σ_2 antagonist, AC927.

SPC formation is believed to be coupled to the observed σ_2 agonist-induced decrease in sphingomyelin (56). Sphingomyelin hydrolysis (removal of the fatty acyl group) would result in SPC formation. Dose dependent [^{3}H]SPC production was demonstrated in detergent extracts of SKBr3 cells which were incubated with [^{3}H]sphingomyelin and CB-184. This would again be consistent with modulation of an enzyme related to spingolipid ceramide N-deacylase, which can both hydrolyze sphingomyelin to form lysosphingolipids and acylate sphingosine to form ceramides (56,59,60).

The exact role of SPC in these cells will need further investigation, but its formation lends additional support to the notion that σ_2 receptors play a role in cell proliferation and survival. Depending on the cell type, SPC has been shown to have proliferative, anti-apoptotic, or pro-apoptotic effects (66). Furthermore, the σ_2 agonist-induced formation of SPC and release of calcium from the endoplasmic reticulum could be coupled since SPC can cause calcium release indirectly by interaction with G protein coupled receptors coupled to phospholipase C and IP_3 formation, or directly by interaction with calcium gating molecules on the endoplasmic reticulum membrane, as described above.

It is interesting to note that transient calcium signals and SPC have both been associated with increases in cell proliferation. In light of the high upregulation of σ_2 receptors in rapidly dividing cells, one could speculate that the true role of this receptor is to mediate trophic or proliferative signals. In this case, acute stimulation of σ_2 receptors might produce growth stimulating or survival signals in the form of SPC and calcium, whereas more chronic stimulation of the receptor could result in ceramide accumulation and apoptosis. It has been proposed that cell life/death decisions are regulated by the balance between the levels of lysosphingolipids and ceramides present, with lysosphingolipids favoring survival and ceramides favoring death (62,63). σ_2 Receptors could act to regulate this balance. Further studies will be needed to explore this possibility.

6. IMPLICATIONS FOR CANCER DIAGNOSIS AND CHEMOTHERAPY

6.1 Efficacy against drug-resistant tumors

The development of drug resistance presents an enormous problem in cancer chemotherapy. Tumor cells develop resistance to antineoplastic agents via several mechanisms (68,69). These include mutations in p53, aberrations in caspase function, and overexpression of drug efflux pumps. A novel approach is to attempt to activate alternative programmed cell death pathways that bypass some of the commonly utilized steps in classical apoptosis (70,71). We have shown that agonist activation of σ_2 receptors results in induction of apoptosis in several breast tumor cell lines which are highly resistant to the apoptotic effects of other antineoplastic agents (33). The mechanism is both caspase- and p53-independent, thus resistant cells remain susceptible. It therefore may be possible to target σ_2 receptors for the development of novel anti-tumor agents effective against a variety of drug-resistant cells.

6.1.1 Tumors with p53 mutations

An important cause of drug resistance in tumors is acquisition of mutations in the tumor suppressor gene, p53 (72,73). p53 is expressed in response to DNA damage, and induces cell cycle arrest and/or apoptosis by mechanisms that include up regulation of pro-apoptotic proteins and stimulation of ubiquitin-mediated degradation of key cellular proteins. Over fifty percent of all tumors examined harbor mutations in the p53 gene, rendering them resistant to apoptosis when DNA is damaged by common antineoplastic agents such as doxorubicin and actinomycin (74,75).

σ_2 Receptor receptor-mediated apoptosis appears to occur independent of involvement of p53. The cytotoxic potency of σ_2 receptor agonists was largely unaffected by the status of p53 across four cell lines examined (33). CB-184 exhibited similar potency in three breast tumor cell lines that harbor various p53 mutations (MCF-7/Adr, SKBr3, and T47D) when compared to a cell line that has wild-type p53 protein (MCF-7). A similar result was observed with the less potent σ_2 agonist, CB-64D. By comparison, doxorubicin was greater than 20-fold less potent in the drug-resistant MCF-7/Adr cells, compared to MCF-7 cells. Therefore, p53 mutations, a common cause of drug resistance, do not confer resistance to σ_2 agonist-induced apoptosis. This indicates that σ_2 receptor agonists would be effective against many drug-resistant tumors that harbor such p53 mutations.

6.1.2   Caspase independence

Caspases are a family of cysteine-aspartyl proteases that are the executioners of apoptotic signals from diverse stimuli, including receptor activation (e.g. Fas ligand and TNFα), DNA-damaging agents, hypoxia, growth factor deprivation, or ionizing radiation (76). The targets of caspases include a vast array of key proteins that are necessary for cell survival. Some include cytoskeletal proteins, cell cycle regulatory proteins, and nuclear matrix proteins. Loss of caspase activity is known to be a mechanism for development of drug-resistance. For example, silencing of caspase-8 mRNA expression, loss of Apaf-1 and caspase-9 regulation, and loss of caspase-3 processing contributes to drug resistance in neuroblastoma, melanoma, and lymphoblastic leukemia, respectively (77,78).

We have demonstrated that σ_2 receptor-mediated apoptosis in breast tumor cell lines is caspase independent. Caspase inhibitors, both broad spectrum (ZVAD-FMK) and selective (DEVD-CHO, capsase-3, YVAD-CHO, caspase-1), failed to block the apoptosis induced by CB-64D and CB-184 in MCF-7 cells (33). However, apoptosis induced by doxorubicin or actinomycin D was blocked by these caspase inhibitors. Therefore, apoptotic cell death induced by σ_2 receptor agonists does not require caspase activation, unlike the mechanisms for these common antineoplastic agents. This would indicate that σ_2 receptor agonists would overcome any drug resistance that results from deficits in caspase function. Further studies will be required to determine if the mechanism is caspase independent in other types of tumor cells.

6.1.3 Effects on P-glycoprotein expression

Tumor cells can also develop drug resistance by overexpressing drug efflux pumps. The multidrug resistance (mdr-1) gene product, P-glycoprotein catalyzes the efflux of hydrophobic compounds that are often toxic to cells, removing them from the internal milieu and thereby reducing or eliminating the effectiveness of some antineoplastic agents (68,69). For example, doxorubicin is a substrate for P-glycoprotein and P-glycoprotein overexpression results in phenotypic resistance to doxorubicin, even in the presence of wild-type p53 (80).

The σ_2 receptor agonists CB-64D and BD737 have been shown to reduce the expression of the MDR gene in human SK-N-SH neuroblastoma and rat C6 glioma cells (81). A 24-hour treatment of C6 glioma cells with BD737 or CB-64D at a concentration of 10 µM (a concentration producing no visible effect on cell morphology or viability) resulted in a 50% reduction in the expression of mdr-1 mRNA as assessed using RT-PCR (81). Treatment of SK-N-SH neuroblastoma cells with 10 µM CB-64D completely inactivated mdr-1 expression. The σ_1 agonist (+)-pentazocine had little or no effect. Neither active compound affected expression of the β-actin gene, which was used as a control for non-specific effects. The σ receptor antagonist, BD1047 (N-[2-(3,4-dichlorophenyl)ethyl]-N-methyl-2-(dimethylamino)ethylamine; 10 µM and 100 µM), blocked the effect of 10 µM CB-64D in SK-N-SH cells. These data suggest that specific activation of σ_2 receptors results in a decrease in mdr-1 gene expression.

As mentioned above, CB-64D and CB-184 were found to have similar cytotoxic potency in MCF-7 and MCF-7/Adr cells (33). In addition to harboring a p53 mutation, MCF-7/Adr cells also overexpress P-glycoprotein compared to MCF-7 cells. This suggests that these σ_2 agonists are not substrates for the drug efflux pumps and will remain potent in cells with highly upregulated P-glycoprotein. It is not clear whether σ_2 agonists will down regulate the high levels of P-glycoprotein in these cells, but results with C6 glioma and SK-N-SH neuroblastoma show promise in this regard. The results also show that σ_2 receptor agonists are efficient at overcoming multiple modes of drug resistance in the same cell.

6.2 Chemo-sensitization

The development of drug resistance in tumors can necessitate using increasing doses of antineoplastic agents in order to maintain clinically effective levels. This results in increased toxic side effects of the antineoplastic agent used. An agent which increases the effectiveness of the anti-tumor drug would be useful in addressing this problem.

σ_2 Receptor agonists when combined with doxorubicin or actinomycin D were found to result in a synergistic increase in cytotoxic potency (33). Importantly, this was observed at doses of σ_2 agonist and antineoplastic agent that were alone minimally toxic. A clear synergistic effect was observed in MCF-7 cells at 24 hours when CB-184 (1 µM) was combined with 10 µM doxorubicin. A similar interaction was observed with drug-resistant MCF-7/Adr cells when 1 µg/ml actinomycin D was combined with CB-184 (1 µM). The same study showed that two clinically available drugs that have σ_2 activity, haloperidol and (±)-pentazocine (Talwin), also displayed the potentiation phenomenon, though not as robustly as CB-184 (33).

The mechanism of this synergistic effect is not clear. The effect of σ_2 agonists on MDR gene expression mentioned above, would also be expected to result in a chemosensitizing effect on those drugs that are substrates for P-glycoprotein in cells that overexpress the MDR gene. However, this is not likely to be the only mechanism since the effect occurred in MCF-7 cells which do not overexpress P-glycoprotein. The simultaneous activation of distinct p53-dependent and p53-independent apoptotic pathways in these cells could synergize at some level in the signaling pathway to produce the enhanced cytotoxicity observed (33). Thus, σ_2 agonists should have useful effects in chemotherapy, even at doses that are not by themselves cytotoxic. Due to the high expression of σ_2 receptors in rapidly dividing tumor cells versus normal cells, σ_2 agonists could increase the specificity of antineoplastic agents for tumor cells leading to less unwanted toxicity.

6.3 Non-invasive tumor imaging

The high expression of σ receptors in many different types of tumor cells and their further upregulation in rapidly proliferating cells, has led to the investigation of σ receptors as targets for diagnostic tumor imaging agents. Several ligands containing iodine or fluorine have been developed which bind with high affinity to σ receptors, and which may be suitable for positron emission tomography (PET) or single photon emission computed tomography (SPECT) imaging of tumors. These include several iodo- or fluoro-benzamides such as IPAB (N-[2-(1'-piperidinyl)ethyl]-4-iodobenzamide), 4-IPB (N-(N-benzylpiperidin-4-yl)-4-iodobenzamide), and PIMBA (N-[2-(1'-piperidinyl) ethyl]-3]iodo-4-methoxybenzamide) (82-88), arylethylenediamines such as IPEMP (N-[2-(4-iodophenyl)ethyl]-N-methyl-2-(1-piperidinyl)ethylamine) (89,90), substituted piperazines such as

SA5845 (1-(4-2'-fluoroethoxy-3-methoxyphenethyl)-4-(3-(4-fluorophenyl) propyl)piperazine) and SA4503 (1-(3,4-dimethoxyphenethyl)-4-(3-phenyl-propyl)piperazine) (91), and arylsulfon-amides (92). These compounds were found to bind with nanomolar affinity to σ receptors of breast tumor, prostate, melanoma, glioma, or lung carcinoma cell lines. In addition to iodine- and fluorine-containing ligands, a Tc99m-based σ radioligand was developed. Tc99m is a widely used imaging isotope because of its high energy, relatively short half-life, and instant availability by generator. [99mTc]BAT-EN6 bound to σ receptors of T47D breast carcinoma cells with K_d = 43.5 nM (93).

Kinetic and biodistribution studies of these radiolabeled compounds in nude mice implanted with tumor xenografts (melanoma, non-small cell lung carcinoma, breast carcinoma, or prostate) have generally shown rapid uptake by the tumor as well as by major organs which possess σ receptors such as liver, kidney, spleen, and brain. However, clearance from normal organs was usually rapid, while tumors tended to retain the radioligand for up to 24 hours.

Clear SPECT images of malignant melanoma xenographs in nude mice were obtained with [^{123}I]PAB (82). At 6 hours after injection of the radiotracer, most organs known to contain σ receptors could be seen in the image, including liver and brain. However, after 24 hours, the tumor implant in the left thigh could be clearly seen, whereas normal organs were clear of labeling. The reason the tumor selectively retains the label is not clear, but could be due to uptake or the high density of receptors compared to normal tissue.

Breast tumor imaging with [^{123}I]PIMBA has been reported in humans (94). SPECT images of 12 patients with mammographically suspicious breast masses were taken 2 hours after injection of [^{123}I]PIMBA. Eight of 10 patients with histologically confirmed lesions were positively identified with [^{123}I]PIMBA. The tumor could be clearly seen relative to surrounding normal tissue, and the tumor-free breast was clear of labeling. This is consistent with a preliminary report which showed that normal breast tissue taken from around a biopsied tumor showed little or no σ receptor binding compared to the tumor itself, which contained abundant σ receptor binding activity (18).

Malignant melanoma has also been successfully imaged in human patients using σ receptor ligands. [^{123}I]IDAB (N-(2-diethylaminoethyl)-4-iodobenzamide) was used to image 110 patients with a history of melanoma (95). Malignant melanoma was detected with 87% accuracy and patients in clinical remission had normal scintography. Metastasis to the lung, bone, and brain could be clearly visualized in the affected patients, where these organs were clear in the normal controls. Although the authors suggested that images might be due to interaction of the radioligand with melanin, it is most likely that the images resulted from binding to σ receptors since IDAB interacts with high affinity to σ_1 receptors. Similar results were obtained with [^{123}I]PIMBA in three melanoma patients (96).

Since tumor cell lines contain high densities of both σ_1 and σ_2 receptors, the question arises as to which subtype should best be targeted for developing diagnostic agents for imaging. It is the σ_2 subtype that is upregulated when cells rapidly divide, whereas σ_1 receptor expression remains fairly constant (22-24). Mach and coworkers have investigated this question using [^{18}F]N-(4'-fluorobenzyl)-4-(3-bromophenyl)acetamide, a radioligand that labels both σ_1 and σ_2 receptors, in nude mice bearing mouse mammary adenocarcinoma line 66 tumor xenografts (97). They showed that masking of σ_1 receptor binding to allow only σ_2 labeling increases the tumor to background ratio. They concluded that σ_2 receptors are thus the best targets for tumor imaging agents, as opposed to σ_1. They also showed in the same study that σ_2 receptor targeting gave better anatomical imaging than the metabolic imaging agent [^{18}F]deoxyglucose and better functional imaging than two DNA precursors [$^{123/124}$I]iododeoxyuridine and [^{11}C]thymidine. These latter agents are used to gauge the proliferative status of tumors by DNA synthesis. In light of these findings, this group has developed a Tc99m-labeled ligand and several benzamide analogs which are highly σ_2 subtype-selective, have high affinity, and which are amenable to imaging by SPECT or PET (98,99). These compounds might be quite useful for imaging σ_2 receptor-expressing tumors and their metastases.

7. SUMMARY

σ Receptors are expressed in the brain, as well as in many other tissues throughout the body. While much is being learned regarding functions of these receptors in the central nervous system, less is known regarding functions in the periphery. The nearly ubiquitous expression across tissues suggests a global function. Upregulation of σ_2 receptors in rapidly proliferating cells, down regulation in quiescent cells, and the ability of σ_2 agonists to inhibit cell proliferation and induce apoptosis suggest a role of σ_2 receptors in cell growth control. This is further supported by the ability of σ_2 receptors to modulate several signaling pathways known to regulate proliferation and cell survival. Current evidence suggests that σ_2 receptors may be useful targets for development of antineoplastic agents that will be effective against drug-resistant tumors. They can also be targeted by noninvasive imaging agents that will be able to detect many types of tumors, monitor their growth, and perhaps detect and treat tumors at metastatic sites. Further exploration with respect to elucidation of σ_2 receptor structure, delineation of endogenous ligands, mechanisms of cell growth control, and development of selective receptor probes as agonists and imaging agents is clearly warranted.

REFERENCES

1. Walker JM, Bowen WD, Walker FO, Matsumoto RR, de Costa BR, Rice KC. Sigma receptors: Biology and function. Pharmacol Rev 1990, 42:355-402.
2. Quirion R, Bowen WD, Itzhak Y, Junien JL, Musacchio JM, Rothma RB, Su TP, Tam SW, Taylor DP. A proposal for the classification of sigma binding sites. Trends Pharmacol Sci 1992, 13: 85-86.
3. Bowen WD. Biochemical pharmacology of sigma receptors. in: Aspects of Synaptic Transmission Vol. II: Acetylcholine, Sigma Receptors, CCK and Eicosanoids, Neurotoxins. (T.W. Stone, ed.), Taylor and Francis, London, U.K., 1993, pp. 113-136.
4. Bowen WD. Sigma receptors: Recent advances and new clinical potentials. Pharmaceutica Acta Helvetiae 2000, 74: 211-218.
5. Su TP, Hayashi T. Understanding the molecular mechanism of sigma-1 receptors: towards a hypothesis that sigma-1 receptors are intracellular amplifiers for signal transduction. Curr Med Chem 2003, 10: 2073-2080.
6. Guitart X, Codony X, Monroy X. Sigma receptors: biology and therapeutic potential. Psychopharmacology 2004, 174: 301-319.
7. Maurice T. Neurosteroids and sigma-1 receptors, biochemical and behavioral relevance. Pharmacopsychiatry 2004, 37(Suppl 3): S171-182.
8. Hellewell SB, Bowen WD. A sigma-like binding site in rat pheochromocytoma (PC12) cells: Decreased affinity for (+)-benzomorphans and lower molecular weight suggest a different sigma receptor form from that in guinea pig brain. Brain Res 1990, 527:244-253.
9. Hanner M, Moebius FF, Flandorfer A, Knaus HG, Striessnig J, Kempner E, Glossmann H. Purification, molecular cloning, and expression of the mammalian sigma$_1$-binding site. Proc Natl Acad Sci USA 1996, 93: 8072-8077.
10. Prasad PD, Li HW, Fei YJ, Ganapathy ME, Fujita T, Plumley LH, Yang-Feng TL, Leibach FH, Ganapathy V. Exon-intron structure, analysis of promoter region, and chromosomal localization of the human type 1 sigma receptor gene. J Neurochem 1998, 70:443-451.
11. Hellewell SB, Bruce A, Feinstein G, Orringer J, Williams W, Bowen WD. Rat liver and kidney contain high densities of sigma-1 and sigma-2 receptors: Characterization by ligand binding and photoaffinity labeling. Eur J Pharmacol Mol Pharmacol Sect 1994, 268: 9-18.
12. Su, TP, London, ED, Jaffe JH. Steroid binding at sigma receptors suggests a link between endocrine, nervous, and immune systems. Science 1988, 240:219-221.
13. Patterson TA, Connor M, Chavkin C. Recent evidence for endogenous substance(s) for sigma receptors. in: The Sigma Receptors, Neuroscience Perspectives Series. (Y. Itzhak, ed.), Academic Press, London, 171-189, 1994.

14. Bowen WD, de Costa BR, Hellewell SB, Walker JM, Rice KC. [^{3}H](+)-Pentazocine: A potent and highly selective benzomorphan-based probe for sigma-1 receptors. Mol Neuropharmacol 1993, 3:117-126.
15. Wolfe Jr SA, Culp SG, De Souza, EB. Sigma receptors in endocrine organs: Identification, characterization, and autoradiographic localization in rat pituitary, adrenal, testis, and ovary. Endocrinology 1989, 124:1160-1172.
16. Thomas GE, Szucs M., Mamone JY, Bem WT, Rush MD, Johnson FE, Coscia CJ. Sigma and opioid receptors in human brain tumors. Life Sci 1990, 46:1279-1286.
17. Bem WT, Thomas GE, Mamone JY, Homan SM, Levy BK, Johnson FE, Coscia CJ. Overexpression of sigma receptors in nonneural human tumors. Cancer Res 1990, 51:6558-6562.
18. John CS, Vilner BJ, Schwartz AM, Bowen WD. Characterization of sigma receptor binding sites in human biopsied solid breast tumors. J Nucl Med 1996, 37:267P.
19. Vilner BJ, Bowen WD. Characterization of sigma-like binding sites of NB41A3, S-20Y, and N1E-115 neuroblastomas, C6 glioma, and NG108-15 neuroblastoma-glioma hybrid cells: Further evidence for sigma-2 receptors. In: *Multiple Sigma and PCP Receptor Ligands: Mechanisms for Neuromodulation and Neuroprotection?* J.-M. Kamenka and E.F. Domino, eds. NPP Books, Ann Arbor, MI, 1992, pp. 341-353.
20. Vilner BJ, John CS, Bowen WD. Sigma-1 and sigma-2 receptors are expressed in a wide variety of human and rodent tumor cell lines. Cancer Res 1995, 55:408-413.
21. Ganapathy ME, Prasad PD, Huang W, Seth P, Leibach FH, Ganapathy V. Molecular and ligand-binding characterization of the sigma-receptor in the Jurkat human T lymphocyte cell line. J Pharmacol Exp Ther 1999, 289:251-260.
22. Mach RH, Smith CR, al-Nabulsi I, Whirrett BR, Childers SR, Wheeler KT. Sigma-2 receptors as potential biomarkers of proliferation in breast cancer. Cancer Res 1997, 57:156-161.
23. Al-Nalbusi I, Mach RH, Wang LM, Wallen CA, Keng PC, Sten K, Childers SR, Wheeler KT. Effect of ploidy, recruitment, environmental factors, and tamoxifen treatment on the expression of sigma-2 receptors in proliferating and quiescent tumor cells. British J Cancer 1999, 81:925-933.
24. Wheeler KT, Wang LM, Wallen CA, Childers SR, Cline JM, Keng PC, Mach RH. Sigma-2 receptors as a biomarker of proliferation in solid tumors. Br J Cancer 2000, 82:1223-1232.
25. Vilner BJ, de Costa BR, Bowen WD. Cytotoxic effects of sigma ligands: Sigma receptor-mediated alterations in cellular morphology and viability. J Neurosci 1995, 15:117-134.
26. de Costa BR, Radesca L, Di Paolo L, Bowen WD. Synthesis, characterization and biological evaluation of a novel class of N-(arylethyl)-N-alkyl-2-(1-pyrrolidinyl)ethylamines: Structural requirements and binding affinity at the sigma receptor. J Med Chem 1992, 35:38-47.
27. Bowen WD, Walker JM, de Costa BR, Wu R, Tolentino PJ, Finn D, Rothman RB, Rice KC. Characterization of the enantiomers of cis-N-[2-(3,4-dichlorophenyl)ethyl]-N-methyl-2-(1-pyrrolidinyl)cyclohexylamine (BD737 and BD738): Novel compounds with high affinity, selectivity, and biological efficacy at sigma receptors. J Pharmacol Exp Ther 1992, 262:32-40.
28. Vilner BJ, Bowen WD. Sigma-2 receptor agonists induce apoptosis in rat cerebellar granule cells and human SK-N-SH neuroblastoma cells. Soc Neurosci Abst 1997, 23:2319, #905.6.

29. Bowen WD, Vilner BJ, Williams W, Bandarage UK, Kuehne ME. Novel ibogaine analogs as selective sigma-2 receptor probes: Ligand binding and functional assays. Soc Neurosci Abst 1997, 23:2319, #905.7.
30. Bowen WD. Sigma receptors and iboga alkaloids. Alkaloids Chem Biol 2001, 56:173-191.
31. Bertha CM, Vilner BJ, Mattson MV, Bowen WD, Becketts K, Xu H, Rothman RB, Flippen-Anderson JL, Rice KC. (E)-8 Benzylidene derivatives of 2-methyl-5-(3-hydroxyphenyl)morphans: Highly selective ligands for the sigma-2 receptor subtype. J Med Chem 1995, 38:4776-4785.
32. Bowen WD. Bertha CM Vilner BJ, Rice KC. CB-64D and CB-184: Ligands with high sigma-2 receptor affinity and subtype selectivity. Eur J Pharmacol 1995, 278:257-260.
33. Crawford KW, Bowen WD. Sigma-2 receptor agonists activate a novel apoptotic pathway and potentiate antineoplastic drugs in breast tumor cell lines. Cancer Res 2002, 62:313-322.
34. Brent PJ, Pang GT. Sigma binding site ligands inhibit cell proliferation in mammary and colon carcinoma cell lines and melanoma cells in culture. Eur J Pharmacol 1995, 278:151-160.
35. Brent PJ, Pang G, Little G, Dosen PJ, Van Helden F. The sigma receptor ligand, reduced haloperidol, induces apoptosis and increases intracellular-free calcium levels $[Ca^{2+}]_i$ in colon and mammary adenocarcinoma cells. Biochem Biophys Res Commun 1996, 219:219-226.
36. Casellas P, Galiegue S, Bourrrie B, Ferrini JB, Jbilo O, Vidal H. SR31747A: a peripheral sigma ligand with potent antitumor activities. Anticancer Drugs 2004, 15:113-118.
37. Renaudo A, Watry V, Chassot AA, Ponzio G, Ehrenfeld J, Soriani O. Inhibition of tumor cell proliferation by sigma ligands is associated with K^+ channel inhibition and p27kip1 accumulation. J Pharmacol Exp Ther 2004, 311:1105-1114.
38. Colabufo NA, Berardi F, Contino M, Niso M, Abate C, Perrone R, Tortorella V. Antiproliferative and cytotoxic effects of some $sigma_2$ agonists and $sigma_1$ antagonists in tumor cell lines. Naunyn Schmiedeberg Arch Pharmacol 2004, 370: 106-113.
39. Spruce BA, Campbell LA, McTavish N, Cooper MA, Appleyard MV, O'Neill M, Howie J, Samson J, Watt S, Murray K, McLean D, Leslie NR, Safrany ST, Ferguson MJ, Peters JA, Prescott AR, Box G, Hayes A, Nutley B, Raynaud F, Downes CP, Lambert JJ, Thompson AM, Eccles S. Small molecule antagonists of the sigma-1 receptor cause selective release of the death program in tumor and self-reliant cells and inhibit tumor growth in vitro and in vivo. Cancer Res 2004, 64:4875-4886.
40. Aydar E, Palmer CP, Djamgoz MB. Sigma receptors and cancer: possible involvement of ion channels. Cancer Res 2004, 64:5029-5035.
41. Berridge MJ, Bootman MD Lipp P. Calcium - a life and death signal. Nature 1998, 395:645-648.
42. McConkey DJ, Orrenius S. The role of calcium in the regulation of apoptosis. J Leuk Biol 1996, 59:775-783.
43. Khal CR, Means AR. Regulation of cell cycle progression by calcium/calmodulin-dependent pathways. Endocrine Rev 2003, 24:719-736.
44. Demaurex N, Distelhorst C. Apoptosis – the calcium connection. Science 2003, 300:65-67.
45. Vilner BJ, Bowen WD. Modulation of cellular calcium by sigma-2 receptors: Release from intracellular stores in human SK-N-SH neuroblastoma cells. J Pharmacol Exp Ther 2000, 292:900-911.

46. Vilner BJ, Bowen WD. Relationship of sigma-2 receptor-mediated increases in intracellular calcium to induction of morphological changes and apoptosis in human SK-N-SH neuroblastoma cells. Soc Neurosci Abst 1998, 24:1594, #627.6.
47. Kolesnick RN, Kronke M. Regulation of ceramide production and apoptosis. Ann Rev Physiol 1998, 60:643-665.
48. Ogretmen B, Hannun YA. Biologically active sphingolipids in cancer pathogenesis and treatment. Nat Rev Cancer 2004, 4:604-616.
49. Obeid LM, Hannun YA. Ceramide: a stress signal and mediator of growth suppression and apoptosis. J Cell Biochem 1995, 58:191-198.
50. Galardi S, Kishikawa K, Kamibayashi C, Mumby MC, Hannun YA. Purification and characterization of ceramide-activated protein phosphatases. Biochem 1998. 37:11232-11238.
51. Chalfant CE, Kishikawa K, Bielawska A, Hannun YA. Analysis of ceramide-activated protein phosphatases. Methods Enzymol 2000, 312:420-428.
52. Zinda MJ, Vlahos CJ, Lai MT. Ceramide induces the dephosphorylation and inhibition of constitutively activated Akt in PTEN negative U87mg cells. Biochem Biophys Res Commun 2001, 280:1107-1115.
53. Ruvolo PP, Clark W, Mumby M., Gao F, May WS. A functional role for the B56 α-subunit of protein phosphatases 2A in ceramide-mediated regulation of Bcl2 phosphorylation status and function. J Biol Chem 2002, 277:22847-22852.
54. Laethem R, Hannun Y, Jayadev S, Sexton CJ, Strum JC, Sundseth R, Smith GK. Increases in neutral, Mg^{2+}-dependent and acidic, Mg^{2+}-independent sphingomyelinase activities precede commitment to apoptosis and are not a consequence of caspase-3-like activity in Molt-4 cells in response to thymidylate synthase inhibition by GW1843. Blood 1998, 91:4350-4360.
55. Mathiasen IS, Lademann U, Jaattela M. Apoptosis induced by vitamin D compounds in breast cancer cells is inhibited by bcl-2 but does not involve known caspases or p53. Cancer Res 1999, 59:4848-4856.
56. Crawford KW, Coop A, Bowen WD. Sigma-2 receptors regulate changes in sphingolipid levels in breast tumor cells. Eur J Pharmacol 2002, 443:207-209.
57. Bowen WD, Crawford KW, Huang S, Walker JW. Activation of sigma-2 receptors causes changes in ceramide levels in neuronal and non-neuronal cell lines. Soc Neurosci Abst 2000, 26:601, #226.11.
58. Bowen WD, Crawford KW, Coop A. Sigma-2 receptors may activate sphingolipid-ceramide N-deacylase (SCDase) as a mechanism to regulate cell growth. Soc Neurosci Abst 2001, 27:948, #364.1.
59. Ito M, Kurita T, Kita K. A novel enzyme that cleaves the N-acyl linkage of ceramides in various glycosphingolipids as well as sphingomyelin to produce their lyso forms. J Biol Chem 1995, 270:24370-24374.
60. Kita K, Kurita T, Ito M. Characterization of the reversible nature of the reaction catalyzed by sphingolipid ceramide N-deacylase: A novel form of reverse hydrolysis reaction. Eur J Biochem 2001, 268:592-602.
61. Stoica BA, Movsesyan VA, Lea PM 4th, Faden AI. Ceramide-induced neuronal apoptosis is associated with dephosphorylation of Akt, BAD, FKHR, GSK-3beta, and induction of the mitochondrial-dependent intrinsic caspase pathway. Mol Cell Neurosci 2003, 22:365-382.
62. Spiegel S, Culliver O, Edsall LC, Kohama T, Menzeleev R, Olah Z, Olivera, A, Pirianov, G, Thomas DM, Tu Z, Van Brocklyn JR, Wang F. Sphingosine-1-phosphate in cell growth and cell death. Ann NY Acad Sci 1998, 845:11-18.

63. Spiegel S, Milstien S. Functions of a new family of sphingosine-1-phosphate receptors. Biochim Biophys Acta 2000, 1484:107-116.
64. Myer zu Heringdorf D, van Koppen, CJ, Jakobs KH. Molecular diversity of sphingolipid signalling. FEBS Lett 1997, 410:34-38.
65. Xu Y. Sphingosylphosphorylcholine and lysophosphatidylcholine: G protein-coupled receptors and receptor-mediated signal transduction. Biochim Biophys Acta 2002, 1582:81-88.
66. Meyer zu Heringdorf D, Himmel HM, Jakobs KH. Sphingosylphosphorylcholine-biological functions and mechanisms of action. Biochim Biophys Acta 2002, 1582:178-189.
67. Mao C, Kim SH, Almenoff JS, Rudner XL, Kearney DM, Kindman LA. Molecular cloning and characterization of SCaMPER, a sphingolipid Ca^{2+} release-mediating protein from endoplasmic reticulum. Proc Natl Acad Sci USA 1996, 93:1993-1996.
68. Stein WD, Bates SE, Fojo T. Intractable cancers: the many faces of multidrug resistance and the many targets it presents for therapeutic attack. Curr Drug Targets 2004, 5:333-346.
69. Longley DB, Johnston PG. Molecular mechanisms of drug resistance. J Pathol 2005, 205:275-292.
70. Mathiasen IS, Jaattela M. Triggering caspase-independent cell death to combat cancer. Trends Mol Med 2002, 8:212-220.
71. Broker LE, Kruyt FA, Giaccone G. Cell death independent of caspases: a review. Clin Cancer Res 2005, 11:3155-3162.
72. Selivanova G. Mutant p53: the loaded gun. Curr Opin Investig Drugs 2001, 2:1136-1141.
73. Gasco M, Crook T. p53 members and chemoresistance in cancer: what we know and what we need to know. Drug Resist Update 2003, 6:323-328.
74. Ryan KM, Vousden KH. Characterization of structural p53 mutants which show selective defects in apoptosis but not cell cycle arrest. Mol Cell Biol 1998, 18:3692-3698.
75. Wallace-Brodeur RR, Lowe SW. Clinical implications of p53 mutations. Cell Mol Life Sci 1999, 55:64-75.
76. Cohen GM. Caspases. The executioners of apoptosis. Biochem J 1997, 326:1-16.
77. Prokop A, Wieder T, Sturm I, Essmann F, Seeger K, Wuchter C, Ludwig WD, Henze G, Dorken B, Daniel PT. Relapse in childhood acute lymphoblastic leukemia is associated with a decrease of the Bax/Bcl-2 ratio and loss of spontaneous caspase-3 processing in vivo. Leukemia 2000, 14:1606-1613.
78. Teitz T, Lahti JM, Kidd VJ. Aggressive childhood neuroblastomas do not express caspase-8: an important component of programmed cell death. J Mol Med 2001, 79:428-436.
79. Baumler C, Duan F, Onel K, Rapaport B, Jahnwar S, Offit K, Elkon KB. Differential recruitment of caspase 8 to cFlip confers sensitivity or resistance to Fas-mediated apoptosis in a subset of familial lymphoma patients. Leuk Res 2003, 27:841-851.
80. Linn SC, Honkoop AH, Hoekman K, van der Valk P, Pinedo HM, Giaccone G. p53 and P-glycoprotein are often co-expressed and are associated with poor prognosis in breast cancer. Br J Cancer 1996, 74:63-68.
81. Bowen WD, Jin B, Blann E, Vilner BJ, Lyn-Cook BD. Sigma receptor ligands modulate expression of the multidrug resistance gene in human and rodent brain tumor cell lines. Proc Am Assoc Cancer Res 1997, 38:479, #3206.

82. John CS, Bowen WD, Saga T, Kinuya S, Vilner BJ, Baumgold J, Paik CH, Reba RC, Neumann RD, Varma VM, McAfee JG. A malignant melanoma imaging agent: Synthesis, characterization, in vitro binding and biodistribution of iodine-125-(2-piperidinylaminoethyl)4-iodobenzamide. J Nucl Med 1993, 34:2169-2175.
83. John CS, Baumgold J, Vilner BJ, McAfee JG, Bowen WD. [^{125}I]N-(2-Piperidinylaminoethyl)4-iodobenzamide and related analogs as sigma receptor imaging agents: High affinity binding to human malignant melanoma and rat C6 glioma cell lines. J Labelled Compd Radiopharm 1994, 33:242-244.
84. John CS, Vilner BJ, Gulden ME, Efange SMN, Langason RB, Moody TW, Bowen WD. Synthesis and pharmacological characterization of 4-[^{125}I]BP: A high affinity sigma receptor ligand for potential imaging of breast cancer. Cancer Res 1995, 55:3022-3027.
85. Dence CS, John CS, Bowen WD, Welch MJ. Synthesis and evaluation of [^{18}F] labeled benzamides: High affinity sigma receptor ligands for PET imaging. Nucl Med Biol 1997, 24:333-340.
86. John CS, Gulden ME, Li JH, Bowen WD, McAfee JG, Thakur ML. Synthesis, in vitro binding, and tissue distribution of radioiodinated 2-[^{125}I]N-(N-benzylpiperidin-4-yl)-2-iodo benzamide, 2-[^{125}I]BP: A potential sigma receptor marker for human prostate tumors. Nucl Med Biol 1998, 25:189-194.
87. John CS, Bowen WD, Fisher SJ, Lim BB, Geyer BC, Vilner BJ, Wahl RL. Synthesis, in vitro pharmacologic characterization, and preclinical evaluation of N-[2-(1'-piperidinyl)ethyl]-3-[^{125}I]iodo-4-methoxybenzamide (P[^{125}I]MBA) for imaging breast cancer. Nucl Med Biol 1999, 26:377-382.
88. John CS, Vilner BJ, Geyer BC, Moody T, Bowen WD. Targeting sigma receptor-binding benzamides as in vivo diagnostic and therapeutic agents for human prostate tumors. Cancer Res 1999, 59:4578-4583.
89. John CS, Gulden ME, Vilner BJ, Bowen WD. Synthesis, in vitro validation and in vivo pharmacokinetics of [^{125}I]N-[2-(4-iodophenyl)ethyl]-N-methyl-2-(1-piperidinyl)ethylamine: A high affinity ligand for imaging sigma receptor positive tumors. Nucl Med Biol 1996, 23:761-766.
90. John CS, Lim BB, Vilner BJ, Bowen WD. Substituted N-(9-benzyl)-N-methyl-2-(1'-piperidinyl)ethylamine (BME) and its analogs as new sigma receptor markers: Synthesis, characterization, and in vivo evaluation. J Labelled Compd Radiopharm 1999, 42(Suppl 1):S411-S413.
91. Van Waarde A, Buursma AR, Hospers GA, Kawamura K., Kobayashi T, Ishii K, Oda K, Ishiwata K, Vaalburg W, Elsinga PH. Tumor imaging with two sigma receptor ligands, ^{18}F-FE-SA5845 and ^{11}C-SA4503: a feasibility study. J Nucl Med 2004, 45:1939-1945.
92. John CS, Lim BB, Vilner BJ, Geyer BC, Bowen WD. Substituted halogenated arylsulfonamides: A new class of sigma receptor binding tumor imaging agents. J Med Chem 1998, 41:2445-2450.
93. John CS, Lim BB, Geyer BC, Vilner BJ, Bowen WD. 99mTc-Labeled sigma-receptor-binding complex: Synthesis, characterization, and specific binding to human ductal breast carcinoma (T47D) cells. Bioconjugate Chem 1997, 8:304-309.
94. Caveliers V, Everaert H, John CS, Lahoutte T, Bossuyt A. Sigma receptor scintigraphy with N-[2-(1'-piperidinyl)ethyl]-3-(123)I-iodo-4-methoxybenzamide of patients with suspected primary breast cancer: first clinical results. J Nucl Med 2002, 43:1647-1649.
95. Michelot JM, Moreau MF, Veyre AJ, Bonafous JF, Bacin FJ, Madelmont JC, Bussiere F, Souteyrand PA, Mauclaire LP, Chossat FM, et al. Phase II scintigraphic clinical trial of malignant melanoma and metastases with iodine-123-N-(2-diethylaminoethyl) 4-iodobenzamide). J Nucl Med 1993, 34:1260-1266.

11. σ₂ receptors

96. Nicholl C, Mohammed A, Hull WE, Bubeck B, Eisenhut M. Pharmacokinetics of iodine-123-IMBA for melanoma imaging. J Nucl Med 1997, 38:127-133.
97. Mach RH, Huang Y, Buchheimer N, Kuhner R, Wu L, Morton TE, Wang L, Ehrenkaufer RL, Wallen CA, Wheeler KT. [^{18}F]N-(4'-fluorobenzyl)-4-(3-bromophenyl) acetamide for imaging the sigma receptor staus of tumors: comparison with [^{18}F]FDG and [^{125}I]IUDR. Nucl Med Biol 2001, 28:451-458.
98. Choi SR, Yang B, Plossl K, Chumpradit S, Wey SP, Acton PD, Wheeler K, Mach RH, Kung HF. Development of a Tc-99m labeled sigma-2 receptor-specific ligand as a potential breast tumor imaging agent. Nucl Med Biol 2001, 28:657-666.
99. Mach RH, Huang Y, Freeman RA, Wu L, Vangveravong S, Luedtke RR. Conformationally-flexible benzamide analogues as dopamine D$_3$ and sigma$_2$ receptor ligands. Bioorg Med Chem Lett 2004, 14:195-202.

Corresponding author: *Dr. Wayne D. Bowen, Mailing address: Department of Molecular Pharmacology, Physiology, & Biotechnology, Box G-B389, Division of Biology and Medicine, Brown University, Providence, RI 02912, USA, Phone: (401) 863-3253, Fax: (401) 863-1595, Electronic mail address: Wayne_Bowen@brown.edu*

Chapter 12

COGNITIVE EFFECTS OF σ RECEPTOR LIGANDS

Tangui Maurice
INSERM Unite 710, University of Montpellier II, c.c. 105, place Eugene Bataillon, 34095 Montpellier Cedex 5, France

1. INTRODUCTION

1.1 Historical perspective

The potential interest of σ receptor ligands in cognitive functions, and notably as anti-amnesic agents, arose from the pioneering article by Earley et al. (1). Based on the observation that selective σ receptor agonists potentiated the N-methyl-D-aspartate (NMDA)-induced neuronal activation of CA3 hippocampal pyramidal neurons, published one year before by Monnet et al. (2), and the importance of NMDA receptor activation in acquisition and learning processes, they tested the effects of systemic administration of a series of reference or selective σ ligands in a passive avoidance task in rats. Curiously, they chose the cholinergic muscarinic antagonist scopolamine, and not an NMDA receptor antagonist, to induce severe learning deficits in rats. Pre-administration of igmesine (JO 1784), (+)-3-(3-hydroxyphenyl)-N-(1-propyl)-piperidine [(+)-3-PPP] or 1,3-di-o-tolylguanidine (DTG) before scopolamine allowed a significant blockade of the drug-induced learning deficit. Rimcazole and (+)-SKF-10,047 (N-allylnormetazocine) were inefficient in their experiment. The authors suggested from these initial observations the potential interest of selective σ compounds like igmesine for the treatment of cognitive dysfunctions, notably associated with aging or dementia (1).

The present chapter will detail the progress obtained during the last ten years concerning the pharmacological potential of σ receptor ligands, interacting mainly with the σ_1 subtype, in learning and memory. We will show, in brief, how the corpus of experiments piled up over time continues to confirm Earley et al. in their conclusions. Even if, to our knowledge, very few compounds are presently in clinical trials for cognitive indications, numerous studies have confirmed and analyzed the behavioral efficacy of σ_1 receptor ligands as anti-amnesic agents in pharmacological and pathological models (for extensive reviews, see 3-6).

1.2 The σ_1 receptor/neuroactive steroid interaction and consequences in memory

One particular therapeutical direction, on which we will dwell, concerns the progressive cognitive impairments resulting from aging and/or dementia. Indeed, the interest of σ_1 receptors against age-related memory impairments has emerged from: (i) its role as one of the targets by which neuroactive steroids exert part of their rapid nongenomic effects; (ii) the well-known decrease of neuroactive steroid levels, in the brain as well as in the periphery, associated with aging or neurodegenerative diseases; and (iii) a series of experiments demonstrating that pharmacological manipulations of neuroactive steroid levels modulated the behavioral efficacy of σ_1 agonists (7,8).

Neuroactive steroids represent steroid hormones synthesized independently in the periphery or *de novo* in the brain, the latter being termed neurosteroids. They act locally to modulate nervous cell activity (9). Neuroactive steroids, i.e. both neuroactive circulating steroids and neurosteroids, modulate neuronal activity rapidly through nongenomic actions and affect learning and memory processes, mood, or depression. Pregnenolone, dehydroepiandrosterone (DHEA) and their sulfate esters are considered as excitatory steroids since they act as negative modulators of γ-aminobutyric type A ($GABA_A$) receptors (10,11) and positive modulators of NMDA receptors (12). Progesterone and its reduced metabolite allopregnanolone act as positive modulators of $GABA_A$ receptor (13,14) and negative modulators of NMDA receptors (15). In turn, neurosteroids play a particular role of efficient neuromodulators regulating the inhibitory/excitatory balance in the brain. In addition, several neuroactive steroids interact with the σ_1 receptor, with *in vitro* affinities for brain [^{3}H](+)-SKF-10,047-labeled σ_1 sites estimated around 300 nM, 1 μM and 3 μM for progesterone, DHEA and pregnenolone, respectively (16,17). Moreover, pregnenolone, DHEA and their sulfate esters also act on physiological tests and behavioral responses as σ_1 receptor agonists while progesterone is a potent σ_1 antagonist (5,6,18,19, see Chapter 1).

Systemic and intracerebral administration of the neuroactive steroids pregnenolone, DHEA, and their sulfate esters, enhanced memory in rodent pharmacological models of amnesia and even elicited pro-mnesic effects in control animals (for detailed review, see 5). These effects have been demonstrated using passive/active avoidance, "Go-No go" visual discrimination, water maze spatial learning or radial arm maze tests. The modulation of GABAergic and/or glutamatergic responses induced by the steroids may undoubtedly be involved in their mnesic effects. Particularly, in the brain structures involved in learning and memory processes —the cerebral cortex, amygdaloid complex or hippocampal formation—, cholinergic and glutamatergic inputs regulate brain activity. However, the interaction with the σ_1 receptor, because of the observed pharmacology, may also contribute to the mnesic effects of neuroactive steroids. In addition, since the endogenous levels of these neurohormones vary considerably with physiological conditions and age, significant consequences on the behavioral efficacy of σ_1 receptor activation are expected.

We will focus here on this σ_1 receptor/neuroactive steroid interaction in learning and memory processes in order to show the particular consequences that could be proposed in terms of therapeutic directions.

1.3 Behavioral models for assessing learning and memory

The study of cognitive functions are, in animals models, mainly restricted to learning and memory processes. Indeed, memory appears as the main physiological function required by animals to interact efficiently with their environment. New information must be acquired and subsequently retrieved to improve their abilities and optimize their behaviors in most of the encountered situations. Memory is a complex physiological process, involving successive and independent phases. Learning, or acquisition, refers to the contact with new environmental situations and thus to new information to acquire. Consolidation refers to physiological changes allowing acquired information to be converted into long-term stable memory traces, and thus converting short-term into long-term memory. Retrieval is involved when animals are again placed in the contextual situation and is highly dependent upon decisional behaviors including attention, motivation, and stimulus selection. Extinction occurs when a memory trace remains unrequested or when stimuli and environmental cues are repeatedly dissociated from the acquired information. Extinction, and putatively reactivation, is a process essential for the selectivity of new learned information, and thus essential for accurate memory discrimination. These different phases of the memory process involve physiological modifications and result in brain plasticity affecting differentially several neurotransmitter pathways and brain structures.

In animal models, different types of learning can be studied (20). Simple learning relies on perception to a particular stimulus that has been previously recognized and is particularly involved during exploration of a new environment. Simple learning could be assessed in alternation tests, for instance: spontaneous or delayed alternation procedures in the Y-maze or T-maze. Animals are free or constraint to explore the three arms of the maze in subsequent alternations and short-term working memory could be evaluated by either the percentage of alternations, the time spent in the novel arm, or the number of errors (21-24).

Associative learning refers to the association between a conditioning stimulus and a systematic behavioral response. Classical conditioning involves a conditioning stimulus followed by an unconditioning stimulus related to an aversive or appetitive response. Passive or active avoidance tests are routinely used to assess such memory. Animals are trained to avoid an aversive stimulus (eg. electric shocks or immersion in water) by learning the position of a safe location, i.e. the white compartment of a step-through type avoidance apparatus (1,25-27) or the wooden platform of a step-down type avoidance apparatus (21-24). Long-term memory is here examined by performing the retention session at least 24 h after the training session and evaluated by the step-through or step-down latency or the number of animals reaching an avoidance criterion. Operant conditioning refers to procedures where conditioned and unconditioned stimuli are associated with facilitation (reward) or diminution (punishment) of the probability of response. Long-term memory is here usually evaluated using the number of session-to-criterion or the evolution of the number of errors over training.

Complex learning is essentially evaluated in spatial learning tasks, where correct acquisition requires learning of the relationship among multiple stimuli and environmental cues in order to optimize the behavioral strategy. Two behavioral tests are routinely used, place learning in the Morris water maze (28-30) or in the eight radial arm maze (31,32). In the water maze task, animals are required to learn the location of a platform immersed below the surface of the water, rendered opaque with powdered milk or lime carbonate. Animals starting from changing departure positions must realize the platform location by using external cues located around the pool. Acquisition requires several training days with 2-4 swimmings per day and progress is evaluated among trials using swimming latency and distance traveled. Place learning involves working memory processes among swimmings each day and reference memory among days. Place learning is usually evaluated after the completion of acquisition in a probe trial, by removing the platform and measuring the place preference of animals in a single swimming. Working memory can also be more selectively evaluated by changing the platform location each day (see 29,30). In the radial arm maze, animals are trained to explore selectively only a few arms, e.g. 4 over 8, randomly chosen, to retrieve food pellets. Acquisition requires several training days to optimize the exploration strategy. Working memory errors (i.e. returns into a previously explored arm) and reference memory errors (i.e. entries into a never baited arm) can be easily determined (31,32). Alternatively, a repeated acquisition procedure can be used, in a three-panel runway apparatus (33). Animals must reach a goal box, baited with food, by passing through four panels, each containing three gates. The selection of possible, non-blocked, gates is chosen randomly. The task requires mainly correct working memory and performance is determined as number of errors. Correct performance in the runway task requires a win-stay strategy and not a win-shift strategy as used in the radial arm maze (33).

Learning and memory can thus be assessed using a large panel of complementary tests in rodents. Analysis of the different performance of animals submitted to either simple learning tasks, associated learning tasks, or complex spatial learning procedures allow us to determine the extent of impairments or the efficacy of pro-mnesic or anti-amnesic drugs. The involvement of σ_1 receptor ligands in learning and memory has been evaluated on tests measuring each kind of learning to demonstrate their potential pharmacological interest.

2. RATIONALE FOR THE INVOLVEMENT OF σ_1 RECEPTORS IN COGNITIVE RESPONSES

As detailed in other chapters of this book, the σ_1 receptor constitutes an unique orphan receptor. It could be considered both as a component of the intracellular trafficking system, and also as a potent neuromodulatory receptor. Indeed, its activation by synthetic and selective agonists, as well as by endogenous steroid hormones or peptides, results not only in its translocation between different organelles of the cell, but also in precise effects at the start and end points of the translocation. In particular, activation of the σ_1 receptor affects the functioning of several major neurotransmitter receptor systems and diverse kinds of ion channels on the plasma membrane that play important roles in the brain plasticity associated with learning and memory formation. We will here remember, in a short survey, the current evidence that σ_1 receptor ligands efficiently modulate the pharmacological systems involved in memory, namely: (i) cholinergic systems, (ii) NMDA-type of glutamatergic receptor, (iii) GABAergic receptors, and (iv) calcium (Ca^{2+}) fluxes (see Chapters 7-11, for details).

2.1 Modulation of cholinergic systems by σ_1 receptor agonists

Acetylcholine and its nicotinic or muscarinic receptors play an important role in memory processes (34). Groups of cholinergic basal forebrain neurons, originating from the nucleus basalis magnocellularis (NBM) and innervating the cerebral cortex, amygdaloid complex, or hippocampal formation, are involved in learning and memory formation (35,36). Patients suffering from Alzheimer's disease, or related dementia, elicit profound degeneration and dysfunction in cortical cholinergic activity, assumed to be partly responsible for the course of memory deficits observed in Alzheimer's disease (37,38). Moreover, administration of muscarinic or nicotinic receptor antagonists such as scopolamine and mecamylamine to control animals led to marked memory deficits, reminiscent of patients with Alzheimer's disease (39,40). Conversely, administration of cholinomimetics appears effective for treating memory impairments in rodents and humans. Acetylcholinesterase inhibitors, such as tacrine or physostigmine, exert some beneficial effects in several animal amnesia models (41-43) and in some Alzheimer's disease patients (44,45). In animal models, these anti-amnesic effects could be observed when drugs were administered either during the acquisition, consolidation, or retention phase of the memory process, indicating the involvement of cholinergic systems at the three different stages.

Cholinergic systems can be efficiently modulated by activation of the σ_1 receptor in the brain. A series of experiments demonstrated that the σ_1 receptor agonists (+)-SKF-10,047, DTG, or igmesine modulated acetylcholine release in different *in vitro* or *in vivo* models. First, the acetylcholine release from superfused slice preparations, from either rat striatum, hippocampus, or ileum longitudinal muscle/myenteric plexus, was modulated by selective σ_1 agonists (46-48). Second, in an *in vivo* microdialysis study, Matsuno et al. (49,50) showed that the σ_1 agonists DTG, (+)-pentazocine, (+)-3-PPP, (+)-SKF-10,047, or 1-(3,4-dimethoxyphenethyl)-4-(3-phenylpropyl)piperazine (SA4503) increased extracellular acetylcholine levels in the rat frontal cortex or hippocampus, but not in the striatum. In turn, σ_1 receptor agonists potentiated cholinergic systems, but differentially according to the brain structure. The lack of potentiation showed by several σ₁ drugs on striatal acetylcholine levels was understood as a lower ability to induce cholinomimetic side effects, an advantage over acetylcholinesterase inhibitors.

The σ_1 receptor agonists DTG, (+)-SKF-10,047 and (+)-pentazocine also inhibited the carbachol-induced phosphatidyl inositol turnover by modulation of muscarinic receptors in the brain. The drugs directly affected muscarinic receptors in the rat cortex, striatum and hippocampus, first, by inhibiting carbachol-induced [^{3}H]inositol 1,4,5-triphosphate (IP$_3$) accumulation and turnover, and, second, by blocking [^{3}H]L-3-quinuclidinyl benzylate binding to its muscarinic site (51,52). The σ_1 receptor ligands (+)-pentazocine, DTG, and haloperidol also inhibited the binding of [^{3}H]methylscopolamine to guinea pig cerebral cortical membranes (53).

Selective σ_1 agonists, by stimulating acetylcholine release or modulating the activation of cholinergic receptors, directly facilitate the activity of cholinergic systems and, in particular, allow the attenuation of disturbances of cholinergic neurotransmission after pharmacologic or pathologic deficits. At the behavioral level, selective σ_1 agonists are thus expected to ameliorate the memory impairments resulting from cholinergic dysfunction or induced by cholinergic blocking agents acting either at the acquisition, consolidation or retention phases.

2.2 NMDA receptor involvement in memory and its modulation by σ_1 agonists

The NMDA receptors mediate the induction of different forms of synaptic plasticity, such as long-term potentiation (LTP) or long-term depression (LTD). These forms of cerebral plasticity are considered to play a critical role in the stabilization/consolidation of synapses in particular brain structures, sustaining learning and memory processes (54,55). For example, induction of LTP in the CA1 hippocampal area involves activation of NMDA receptors and subsequent Ca^{2+} influx (56,57). Activation of NMDA receptors is in turn necessary to learning and memory encoding. In particular, recent studies using transgenic mice, with selective deletion of the NMDA receptor in the CA1 pyramidal cells of the hippocampus, demonstrated that acquisition and storage of spatial memory need NMDA-dependent strengthening of CA1 synapses (58). Competitive or noncompetitive NMDA antagonists delivered into the brain or administered systemically impair both hippocampal LTP and several kinds of learning processes, such as spatial learning in the water maze or nonspatial passive avoidance learning (28,59,60).

The modulation exerted by σ_1 agonists on several responses induced by NMDA receptor activation, particularly in the hippocampus, is now well documented, although the exact mechanism remains to be characterized (see Chapter 10). The σ_1 agonists modulate different responses to NMDA in a stimulatory or inhibitory manner, whereas σ_1 antagonists are devoid of effect alone, but block the agonist-induced modulation. Some differences appear, however, depending on the brain structure or the physiologic response involved. Low doses of σ_1 agonists enhanced, for instance, dose dependently but in a bell-shaped manner the NMDA-induced [^{3}H]norepinephrine release from rat hippocampal slices, the [^{3}H]dopamine release from rat striatum slices *in vitro*, and the excitatory response of CA3 pyramidal neurons in the rat hippocampus *in vivo* or in slice preparations from pyramidal cells of rat prefrontal cortex (2,61-63).

Monnet et al. (18) demonstrated the neurophysiological relevance of the interaction between neuroactive steroids and the σ_1 receptor using the σ_1 receptor-mediated modulation of NMDA responses. They observed that DHEA sulfate potentiated whereas pregnenolone sulfate inhibited the NMDA-induced [^{3}H]norepinephrine release from rat hippocampal slices. The σ_1 receptor antagonists haloperidol and 1[2-(3,4-dichlorophenyl)ethyl]-4-methylpiperazine (BD1063) prevented the effects of DHEA sulfate and pregnenolone sulfate, and conversely, the inhibitory effects of DTG and pregnenolone sulfate were blocked by progesterone (18). DHEA sulfate thus induced a σ_1-like potentiation of the NMDA-induced response, whereas pregnenolone sulfate exerts two opposite effects on the NMDA receptor: a direct potentiation, possibly by binding to the NMDA receptor complex, and an indirect, σ_1-like inhibition. Furthermore, the σ_1 antagonist activity of progesterone was supported by *in vivo* experiments showing that centrally administered progesterone, inactive on the NMDA-induced neuronal activation, counteracted the DTG-induced modulation of the responses of CA3 hippocampal pyramidal neurons to NMDA (19). DHEA potentiated the NMDA-evoked electric activity of CA3 pyramidal neurons, as did σ_1 receptor agonists, an effect that could be blocked by the σ_1 antagonists haloperidol, N,N-dipropyl-2-[4-methoxy-3-(2-phenylethoxy)phenyl]ethyl-amine (NE-100), and progesterone. However, neither pregnenolone nor its sulfate affected this NMDA-induced response *in vivo* (19). These studies indicated a clear crossed pharmacology between neuroactive steroids and σ_1 receptor ligands on the modulation of NMDA-dependent responses, although some discrepancies appeared depending on the model.

A preliminary report indicated that σ_1 agonists, including DTG, (+)-pentazocine, igmesine, and pregnenolone sulfate modulated LTP components, i.e. both the magnitude and slope of population spikes and field excitatory postsynaptic potentials in the rat hippocampus (64). This observation needs to be replicated, but it strengthened the hypothesis of a putative effect of σ_1 agonists on NMDA receptor-mediated learning and memory processes.

2.3 Modulation of GABAergic inhibitory responses by σ_1 agonists

GABAergic systems play an effective role of neuromodulatory systems in learning and memory processes. Indeed, GABAergic neurons influence the activity of cholinergic neurons, particularly in the hippocampus or NBM, and the mnemonic effects basal forebrain infusions of GABAergic agents, including $GABA_A$ agonists like muscimol, inverse agonists like β-carbolines, or benzodiazepines like diazepam, have been widely tested (65).

Very few studies addressed the effects of σ_1 ligands on GABAergic neurotransmission. The only reports showed that (+)-SKF-10,047 inhibited potassium-induced [^{3}H]GABA efflux in slices of the rat substantia nigra (66); dexoxadrol, a nonselective σ_1 agonist, potentiated GABA-induced inhibition in cerebellar Purkinje neurons of anesthetized rats (67), and DTG inhibited the firing rate of GABA-containing midbrain interneurons in anesthetized rats (68). These few reports tended to suggest that σ_1 receptor agonists, similarly as pregnenolone or DHEA, may behave as negative modulators of $GABA_A$ receptors. Indeed, numerous evidences showed that the neuroactive steroids pregnenolone and DHEA sulfates act as negative modulators of $GABA_A$ receptors. DHEA sulfate has been reported to antagonize the $GABA_A$ receptor by interacting with the barbiturate site, i.e. to decrease the potency of pentobarbital to potentiate [^{3}H]flunitrazepam binding, and to inhibit GABA-induced currents in neurons (69). DHEA sulfate also produced a concentration-dependent blockade of GABA-induced currents in cultured neurons from ventral mesencephalon (70). In addition, DHEA sulfate blocked the $GABA_A$ receptor in primary cultures of ventral midbrain neurons of fetal rats by accelerating desensitization and not by acting on the conductance of the chloride channel (71). DHEA sulfate thus acts as a noncompetitive modulator of the $GABA_A$ receptor. Pregnenolone sulfate also behaves as an endogenous allosteric antagonist of the $GABA_A$ receptor. The steroid, for instance, reversibly inhibited GABA-induced currents in rat isolated cerebral cortical neurons and inhibited GABA and glycine-mediated chloride currents in cultured chick spinal cord neurons (10,72).

Effects on GABAergic neurons may thus constitute a major pharmacological difference between neuroactive steroids and σ_1 receptor ligands. It is usually considered that σ_1 receptor ligands do not affect the GABAergic systems, and consequently, that the pharmacological profile of neuroactive steroids interacting with GABAergic receptors is unrelated from the σ_1 receptor component. Noteworthy, allopregnanolone, the most efficient positive modulator of $GABA_A$ receptors, is devoid of affinity for σ_1 receptors. However, more detailed studies are crucially needed to address this issue. Very recent data may contradict this conclusion. Mtchedlishvili and Kapur (73) reported that low concentrations of pregnenolone sulfate inhibited presynaptic GABA release and the resulting miniature inhibitory presynaptic currents in hippocampal neurons cultures. This effect was abolished by the σ_1 receptor antagonist BD1063, suggesting that σ_1 receptor activation modulates GABA release and neurotransmission. This effect deserves further studies using selective σ_1 ligands. Until convincing arguments become evident, we will here consider that σ_1 agonists fail to efficiently modulate GABAergic neurotransmission, contrarily to neuroactive steroids, and thus, that this component is absent from the σ_1 receptor-mediated effects on learning and memory.

2.4 Modulation of calcium homeostasis by activation of the σ_1 receptor

A major role of σ_1 receptors is related to regulation of Ca^{2+} mobilization. This was first forewarned through the intracellular localization of the receptor. Recent studies using a specific antibody showed that the σ_1 receptor is mainly present on the endoplasmic reticulum, and on the mitochondrial, nuclear, and plasma membranes (30,74-76). Second, different physiological experiments showed direct effects of σ_1 agonists on Ca^{2+} fluxes. The σ_1 agonists increased contractility, beating rate, and Ca^{2+} influx in cultured cardiac myocytes (77), and increased intracellular levels of IP_3 in cultured myocytes (78). In NG108 cells, σ_1 agonists such as (+)-pentazocine, 2-(4-morpholino)ethyl-1-phenylcyclo hexane-1-carboxylate (PRE-084), or pregnenolone sulfate, enhanced bradykinin-induced increases in $[Ca^{2+}]_i$, suggesting that the σ_1 receptor facilitated IP_3-induced Ca^{2+} release (75). (+)-Pentazocine or pregnenolone sulfate inhibited, whereas PRE-084 potentiated, the KCl-induced increase in $[Ca^{2+}]_i$ in a pertussis toxin sensitive manner, suggesting that σ_1 ligands affected the depolarization-induced $[Ca^{2+}]_i$ influx at the plasma membrane by a mechanism implying $G_{i/o}$ proteins (75). The σ_1 receptor thus act as a sensor/modulator of Ca^{2+} signaling, acting both at the endoplasmic reticulum by modulating intracellular IP_3 receptor-gated pools and at the plasma membrane by modulating extracellular Ca^{2+} entries through voltage-dependent Ca^{2+} channels (VDCC) and ionotropic neurotransmitter receptors (eg. NMDA, nicotinic receptors). Such effects may either constitute an unifying track to explain the wide-range neuromodulatory effects induced by selective σ_1 agonists or be an additional argument to suggest their involvement in learning and memory processes. Indeed, induction of LTP in hippocampal pyramidal neurons requires a rise in postsynaptic Ca^{2+} (79). The Ca^{2+} entry is mainly due to the activation of NMDA receptors. However, another source for Ca^{2+} could be VDCC; the dendrites of CA1 pyramidal neurons contain both N-type and L-type VDCC (80,81), and repetitive activation of L-type VDCC generates a long lasting enhancement that mimics LTP (82). The implication of Ca^{2+} fluxes in memory processes is suggested by several direct experiments. Acute administration of nifedipine or nimodipine, dihydropyridine VDCC blockers, impaired retention of a visual discrimination task in chicks, two-way active avoidance behavior in rats, or spontaneous alternation behavior, step-down passive avoidance and place learning in a water maze in mice (83-85).

3. PRO-MNESIC EFFECTS OF σ_1 LIGANDS

In most of the studies examining the anti-amnesic potential of σ_1 agonists in either pharmacological or pathological models of amnesia (summarized in Tables 12-1 and 12-2), a putative pro-mnesic effect was examined by injecting each drug alone to control animals. The literature thus reports numerous experiments using (+)-SKF-10,047, (+)-pentazocine, PRE-084, igmesine, SA4503, or DTG administered to animals submitted to short-term or long-term memory tests. In a convergent manner, none of these compounds, tested in large dose range, facilitated learning in control animals (see for instance, 22,33,86-88), indicating that activation of the σ_1 receptor failed to improve the quality of learning in memory tests. Conversely, σ_1 antagonists failed to impede learning ability in control animals. This was observed for α-(4-fluorophenyl)-4-(5-fluoro-2-pyrimidinyl)-1-piperazine-butanol (BMY 14,802), haloperidol, or NE-100. Blockade of the σ_1 receptor did not result in learning impairments. Interestingly, the down regulation of σ_1 receptor expression using *in vivo* antisense strategies (86,87) also failed to affect the learning ability of mice submitted to a passive avoidance test, confirming the lack of involvement of the receptor in normal learning processes. This is presently understood as a consequence of the precise neuromodulatory role of σ_1 receptors. As usually observed in most of the physiological or pharmacological tests used to evidence σ_1 receptor pharmacology, compounds are devoid of effect alone, but exert some action only when the transmission is perturbed. The absence of important promnesic effects of σ_1 agonists could thus represent the behavioral consequence of the strict modulatory role of σ_1 receptors.

Further experiments could however examine a putative direct mnesic effect of σ_1 agonists, particularly in cases of moderate learning. The previously reported results referred indeed to behavioral tests where learning level was already considerably high for control animals, since procedures were set to easily observe drug-induced impairments. In addition, most of the studies focused on acquisition of memory and σ_1 agonists could affect other phases of the mnesic process. For example, we performed a preliminary study examining the effect of PRE-084 in the different phases of passive avoidance learning in control Swiss mice (Figure 12-1). Animals were trained repeatedly to stay on a wooden platform, without stepping down on the gridfloor of the cage and thus avoiding electric shocks (Figure 12-1A). When PRE-084 was injected before acquisition, it failed to affect the number of session-to-criterion (Figure 12-1B) or the retention latency on Day 2 (Figure 12-1C). Similarly, PRE-084 was without effect when injected before retention (Figure 12-1C). Following retention, the avoidance response was extinguished by placing the animal again on the platform every day during 10 days. When it stepped down, no shock was applied. Avoidance was progressively extinguished but could be reactivated by subsequent reinstallation of the footshock on Day 12. When PRE-084 was administered every day during the 10 day extinction period, a significant facilitation of extinction was measured (Figure 12-1D). Furthermore, when the shock was reinstalled, PRE-084-treated animals showed a higher reactivation latency (Figure 12-1D). This experiment suggested that activation of the σ_1 receptor may facilitate the extinction/reactivation phase of memory in control animals and thus the discrimination with new learned information.

12. Cognitive effects 249

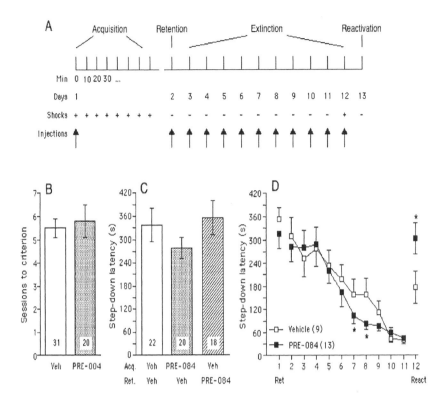

Figure 12-1. Effect of the selective σ_1 agonist PRE-084 on learning processes in control animals. (A) Experimental procedure. Naïve, male, Swiss mice (28-30 g) were submitted to a step-down type passive avoidance training, as previously described (24). Animals were placed on a wooden platform set at the center of a cage with a grid floor. When the animal stepped-down, it received an aversive, electric shock. Training was repeated every 10 min, until the animal remained for 60 s on the platform without stepping down (i.e. acquisition criterion). Retention (Ret) was checked after 24 h, by placing the mouse on the platform and measuring its step-down latency up to 600 s. The behavior was extinguished during the following 10 days by allowing the mouse to stay for 30 s on the grid floor without receiving footshock. After complete extinction, the passive avoidance was reactivated on Day 12 by reinstalling the footshock in a new session and the extent of reactivation (React) was measured after 24 h. PRE-084 (0.5 mg/kg, i.p.) was administered either before the first acquisition session on Day 1, or before the retention test on Day 2, or before each extinction session between Day 2 to 12. Reactivation on Day 13 was checked without drug injection. PRE-084 administered before the first training session failed to affect acquisition, in terms of number of sessions to criterion (B), or step-down latency during the retention session (C). PRE-084 administered before the retention session failed to affect retrieval (C). However, PRE-084 repeated treatment facilitated extinction, with significant differences on Days 7 and 8 (D). Reactivation was significantly enhanced for PRE-084-treated animals (D). The number of animals is indicated within the columns (B and C) and in the caption (D). *$P < 0.05$ vs. the vehicle-treated group (Student's t-test after a two-way ANOVA).

Table 12-1. Anti-amnesic properties of σ_1 receptor agonists and related drugs in pharmacological models of amnesia: interaction with cholinergic and glutamatergic systems

Amnesia Models	Species	σ_1 Agonists	Related Drugs	σ_1 Antagonists	Behavioral Tests	References
(A) Cholinergic Dysfunction Models						
Scopolamine	Rat	Igmesine (+)-3-PPP DTG			Passive avoidance	1
Scopolamine	Rat	(+)-SKF-10,047 SA4503			Passive avoidance	25
Scopolamine	Rat	OPC-14523			Passive avoidance	119
p-Chloroamphetamine	Mouse	(+)-SKF-10,047			Passive avoidance	25
		(+)-Pentazocine (+)-3-PPP DTG				26
Scopolamine	Mouse	(+)-SKF-10,047 (+)-3-PPP DTG		Haloperidol NE-100	Passive avoidance	89
Scopolamine	Mouse		DHEA sulfate Pregnenolone sulfate	NE-100 Progesterone	Spontaneous alternation Water maze spatial learning	93
Scopolamine	Mouse	SA4503		Antisense	Spontaneous alternation Passive avoidance	86

Table 12-1. Anti-amnesic properties of σ_1 receptor agonists and related drugs in pharmacological models of amnesia: interaction with cholinergic and glutamatergic systems (cont.)

Amnesia Models	Species	σ_1 Agonists	Related Drugs	σ_1 Antagonists	Behavioral Tests	References
(B) Glutamatergic Dysfunction Models						
Dizocilpine	Mouse	DTG		BMY 14802	Spontaneous alternation	21
		(+)-SKF-10,047		NE-100	Passive avoidance	22
		(+)-Pentazocine		Haloperidol	Elevated plus maze	24
		PRE-084		Antisense #1		86
				Antisense #2		87
Dizocilpine	Mouse		NPY	BMY-14802	Passive avoidance	94
			[Leu^{31}Pro^{34}]NPY			
			PYY			
			CGRP			
			[Cys(ACM)$^{2-7}$] hCGRFα			
Dizocilpine	Mouse	SA4503	DHEA sulfate	Haloperidol	Spontaneous alternation	88
L-NAME			Pregnenolone sulfate	Pregnenolone	Passive avoidance	98
				BMY 14802		87
				Antisense #2		
Dizocilpine	Rat	(+)-SKF-10,047			Three-panel runaway maze	33
Dizocilpine	Rat	(+)-SKF-10,047	DHEA sulfate	NE-100	Radial arm maze	31
Delay btw choices		SA4503	Pregnenolone sulfate	Progesterone		32
		BMY 14802				

4. AMNESIA RESULTING FROM BLOCKADE OF CHOLINERGIC SYSTEMS

As summarized in Table 12-1A, numerous selective σ_1 agonists were tested in amnesia models involving blockade of cholinergic systems. In line with the study of Earley et al. (1), several groups, and notably the group of Matsuno (25-27,89), showed that several σ_1 receptor ligands reversed the amnesia induced by scopolamine or by the serotonin depleter p-chloroamphetamine in mice and rats. The prototypic σ ligands (+)-SKF-10,047, (±)-pentazocine, DTG, (+)-3-PPP, the selective σ_1 agonists igmesine or SA4503, and the mixed σ_1/5-HT$_{1A}$ agonist 1-[3-[4-(3-chlorophenyl)-1-piperazinyl]propyl]-5-methoxy-3,4-dihydro-2-quinolinone monomethamesulfonate (OPC-14523) prevented the scopolamine-induced amnesia in the step-through type passive avoidance task (Table 12-1). Furthermore, the effect on p-chloroamphetamine-induced amnesia was also observed when the σ_1 agonists were administered before training or before retention, confirming that they could improve the cholinergic-dependent memory processes either during the acquisition, consolidation, or retention phase (26). In addition, we reported that DTG or PRE-084 attenuated the learning impairments induced by the nicotinic cholinergic antagonist mecamylamine in mice (22,24), in accordance with the described neuromodulatory effect exerted by σ_1 receptor ligands on nicotinic receptor activation. Due to the recent cloning of the protein and the availability of molecular biology tools, the implication of the σ_1 receptor in learning and memory processes could be firmly established. A phosphorothioate-modified antisense oligodeoxynucleotide, targeting the σ_1 receptor mRNA, was administered intracerebroventricularly during 3 days in mice. The anti-amnesic effects of SA4503, observed against the learning impairments induced by scopolamine, were blocked after σ_1 antisense oligodeoxynucleotide administration, but not after a saline or a control σ_1 mismatch oligodeoxynucleotide treatment. These observations clearly brought a molecular basis to the implication of the cloned σ_1 receptors in memory processes (86).

Exogenous administration of the neuroactive steroids pregnenolone and DHEA sulfate also reversed the amnesic effects of scopolamine in mice submitted to a footshock active avoidance test (90), a step-through type passive avoidance test (91), or a "Go-No go" visual discrimination task (92). We observed that peripheral administration of DHEA sulfate or pregnenolone sulfate, but not progesterone, counteracted the scopolamine-induced deficits of either spontaneous alternation behavior or place learning in a water maze in mice (93, Table 12-1A). The involvement of the σ_1 receptor in the anti-amnesic effect induced by the steroids was indicated by: (i) the blockade of both DHEA sulfate and pregnenolone sulfate effects by the selective σ_1 antagonist NE-100, and (ii) the observation that progesterone behaved as a clear antagonist of the steroid effects.

In conclusion, σ_1 receptor agonists exert a neurophysiological modulation of cholinergic systems that appeared particularly effective in cases of cholinergic deficits. At the behavioral level, σ_1 agonists constitute potent anti-amnesic drugs against amnesia resulting from cholinergic dysfunction. Activation of the σ_1 receptor has been pharmacologically demonstrated to be involved in the anti-amnesic effects induced by neuroactive steroids, although direct modulation of glutamatergic and putative GABAergic inputs on cholinergic neurons may also be involved.

5. LEARNING IMPAIRMENTS RESULTING FROM BLOCKADE OF NMDA RECEPTORS

Since selective σ_1 receptor agonists were shown to potentiate several NMDA receptor-mediated physiological responses, it was expected as a behavioral consequence, that they could produce either a facilitation of memory acquisition, or at least anti-amnesic effects against the deficits induced by NMDA receptor blockade. As summarized in Table 12-1B, synthetic σ_1 receptor agonists showed potent anti-amnesic effects against the impairments induced by pre-acquisition injection of the noncompetitive NMDA receptor antagonist dizocilpine. Numerous compounds were tested, including the prototypic σ_1 agonist (+)-SKF-10,047, (+)-pentazocine, DTG or more selective and efficient compounds like PRE-084 or SA4503. These effects were stereoselective, blocked by BMY 14802 or NE-100, after subchronic treatment with haloperidol, in accordance with binding studies, or using *in vivo* antisense strategies (86,87). Most of the studies by our group were performed using mice submitted to spontaneous alternation, passive avoidance or water maze tests, i.e. spatial or nonspatial tests assessing working or reference memory. In rats, Ohno and Watanabe (33) reported that intra-hippocampal administration of (+)-SKF-10,047, ineffective by itself, prevented the dizocilpine-induced increase in the number of working errors in a three-panel runway task in rats. Zou et al. (31,32) observed an ameliorating effect of (+)-SKF-10,047 or SA4503 against the dizocilpine-induced working memory deficits assessed using the radial arm maze task.

The interaction of neuropeptide Y (NPY) or calcitonin gene-related peptide (CGRP) was demonstrated *in vivo* using the dizocilpine-induced amnesia model (94, Table 12-1). Intracerebroventricular administration of NPY, [Leu^{31},Pro^{34}]NPY, peptide YY, CGRP, or [Cys(ACM)$^{2-7}$]hCGRPα attenuated the dizocilpine-induced passive avoidance deficits in Swiss mice. These effects were blocked by BMY 14,802 and NPY[13-36] and CGRP[8-37] were inefficient, in a consistent manner as the structure-activity relationship previously observed at σ_1 receptors (95-97). Thus, the anti-amnesic effect of NPY-related peptides, already described on several other amnesia models, may in part involve the controversial, but effective *in vivo*, interaction with σ_1 receptors.

The interaction of neuroactive steroids was also demonstrated using dizocilpine-induced amnesia models (Table 12-1). First, DHEA sulfate attenuated the dizocilpine-induced learning deficits. In mice, its anti-amnesic effects on spontaneous alternation or passive avoidance were blocked by preadministration of BMY 14,802 or using an *in vivo* antisense strategy (87,98). In rats, the working memory and reference memory errors induced by dizocilpine in the radial arm maze were attenuated by DHEA sulfate and these effects were blocked by the σ_1 antagonist NE-100 (32). Second, pregnenolone sulfate also induced anti-amnesic effects against the dizocilpine-induced deficits. However, although NE-100 allowed a blockade of the working and reference memory errors in the radial arm maze (32), the *in vivo* antisense strategy failed to affect the beneficial effect of pregnenolone in mice submitted to a passive avoidance test (87). This observation suggests that the steroid interacts more directly and efficiently with the GABA$_A$ and/or NMDA systems to mediate its behavioral effects. Third, progesterone behaved consistently as an antagonist. The anti-amnesic effects of SA4503 against the learning impairments induced in mice by dizocilpine or N-ω-nitro-L-arginine methyl ester (L-NAME), the nitric oxide synthase inhibitor, could be blocked by progesterone, in a similar manner to haloperidol (88). In rats, the anti-amnesic effects of SA4505, DHEA sulfate or pregnenolone sulfate in the radial arm maze were also blocked by progesterone in a similar manner as with NE-100 (32). In turn, exogenous administration of neuroactive steroids modulates the pharmacological amnesia induced by blockade of NMDA receptors. The crossed pharmacology studies as well as the relative activity of each steroid, i.e. DHEA sulfate and pregnenolone sulfate act as agonists while progesterone is an antagonist, confirmed the involvement of the σ_1 receptor in the cognitive effects of the steroids. A major consequence of these observations is that neurosteroid levels, i.e. the brain levels of steroids and their physiological variations, will directly affect either learning or memory processes, or the behavioral efficacy of selective σ_1 agonists. This latter point was in particular directly addressed in a study using adrenalectomy/castration and subsequent treatment with trilostane, a 3β-hydroxysteroid dehydrogenase inhibitor, and finasteride, a 5α-reductase inhibitor, in order to manipulate endogenous levels in neurosteroids (7). The *in vivo* binding parameters to σ_1 sites in the mouse brain and the σ_1 receptor-mediated anti-amnesic effects appeared markedly sensitive to endocrine conditions. The binding levels and intensity of the PRE-084 anti-amnesic effects against dizocilpine-induced deficits were increased after surgery and furthermore after the treatment with trilostane, and conversely diminished after the treatment with finasteride. The modulation of brain levels in progesterone appeared particularly coherent with such effects, confirming at a behavioral level, that neurosteroids are potent endogenous modulators of the σ_1 receptor (7).

6. LEARNING IMPAIRMENTS INDUCED BY OTHER PHARMACOLOGICAL AGENTS

6.1 Amnesic effect of the benzodiazepine diazepam

Administration of the typical benzodiazepine, diazepam, is known to impair both acquisition and retention of memory and to affect short-term memory (65). For instance, the drug blocked both spontaneous alternation behavior and passive avoidance learning in mice (22,65). Until now, only one study examined the putative anti-amnesic effect of a σ_1 agonist against the diazepam-induced learning deficits (22). The nonselective σ_1/σ_2 agonist DTG failed to affect the drug-induced impairment of alternation behavior. However, further studies with more selective σ_1 compounds are required to exclude an effect of σ_1 agonists on the behavioral, and particularly mnemonic, consequences of GABAergic hyperactivation.

6.2 Learning impairments induced by the L-type VDCC blocker nimodipine

Activation of σ_1 receptors results in modulation of Ca^{2+} mobilization from intracellular pools and effects on Ca^{2+} entries either through modulation of ionotropic neurotransmitter receptors or VDCC (75,76). A direct behavioral relevance for this hypothesized mechanism of action was investigated in an experiment examining the effect of PRE-084 against the learning impairments induced by acute administration of a low dose of nimodipine in mice. The last time σ_1 agonist alleviated the VDCC blocker-induced deficits in both short-term and long-term memory tests, namely the spontaneous alternation in the Y-maze, the step-down type passive avoidance test and place learning in a water maze (99). Therefore, in addition to the modulation of neurotransmitter responses, a direct effect on activation of VDCC may be part of the mechanism of action of selective σ_1 agonists in learning and memory processes.

7. ANTI-AMNESIC EFFECTS IN PATHOLOGICAL MODELS OF COGNITIVE DEFICITS

7.1 Hypoxic insults - σ_2 vs. σ_1 receptor involvement in memory

Couture and Debonnel (100) reported that, similarly to σ_1 receptor ligands, σ_2 agonists, such as 1'-[4-[1-(4-fluorophenyl)-1H-indol-3-yl]-1-butyl]spiro[iso benzofuran-1(3H),4'-piperidine (Lu 28-1798), N-[2-(3,4-dichlorophenyl)ethyl]-N-methyl-2-(1-pyrrolidinyl)ethylamine (BD1008) or ibogaine, are also able to potentiate the neuronal response to NMDA in the CA3 region of the rat dorsal hippocampus. These effects were insensitive to the σ_1 antagonists haloperidol, NE-100, and progesterone. Since the anti-amnesic effects of σ_1 receptor agonists could be considered as relevant to this cellular physiological mechanism, such results questioned the putative anti-amnesic effects of σ_2 agonists. As regards the different pharmacological models of amnesia, the major involvement of the σ_1 receptor is now widely accepted, mainly because several σ_1 antagonists efficiently blocked the anti-amnesic effects of the agonists (Table 12-1).

We extended the study of the σ pharmacology to two lesional models of amnesia, different from the previously described pharmacological models (101). First, the successive exposure of mice to carbon monoxide gas induces, after 5 to 7 days, delayed amnesia observable in the Y-maze and passive avoidance tests, and delayed neuronal death that remains restricted to the CA1 area of the hippocampal formation (23,102). Second, administration of the neurotoxicant trimethyltin produces in rats or mice important damage in selective neuronal populations from the limbic structures in the brain (103-105).

Table 12-2. Anti-amnesic properties of σ_1 receptor agonists and related drugs in other pharmacological models of amnesia and age- and Alzheimer's Disease-related models of cognitive deficits

Amnesia Models	Species	σ_1 Agonists	Related Drugs	σ_1 Antagonists	Behavioral Tests	References
(A) Other Pharmacological Amnesia Models						
Nimodipine	Mouse	PRE-084		BMY 14802	Spontaneous alternation	99
					Passive avoidance	
CDEP	Mouse	(+)-SKF-10,047			Passive avoidance	121
		(+)-3-PPP				
		DTG				
	Mouse		DHEA sulfate	Haloperidol	Passive avoidance	122
			Pregnenolone sulfate			
(B) Pathological Amnesia Models						
CO intoxication	Mouse	DTG		BMY 14802	Spontaneous alternation	23
Trimethyltin intoxication	Mouse	(+)-SKF-10,047		NE-100	Passive avoidance	101
		PRE-084				
		BD1008 (σ_2)				
		Haloperidol (σ_2)				
β_{25-35}-Amyloid peptide	Mouse	(+)-Pentazocine	DHEA sulfate	BMY 14802	Spontaneous alternation	110
		PRE-084	DHEA	Haloperidol	Passive avoidance	86
		SA4503	Pregnenolone	Progesterone		
			Pregnenolone sulfate			

Table 12-2. Anti-amnesic properties of σ_1 receptor agonists and related drugs in other pharmacological models of amnesia and age- and Alzheimer's Disease-related models of cognitive deficits (cont.)

Amnesia Models	Species	σ_1 Agonists	Related Drugs	σ_1 Antagonists	Behavioral Tests	References
(B) Pathological Amnesia Models (cont.)						
Basal forebrain lesion	Rat	SA4305			Passive avoidance	108
					Water maze spatial learning	109
(C) Aging-Related Deficit Models						
SAM	Mouse	PRE-084		BMY 14802	Spontaneous alternation	118
		Igmesine			Passive avoidance	
Aged (C57BL/6, 24 m.o.)	Mouse	PRE-084			Water maze spatial learning	30
Aged (Wistar, 24 m.o.)	Rat	PRE-084			Water maze spatial learning	29
Aged (Fisher 344, 22 m.o.)	Rat	OPC-14523			Conditioned active avoidance	119
					Water maze spatial learning	

CO = carbon monoxide; m.o. = month old; SAM = senescence accelerated mice

The selective σ_1 ligand PRE-084 or the nonselective σ_1/σ_2 compounds DTG, BD1008, and haloperidol reversed the spontaneous alternation deficits observed 7 days after exposure to carbon monoxide or 14 days after intoxication with trimethyltin. The selective σ_1 antagonist NE-100 was ineffective by itself, but blocked completely the PRE-084 effect, partially the DTG effect, and did not affect the effect induced by BD1008 or haloperidol. A similar pharmacological profile was observed in the step-down type passive avoidance test performed 8 days after exposure to carbon monoxide. In contrast to the previously reported amnesia models, the impairments induced after exposure to carbon monoxide or intoxication with trimethyltin could be alleviated not only by σ_1 receptor agonists but also by σ_2 agonists. The exact nature of this σ_2 site remains to be clarified. Matsumoto et al. (106) showed that it is involved in motor and posture control. Our results reveal that it may also be involved in cognitive functions. It remains to determine why the σ_2 site ligands show such anti-amnesic effect in these particular models as compared to pharmacological amnesia models. The particular pattern of neurodegeneration, almost similar between carbon monoxide exposure and trimethyltin intoxication, may be involved.

7.2 Nontransgenic Alzheimer's disease-related amnesia models

Based on the observation that selective σ_1 agonists allowed to potently attenuated scopolamine-induced amnesia (Table 12-1), usually considered as a primitive pharmacological model of Alzheimer's disease, several authors extended the study to other nontransgenic models of Alzheimer's disease - related cognitive deficits (Table 12-2). First, lesion of the basal forebrain area is considered to mimic degeneration of the nucleus basalis of Meynert in humans, a common physiopathological feature observed in post-mortem Alzheimer's disease patient brains (38). Second, central administration of the aggregated form of β[25-35] peptide fragment of the β-amyloid protein into the mouse or rat brain induced histological and biochemical changes and behavioral deficits (43). These changes appeared reminiscent of the deposition of amyloid plaques, endogenously constituted in human from β[1-40] and β[1-42]amyloid proteins, but not β[25-35], which extent correlates with the progressive cognitive deficits and memory impairments observed in Alzheimer's disease patients (107).

Senda et al. (108,109) reported that the passive avoidance impairments or place learning deficits in the water maze, observed after basal forebrain lesions in rats, could be attenuated by SA4503. The σ_1 agonist was as efficient as the cholinesterase inhibitor tacrine, but its effect did not involve modulation of choline acetyltransferase or cholinesterase activity notably in the frontal cortex, a structure where enzyme activities were decreased in lesioned rats (109). The anti-amnesic potencies of σ_1 agonists and neuroactive steroids were also evaluated in mice centrally administered with the β[25-35]amyloid related peptide (110). The σ_1 agonists (+)-pentazocine, PRE-084, or SA4503 attenuated, in a dose dependent but bell-shaped manner, the deficits of spontaneous alternation and step-down passive avoidance observed in β[25-35]-treated mice (110). In parallel, DHEA, pregnenolone and their sulfate esters, also reduced the β[25-35]-induced deficits. Progesterone behaved as an antagonist, blocking the effects of both active steroids and σ_1 agonists. Conversely, haloperidol blocked the effects of active steroids, confirming the crossed pharmacology between σ_1 ligands and neuroactivesteroids (110). It appears thus that σ_1 agonists may be efficient in improving the cognitive deficits observed in Alzheimer's disease, although no study presently addressed the efficacy of σ_1 agonists against the memory deficits observed in transgenic rodent models of Alzheimer's disease (111).

Noteworthy, a recent study examined the behavioral despair response of β[25-35]amyloid peptide-treated animals (112). The antidepressant effect of the σ_1 agonists igmesine or PRE-084 was increased in β[25-35]-treated animals. The tricyclic antidepressant desipramine reduced the immobility duration similarly among groups and the selective serotonin reuptake inhibitor fluoxetine appeared less potent in β[25-35]-treated animals. Since β[25-35]-treated animals exhibited decreased progesterone levels in the hippocampus (-47%), the enhanced efficacy of σ_1 agonists could be explained by a decreased level of the steroid acting as endogenous antagonist at σ_1 receptors. In turn, σ_1 agonists appear as a promising therapeutic issue in Alzheimer's disease, that may allow alleviation of not only the memory deficits, but also the depressive symptoms frequently associated with Alzheimer's disease (111).

8. AGING-RELATED COGNITIVE DEFICITS

Aging, even in non-pathological conditions, leads to a progressive decrease of efficacy of central neuromodulatory mechanisms, resulting in loss of brain plasticity and neurotransmission deficits. Present strategies aimed at attenuating the extent of age-related cognitive deficits focus mainly on sustaining residual neurotransmitter functions. Acting through the σ_1 receptor, and thus exerting a wide-range facilitation of the residual neurotransmission activity simultaneously on different systems may also offer a unique way to maintain effective cognitive functions during aging.

Age-associated learning impairments result from multifactorial origins. First, both central cholinergic and glutamatergic dysfunctions were evidenced. Second, the neuromodulatory tonus exerted by different neuroendocrine systems is affected. Third, disturbances in intraneuronal Ca^{2+} homeostasis were observed, due to direct dysfunction of Ca^{2+} mobilization from intracellular pools, of VDCC activation, or as a consequence of ionotropic neurotransmitter receptors dysfunctions. Thus, σ_1 agonists, because of their previously described efficacy in the cholinergic or glutamatergic pharmacological models of amnesia, were expected to be efficient in aging-related models. In addition, supplementation in neuroactive steroids, and notably in pregnenolone or DHEA sulfate, was reported to improve learning and memory in aged rodents. For instance, post-training administration of DHEA sulfate improved retention in an active avoidance test in aged (24-month old) C57BL/6 mice (113). Vallée et al. (114) showed that individual differences in learning ability could be observed among aged Sprague Dawley rats submitted to a delayed alternation test in the Y-maze or place learning in the water maze. These differences correlated with pregnenolone sulfate level variations. In addition, systemic or central administration of pregnenolone restored correct learning ability in aged animals by stimulating acetylcholine release in the hippocampus (114). The involvement of σ_1 receptor activation in these effects of neuroactive steroids was not examined. However, it was of interest to determine whether selective σ_1 agonists share this beneficial effect against age-related cognitive deficits.

The first parameter to check was the expression of σ_1 receptors during aging. In the aged C57BL/6 mouse brain, σ_1 receptor mRNA quantification, σ_1 protein immunohistolabeling and σ_1 receptor binding measures showed a remarkable preservation of the σ_1 receptor expression (30). The anatomical distribution was carefully examined. The brain structures involved in cognitive functions, notably the hippocampus, cortex, basal forebrain, and hypothalamus, showed high levels of σ_1 receptors, highly comparable to the expression measured in young mice (30). The receptor ontogeny has not yet been published, but immunohistochemical observations from our laboratory revealed that it is already expressed at the embryonic stage (N König, unpublished result). The anatomical distribution observed in adult is found similar to that observed at postnatal day 15 (VL Phan, unpublished result). The σ_1 receptor expression is thus remarkably stable during the course of late development, adulthood, and aging. Recent studies addressed selectively the σ_1 receptor binding levels in aged rodents. Wallace et al. (115) reported a mild but significant increase of haloperidol-sensitive [^{3}H]DTG binding sites in the striatum of 24-month old Fisher 344 rats, a result recently extended by Ishikawa et al. (116). In the same animal, these authors reported increased binding levels of σ_1 sites labeled with either [^{3}H]SA4503, [^{3}H](+)-pentazocine, or [^{3}H]DTG in whole brain homogenates. In addition, the *in vivo* positron emission tomography binding parameters of [^{11}C]SA4503 to σ_1 receptors were evaluated in the monkey brain during aging (117). Although the quantity of radiolabeled tracer crossing the blood-brain barrier and reaching the brain was diminished in all structures in the aged monkey brain, the binding ability (estimated in terms of ratio between association and dissociation rate) was increased in all structures examined. These coherent results confirmed that the σ_1 receptor bioavailability during aging is at least preserved if not increased. These observations strengthened the validity of targeting the σ_1 receptor during aging, and confirmed its lack of involvement in the etiology of cognitive dysfunctions (i.e. although its expression is preserved with age, cognitive deficits progressively develop).

At the behavioral level (see Table 12-2), several studies addressed the efficacy of σ_1 agonists to reverse the memory deficits in aged mice (30,118) and rats (29,119). One study addressed the anti-amnesic efficacy of igmesine and PRE-084 in a particular model of aging, the senescence-accelerated SAM mouse (118); others used classical models of aged mice, 24-month old C57BL/6 (30), or 24-month old Wistar, or 22-month old Fisher 344 rats (29,119). Animals were submitted to passive avoidance, conditioned active avoidance, or place learning in the water maze and treatment with the selective σ_1 agonist PRE-084 or the mixed σ_1/5-HT$_{1A}$ agonist OPC-14523 restored correct learning abilities in aged rodents (Table 12-2).

The σ_1 receptor is not only highly expressed in the aged brain, but allows marked pro-mnesic effects, contrarily to what was observed in younger animals (3). The mechanism of this efficiency remains to be determined, but likely involved stimulation of several cellular processes (Ca^{2+} homeostasis and mobilization, ionotropic receptor-mediated responses, restoration of the modulatory actions exerted by neurohormones, intracellular trafficking, etc.) that are significantly affected during aging.

9. PERSPECTIVES FOR THE USE OF σ_1 LIGANDS IN COGNITIVE INDICATIONS

As seen, although activation of the σ_1 receptor seems poorly involved in mnemonic functions in control animals, selective σ_1 receptor agonists showed an efficient anti-amnesic ability in numerous animal models of amnesia. This property was observed not only in pharmacological models of amnesia, deficits being induced by systemic administration of selective neurotransmitter receptor antagonists, but also in lesional and pathological models of learning impairments. The neurophysiological mode of action underlying such apparently nonselective behavioral effects is currently under extensive study. Recent studies demonstrated that the σ_1 receptor regulates several neuronal functions like intracellular Ca^{2+} mobilization, Ca^{2+} channel activity, neurotransmitter release and neuronal firing. In terms of behavioral relevance, σ_1 receptors were reported to affect not only learning and memory processes, but also mood, depression, response to stress, psychotomimetic states, and even acquisition of drug addiction. Several σ_1 agonists also present interesting neuroprotective activities. In turn, convergent amounts of data argue in favor of a major role of σ_1 receptors in neuronal plasticity. It remains thus intriguing that acute or chronic administration of σ_1 ligands, agonists as well as antagonists, failed to induce any effect in control animals. In learning and memory tests, similarly as reported on numerous physiological tests, σ_1 receptors act as pure neuromodulatory receptors, devoid of effect in control conditions but highly effective as soon as a pharmacological or pathological unbalanced state is induced. The σ_1 receptor thus appears as a very promising target, allowing to finely restore behavioral functions but devoid of undesirable side effects.

Major unresolved questions remain, such as the identification of the endogenous σ_1 receptor ligand or the determination of the physiological regulation of σ_1 receptor expression and function. In that matter, the validity and physiologically relevant interaction between neuroactive steroids and σ_1 receptors is particularly important. Pharmacological and neuroendocrine studies demonstrated that at the behavioral level, DHEA and at a lesser extent pregnenolone, as well as their sulfate esters, exert part of their rapid nongenomic effect through their σ_1 receptor agonist activity. Conversely, progesterone acts endogenously as a potent σ_1 antagonist. As a consequence, it was observed that endogenous progesterone levels directly affected the behavioral efficacy of synthetic σ_1 agonists. In cases of physiological or pathological diminutions of steroid levels in the brain, σ_1 agonists became particularly effective (7). Indeed, recent studies showed, for instance, the efficacy of σ_1 agonists against age-related learning deficits. In senescence-accelerated (SAM) mice or aged mice or rats, σ_1 agonists showed pro-mnesic ability (29,30,118,119). Brain levels of steroids, and particularly progesterone, are significantly decreased during aging: -65% in 18-month old senescent SAMP/8 as compared to age-matched controls SAMR/1; -55% in 24-month old C57BL/6 mice as compared to 2-month old mice (30,120). A current hypothesis suggests that neurosteroids could represent the endogenous ligand system for σ_1 receptors. This idea raises, however, several concerns. First, among steroids, progesterone shows the highest affinity and seems to present the most important influence on σ_1 receptor activity. The steroid behaves, however, as an antagonist contrarily to what is expected for a prototypic endogenous ligand. Due to the intense variations in progesterone, little could be understood concerning the role of other steroids, notably acting as agonists, i.e. DHEA, pregnenolone and their sulfate esters. Second, steroid syntheses vary importantly according to age and physiological conditions, such as stress or pregnancy. What has been described so far concerning the σ_1 receptor expression suggests that, on the contrary, the receptor is particularly stable, notably during development and throughout aging. Third, steroid hormone tonus is considered as a major factor of interindividual differences and sensitivity to particular endocrine and physiological responses. For instance, glucocorticoids have a major role in individual vulnerability to drug addiction. A comparative study between different mouse strains showed indeed marked genetic differences in brain neurosteroid levels, resulting in marked differences in the behavioral effects of selective σ_1 agonists, notably in learning and memory tests (121). However, the σ_1 expression, examined at the mRNA, protein, and receptor binding level, was unchanged among mouse strains, evidencing a discrepancy between the expression of the endogenous effector and expected target. It is highly speculative from the data obtained until now to suggest that the σ_1 receptor may play a role in individual vulnerability to cognitive disorders, but its role as a target for neuroactive steroids undoubtedly will direct clinical practice towards cognitive deficits associated with hormonal depletion, as seen for neurodegenerative diseases or during normal or pathological aging.

In conclusion, selective σ_1 receptor agonists are particularly interesting compounds for alleviating learning and memory dysfunctions. Pharmacological and pathological animal models showed that σ_1 agonists are effective after systemic administration. Precise localization of the effect among brain structures involved in learning and memory processes remains however to be performed. Several synthetic σ_1 agonists show the efficacy and innocuity allowing their use in clinical trials. These compounds are now expected to give promising results in therapeutical indications, where active drugs are still crucially needed.

ACKNOWLEDGEMENTS

The author thanks Dr. S. Gaillet and P. Romieu (Montpellier) for helpful comments on the manuscript. The work of our team was supported by INSERM, CNRS and Pfizer GRD (Fresnes).

REFERENCES

1. Earley B, Burke M, Leonard BE, Gouret CJ, Junien JL. Evidence for an anti-amnesic effect of JO1784 in the rat: a potent and selective ligand for the σ receptor. Brain Res 1991, 546:282-286.
2. Monnet FP, Debonnel G, Junien JL, De Montigny C. N-methyl-D-aspartate-induced neuronal activation is selectively modulated by σ receptors. Eur J Pharmacol 1990, 179:441-445.
3. Maurice T, Lockhart BP. Neuroprotective and anti-amnesic potentials of sigma (σ) receptor ligands. Prog Neuro-Psychopharmacol Biol Psychiatry 1997, 21:69-102.
4. Matsuno K, Mita S. SA4503: A novel σ_1 receptor agonist. CNS Drug Rev 1998, 4:1-24.
5. Maurice T, Phan VL, Urani A, Kamei H, Noda Y, Nabeshima T. Neuroactive neurosteroids as endogenous effectors for the sigma-1 (σ_1) receptor: Pharmacological evidence and therapeutic opportunities. Jpn J Pharmacol 1999, 81:125-155.
6. Maurice T, Urani A, Phan VL, Romieu P. The interaction between neuroactive steroids and the sigma-1 (σ_1) receptor function: Behavioral consequences and therapeutic opportunities. Brain Res Rev 2001, 37:116-132.
7. Phan VL, Su TP, Privat A, Maurice T. Modulation of steroidal levels by adrenalectomy/castration and inhibition of neurosteroid synthesis enzymes affect the σ_1 receptor-mediated behaviour in mice. Eur J Neurosci 1999, 11:2385-2396.
8. Urani A, Roman FJ, Phan VL, Su TP, Maurice T. The antidepressant-like effect induced by sigma$_1$ (σ_1) receptor agonists and neuroactive steroids in mice submitted to the forced swimming test. J Pharmacol Exp Ther 2001, 298:1269-1279.
9. Baulieu EE. Steroid hormones in the brain: Several mechanisms ? In *Steroid Hormone Regulation of the Brain*. Fuxe K, Gustafson JA, Wettenberg L., eds. Oxford: Pergamon Press, 1981, pp 3-14.
10. Majewska MD, Mienville JM, Vicini S. Neurosteroid pregnenolone sulfate antagonizes electrophysiological responses to GABA in neurons. Neurosci Lett 1988, 90:279-284.
11. Majewska MD, Demirgören S, Spivak CE, London ED. The neurosteroid dehydroepiandrosterone sulfate is an allosteric antagonist of the GABA$_A$ receptor. Brain Res 1990, 526:143-146.
12. Wu FS, Gibbs TT, Farb DH. Pregnenolone sulfate: A positive allosteric modulator at the N-methyl-D-aspartate receptor. Mol Pharmacol 1991, 40:333-336.
13. Schumacher M, McEwen BS. Steroid and barbiturate modulation of the GABA$_A$ receptor. Mol Neurobiol 1989, 3:275-280.
14. Wu FS, Gibbs TT, Farb DH. Inverse modulation of gamma-aminobutyric acid- and glycine-induced currents by progesterone. Mol Pharmacol 1990, 37:597-602.
15. Smith SS. Progesterone administration attenuates excitatory amino acid responses of cerebellar Purkinje cells. Neuroscience 1991, 42:309-320.

16. Su TP, London ED, Jaffe JH. Steroid binding at sigma receptors suggests a link between endocrine, nervous, and immune systems. Science 1988, 240: 219-221.
17. Maurice T, Roman FJ, Privat A. Modulation by neurosteroids of the *in vivo* (+)-[^{3}H]SKF-10,047 binding to σ_1 receptors in the mouse forebrain. J Neurosci Res 1996, 46:734-743.
18. Monnet FP, Mahé V, Robel P, Baulieu EE. Neurosteroids, via σ receptors, modulate the [^{3}H] norepinephrine release evoked by N-methyl-D-aspartate in the rat hippocampus. Proc Natl Acad Sci USA 1995, 92: 3774-3778.
19. Bergeron R, de Montigny C, Debonnel G. Potentiation of neuronal NMDA response induced by dehydroepiandrosterone and its suppression by progesterone: effects mediated via sigma receptors. J Neurosci 1996, 16:1193-1202.
20. Vallée M, Mayo W, Koob GF, Le Moal M. Neurosteroids in learning and memory processes. Int Rev Neurobiol 2001, 46:273-320.
21. Maurice T, Hiramatsu M, Itoh J, Kameyama T, Hasegawa T, Nabeshima T. Low dose of 1,3-di(2-tolyl)guanidine (DTG) attenuates MK-801-induced spatial working memory impairment in mice. Psychopharmacology 1994, 114:520-522.
22. Maurice T, Hiramatsu M, Itoh J, Kameyama T, Hasegawa T, Nabeshima T. Behavioral evidence for a modulating role of σ ligands in memory processes. I. Attenuation of dizocilpine (MK-801)-induced amnesia. Brain Res 1994, 647:44-56.
23. Maurice T, Hiramatsu M, Kameyama T, Hasegawa T, Nabeshima T. Behavioral evidence for a modulating role of σ ligands in memory processes. II. Reversion of carbon monoxide-induced amnesia. Brain Res 1994, 647:57-64.
24. Maurice T, Su TP, Parish DW, Nabeshima T, Privat A. PRE-084, a σ selective PCP derivative, attenuates MK-801-induced impairment of learning in mice. Pharmacol Biochem Behav 1994, 49:859-869.
25. Matsuno K, Senda T, Matsunaga K, Mita S, Kaneto H. Similar ameliorating effects of benzomorphans and 5-HT$_2$ antagonists on drug-induced impairment of passive avoidance response in mice: Comparison with acetylcholinesterase inhibitors. Psychopharmacology 1993, 112:134-141.
26. Matsuno K, Senda T, Matsunaga K, Mita S. Ameliorating effects of σ receptor ligands on the impairment of passive avoidance tasks in mice: Involvement in the central acetylcholinergic system. Eur J Pharmacol 1994, 261:43-51.
27. Matsuno K, Senda T, Kobayashi T, Okamoto K, Nakata K, Mita S. SA4503, a novel cognitive enhancer, with σ_1 receptor agonistic properties. Behav Brain Res 1997, 83:221-224.
28. Morris RG, Anderson E, Lynch GS, Baudry M. Selective impairment of learning and blockade of long-term potentiation by an N-methyl-D-aspartate receptor antagonist, AP5. Nature 1986, 319:774-776.
29. Maurice T. Beneficial effect of the σ_1 receptor agonist PRE-084 against spatial learning deficits in aged rats. Eur J Pharmacol 2001, 431:223-227.
30. Phan VL, Urani A, Sandillon F, Privat A, Maurice T. Preserved sigma-1 (σ_1) receptor expression and behavioral efficacy in the aged C57BL/6 mouse. Neurobiol Aging 2003, 24:865-881.
31. Zou LB, Yamada K, Nabeshima T. Sigma receptor ligands (+)-SKF10,047 and SA4503 improve dizocilpine- induced spatial memory deficits in rats. Eur J Pharmacol 1998, 355:1-10.
32. Zou LB, Yamada K, Sasa M, Nakata Y, Nabeshima T. Effect of σ_1 receptor agonist SA4503 and neuroactive steroids on performance in a radial arm maze task in rats. Neuropharmacology 2000, 39:1617-1627.

33. Ohno M, Watanabe S. Intrahippocampal administration of (+)-SKF 10,047, a σ ligand, reverses MK-801-induced impairment of working memory in rats. Brain Res 1995, 684:237-242.
34. Björklund A, Dunnett SB. Cognitive function. Acetylcholine revisited. Nature 1995, 375:446.
35. Aigner TG. Pharmacology of memory: Cholinergic-glutamatergic interactions. Curr Opin Neurobiol 1995, 5:155-160.
36. Gallagher M, Colombo PJ. Ageing: The cholinergic hypothesis of cognitive decline. Curr Opin Neurobiol 1995, 5:161-168.
37. Davies P, Maloney AJ. Selective loss of central cholinergic neurons in Alzheimer's disease. Lancet 1976, 2:1403.
38. Bartus RT, Dean RL, Beer B, Lippa AS. The cholinergic hypothesis of geriatric memory dysfunction. Science 1982, 217:408-414.
39. Glick SD, Zimmerberg B. Amnesic effects of scopolamine. Behav Biol 1972; 7:245-54.
40. Smith CM, Swash M. Possible biochemical basis of memory disorder in Alzheimer disease. Ann Neurol 1978, 3:471-473.
41. McGaugh JL. Drug facilitation of learning and memory. Ann Rev Pharmacol 1973, 13:229-241.
42. Dokla CP, Parker SC, Thal LJ. Tetrahydroaminoacridine facilitates passive avoidance learning in rats with nucleus basalis magnocellularis lesions. Neuropharmacology 1989, 28:1279-1282.
43. Maurice T, Lockhart BP, Privat A. Amnesia induced in mice by centrally administered beta-amyloid peptides involves cholinergic dysfunction. Brain Res 1996, 706:181-193.
44. Summers WK, Majovski LV, Marsh GM, Tachiki K, Kling A. Oral tetrahydroaminoacridine in long-term treatment of senile dementia, Alzheimer type. N Engl J Med 1986, 315:1241-1245.
45. Farlow M, Gracon SI, Hershey LA, Lewis KW, Sadowsky CH, Dolan-Ureno J. A controlled trial of tacrine in Alzheimer's disease. The Tacrine Study Group. J Am Med Assoc 1992, 268:2523-2529.
46. Leventer SM, Johnson KM. Phencyclidine-induced inhibition of striatal acetylcholine release: comparisons with mu, kappa, and sigma opiate agonists. Life Sci 1984, 34:793-801.
47. Cambell BG, Keana JF, Weber E. Sigma receptor ligand N,N'-di-(ortho-tolyl)guanidine inhibits release of acetylcholine in the guinea pig ileum. Eur J Pharmacol 1991, 205:219-223.
48. Junien JL, Roman FJ, Brunelle G, Pascaud X. JO1784, a novel sigma ligand, potentiates [^{3}H]acetylcholine release from rat hippocampal slices. Eur J Pharmacol 1991, 200:343-345.
49. Matsuno K, Matsunaga K, Mita S. Increase of extracellular acetylcholine level in rat frontal cortex induced by (+)N-allylnormetazocine as measured by brain microdialysis. Brain Res 1992, 575:315-319.
50. Matsuno K, Matsunaga K, Senda T, Mita S. Increase in extracellular acetylcholine level by σ ligands in rat frontal cortex. J Pharmacol Exp Ther 1993, 265:851-859.
51. Bowen WD, Kirschner BN, Newman AH, Rice KC. Sigma receptors negatively modulate agonist-stimulated phosphoinositide metabolism in rat brain. Eur J Pharmacol 1988, 149:399-400.
52. Candura SM, Coccini T, Manzo L, Costa LG. Interaction of σ-compounds with receptor-stimulated phosphoinositide metabolism in the rat brain. J Neurochem 1990, 55:1741-1748.

53. Vargas HM, Pechnick RN. Binding affinity and antimuscarinic activity of sigma and phencyclidine receptor ligands. Eur J Pharmacol 1991, 195:151-156.
54. Izquierdo I. Role of NMDA receptors in memory. Trends Pharmacol Sci 1991; 12:128-9.
55. McEntee WJ, Crook TH. Glutamate: Its role in learning, memory, and the aging brain. Psychopharmacology 1993, 111:391-401.
56. Bliss TV, Collingridge GL. A synaptic model of memory: Long-term potentiation in the hippocampus. Nature 1993, 361:31-39.
57. Hawkins RD, Kandel ER, Siegelbaum SA. Learning to modulate transmitter release: Themes and variations in synaptic plasticity. Ann Rev Neurosci 1993, 16:625-665.
58. Tsien JZ, Huerta PT, Tonegawa S. The essential role of hippocampal CA1 NMDA receptor-dependent synaptic plasticity in spatial memory. Cell 1996, 87:1327-1338.
59. Parada-Turska J, Turski WA. Excitatory amino-acid antagonists and memory: Effect of drugs acting at N-methyl-D-aspartate receptors in learning and memory tasks. Neuropharmacology 1990, 29:1111-1116.
60. Venable N, Kelly PH. Effects of NMDA receptor antagonists on passive avoidance learning and retrieval in rats and mice. Psychopharmacology 1990, 100:215-221.
61. Monnet FP, Blier P, Debonnel G, de Montigny C. Modulation by σ ligands of N-methyl-D-aspartate-induced [^{3}H]noradrenaline release in the rat hippocampus: G-protein dependency. Naunyn Schmiedeberg Arch Pharmacol 1992, 346:32-39.
62. Gonzalez-Alvear GM, Werling LL. Regulation of [^{3}H]dopamine release from rat striatal slices by σ receptor ligands. J Pharmacol Exp Ther 1994, 271:212-219.
63. Liang X, Wang RY. Biphasic modulatory action of the selective σ receptor ligand SR31742A on N-methyl-D-aspartate-induced neuronal responses in the frontal cortex. Brain Res 1998, 807:208-213.
64. Monnet FP. Sigma receptors and intracellular signaling: Impact on synaptic plasticity? XXIII CINP Meeting Abstr. 2002, S.29.3
65. Venault P, Charpouthier G, Prado de Carvalho L, Simiand J, Morre M, Dodd RH, Rossier J. Benzodiazepine impairs and beta-carboline enhances performance in learning and memory tasks. Nature 1986, 321:864-866.
66. Starr MS. Multiple opiate receptors may be involved in suppressing gamma-aminobutyrate release in substantia nigra. Life Sci 1985, 37:2249-2255.
67. Wang Y, Lee HK. Facilitation of gamma-aminobutyric acid-induced depression by (+)PCMP and dexoxadrol in the cerebellar Purkinje neurons of the rat. Neuropharmacology 1989, 28:343-350.
68. Zhang J, Chiodo LA, Freeman AS. Effects of phencyclidine, MK-801 and 1,3-di(2-tolyl) guanidine on non-dopaminergic midbrain neurons. Eur J Pharmacol 1993, 230:371-374.
69. Majewska MD. Neuronal actions of dehydroepiandrosterone. Possible roles in brain development, aging, memory, and affect. Ann NY Acad Sci 1995, 774:111-120.
70. Demirgören S, Majewska MD, Spivak CE, London ED. Receptor binding and electrophysiological effects of dehydroepiandrosterone sulfate, an antagonist of the GABA$_A$ receptor. Neuroscience 1991, 45:127-135.
71. Spivak CE. Desensitization and noncompetitive blockade of GABA$_A$ receptors in ventral midbrain neurons by a neurosteroid dehydroepiandrosterone sulfate. Synapse 1994, 16:113-122.
72. Wu FS, Chen SC, Tsai JJ. Competitive inhibition of the glycine-induced current by pregnenolone sulfate in cultured chick spinal cord neurons. Brain Res 1997, 750:318-320.
73. Mtchedlishvili Z, Kapur J. A presynaptic action of the neurosteroid pregnenolone sulfate on GABAergic synaptic transmission. Mol Pharmacol 2003, 64:857-864.

74. Alonso G, Phan VL, Guillemain I, Saunier M, Legrand A, Anoal M, Maurice T. Immunocytochemical localization of the sigma$_1$ (σ_1) receptor in the adult rat central nervous system. Neuroscience 2000, 97:155-170.
75. Hayashi T, Maurice T, Su TP. Ca^{2+} signaling via σ_1-receptors: Novel regulatory mechanism affecting intracellular Ca^{2+} concentration. J Pharmacol Exp Ther 2000, 293:788-798.
76. Hayashi T, Su TP. Regulating ankyrin dynamics: roles of σ_1 receptors. Proc Natl Acad Sci USA 2001, 98:491-496.
77. Ela C, Barg J, Vogel Z, Hasin Y, Eilam Y. Sigma receptor ligands modulate contractility, Ca^{2+} influx and beating rate in cultured cardiac myocytes. J Pharmacol Exp Ther 1994, 269:1300-1309.
78. Novakova M, Ela C, Bowen WD, Hasin Y, Eilam Y. Highly selective sigma receptor ligands elevate inositol 1,4,5-trisphosphate production in rat cardiac myocytes. Eur J Pharmacol 1998, 353:315-317.
79. Nicoll RA, Kauer JA, Malenka RC. The current excitement in long-term potentiation. Neuron 1988, 1:97-103.
80. Jones OT, Kunze DL, Angelides KJ. Localization and mobility of ω-conotoxin-sensitive Ca^{2+} channels in hippocampal CA1 neurons. Science 1989, 244:1189-1193.
81. Westenbroek RE, Ahlijanian MK, Catterall WA. Clustering of L-type Ca^{2+} channels at the base of major dendrites in hippocampal pyramidal neurons. Nature 1990, 347:281-284.
82. Grover LM, Teyler TJ. Two components of long-term potentiation induced by different patterns of afferent activation. Nature 1990, 347:477-479.
83. Deyo RA, Nix DA, Parker TW. Nifedipine blocks retention of a visual discrimination task in chicks. Behav Neural Biol 1992, 57:260-262.
84. Nikolaev E, Kaczmarek L. Disruption of two-way active avoidance behavior by nimodipine. Pharmacol Biochem Behav 1994, 47:757-759.
85. Maurice T, Bayle J, Privat A. Learning impairment following acute administration of the calcium channel antagonist nimodipine in mice. Behav Pharmacol 1995, 6:167-175.
86. Maurice T, Phan VL, Privat A. The anti-amnesic effects of sigma$_1$ (σ_1) receptor agonists confirmed by in vivo antisense strategy in the mouse. Brain Res 2001a, 898:113-121.
87. Maurice T, Phan VL, Urani A, Guillemain I. Differential involvement of the sigma$_1$ (σ_1) receptor in the anti-amnesic effect of neuroactive steroids, as demonstrated using an in vivo antisense oligodeoxynucleotide strategy in the mouse. Br J Pharmacol 2001, 134:1731-1741.
88. Maurice T, Privat A. SA4503, a novel cognitive enhancer with σ_1 receptor agonist properties, facilitates NMDA receptor-dependent learning in mice. Eur J Pharmacol 1997, 328:9-18.
89. Senda T, Matsuno K, Kobayashi T, Mita S. Reduction of the scopolamine-induced impairment of passive avoidance performance by σ receptor agonist in mice. Physiol Behav 1997, 61:257-264.
90. Flood JF, Smith GE, Roberts E. Dehydroepiandrosterone and its sulphate enhance memory retention in mice. Brain Res 1988, 447:269-278.
91. Li PK, Rhodes ME, Jagannathan S, Johnson DA. Reversal of scopolamine-induced amnesia in rats by the steroid sulfatase inhibitor estrone-3-O-sulfamate. Cog Brain Res 1995, 2:251-254.
92. Meziane H, Mathis C, Paul SM, Ungerer A. The neurosteroid pregnenolone sulfate reduces learning deficits induced by scopolamine and has promnesic effects in mice performing an appetitive learning task. Psychopharmacology 1996, 126:323-330.

93. Urani A, Privat A, Maurice T. The modulation by neurosteroids of the scopolamine-induced learning impairment in mice involves an interaction with sigma-1 (σ_1) receptors. Brain Res 1998, 799:64-77.
94. Bouchard P, Maurice T, St-Pierre S, Privat A, Quirion R. Neuropeptide Y and calcitonin gene-related peptide attenuate learning impairments induced by MK-801 via a σ receptor-related mechanism. Eur J Neurosci 1997, 9:2142-2151.
95. Bouchard P, Dumont Y, Fournier A, St-Pierre S, Quirion, R. Evidence for in vivo interactions between neuropeptide Y-related peptides and σ receptors in the mouse hippocampal formation. J Neurosci 1993, 13:3926-3931.
96. Bouchard P, Monnet F, Bergeron R, Roman F, Junien JL, de Montigny C, Debonnel G, Quirion R. In vivo modulation of σ receptor sites by calcitonin gene-related peptide in the mouse and rat hippocampal formation: Radioligand binding and electrophysiological studies. Eur J Neurosci 1995, 7:1952-1962.
97. Bouchard P, Roman F, Junien JL, Quirion R. Autoradiographic evidence for the modulation of in vivo σ receptor labelling by neuropeptide Y and calcitonin gene-related peptide in the mouse brain. J Pharmacol Exp Ther 1996, 276:223-230.
98. Maurice T, Junien JL, Privat A. Dehydroepiandrosterone sulfate attenuates dizocilpine induced learning impairment in mice via σ_1-receptors. Behav Brain Res 1997, 83:159-164.
99. Maurice T, Su TP, Parish DW, Privat A. Prevention of nimodipine-induced impairment of learning by the selective σ ligand PRE-084. J Neural Transm 1995, 102:1-18.
100. Couture S, Debonnel G. Modulation of the neuronal response to N-methyl-D-aspartate by selective σ_1 ligands. Synapse 1998, 29:62-71.
101. Maurice T, Phan VL, Noda Y, Yamada K, Privat A, Nabeshima T. The attenuation of learning impairments induced after exposure to CO or trimethyltin in mice by sigma (sigma) receptor involves both σ_1 and σ_2 sites. Br J Pharmacol 1999a, 127:335-342.
102. Nabeshima K, Katoh A, Ishimaru I, Yoneda Y, Ogita K, Murase K, Ohtsuka H, Inari K, Fukuta T, Kameyama T. Carbon monoxide-induced delayed amnesia, delayed neuronal death and change in acetylcholine concentration in mice. J Pharmacol Exp Ther 1991, 256:378-384.
103. Brown AW, Aldridge WN, Street BW, Verschoyle RD. The behavioral and neuropathologic sequelae of intoxication by trimethyltin compounds in the rat. Am J Pathol 1979, 97:59-82.
104. Chang LW, Dyer RS. A time-course study of trimethyltin induced neuropathology in rats. Neurobehav Toxicol Teratol 1983, 5:443-459.
105. Hagan JJ, Jansen JHM, Broekkamp CLE. Selective behavioural impairment after acute intoxication with trimethyltin (TMT) in rats. Neurotoxicology 1988, 9:53-74.
106. Matsumoto RR, Hemstreet MK, Lai NL, Thurkauf A, de Costa BR, Rice KC, Hellewell SB, Bowen WD, Walker JM. Drug specificity of pharmacological dystonia. Pharmacol Biochem Behav 1990, 36:151-155.
107. Ishikawa K, Kobayashi T, Kawamura K, Matsuno K. Age-related changes of the binding of [^{3}H]SA4503 to σ_1 receptors in the rat brain. Ann Nucl Med 2003, 17:73-77.
108. Selkoe DJ. The molecular pathology of Alzheimer's disease. Neuron 1991, 6:487-498.
109. Senda T, Matsuno K, Okamoto K, Kobayashi T, Nakata K, Mita S. Ameliorating effect of SA4503, a novel σ_1 receptor agonist, on memory impairments induced by cholinergic dysfunction in rats. Eur J Pharmacol 1996, 315:1-10.
110. Senda T, Matsuno K, Kobayashi T, Nakazawa N, Nakata K, Mita S. Ameliorative effect of SA4503, a novel cognitive enhancer, on the basal forebrain lesion-induced impairment of the spatial learning performance in rats. Pharmacol Biochem Behav 1998, 59:129-134.

111. Maurice T, Su TP, Privat A. Sigma$_1$ (σ_1) receptor agonists and neurosteroids attenuate β-25-35-amyloid peptide-induced amnesia in mice through a common mechanism. Neuroscience 1998, 83:413-428.
112. Maurice T. Improving Alzheimer's disease-related cognitive deficits with sigma$_1$ (σ_1) receptor agonists. Drugs News Perspect 2002, 15:617-625.
113. Urani A, Romieu P, Roman FJ, Maurice T. Enhanced antidepressant effect of sigma$_1$ (σ_1) receptor agonists in β-25-35-amyloid peptide-treated mice. Behav Brain Res 2002, 134:239-247.
114. Flood JF, Roberts E. Dehydroepiandrosterone sulfate improves memory in aging mice. Brain Res 1988, 448:178-183.
115. Vallée M, Mayo W, Darnaudéry M, Corpéchot C, Young J, Koehl M, Le Moal M. Baulieu EE, Robel P, Simon H. Neurosteroids: Deficient cognitive performance in aged rats depends on low pregnenolone sulfate levels in the hippocampus. Proc Natl Acad Sci USA 1997, 94:14865-14870.
116. Wallace DR, Mactutus CF, Booze RM. Sigma binding sites identified by [^{3}H]DTG are elevated in aged Fisher-344 x Brown Norway (F1) rats. Synapse 2000, 35:311-3.
117. Kawamura K, Kimura Y, Tsukada H, Kobayashi T, Nishiyama S, Kakiuchi T, Ohba H, Harada N, Matsuno K, Ishii K, Ishiwata K. An increase of σ_1 receptors in the aged monkey brain. Neurobiol Aging 2003, 24:745-752.
118. Maurice T, Roman FJ, Su TP, Privat A. Beneficial effects of σ agonists on the age-related learning impairment in senescence-accelerated mouse (SAM). Brain Res 1996, 733:219-230.
119. Tottori K, Nakai M, Uwahodo Y, Miwa T, Yamada S, Oshiro Y, Kikuchi T, Altar A. Attenuation of scopolamine-induced and age-associated memory impairments by the σ and 5-hydroxytryptamine-1A receptor agonist OPC-14523 (1-(3[4-(3-chlorophenyl)-1-piperazinyl]propyl)-5-methoxy-3,4-dihydro-2[1*H*]-quinolinone monomethanesulfonate). J Pharmacol Exp Ther 2002, 301:249-257.
120. Phan VL, Miyamoto Y, Nabeshima T, Maurice T. Age-related expression and antidepressant efficacy of sigma1 (σ1) receptors in the senescence-accelerated (SAM) mouse. J Neurosci Res 2005, 79:561-572.
121. Phan VL, Urani A, Romieu P, Maurice T. Strain differences in σ_1 receptor-mediated behaviours are related to neurosteroid levels. Eur J Neurosci 2002, 15:1423-1434.
122. Matsuno K, Senda T, Kobayashi T, Murai M, Mita S. Reduction of 4-cyclohexyl-1-[(1R)-1,2-diphenylethyl]-piperazine-induced memory impairment of passive avoidance performance by σ_1 receptor agonists in mice. Meth Find Exp Clin Pharmacol 1998, 20:575-580.
123. Reddy DS, Kulkarni SK. The effects of neurosteroids on acquisition and retention of a modified passive-avoidance learning task in mice. Brain Res 1998, 791:108-116.

Corresponding author: Dr. Tangui Maurice, Mailing address: University of Montpellier II, INSERM Unite 710, c.c. 105, place Eugène Bataillon, 3409 Montpellier cedex 5, France, Phone: 33/0 4 67 14 36 23, Fax : 33/0 4 67 14 33 86, Electronic mail address: maurice@univ-montp2.fr

Chapter 13

σ RECEPTORS AND SCHIZOPHRENIA

Xavier Guitart
Prous Science, Provenza, 388, Barcelona 08025, Spain

1. INTRODUCTION

Schizophrenia is a severe psychiatric disorder that is characterized by both disturbed form and content of thought. Schizophrenic patients suffer from delusions, hallucinations, and social withdrawal, and are therefore unable to fully participate in society. The disease affects almost 1% of the population, and only one third of these patients have a good outcome. In addition to its devastating effects on schizophrenic individuals and their families, the disease represents a major economic issue, costing millions of dollars annually. The cause of schizophrenia remains unknown.

The "dopamine hypothesis" of schizophrenia was first proposed many years ago. This hypothesis was originally based on several lines of evidence suggesting that the neurotransmitter dopamine plays an important role in the disease, including:

a) symptoms similar to those of schizophrenia may be induced by either dopamine agonists or by certain stimulants, such as amphetamine, which cause large amounts of dopamine to be released (1);

b) there is a correlation between the *in vitro* dopamine receptor binding affinity of antipsychotic drugs and their clinical efficacy (2,3); and

c) post-mortem studies showed that brain tissue from schizophrenic patients exhibited increased numbers of dopamine D_2 receptors as compared to non-schizophrenics, although it is currently accepted that this increase reflects a response to chronic antipsychotic treatment (4).

Based on findings obtained in the last decade, it seems clear that the original dopamine hypothesis needs some revision. Our knowledge about dopamine neurotransmitter systems has increased considerably over this period. Furthermore, novel antipsychotic drugs have been developed, with fewer effects on the dopaminergic system. Finally, certain symptoms of schizophrenia, primarily the so-called "negative symptoms" (blunted affect, anhedonia, and attentional impairment, among others) and the cognitive deficits, are less responsive to treatment with dopamine antagonists than are the positive symptoms of the illness (hallucinations, delusions, disorganized speech and behavior) (5). This suggests that some of the central symptoms of schizophrenia may be less related or unrelated to an increase in dopaminergic function.

2. PATHOPHYSIOLOGY OF SCHIZOPHRENIA

Although the dopamine hypothesis cannot be supplanted at this time, other hypotheses suggest that serotonin and glutamate may play a significant role in the pathophysiology of schizophrenia (6-10). Moreover, σ receptors (and the existence of the σ_1 and σ_2 subtypes) have generated a great deal of interest because of their possible roles in various behaviors, including depression, anxiety, learning processes, and psychosis (see other chapters in this book for a detailed review of σ receptors). The potential involvement of σ receptors in the pathophysiology of schizophrenia was originally proposed when it was discovered that the first synthetic σ ligands exhibited psychotomimetic properties and because several neuroleptic drugs have high affinity for σ receptors.

It is important to note that there is an intricate relationship between σ receptors and the dopamine and glutamate neurotransmitter systems. The role of N-methyl-D-aspartate (NMDA) in the pathogenesis of schizophrenia has generated considerable interest. Phencyclidine (PCP), which both binds to σ receptors and blocks NMDA receptor channels with different affinity, can induce a schizophrenia-like psychosis in humans. Several postmortem studies have shown that schizophrenic brains exhibit abnormalities in glutamate levels and glutamate receptor densities (11-14). Moreover, certain biochemical and behavioral studies have suggested that antipsychotics may alter the function of NMDA receptors, possibly via σ receptors. Among the various effects attributed to σ_1 receptors, it is of interest to note that these receptors may regulate cognitive processes by modulating the activity of glutamatergic transmission at the NMDA receptor complex (15-17). In a recent study, using 4-IBP (N-(N-benzylpiperidin-4-yl)-4-iodobenzamide), it has been suggested the existence of two different pathways (mediated by two different σ_1 receptor subtypes) modulating glutamatergic neurotransmission in the hippocampus (18). Interestingly, σ_1 receptors are abundant in both dopamine-rich brain regions, such as different areas of the striatum, and in glutamate-rich brain regions, such as the hippocampus. The mesocortical, mesolimbic, and nigrostriatal dopaminergic systems, which have been implicated in the pathophysiology of schizophrenia, receive large glutamatergic imputs. Therefore, the role of σ_1 receptors in the regulation of dopaminergic activity could play a role in determining whether these sites modulate the glutamate sensitivity of dopaminergic neurons, via interaction with NMDA receptors (19). Currently, most clinicians believe that schizophrenia results from a complicated imbalance that includes several neurotransmitters and their receptors, and that these systems can modulate and interact with one another.

3. TYPICAL AND ATYPICAL ANTIPSYCHOTICS

Given the close association of dopamine with schizophrenia, it is not surprising that the first medications used to treat this disease were based on their interactions with dopaminergic neurotransmission. These antipsychotic medications (primarily drugs with dopamine D_2 antagonistic activity) markedly reduced the positive symptoms of the illness, but were much less helpful for the negative symptoms. Unfortunately, the first generation of antipsychotic drugs, known as typical antipsychotics, also exhibited several disturbing side effects, mainly related to involuntary movements (20), such as akathisia, akinesia, rigidity, tremor, dystonia, and tardive dyskinesia. In an attempt to avoid these deleterious side effects, as well as to treat the negative symptoms of schizophrenia, a second generation of drugs, designated as atypical antipsychotics, was eventually developed. The atypical antipsychotics act on other neurotransmitter systems in addition to the dopaminergic system, and they are currently first-line choices for treatment. These drugs can improve both the positive and negative symptoms of schizophrenia, and are less likely to induce undesirable movement disorders.

Since the development of second generation (atypical) antipsychotics, other novel pharmacological strategies for developing novel antipsychotic agents have arisen. It has been suggested that serotonergic neurotransmission may also be altered in schizophrenia (6,7), and antagonism of the serotonin 5-HT$_{2A}$ receptor seems to improve the manifestation of negative symptoms in schizophrenic patients. On the other hand, several atypical antipsychotics have relatively potent affinities for the serotonin 5-HT$_{2C}$ receptor (21). It has also been proposed that serotonin 5-HT$_6$ and 5-HT$_7$ receptors may play a role in schizophrenia (22). As discussed previously, glutamate has also been implicated in schizophrenia, and a decrease in NMDA-mediated synaptic transmission could contribute to certain schizophrenic symptoms. Several antipsychotics can modulate NMDA receptor channel function, and it has recently been demonstrated that selective agonists of the metabotropic glutamate receptor (group II) have potent antipsychotic effects in animal models that mimic schizophrenia (23). However, side effects remain problematic for these compounds. Thus, the development of new antipsychotic drugs remains a hot field of research in neuropharmacology (see 24-27 for recent reviews).

4. POTENTIAL TREATMENT FOR SCHIZOPHRENIA

σ Receptors have been extensively reviewed in other chapters of this book. In brief, the characteristics of the 223-amino acid σ_1 type receptor are quite unique; this protein is unrelated to any other known mammalian protein (28-29). While a σ_2 type receptor has also been described, it has not yet been cloned, and therefore much less is known about its molecular structure. In addition to their high affinity for dopamine D_2 receptors, many typical antipsychotics also act on σ receptors (30,31). This has led to speculation that σ receptors could also play a role in the antipsychotic actions of these drugs (32). However, no correlation between the affinity of neuroleptics for σ receptors and their therapeutic efficacy has been found. Thus, it has been suggested that σ receptors may instead mediate the motor side effects of antipsychotic drugs (33,34). Recently, the relationship between the ability of neuroleptics to interact with σ_1 and σ_2 receptors and their tendency to induce dystonic reactions in humans was evaluated (35). The findings suggested that the motor side effects induced by neuroleptics could be mediated through both σ_1 and σ_2 receptors. Recently, a σ_1 receptor knockout mouse has been produced (36). The study of these animals could help to clarify the role of σ_1 receptors in different pathophysiological processes (including schizophrenia), where these receptors have been implicated.

σ Receptors are currently considered to be a compelling target for the development of atypical antipsychotics (37-39). Certain novel compounds that exhibit antipsychotic activity in preclinical experiments appear to have σ binding affinity (40-45). σ Receptors have been found in cortical and limbic structures in human postmortem brain (46), and selective loss of σ sites has been reported in schizophrenia (47). However, studies of polymorphisms of the σ_1 receptor gene in schizophrenia or drug-induced psychosis have yielded contradictory results (48-52). Although no selective

13. Schizophrenia

σ ligand has yet been marketed as an antipsychotic drug, the putative antipsychotic activity of new σ receptor ligands has been frequently described (45,53,54). Figure 13-1 depicts the chemical structure of certain σ_1 receptor ligands that have been described as antipsychotics or putative antipsychotic drugs. It is interesting to note that there is considerable heterogeneity of the chemical structures among the different compounds. Table 13-1 shows the binding affinity of some of these molecules for the σ_1 receptor.

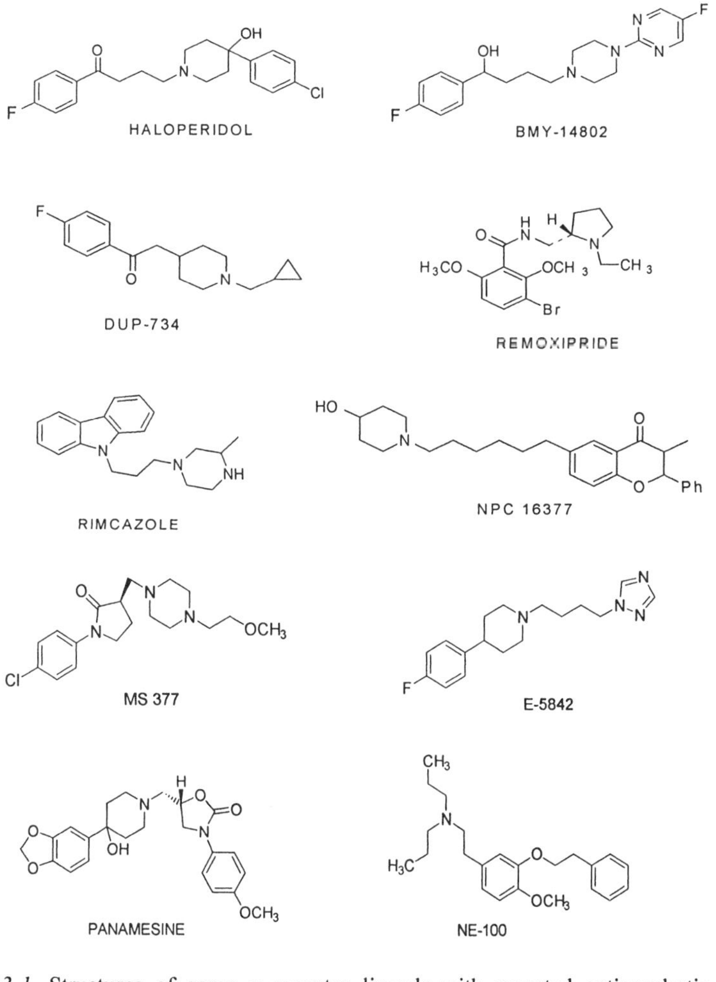

Figure 13-1. Structures of some σ receptor ligands with reported antipsychotic activity. Structures correspond to the compounds in Table 13-1. Note the diversity of chemical structural classes.

Table 13-1. Binding affinity for the σ_1 receptor of some selected compounds with clinical or preclinical activity as antipsychotics

Compound	σ_1 receptor affinity (IC$_{50}$ in nM)	Reference
Haloperidol	2.3	45
BMY 14,802	237	55
Remoxipride	60	56
Rimcazole	500	57
NPC 163777	36	41
MS-377	75 (K_i)	54
E-5842	5	44
Panamesine	6	58
NE-100	1.6	45

4.1 σ Receptors and schizophrenia

It is widely accepted that the therapeutic actions of antipsychotic drugs are mediated within mesolimbic and mesocortical dopaminergic projections of the brain. Dopaminergic cell bodies are mainly located in the midbrain and the diencephalon. Three major nuclei in the brain contain dopaminergic neurons: the ventral tegmental area (also known as A10), the pars compacta of the substantia nigra (also known as A9), and the arcuate nucleus of the hypothalamus. The neurons of the substantia nigra project mainly to the neostriatum, and this is thought to be the biological substrate for the extrapyramidal side effects produced by typical antipsychotics. It has already been stated below the possible relationship between σ_1 and σ_2 receptors and the appearance of motor side effects after treatment with neuroleptics. The neurons of the ventral tegmental area project primarily to limbic and cerebral cortical structures (eg. the nucleus accumbens, prefrontal cortex, and cingulate cortex).

It has been proposed that antipsychotic drugs produce their antipsychotic effects by antagonizing the actions of dopamine in this neuronal system. However, this hypothesis remains unproven, and it is unlikely that dopamine receptor antagonism is the sole mechanism by which prolonged antipsychotic treatment induces adaptations in the brain. Many atypical antipsychotics are thought to act through other receptor types. For example, clozapine, the prototypical atypical antipsychotic, is a weak dopamine D_2 antagonist, but has high affinity for the dopamine D_4 receptor. Thus, it does not appear that a single ligand-receptor interaction mediates all of the observed clinical properties of antipsychotic drugs.

The mechanism(s) of action of the antipsychotic drugs is undoubtedly quite complex, and drugs that initially act on distinct receptors types or intracellular signal transduction pathways could converge downstream to mediate the same types of neural adaptations. In view of this complex landscape, it is clear that more work needs to be done to develop new pharmacological avenues to treat schizophrenia. Increasing evidence suggests that the σ receptor may be a valuable target for antipsychotic drug development.

Receptor binding studies using membrane homogenates and quantitative autoradiography have been used to characterize the distribution of σ_1 and σ_2 binding sites in the brain of several species including mouse, rat, guinea pig, cat, monkey and human (45,59-64). The anatomical distribution of the expression of the σ_1 receptor and its mRNA in the mammalian brain has been studied by in situ hybridization (65,66) and immunohistochemistry (67,68). These studies showed a widespread distribution of the σ_1 receptor or its message through the brain, with more intense labeling in some specific areas, such as the granular layer of the olfactory bulb, pyramidal layers of the hippocampus, several hypothalamic nuclei, and the septum, among others. Recently, PET-imaging studies with fluorinated analogs of σ ligands have been performed in conscious monkeys. These studies showed a characteristic binding of the radioligands in different brain regions and have potential for mapping σ receptors in the human brain (69).

σ Receptors have been found in cortical and limbic structures in human postmortem brain (46,70,71). Selective loss of σ binding sites in schizophrenia has been reported as a decrease in maximal receptor densities (B_{max}), but without a significant difference in receptor affinity (K_d) in the temporal cortex (47). However, certain pharmacological studies have reported a lack of correlation between behavioral effects of antipsychotic drugs in animal models of schizophrenia and affinity for the σ receptor (72). Although no pure "σ compounds" have yet been marketed for the treatment of schizophrenia, it is quite possible that σ receptors may play a pivotal role in the pathophysiology of this illness. Clearly, further research is needed to determine the role of σ receptors in schizophrenia and their potential role in treatment for this disease.

4.2 σ Receptor ligands and putative antipsychotic compounds

Over the past few years, several σ receptor compounds have been extensively studied and brought to different developmental clinical phases. Table 13-2 shows the activity of commercially available antipsychotics and putative antipsychotic compounds with remarkable affinity for the σ_1 receptor in different preclinical tests which predict potential antipsychotic activity. Additional details about each of these compounds are summarized below.

4.2.1 Rimcazole

Rimcazole (9-[3-(cis-3,5-dimethyl-1-piperzinyl)propyl]carbazole), although a relatively weak ligand for σ receptors, was one of the first compounds to be proposed as a treatment for schizophrenia. Rimcazole was effective in open clinical studies (73,74), improving anxiety, depression, and thought disorders, but the double blind trial was discontinued, due to induction of seizures.

4.2.2 BMY 14802

BMY 14802 (α-(4-fluorophenyl)-4-(5-fluoro-2-pyrimidiny)-1-piperazine butanol) is a compound with significant affinity for σ, serotonin $5HT_{1A}$, and serotonin $5HT_2$ receptors and clearly weak affinity for dopamine and other neurotransmitter receptors (75). This σ antagonist was developed in an effort to identify new compounds with antipsychotic activity that could overcome the negative side effects caused by prolonged antagonism of the dopamine D_2 receptor, and was identified as a candidate antipsychotic drug based on classical neuropharmacological tests. Thus, BMY 14802 can inhibit apomorphine-induced stereotypy in rats, and apomorphine-induced climbing in mice (76), with a potency clearly lower than that of haloperidol, but similar to that of clozapine. BMY 14802 reversed amphetamine-induced excitations of single-unit activity in the neostriatum of freely moving rats (77), and attenuated the locomotor effects of amphetamine. In another electrophysiological study, acute administration of BMY 14802 reversed the rate suppressant effect of the dopamine agonist apomorphine on dopaminergic neurons of both the ventral tegmental area and the substantia nigra. Repeated administration of BMY 14802 reduced the number of spontaneously active dopaminergic ventral tegmental neurons without affecting nigral dopaminergic neurons. This effect was partially reversed by apomorphine. These observations suggest that this σ compound differs from typical antipsychotics in its effects on depolarization block, and that it may influence dopaminergic neurotransmission via a non-dopaminergic mechanism.

In another set of experiments, BMY 14802 and haloperidol (which also acts as a non-specific σ receptor ligand) have been shown to increase levels of neurotensin within specific brain regions, namely the nucleus accumbens and caudate, in response to acute and chronic treatment (78). Neurotensin is a peptide localized in dopaminergic neurons, and most of the antipsychotic drugs identified to date are known to regulate neurotensin. In these experiments it was shown that the dopamine D_2 receptor antagonist sulpiride also increased levels of neurotensin, and that this effect was blocked by concomitant administration of SCH 23390 [(R)-(+)-7-chloro-8-hydroxy-3-methyl-1-phenyl-2,3,4,5-tetrahydro-1H-3-benzazepine], a dopamine D_1 receptor antagonist (D_1 antagonists may attenuate some D_2 antagonists effects). The effects of BMY 14802 and haloperidol were not antagonized by SCH 23390, suggesting that BMY 14802 and haloperidol may regulate neurotensin by a different mechanism than that of a pure dopamine D_2 antagonist such as sulpiride.

BMY 14802 is moderately active in the rat conditioned avoidance response test and in apomorphine-induced pole climbing in mice (76). BMY 14802 also blocks methamphetamine sensitization (79), and the development of cocaine sensitization (80), which is used as a pharmacological model of schizophrenia.

Taken together, this favorable preclinical profile suggests that BMY 14802 could mediate antipsychotic effects in humans without the extrapyramidal side effects associated with standard neuroleptics. However, clinical trials with BMY 14802 showed no efficacy or significant improvement in psychiatric symptoms (81). In the same study, no changes in involuntary movements or in extrapyramidal symptoms were found.

Table 13-2. Activity of commercially available and putative antipsychotic compounds with remarkable affinity for the σ_1 receptor in different preclinical tests which predict potential antipsychotic activity

	Inhibition of APO Induced Climbing (Mice)	Reversion of d-AMPH Induced Hyperactivity (Rats)	Inhibition of Conditioned Avoidance Response (Rats)	Induction of Catalepsy (Rats)	Reference(s)
Haloperidol	0.13	0.2	0.4	0.15	44
Clozapine	14.50	4.1	10.4	>80	44
E-5842	7.70	3.8	13.3	>80	44
NE-100	>10*	>10 (mice)	n.f.	>10	99
Panamesine	0.63	3.0	n.f.	>20	42
MS-377	24.00	n.f.	n.f.	>80	107
BMY 14802	60.00	11.0	26.0	>80	112,113

Results express ED_{50} values (mg/kg). *Results refer to the no-effect of NE-100 on apomorphine (APO)-induced stereotyped behavior in rats (99). APO, apomorphine; AMPH, amphetamine; n.f., not found.

4.2.3 E-5842

E-5842 (4-[4-fluorophenyl]-1,2,3,6-tetrahydro-1-[4-[1,-2,4-triazol-1-il] butyl]pyridine) is a compound with very high affinity for the σ_1 receptor (K_i = 4 nM), and moderate affinity for other central nervous system receptors, including several dopamine, serotonin and glutamate receptors (44). E-5842 has been widely tested for antipsychotic activity at the preclinical level, where its profile was clearly consistent with that of an atypical antipsychotic (44). E-5842 inhibits apomorphine-induced climbing in mice and inhibits the conditioned avoidance response in rats with ED_{50} values similar to those of clozapine and completely different from those of typical dopamine D_2 antagonists, such as haloperidol and risperidone (44). E-5842 also inhibits amphetamine-induced locomotor activity in rats in a dose-dependent manner. However, unlike typical dopamine D_2 receptor antagonists, high doses of this σ_1 compound did not induce catalepsy in rats (44). It has been widely argued that antagonism of the serotonin 5-HT_2 receptor could account for the lack of cataleptogenic activity of certain atypical antipsychotics. The fact that E-5842 also failed to induce catalepsy seems to contradict this hypothesis, given its low affinity for serotonin 5-HT_2 receptors. However, it is important to note that the 5-HT_2 binding properties of this compound are based on data from *in vitro* binding studies, which may not accurately reflect its *in vivo* binding. Pre-pulse inhibition of the acoustic startle response has been used as an animal model for the sensorimotor gating defects observed in schizophrenia (82,83). Pre-pulse inhibition can be pharmacologically impaired by apomorphine or NMDA receptor antagonists (84,85), and this disruption can be blocked by several typical and atypical antipsychotics (86). E-5842 can also block apomorphine-induced disruption of pre-pulse inhibition (44), although it has no effect on NMDA receptor antagonist-induced disruption of pre-pulse inhibition.

Similar to BMY 14802, E-5842 may also alter the electrophysiological characteristics of dopaminergic neurons. Chronic administration of E-5842 decreased the number of spontaneously active ventral tegmental, but not nigral dopamine neurons (87). This effect, which was reversed by apomorphine, indicates a possible depolarization inactivation phenomenon, similar to that observed with atypical antipsychotics (88,89). Acute administration of E-5842 also increases the levels of Fos protein expressed in selected regions of the rat forebrain (90), similar to the effects of panamesine, a selective σ receptor ligand. In the case of E-5842, Fos levels were increased in the prefrontal cortex and the nucleus accumbens, with little effect in the dorsolateral striatum. These findings are consistent with the possibility that this σ compound may mediate antipsychotic activity.

The intracellular signaling systems regulated by σ ligands and their effects on other classical neurotransmitter systems have been extensively addressed in other chapters of this book. When given acutely, E-5842 increases phospholipase C (PLC) activity in the striatum and hippocampus. Repeated administration of E-5842 increases PLC activity in the frontal cortex and the striatum. In the frontal cortex, the increase in enzymatic activity is accompanied by an increase in immunoreactivity levels of PLCβ and the associated α subunit of the $G_{q/11}$ GTP-binding protein (91). It has been previously suggested that intracellular $σ_1$ receptors may modulate PLC and protein kinase C activities in the brainstem (92). The $σ_1$ receptor ligand E-5842 can also alter intracellular cyclic AMP levels. Repeated administration of E-5842 significantly decreased adenylyl cyclase type I immunoreactivity and adenylyl cyclase enzyme activity in membranes from rat frontal cortex, with a similar tendency observed in the striatum (93). In addition, administration of E-5842 induced changes in levels of fibroblast growth factor (FGF)-2 and brain-derived neurotrophic factor (BDNF) in different brain regions (protein and mRNA) (94,95). Interestingly, repeated administration of E-5842 altered immunoreactivity levels of various subunits of the ionotropic glutamate receptors (96). A similar effect has been previously described for other antipsychotics (97,98) but not for any $σ_1$ receptor ligands. It is not known if this effect is related to a possible dysfunction in the excitatory amino acid system involved in the pathophysiology of schizophrenia (8-10).

4.2.4 NE-100

NE-100 (N,N-dipropyl-2-[4-methoxy-3-(2-phenylethoxy)phenyl]ethylamine) is another putative antipsychotic drug with high affinity for the $σ_1$ receptor, and a lower affinity for the $σ_2$, dopamine D_2, serotonin 5-HT_2 and PCP receptors (45,99). NE-100 can antagonize PCP-induced behavior, whereas it has no apparent effect on methamphetamine- or apomorphine-induced stereotyped behavior or hyperactivity, and does not induce catalepsy in rats (99,100). NE-100 and other $σ_1$ antagonists have been shown to regulate dopamine release through the NMDA receptor (101) and/or to modulate NMDA-induced currents in dopaminergic neurons of the ventral tegmental area (102). Taken together with results from studies of BMY 14802 and E-5842 (44), these data clearly point to a role for σ compounds in both dopaminergic and glutamatergic neurotransmission. This topic is discussed in extensive detail in another chapter of this book.

4.2.5 Remoxipride

In the last decade, other compounds with high affinity for the σ receptor have been studied as possible treatments for schizophrenia. Remoxipride has considerable binding affinity for the σ receptor and was found to exhibit atypical antipsychotic activity in humans. Despite the several-fold difference in binding affinity for dopamine D_2 receptors versus σ receptors, remoxipride appears to be a more potent drug at dopamine D_2 receptors than was originally established in preclinical studies (103). However, subchronic administration of remoxipride did not affect either the density or the affinity of σ receptors in the rat brain (104), as determined by using $[^3H](+)$-3-PPP radioligand binding. This finding was unexpected, in that haloperidol, which is also assumed to be an antagonist of the $σ_1$ receptor, has been reported to decrease $σ_1$ binding without affecting $σ_1$ receptor mRNA levels in the brains of rats and guinea pigs (105). Furthermore, repeated administration of E-5842 has been shown to increase $σ_1$ receptor mRNA levels in several brain areas (65).

4.2.6 NPC 16377

NPC 16377 (6-[6-(4-hydroxypiperidinyl)hexyloxy]-3]methylflavone) has been described as another potent and selective σ ligand with antipsychotic potential. NPC 16377 reversed amphetamine-induced hyperactivity and apomorphine-induced climbing in mice, and failed to induce catalepsy, but had no effect on dopamine turnover in various brain areas related to schizophrenia in humans (41,106).

4.2.7 MS-377

MS-377 [(R)-(+)-1-(4-chlorophenyl)-3-[4-(2-methoxyethyl)piperazin-1-yl]methyl-2-pyrrolidinone] is another $σ_1$ receptor antagonist that has recently been reported to have antipsychotic activity (107) and to potentiate the inhibitory effects of haloperidol on apomorphine-induced climbing behavior in a dose-dependent manner (108). In contrast, MS-377 did not antagonize the apomorphine-induced disruption of pre-pulse inhibition of acoustic startle, but did antagonize PCP-induced pre-pulse inhibition (109).

13. Schizophrenia

4.2.8 Panamesine (EMD 57445)

Panamesine (EMD 57445) (42) also exhibits high affinity for σ receptors and potent antidopaminergic activity, despite its lack of affinity for any other receptors tested, including dopamine receptors (42,48). Panamesine reduces apomorphine-induced stereotypy and climbing, inhibits avoidance behavior, and shows only weak cataleptic activity at very high doses. Panamesine also antagonizes amphetamine-induced locomotor hyperactivity and stereotypy and can be clearly classified as an atypical antipsychotic, although it has been described that a metabolite of panamesine also binds to the dopamine D_2 receptor (58). Similar to other σ compounds, panamesine has entered clinical trials for the treatment of schizophrenia and has shown favorable efficacy and safety in open studies (110,111).

5. CONCLUDING REMARKS

σ Receptors ligands can influence various neurotransmitter systems, including the dopaminergic and glutamatergic systems, which are widely considered to play a role in the pathophysiology of schizophrenia. While the initial σ ligands were not highly selective, much more selective compounds have been recently developed. Some of these highly selective σ receptor antagonists have been reported to exhibit antipsychotic activity in animal models, although a pure σ compound has not yet been marketed for the treatment of schizophrenia. These compounds do not exhibit remarkable binding affinities for neurotransmitter receptors other than the σ receptor; nevertheless, they have been shown to modulate dopaminergic and glutamatergic neurotransmission. One interesting property of σ receptors in drug discovery and development is that drugs from different pharmacological classes bind to these receptors, but not all drugs within a single class bind to σ receptors with the same affinity. Future behavioral and/or biochemical studies of σ receptor function are clearly needed to address the modulatory role of σ sites and its importance in the manifestation of the various symptoms of schizophrenia. The potential utility of σ ligands for the treatment of schizophrenia represents an important direction in neuropharmacological research. Such studies promise to facilitate our understanding of the role of σ receptors in schizophrenia and the more general roles played by these receptors in the normal physiology of the central nervous system as well as in various brain disorders.

REFERENCES

1. Randrup A, Munkvad I. Special antagonism of amphetamine-induced abnormal behaviour. Inhibition of stereotyped activity with increase of some normal activities. Psychopharmacologia 1965, 7:416-422.
2. Carlsson A, Lindquist M. Effect of chlorpromazine or haloperidol on formation of 3-methoxythyramine and normethanephrine in mouse brain. Acta Pharmacol Toxicol 1963, 20:140-144.
3. Creese I, Burt DR, Snyder SH. Dopamine receptor binding predicts clinical and pharmacological potencies of antischizophrenic drugs. Science 1976, 192:481-483.
4. Farde L, Wiesel FA, Hall H, Halldin C, Stone-Elander S, Sedvall G. No D_2 receptor increase in PET study of schizophrenia. Arch Gen Psychiatry 1987, 44:671-672.
5. Andreasen NC, Olson S. Negative versus positive schizophrenia: definition and validation. Arch Gen Psychiatry 1982, 39:789-794.
6. Meltzer HY. Role of serotonin in the action of atypical antipsychotic drugs. Clin Neurosci 1995, 3:64-75.
7. Meltzer HY. The role of serotonin in antipsychotic drug action. Neuropsychopharmacology 1999, 21(Suppl):106S-115S.
8. Carlsson M, Carlsson A. Interactions between glutamatergic and monoaminergic systems within the basal ganglia - Implications for schizophrenia and Parkinson's disease. Trends Neurosci 1990, 13:272-276.
9. Moghaddam B. Recent basic findings in support of excitatory amino acid hypotheses of schizophrenia. Prog Neuropsychopharmacol Biol Psychiatry 1994, 18:859-870.
10. Tamminga CA. Schizophrenia and glutamatergic transmission. Crit Rev Neurobiol 1995, 12:21-36.
11. Meador-Woodruff JH, Healy DJ. Glutamate receptor expression in schizophrenic brain. Brain Res Brain Res Rev 2000, 31:288-294.
12. Gao XM, Sakai K, Roberts RC, Conley RR, Dean B, Tamminga CA. Ionotropic glutamate receptors and expression of N-methyl-D-aspartate receptor subunits in subregions of human hippocampus: effects of schizophrenia. Am J Psychiatry 2000, 157:1141-1149.
13. Noga JT, Wang H. Further postmortem autoradiographic studies of AMPA receptor binding in schizophrenia. Synapse 2002, 45:250-258.
14. Zavitsanou K, Ward PB, Huang XF. Selective alterations in ionotropic glutamate receptors in the anterior cingulated cortex in schizophrenia. Neuropsychopharmacology 2002, 27:826-833.
15. Monnet FP, Debonnel G, deMontigny C. In vivo electrophysiological evidence for a selective modulation of N-methyl-D-aspartate-induced neuronal activation in rat CA3 dorsal hippocampus by σ ligands. J Pharmacol Exp Ther 1992, 261:123-130.
16. Zou LB, Yamada K, Nabeshima T. Sigma receptor ligands (+)-SKF 10,047 and SA4503 improve dizocilpine-induced spatial memory deficits in rats. Eur J Pharmacol 1998, 355:1-10.
17. Kitaichi K, Chabot J-G, Moebius FF, Flandorfer A, Glossmann H, Quirion R. Expression of the purported sigma$_1$ (σ_1) receptor in the mammalian brain and its possible relevance in deficits induced by antagonism of the NMDA receptor complex as revealed using an antisense strategy. J Chem Neuroanat 2000, 20:375-387.
18. Bermack JE, Debonnel G. Distinct modulatory roles of sigma receptor subtypes on glutamatergic responses in the dorsal hippocampus. Synapse 2005, 55:37-44.

19. Gronier B, Debonnel G. Involvement of σ receptors in the modulation of the glutamatergic/NMDA neurotransmission in the dopaminergic systems. Eur J Pharmacol 1999, 368:183-196.
20. Kane JM. Schizophrenia. The New England Journal of Medicine 1996, 334:34-41.
21. Rauser L, Savage JE, Meltzer HY, Roth BL. Inverse agonist actions of typical and atypical antipsychotic drugs at the human 5-hydroxytryptamine/2C receptor. J Pharmacol Exp Ther 2001, 299:83-89.
22. Roth BL, Craigo SC, Choudhary MS, Uluer A, Monsma FJ Jr, Shen Y, Meltzer HY, Sibley DR. Binding of typical and atypical antipsychotic agents to 5-hydroxytryptamine-6 and 5-hydroxytryptamine-7 receptors. J. Pharmacol Exp Ther 1994, 268:1403-1410.
23. Moghaddam B, Adams BW. Reversal of phencyclidine effects by a group II metabotropic glutamate receptor agonist in rats. Science 1998, 281:1349-1352.
24. Nakazato A, Okuyama S. Recent advances in novel atypical antipsychotic agents: potential therapeutic agents for the treatment of schizophrenia. Exp Opin Ther Patents 2000, 10:75-98.
25. Volz HP, Stoll KD. Clinical trials with sigma ligands. Pharmacopsychiatry 2004, 37:S214-220.
26. Skuza G, Wedzony K. Behavioral pharmacology of sigma-ligands. Pharmacopsychiatry 2004, 37:183-188.
27. Hayashi T, Su TP. Sigma-1 receptor ligands: potential in the treatment of neuropsychiatric disorders. CNS Drugs 2004, 18:269-284.
28. Hanner M, Moebius FF, Flandorfer A, Knaus HG, Striessnig J, Kempner E, Glossmann H. Purification, molecular cloning, and expression of the mammalian σ_1-binding site. Proc Natl Acad Sci USA 1996, 93:8072-8077.
29. Seth P, Fei Y-J, Li HW, Huang W, Leibach FH, Ganapathy V. Cloning and functional characterization of a σ receptor from rat brain. J Neurochem 1998, 70:922-931.
30. Tam SW, Cook L. Sigma opiates and certain antipsychotic drugs mutually inhibit (+)[^{3}H]SKF 10,047 and [^{3}H]haloperidol binding in guinea pig brain membranes. Proc Natl Acad Sci USA 1984, 81:5618-5621.
31. Largent BL, Wilkstrom H, Snowman AM, Snyder SH. Novel antipsychotic drugs share high affinity for σ receptors. Eur J Pharmacol 1988, 155:345-347.
32. Gillgan PJ, Tam SW. Sigma receptor ligands: potential drugs for the treatment of CNS disorders? Drug News Perspectives 1994, 7:13-18.
33. Walker JM, Bowen WD, Walker FO, Matsumoto RR, de Costa BR, Rice KC. Sigma receptors: biology and function. Pharmacol Rev 1990, 42:355-402.
34. Walker JM, Martin WJ, Hohmann AG, Hemstreet MK, Roth JS, Leitner ML, Weiser SD, Patrick SL, Patrick RL, Matsumoto RR. Role of σ receptors in brain mechanisms of movement. In *Sigma Receptors*. Yossef Itzhak, ed., Miami, FL: Academic Press, 1994.
35. Matsumoto RR, Pouw B. Correlation between neuroleptic binding to σ_1 and σ_2 receptors and acute dystonic reactions. Eur J Pharmacol. 2000, 401:155-160.
36. Langa F, Codony X, Tovar V, Lavado A, Gimenez E, Cozar P, Cantero M, Dordal A, Hernandez E, Perez R, Monroy X, Zamanillo D, Guitart X, Montoliu L. Generation and phenotypic analysis of sigma receptor type I (σ1) knockout mice. Eur J Neurosci 2003, 18:2188-2196.
37. Ferris CD, Hirsch DJ, Brooks BP, Snyder SH. Sigma receptors: From molecule to man. J Neurochem 1991, 57:1207-1212.
38. Su TP. Sigma receptors: putative links between nervous, endocrine and immune systems. Eur J Biochem 1991, 200:633-642.

39. Tam SW. Potential therapeutic application of σ receptor antagonists. In *Sigma Receptors*. Yossef Itzhak, ed. Miami, FL: Academic Press, 1994.
40. Akunne HC, Whetzel SZ, Wiley JN, Corbin AE, Ninteman FW, Tecle H, Pei Y, Pugsley TA, Heffner TG. The pharmacology of the novel and selective σ ligand, PD 144418. Neuropharmacology 1997, 36:51-62.
41. Karbon EW, Abreu ME, Erickson RH, Kaiser C, Natalie KJ Jr, Clissold DB, Borosky S, Bailey M, Martin LA, Pontecorvo MJ, Enna SJ, Feranaky JW. NPC 16377, a potent and selective σ ligand. I. Receptor binding, neurochemical and neuroendocrine profile. J Pharmacol Exp Ther 1993, 265:866-875.
42. Maj J, Rogóz Z, Skuza G, Mazela H. Neuropharmacological profile of EMD57445, a σ-receptor ligand with potential antipsychotic activity. Eur J Pharmacol 1996, 315:235-243.
43. Tam SW, Steinfels GF, Gilligan PJ, Schimdt WK, Cook L. DuP 724 [1-(cyclopropylmethyl)-4-(2'(4''-fluorophenyl)-2'-oxoethyl)-piperidine HBr], a σ- and 5-hydroxytryptamine$_2$ receptor antagonist; receptor-binding, electrophysiological and neuropharmacological profiles. J Pharmacol Exp Ther 1992, 263:1167-1174.
44. Guitart X, Codony X, Ballarín M, Dordal A, Farré AJ. E-5842: a new potent and preferential sigma ligand. Preclinical pharmacological profile. CNS Drug Rev 1998, 4:201-224.
45. Chaki S, Tanaka M, Muramatsu M, Otomo S. NE-100, a novel potent σ ligand, preferentially binds to σ$_1$ binding sites in guinea pig brain. Eur J Pharmacol 1994, 251:R1-R2.
46. Weissman AD, Su TP, Hedreen JC, London DE. Sigma receptors in post-mortem human brains. J Pharmacol Exp Ther 1988, 247:29-33.
47. Weissman AD, Casanova MF, Kleinman JE, London ED, De Souza EB. Selective loss of cerebral cortical σ, but not PCP-binding sites in schizophrenia. Biol Psychiatry 1991, 29:41-54.
48. Ishiguro H, Ohtsuki T, Toru M, Itokawa M, Aoki J, Shibuya H, Kurumaji A, Okubo Y, Iwawaki A, Ota K, Shimizu H, Hamaguchi H, Arinami T. Association between polymorphisms in the type 1 σ receptor gene and schizophrenia. Neurosci Lett 1998, 257:45-48.
49. Ohmori O, Shinkai T, Suzuki T, Okano C, Kojima H, Terao T, Nakamura J. Polymorphisms of the σ$_1$ receptor gene in schizophrenia: an association study. Am J Med Genet 2000, 96:118-122.
50. Uchida N, Ujike H, Nakata K, Takaki M, Nomura A, Katsu T, Tanaka Y, Imamura T, Sakai A, Kuroda S. No association between the sigma receptor type 1 gene and schizophrenia: results on analysis and meta-analysis of case-control studies. BMC Psychiatry 2003, 21:13.
51. Inada T, Iijima Y, Uchida N, Maeda T, Iwashita S, Ozaki N, Harano M, Komiyama T, Yamada M, Sekine Y, Iyo M, Sora I, Ujike H. No association found between the type 1 sigma receptor gene polymorphisms and methamphetamine abuse in the Japanese population: a collaborative study by the Japanese Genetics Initiative for Drug Abuse. Ann NY Acad Sci USA 2004, 1025:27-33.
52. Satoh F, Miyatake R, Furukawa A, Suwaki H. Lack of association between sigma receptor gene variants and schizophrenia. Psychiatry Clin Neurosci 2004, 58:359-363.
53. Bartoszky GD, Bender HM, Hellman J, Schnorr C, Seyfred CA. EMD 57445: a selective σ ligand with the profile of an atypical neuroleptic. CNS Drug Rev 1996, 2:175-194.

54. Tahashi S, Sonehara K, Takagi K, Miwa T, Horikomi K, Mita N, Nagase H, Iizuka K, Sakai K. Pharmacological profile of MS-377, a novel selective antipsychotic agent with selective affinity for σ receptors. Psychopharmacology 1999, 145:295-302.
55. Koe BK, Burkhart CA, Lebel LA. (+)-[^{3}H]3-(3-hydroxyphenyl)-N-(1-propyl)-piperidine binding to σ receptors in mouse brain in vivo. Eur J Pharmacol 1989, 161:263-266.
56. Wadworth AN, Heel RC. Remoxipride. A review of its pharmacodynamic and pharmacokinetic properties, and therapeutic potential in schizophrenia. Drugs 1990, 40:863-879.
57. Bowen WD, Walker JM, de Costa BR, Wu R, Tolentino PJ, Finn D, Rothman RB, Rice KC. Characterization of the enantiomers of cis-N-[2-(3,4-dichlorophenyl)ethyl]-N-methyl-2-(1-pyrrolidinyl)cyclohexylamine (BD737 and BD738): novel compounds with high affinity, selectivity and biological efficacy at σ receptors. J Pharmacol Exp Ther 1992, 262:32-40.
58. Grunder G, Muller MJ, Andreas J, Heydari N, Wetzel H, Schlosser R, Schlegel R, Nickel O, Eissner D, Benkert O. Occupancy of striatal D$_2$-like dopamine receptors after treatment with the σ ligand EMD 57445, a putative atypical antipsychotic. Psychopharmacology 1999, 146:81-86.
59. Jansen KLR, Faull RLM, Dragunow M, Leslie RA. Autoradiographic distribution of σ receptors in human neocortex, hippocampus, basal ganglia, cerebellum, pineal and pituitary gland. Brain Res 1991, 559:172-177.
60. Mash DC, Zabetian CP. Sigma receptors are associated with cortical limbic areas in the primate brain. Synapse 1992, 12:195-205.
61. Su T-P, Junien J-L. Sigma receptors in the central nervous system and the periphery. In *Sigma Receptors.* Yossef Itzhak, ed. Miami, FL: Academic Press, 1994.
62. Gonzalcz-Alvcar GM, Thompson-Montgomery D, Deben SE, Werling LL. Functional and binding properties of σ receptors in rat cerebellum. J Neurochem 1995, 65:2509-2516.
63. Okuyama S, Chaki S, Yae T, Nakazato A, Muramatsu M. Autoradiographic characterization of binding sites for [^{3}H]NE-100 in guinea pig brain. Life Sci 1995, 57:L333-L337.
64. Bouchard P, Quirion R. [^{3}H]1,3-di(2-tolyl)guanidine and [^{3}H](+)pentazocine binding sites in the rat brain: autoradiographic visualization of the putative sigma1 and σ$_2$ receptor subtypes. Neuroscience 1997, 76:467-477.
65. Zamanillo D, Andreu F, Ovalle S, Pérez MP, Romero G, Farré AJ, Guitart X. Up-regulation of σ$_1$ receptor mRNA in rat brain by a putative atypical antipsychotic and sigma receptor ligand. Neurosci Lett 2000, 282:169-172.
66. Kitaichi K, Chabot J-G, Moebius F, Flandorfer A, Glossmann H, Quirion R. Expression of the purported sigma$_1$ (σ$_1$) receptor in the mammalian brain and its possible relevance in deficits induced by antagonism of the NMDA receptor complex as revealed using an antisense strategy. J Chem Neuroanat 2000, 20:375-387.
67. Alonso G, Phan V, Guillemain I, Saunier M, Legrand A, Anoal M, Maurice T. Immunocytochemical localization of the σ$_1$ receptor in the adult rat central nervous system. Neuroscience 2000; 97:155-170.
68. Phan VL, Urani A, Romieu P, Maurice T. Strain differences in σ$_1$ receptor-mediated behaviours are related to neurosteroid levels. Eur J Neurosci 2002, 15:1523-1534.
69. Elsinga PH, Tsukada H, Harada N, Kakiuchi T, Kawamura K, Vaalburg W, Kimura Y, Kobayashi T, Ishiwata K. Evaluation of [^{18}F]fluorinated sigma receptor ligands in the conscious monkey brain. Synapse 2004, 52:29-37.
70. Largent BL, Gundlach AL, Snyder SH. Pharmacological and autoradiographic discrimination of sigma and phencyclidine receptor binding sites in brain with (+)-[^{3}H]

SKF 10,047, (+)-[^{3}H]-3-[3-hydroxyphenyl]-N-(1-propyl)piperidine and [^{3}H]-1-[1-(2-thienyl)cyclohexyl]piperidine. J Pharmacol Exp Ther 1986, 238:739-748.
71. Shibuya H, Mori H, Toru M. Sigma receptors in schizophrenic cerebral cortices. Neurochem Res 1992, 17:983-990.
72. Lang A, Soosaar A, Koks S, Volke V, Bourin M, Bradwejn J, Vasar E. Pharmacological comparison of antipsychotic drugs and σ-antagonists in rodents. Pharmacol Toxicol 1994, 75:222-227.
73. Davidson J, Miller R, Wingfield M, Zung W, Dren AT. The first clinical study of BW-234U in schizophrenia. Psychopharmacol Bull 1982, 18:1159-1166.
74. Chouinard G, Annable L. An early phase II clinical trial of BW-234U in the treatment of acute schizophrenia in newly admitted patients. Psychopharmacology 1984, 84:282-284.
75. Taylor DP, Dekleva J. The potential antipsychotic BMY 14802 selectively binds to σ sites. Fed Proc 1987, 46:1304.
76. Taylor DP, Eison MS, Moon SL, Schlemmer RF Jr, Shukla UA, VanderMaelen CP, Yocca FD, Gallant DJ, Behling SH, Boissard CG, Braselton JP, Davis HH Jr, Duquette MN, Lamy RC, Libera JM, Ryan E, Wright RN. A role for σ binding in the antipsychotic profile of BMY 14802? In *Sigma, PCP, and NMDA Receptors*. EB De Souza, D Clouet, E London, eds. Rockville, MD: NIDA Research Monograph Series, 1993.
77. Wang Z, Haracz JL, Rebec GV. BMY-14802, a σ ligand and potential antipsychotic drug, reverses amphetamine-induced changes in neostriatal single-unit activity in freely moving rats. Synapse 1992, 12:312-321.
78. Levant B, Nemeroff CB. Further studies on the modulation of regional brain neurotensin concentrations by antipsychotic drugs: focus on haloperidol and BMY 14802. J Pharmacol Exp Ther 1992, 262:348-355.
79. Ujike H, Kanzaki A, Okumura K, Akiyama K, Otsuki S. Sigma antagonist BMY 14802 prevents methamphetamine-induced sensitization. Life Sci 1992, 50:PL129-PL134.
80. Ujike H, Kuroda S, Otsuki S. Sigma receptor antagonists block the development of sensitization to cocaine. Eur J Pharmacol 1996, 296:123-128.
81. Gewirtz GR, Gorman JM, Volavka J, Macaluso J, Gribkoff G, Taylor DP, Borison R. BMY 14802, a σ receptor ligand for the treatment of schizophrenia. Neuropsychopharmacology 1994, 10:37-40.
82. Braff DL, Grillon C, Geyer M. Gating and habituation of the startle reflex in schizophrenic patients. Arch Gen Psychiatry 1992, 49:206-214.
83. Braff DL, Swerdlow NR, Geyer MA. Symptom correlates of prepulse inhibition deficits in male schizophrenic patients. Am J Psychiatry 1999, 156:596-602.
84. Canal NM, Gourevitch R, Sandner G. Non-monotonic dependency of PPI on temporal parameters: differential alteration by ketamine and MK-801 as opposed to apomorphine and DOI. Psychopharmacology 2001, 156:169-176.
85. Breese GR, Knapp DJ, Moy SS. Integrative role for serotonergic and glutamatergic receptor mechanisms in the action of NMDA antagonists: potential relationship to antipsychotic drug actions on NMDA antagonist responsiveness. Neurosci Biobehav Rev 2002, 26:441-455.
86. Kumari V, Sharma T. Effects of typical and atypical antipsychotics on prepulse inhibition in schizophrenia: a critical evaluation of current evidence and directions for future research. Psychopharmacology 2002, 162:97-101.
87. Sánchez-Arroyos R, Guitart X. Electrophysiological effects of E-5842, a σ_1 receptor ligand and potential atypical antipsychotic, on A9 and A10 dopamine neurons. Eur J Pharmacol 1999, 378:31-37.
88. Goldstein JM, Litwin LC, Sutton EB, Malick JB. Seroquel: electrophysiological profile of a potential atypical antipsychotic. Psychopharmacology 1993, 112:293-298.

89. Stockton ME, Rasmussen K. Electrophysiological effects of olanzapine, a novel atypical antipsychotic, on A9 and A10 dopamine neurons. Neuropsychopharmacology 1996, 14:97-104.
90. Guitart X, Farré AJ. The effect of E-5842, a s receptor ligand and potential atypical antipsychotic, on Fos expression in rat forebrain. Eur J Pharmacol 1998, 363:127-130.
91. Romero G, Pérez MP, Carceller A, Monroy X, Farré AJ, Guitart X. Changes in phosphoinositide signalling activity and levels of the α subunit of $G_{q/11}$ protein in rat brain induced by E-5842, a σ_1 receptor ligand and potential atypical antipsychotic. Neurosci Lett 2000, 290:189-192.
92. Morin-Surun MP, Collin T, Denavit-Saubié M, Baulieu E-E, Monnet FP. Intracellular σ_1 receptor modulates phospholipase C and protein kinase C activities in the brainstem. Proc Natl Acad Sci USA 1999, 96:8196-8199.
93. Monroy X, Romero G, Pérez MP, Farré AJ, Guitart X. Decrease of adenylyl cyclase activity and expression by a σ_1 receptor ligand and putative atypical antipsychotic. Neuroreport 2001, 12:1989-1992.
94. Ovalle S, Zamanillo D, Andreu F, Farre AJ, Guitart X. Fibroblast growth factor-2 is selectively modulated in the rat brain by E-5842, a preferential sigma-1 receptor ligand and putative atypical antipsychotic. Eur J Neurosci 2001, 13:909-915.
95. Ovalle S, Andreu F, Perez MP, Zamanillo D, Guitart X. Effect of the novel sigma 1 receptor ligand and putative atypical antipsychotic E-5842 on BDNF mRNA expression in the rat brain. Neuroreport 2002, 13:2345-2348.
96. Guitart X, Méndez R, Ovalle S, Andreu F, Carceller A, Farré AJ, Zamanillo D. Regulation of ionotropic glutamate receptor subunits in different rat brain areas by a preferential σ_1 receptor ligand and potential atypical antipsychotic. Neuropsychopharmacology 2000, 23:539-546.
97. Fitzgerald LW, Deutch AY, Gasic G, Heinemann SF, Nestler EJ. Regulation of cortical and subcortical glutamate receptor subunit expression by antipsychotic drugs. J Neurosci 1995, 15:2453-2461.
98. Tascedda F, Lovati E, Blom JMC, Muzzioli P, Brunello N, Racagni G, Riva MA. Regulation of ionotropic glutamate receptors in the rat brain in response to the atypical antipsychotic seroquel (quetiapine fumarate). Neuropsychopharmacology 1999, 21:211-217.
99. Okuyama S, Imagawa Y, Ogawa S, Araki H, Ajima A, Tanaka M, Muramatsu M, Nakazato A, Yamaguchi K, Yoshida M, Otomo S. NE-100, a novel σ receptor ligand: in vivo tests. Life Sci 1993, 53:285-290.
100. Okuyama S, Ogawa S, Nakazato A, Tomizawa K. Effect of NE-100, a novel σ receptor ligand, on phencyclidine-induced delayed cognitive dysfunction in rats. Neurosci Lett 1995, 189:60-62.
101. Chaki S, Okuyama S, Ogawa S, Tomisawa K. Regulation of NMDA-induced [^{3}H]dopamine release from rat hippocampal slices through σ_1 binding sites. Neurochem Int 1998, 33:29-34.
102. Yamazaki Y, Ishioka M, Matsubayashi H, Amano T, Sasa M. Inhibition by sigma receptor ligand, MS-377, of N-methyl- D-aspartate-induced currents in dopamine neurons of the rat ventral tegmental area. Psychopharmacology 2002, 161:64-69.
103. Nadal R. Pharmacology of the atypical antipsychotic remoxipride, a dopamine D_2 receptor antagonist. CNS Drug Rev 2001, 7:265-282.
104. Ericson H, Ross SB. Subchronic treatment of rats with remoxipride fails to modify σ binding sites in the brain. Eur J Pharmacol 1992, 226:157-161.

105. Inoue A, Sugita S, Shoji H, Ichimoto H, Hide I, Nakata Y. Repeated haloperidol treatment decreases σ_1 receptor binding but does not affect it mRNA levels in the guinea pig or rat brain. Eur J Pharmacol 2000, 401:307-316.
106. Clissold DB, Pontecorvo MJ, Jones BE, Abreu ME, Karbon EW, Erickson RH, Natalie KJ Jr, Borosky S, Hartman T, Mansbach RS, Balster RL, Fernaky JW, Enna SJ. NPC 16377, a potent and selective σ-ligand. II. Behavioral and neuroprotective profile. J Pharmacol Exp Ther 1993, 265:876-886.
107. Takahashi S, Sonehara K, Takagi K, Miwa T, Horikomi K, Mita N, Nagase H, Iizuka K, Sakai K. Pharmacological profile of MS-377, a novel antipsychotic agent with selective affinity for σ receptors. Psychopharmacology 1999, 145:295-302.
108. Karasawa J, Takahashi S, Takagi K, Horikomi K. Effects of σ_1 receptor ligand MS-377 on D_2 antagonists-induced behaviors. Pharmacol Biochem Behav 2002, 73:505-510.
109. Yamada S, Yamauchi K, Hisatomi S, Annoh N, Tanaka M. Effects of σ_1 receptor ligand, MS-377 on apomorphine- or phencyclidine-induced disruption of prepulse inhibition of acoustic startle in rats. Eur J Pharmacol 2000, 402:251-254.
110. Frieboes RM, Murck H, Wiedemann K, Holsboer F, Steiger A. Open clinical trial of the σ ligand panamesine in patients with schizophrenia. Psychopharmacology 1997, 132:82-88.
111. Huber MT, Gotthardt U, Schreiber W, Krieg JC. Efficacy and safety of the sigma receptor ligand EMD 57445 (panamesine) in patients with schizophrenia: an open clinical trial. Pharmacopsychiatry 1999, 32:68-72.
112. Taylor Duncan P, Eison MS, Moon SL, Yocca FD. BMY 14802, a potential antipsychotic with selective affinity for σ–binding sites. In *Advances in Neuropsychiatry and Psychopharmacology, Vol 1: Schizophrenia Research*. CA Tamminga, SC Schulz, eds. New York, NY: Raven Press, 1991.
113. Bristow LJ, Baucutt L, Thorn L, Hutson PH, Noble A, Beer M, Middlemiss DN, Tricklebank MD. Behavioral and biochemical evidence of the interaction of the putative antipsychotic agent, BMY 14802 with the $5-HT_{1A}$ receptor. Eur J Pharmacol 1991, 204:21-28.

Corresponding author: *Dr. Xavier Guitart, Mailing address: Prous Science, Provenza 388, Barcelona 08025, Spain, Phone: 34934592220 Fax: 34934581535, Electronic mail address: xguitart@prous.com*

Chapter 14

POTENTIAL ROLE OF σ LIGANDS AND NEUROSTEROIDS IN MAJOR DEPRESSION

Guy Debonnel, Malika Robichaud and Jordanna Bermack
Department of Psychiatry, McGill University, Montreal, Quebec, Canada H3A 1A1

1. σ RECEPTORS

Since their discovery in 1976 by the group of Martin (1), σ receptors have always been a subject of controversy, regarding their existence, their nature and their role. Initially classified as opiate receptors, their appartenance to this class was rapidly refuted as several of the effects of the prototypal σ ligand SKF-10,047 (N-allylnormetazocine) were not reversed by opiate antagonists. The existence of two different subtypes of σ receptors is officially acknowledged (2) but numerous evidences suggest that there are many more than that. Nearly 30 years after their discovery, at least one subtype has been cloned (3) but no endogenous ligand has been identified. For many years, σ receptors have been suggested to be associated with neuropsychiatric disorders, but this has also been controversial.

2. σ RECEPTORS AND PSYCHOSIS

It was initially proposed that σ receptors may play an important role in the pathophysiology or the treatment of psychosis. This assumption was based on two evidences: first, the psychotomimetic effects induced by SKF-10,047 and early σ ligands, and second, the high affinity of the classical antipsychotic medication haloperidol for σ receptors (4,5). However, 20 years later, the role of σ receptors in psychosis remains an open and controversial question. Indeed, it was rapidly demonstrated that the psychotomimetic effects of SKF-10,047 were in fact due to its relatively high affinity for phencyclidine (PCP) receptors (6) and that new σ ligands, devoid of affinity for the PCP site, did not produce any psychotomimetic effects. Moreover, the reality of an antipsychotic effect of σ ligands is still unclear. If some σ ligands like haloperidol, rimcazole or remoxipride are effective antipsychotic compounds, they also present, besides their affinity for σ receptors, a high affinity for dopaminergic D_2 receptors which have been considered for long, as responsible for the effects of all classical antipsychotic drugs (7-9). Even drugs presented as more selective σ ligands like BMY 14802 (α-(4-fluorophenyl)-4-(5-fluoro-2-pyrimidinyl)-1-piperazine-butanol) which demonstrated some antipsychotic effects (10,11) also present some affinity for D_2 receptors through the activity of their metabolites. As of now, despite several attempts, no selective σ antagonist has yet demonstrated a clear antipsychotic effect. Therefore, the potential antipsychotic effect of a selective σ ligand remains to be established.

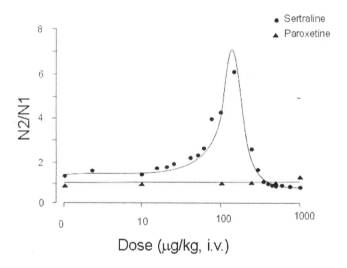

Figure 14-1. Dose-response curve of the effects of intravenously administered sertraline and paroxetine on the neuronal activation of CA3 dorsal hippocampal pyramidal neurons induced by microiontophoretic application of N-methyl-D-aspartate (NMDA). The effects of these drugs were assessed by determining the ratio (N_2/N_1) of the number of spikes generated/nC of NMDA before (N_1) and after (N_2) the injection of the drug. One dot represents the effect of one dose of drug administered to one rat while recording from one neuron. Adapted from ref. (14).

3. σ RECEPTORS AND DEPRESSION

More recently, it has also been proposed that σ ligands may act as antidepressants or that σ receptors may play a role in the pathophysiology of depression. This hypothesis is based on several preclinical lines of evidence and limited clinical data.

The first observation was that several antidepressant medications such as sertraline (a serotonin reuptake blocker, SSRI) (12) and clorgyline (a monoamine oxidase inhibitor, MAOI) (13), present a high affinity for σ receptors and act as σ agonists, and induce the same bell-shaped dose-response curves, classically observed with selective σ compounds (14) (Figure 14-1). However, this does not constitute a very strong argument, given the extreme variety of compounds displaying a relatively high affinity for σ receptors.

More convincing are the results obtained in behavioral studies, which have shown that various σ agonists produce antidepressant-like effects in several animal models of depression. For example, the selective σ ligands igmesine (JO 1784) (15), SA4503 (1-(3,4-dimethoxyphenethyl)-4-(3-phenylpropyl)piperidine) (16), SKF-10,047, DTG (di-o-tolylguanidine) (17) and (+)-pentazocine (18) decrease the immobility time, in the forced swimming test, an effect reversed by the σ antagonists BD1047 (N-[2-(3,4-dichlorophenyl)ethyl]-N-methyl-2-(dimethylamino)ethylamine) (19) or NE-100 (N,N-dipropyl-2-[4-methoxy-3-(2-phenethoxy)phenyl]ethylamine) (20-22) (Figure 14-2). A similar effect, in the same paradigm was found with OPC-14523 (1-[3-[4-(3-chlorophenyl)-1-piperazinyl]propyl]-5-methoxy-3,4-dihydro-2-quinolinone monomethanesulfonate), a σ_1 agonist with a high affinity for 5-HT_{1A} receptors. In these experiments, the effects of OPC-14523 were also reversed by the selective σ_1 antagonist NE-100 (23). Interestingly the antidepressant-like effect of SA4503 in the forced swim test, is potentiated by the non-competitive N-methyl-D-aspartate (NMDA) antagonist amantadine (24). SA4503 also produces antidepressant-like effects in another animal model of depression, the tail-suspension test in mice. This effect as well as those of two other σ agonists, (+)-pentazocine and DTG, were again reversed by the σ_1 antagonist NE-100 (25). Less direct evidence were also reported with SKF-10,047 and dextromethorphan. These σ ligands reverse the motor suppression, in response to chronic stress, which is believed to be implicated in the etiology of affective disorders (26,27). The antidepressant properties do not appear to be exclusive for σ_1 ligands, since the selective σ_2 ligand Lu-28,179 (1'-[4-[1-(4-fluorophenyl)-1H-indol-3-yl]-1-butyl]spiro[isobenzofuran-1(3H)4'-piperidine) also demonstrates an antidepressant-like profile in a chronic, mild stress model of depression (28).

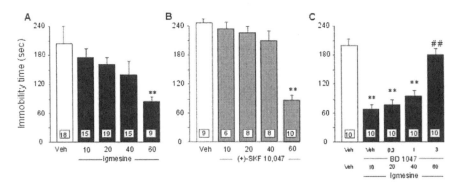

Figure 14-2. Antidepressant-like effect of σ_1 receptor ligands in the forced swim test in Swiss mice. Dose-response effect of igmesine (A), (+)-SKF-10,047 (B), and antagonism by BD1047 of the igmesine-induced effect (C). Drugs were injected i.p. 30 min before the session on Day 2. BD1047 was administered i.p. 15 min before igmesine, which was given 30 min before the session on Day 2. The duration of immobility was recorded for the last 5 min over a 6-min session. Values are expressed as mean ± S.E.M. of the number of animals indicated inside each column. **$P<0.01$ versus the vehicle-treated group (Veh); ##$P<0.01$ versus the igmesine-treated group (Dunnett's test). Adapted from ref. (22).

Very few clinical data are available regarding the effect of σ ligands in depressed patients. In one open study with severely depressed patients, igmesine was administered at doses ranging between 10 and 300 mg per day during four weeks. The intensity of the depression measured by the scores on the Hamilton Rating Scale for Depression, dropped from 31 to 16 within two weeks, and to 13 after four weeks of treatment; six patients out of 18 showed a complete remission. Interestingly (see below), the patients having received a lower dose presented better results than those receiving higher doses (JL Junien and F Roman, personal communication). For sake of comparison, two other new antidepressant medications, venlafaxine and duloxetine were shown, in open studies with groups of 20 to 30 patients, to induce an improvement of less than 35% after two weeks of treatment and of around 50% after four weeks of treatment (29-31). Moreover, an interim analysis of a double-blind placebo controlled study, conducted in 60 patients, demonstrated that a dose of 20 mg/day of igmesine was superior to placebo and to a daily dose of 20 mg of fluoxetine. However, the daily dose of 100 mg of igmesine was not superior to placebo (32).

Therefore, the few clinical data obtained thus far, also support the hypothesis that σ agonists could represent a new class of effective antidepressant medications. However, the precise mechanisms of action, by which these compounds could act as antidepressants, are not obvious.

4. MAJOR DEPRESSION

Major depression is a severe psychiatric disorder, with an incidence higher than 4% of the general population (33). Susceptibility to affective disorders differs between genders and women are twice as likely to suffer from major depression, than men (34). In addition, women can experience dramatic swings in affective states during and following pregnancy (35-42). Typically, during the last two trimesters of pregnancy, women report feelings of elation and tranquility and they present a lower risk of developing emotional disturbances. In contrast, during the post-partum period, there is a higher incidence of psychiatric disorders (43).

The biological mechanisms underlying major depression are not yet fully understood, but it is now generally accepted that, at least, the two neurotransmitters serotonin and norepinephrine are involved in its pathophysiology (for reviews see 44,45).

5. SEROTONIN AND DEPRESSION

Serotonin (5 HT) is a neurotransmitter involved in numerous aspects of body function and was first hypothesized to be implicated in the pathophysiology of depression in the late 1960s (46,47). An enhancement in 5-HT neurotransmission is believed to be responsible for the therapeutic effect of antidepressant medications (48,49).

5-HT acts upon several subtypes of receptors which have been identified, up to now, as being as many as 17 (50,51). The 5-HT$_{1A}$ receptors are of particular importance in the regulation of 5-HT neuronal activity. Activation of these receptors triggers the opening of potassium channels which induces a hyperpolarization of the neuron and decreases its firing activity. Therefore, when activated, the 5-HT$_{1A}$ somatodendritic autoreceptor, located on the soma of 5-HT neurons located in the dorsal raphe nucleus induces a negative feedback mechanism which reduces the firing activity of 5-HT neurons. Thus, the acute administration of antidepressants like the SSRIs, which induces an initial moderate increase of 5-HT concentration in the synaptic cleft, as well as in the vicinity of 5-HT cell bodies, triggers an activation of somatodendritic and terminal autoreceptors and a reduction of firing activity of the 5-HT neurons (52-55). However, in the course of the long-term administration of SSRIs, 5-HT neurons recover their normal firing activity, as a consequence of the desensitization of these autoreceptors (48). This mechanism explains the three to four weeks delay observed in depressed patients before the onset of action of antidepressant treatments.

6. σ RECEPTORS AND SEROTONIN

In order to obtain a better understanding of the mechanisms which could explain the antidepressant-like properties of σ ligands, it deemed interesting to assess their effects on serotonergic neurotransmission. We used an electrophysiological model of unitary recordings of 5-HT neurons, from the dorsal raphe nucleus (DRN), in rats anesthetized with chloral hydrate (400 μg/kg, i.p.) and mounted in a stereotaxic apparatus. Glass micropipettes, pulled and broken back under microscopic control to obtain a tip of about 5 to 12 μm, were filled with a 2M NaCl solution and used to obtain the extracellular recordings and measure the basal firing activity of these 5-HT neurons.

The effects observed were highly variable from one σ ligand to another. Specifically, short-term treatments with igmesine and the selective σ_1 agonist PRE-084 (2-(4-morpholino)ethyl-1-phenylcyclohexane-1-carboxylate) (56) (2 mg/kg/day) produced no change in the firing activity of 5-HT neurons. However, short-term treatments with the σ ligands 4-IBP (N-(N-benzyl-piperidin-4-yl)-4-iodobenzamide) (57), (2 mg/kg/day) or (+)-pentazocine (2 mg/kg/day) for 2 days induced a significant increase (35%) in firing activity of 5-HT neurons when compared to saline-treated controls. The co-administration of NE-100 (10 mg/kg/day), did not modify the effect of 4-IBP on serotonergic firing activity. However, it completely prevented the increase in firing induced by (+)-pentazocine treatments (Figure 14-3). Following long-term (21 days) treatments, both 4-IBP and (+)-pentazocine maintained their increase in the firing activity of 5-HT neurons. As for the short-term treatments, the co-administration of NE-100 prevented the increase induced by (+)-pentazocine but not that induced by 4-IBP. In both cases however, the effects produced by 4-IBP were however, blocked by the co-administration of the non-selective $\sigma_{1/2}$ antagonist haloperidol (58).

DTG, a $\sigma_{1/2}$ agonist, also potentiated the 5-HT firing activity after short- and long-term treatments. Similarly to what was observed with (+)-pentazocine, its effect was blocked by the co-administration of NE-100 suggesting that this effect is mediated via σ_1 receptors (58).

14. Major depression

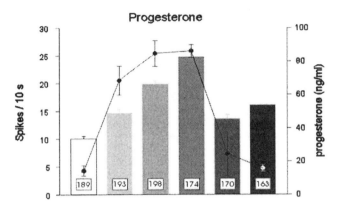

Figure 14-3. Effect of pregnancy on the spontaneous firing activity of dorsal raphe nucleus 5-HT neurons. Mean firing rate of dorsal raphe 5-HT neurons expressed in spikes per 10 s (spk/10 s, mean ± S.E.M.) for each experimental group (CF, P11, P14, P17, P21 and PP). Significance compared to CF is indicated by an asterisk (P<0.001). Adapted from ref. (84).

As stated above, the significant increase of the firing rate of 5-HT neurons after only 2 days of treatment contrasts with what has been shown, up to now, in electrophysiological studies assessing the effects of antidepressants. More specifically, short-term treatments with SSRIs or MAOIs lead to an initial decrease of the firing activity of 5-HT neurons of the DRN. As the treatment continues, the 5-HT neurons regain a normal firing activity due to a desensitization of 5-HT_{1A} receptors (59-62). We have thus examined the potential modifications of the 5-HT_{1A} mediated response using 8-OH-DPAT [(±)-8-hydroxy-2-dipropylaminotetralin], a 5-HT_{1A} agonist. But neither igmesine nor (+)-pentazocine showed any significant effect on the response to intravenous administration of 8-OH-DPAT (4 mg/kg) following either acute, short- or long-term treatments (63), suggesting that the increase in the firing activity of 5-HT neurons is due to a different mechanism which remains to be determined.

7. σ RECEPTORS AND STEROIDS

The finding that progesterone and testosterone act as competitive inhibitors of $[^3\text{H}](+)$-SKF-10,047 binding to σ receptors (64,65) raised the possibility that some neurosteroids may function as endogenous ligands for the σ receptors. Indeed, the σ_1 subtype of receptor has been cloned and the protein shares homology with fungal proteins involved in sterol synthesis (3).

In mammals, all steroid hormones derive from cholesterol and are mainly synthesized in various endocrine organs such as adrenal glands and gonads. Due to their high lipophilic properties, steroids can easily cross the blood-brain barrier. Most of them, like 17-β-estradiol and progesterone, diffuse through the cell membrane to bind their respective receptors mainly located in the cell nucleus (66) where they act as transcription factors in the regulation of gene expression (66). However, in the past decade, considerable evidence has emerged, indicating that several steroids alter neuronal excitability via interactions with specific neurotransmitter receptors. These steroids were designated as "neuro-active steroids" (67). The term "neurosteroid" describes a family of steroid compounds like progesterone or dehydroepiandrosterone (DHEA) which are not only present in the central nervous system (CNS), where they act as neurotransmitters or neuromodulators, but for which there is strong evidence of local synthesis in the CNS (68,69).

Our understanding of the role of neuroactive steroids changed dramatically when it was first realized that it was not limited to their genomic effects. Fifteen years ago, Majewska's group was the first to report an interaction between neurosteroids and γ-aminobutyric acid (GABA), one of the major inhibitory neurotransmitters.

Neurosteroids are also modulators of other types of receptors including NMDA, AMPA (α-amino-3-hydroxy-5-methylisoxazole-4-propionate), kainite, and glycine (67,70). Finally, neurosteroids have also been shown to exert potent effects on the 5-HT system. They markedly increase the gene and the protein expression of tryptophan hydroxylase (the rate-limiting enzyme in the synthesis of 5-HT) in the DRN (66,71,72). In this locus, progesterone and 17-β-estradiol treatments have been shown not only to modify 5-HT transporter mRNA levels and its efficacy (73,74), but also the 5-HT$_{1A}$ autoreceptor gene expression (72). Furthermore, in ovariectomized rats, 17-β-estradiol treatments reduces the response of the 5-HT$_{1A}$ autoreceptor (75) whereas the response might be increased or decreased at the postsynaptic level, depending on the region (71,76). Finally, modifications of other receptors such as the 5-HT$_{2A}$ or the 5-HT$_3$ have also been reported (77-79).

As stated above, it is well known that the risk for mood disorders varies markedly with pregnancy. During this period, the levels of progesterone, as well as those of other neuroactive steroids, are the highest ever in a woman's life (80). The plasma levels of progesterone in pregnant rats are increased by six-fold in late pregnancy, and these levels decrease very rapidly just before delivery (81).

14. Major depression

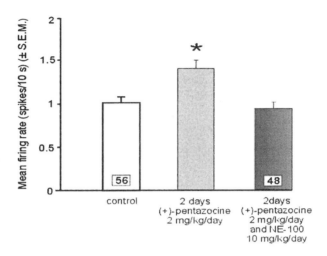

Figure 14-4. Mean firing activity expressed as spikes/10 seconds (mean + S.E.M.) of dorsal raphe nucleus 5-HT neurons measured in rats treated with saline (control), (+)-pentazocine (2 mg/kg/day), or coadministered (+)-pentazocine (2 mg/kg/day) and NE-100 (10 mg/kg/day) for 2 days. *P<0.05. Adapted from ref. (58).

Interestingly, estrogen and progesterone receptors are present in the DRN, but female DRN contain 30% more progesterone receptors than males, while the amount of estrogen receptors is the same between sexes, suggesting the possibility of a greater sensitivity of the female serotonergic system to progesterone modulation (82). A similar gender difference has also been reported for 5-HT_{1A} and 5-HT_{2A} receptors (83). Because of the evidence suggesting an interaction between neurosteroids and 5-HT neurotransmission, we assessed their effect in our electrophysiological model.

We first investigated potential variations in the function of DRN 5-HT neurons with gender and pregnancy. A gender difference in the spontaneous firing rate of 5-HT neurons was found with male 5-HT neurons having a 50% higher firing frequency than the female ones (84). Conversely, during pregnancy, there was a striking correlation between the firing rate of 5-HT neurons and progesterone levels. Indeed, the 5-HT neuronal firing activity increased in parallel with progesterone; it peaked at the 17th day of pregnancy (Figure 14-4) and then declined just before parturition. These results strongly suggested a role for progesterone in modulating the spontaneous firing activity of female 5-HT neurons during pregnancy (84).

However, short- and long-term treatments with progesterone up to a dose of 1000 μg/kg/day failed to modify the basal firing activity of 5-HT neurons in female rats, suggesting that the increased firing activity observed during pregnancy was not due directly to progesterone, that it could be induced by some of its metabolites or other neurosteroids. Indeed, several progesterone metabolites like 3-α-dihydroxy-5-α-pregnane-20-one and 3-β-5-α-pregnane-3-α-20-one administered at the dose of 50 μg/kg, i.c.v. for one week increase markedly the firing activity of 5-HT neurons. A similar effect was observed with other neurosteroids like 17-β-estradiol, testosterone or dehydroepiandrosterone (DHEA). Interestingly, some of these effects like those of DHEA were reversed by NE-100 whereas others were not (85).

In pregnant rats, the dose-response curve assessing the response of the 5-HT neurons to the intravenous administration of lysergic acid diethylamide (LSD) indicated that the 5-HT$_{1A}$ autoreceptor was partly desensitized. This could explain the enhanced 5-HT neuronal firing rate observed during pregnancy (86).

Because some progesterone metabolites are potent GABA$_A$ receptor modulators, we also compared the tonic GABAergic inhibition of 5-HT neurons between females, males and pregnant females (at the 17th day of gestation). We found that the tonic inhibitory GABAergic input onto 5-HT neurons was drastically reduced at Day 17 in pregnant rats, whereas in male, a decreased sensitivity of GABA$_A$ receptors was found as compared to females (86).

Obviously, the modulation of 5-HT neurotransmission by σ ligands and neurosteroids could constitute an important part in the mediation of their antidepressant-like effects. However, this could not constitute the only aspect, as some σ ligands like igmesine did not appear to modify 5-HT function.

8.   σ RECEPTORS AND NMDA RECEPTORS

We had previously designed another electrophysiological model to differentiate σ agonists and antagonists. Specifically, this model involves extracellular recordings of CA3 dorsal hippocampus pyramidal neuronal response to microiontophoretic applications of NMDA, in urethane anesthetized male rats. In this model, σ agonists (e.g. DTG, igmesine, (+)-pentazocine) potentiate the response to microiontophoretic applications of NMDA while σ antagonists (e.g. NE-100, BD1047, haloperidol) have no effect on their own but block the effect of the σ agonists (87-89). The effects of the agonists demonstrate bell-shaped dose-response curves, such that low doses (0.5-50 µg/kg) potentiate the NMDA response whereas the effects of higher doses (>500 µg/kg) progressively decrease and disappear (Figure 14-1). At high doses, σ "agonists" can even act as "antagonists" as they reverse or prevent the potentiation induced by other agonists (90,93). Bell-shaped dose response curves have also been demonstrated with σ ligands in several other models (94-97,20,92,98-100,101).

Interestingly, in this electrophysiological model also, σ ligands and neurosteroids have similar properties. Progesterone and testosterone have no effect on the NMDA response when administered intravenously (1-1000 µg/kg) but suppress, at low doses (20 µg/kg, i.v.), the potentiation induced by selective σ agonists such as DTG (89,102). Conversely the neurosteroid DHEA is acting as a σ agonist. It potentiates selectively, in a dose-dependent manner, the neuronal response to NMDA. This potentiation is reversed by low doses of progesterone, testosterone, NE-100 and haloperidol. Pregnenolone and pregnenolone sulphate, which have very low affinity for σ receptors (64) are ineffective in potentiating the NMDA response as well as in reversing the potentiation induced by DTG and DHEA. Furthermore in female rats, two weeks following ovariectomy, the degree of the potentiation of the NMDA response induced by DTG is significantly higher than in control female rats, suggesting that receptors might be tonically inhibited by endogenous progesterone (102,103).

9. NMDA RECEPTORS AND DEPRESSION

A number of studies have demonstrated interactions between several antidepressant drugs and the NMDA receptor complex. For example, the acute administration of desipramine, imipramine and nortriptyline drastically reduce NMDA-induced toxicity and seizures (104,105). Several classes of antidepressant also potentiate the behaviors induced by the NMDA antagonist, an effect which is blocked by haloperidol, but not by the D_1 and the D_2 selective antagonists SCH 23390 [(R)-(+)-7-chloro-8-hydroxy-3-methyl-1-phenyl-2,3,4,5-tetrahydro-1H-3-benzazepine] or sulpiride, respectively (106,107), thus suggesting that this effect of haloperidol is mediated by receptors whereas long-term treatments with tricyclics inhibit the binding of [^{3}H]dizocilpine (108). In radioligand binding studies, desipramine behaves as a non-competitive NMDA antagonist, acting on a site different from the glycine and the PCP binding sites (109,110). Moreover, it has also been reported that the acute administration of 5-HT$_{1A}$ agonists suppresses the behavioral syndromes induced by NMDA antagonists (111).

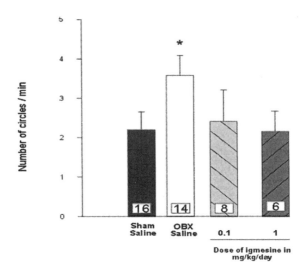

Figure 14-5. Effect of surgery and treatment with dizocilpine-induced behavior. Black column: Sham surgery + saline; White column: OBX surgery + saline; Light gray column: OBX + igmesine 100 µg/kg/day (21 days); Dark gray column: OBX + igmesine 1000 µg/kg/day (7 days). Adapted from ref. (126).

Conversely, NMDA antagonists have been reported to display some "antidepressant-like" characteristics in the forced swim test and the tail suspension test (112,113).

Olfactory bulbectomy (OBX) is currently recognized as one of the most valuable animal models of major depression (114-116). In rodents, OBX provokes a variety of neurochemical and behavioral alterations, which are not related to anosmia (115-118). The behavioral changes, appearing two to three weeks after the surgery, are characterized by hyperactivity, irritability, deficits in learning avoidance responses and spatial memory, disruption of sexual behavior, and sleep disturbances (119,120). Plasma levels of corticosterone are increased, as in depressed patients (114,118,121). Uptake of [^{3}H]5-HT into platelets, which is reduced during depression, and normalizes in patients responding to antidepressant treatment, is also reduced in OBX rats. Moreover, OBX results in widespread reductions of 5-HT turnover, in particular in the frontal cortex, the nucleus accumbens, the hippocampus and the striatum (115) whereas OBX induces an increase of 5-HT synthesis in the dorsal raphe nuclei and a decrease in the terminal regions (122). In opposition to other animal models of depression like the forced swim test, the time-frame of the effects of antidepressants in the OBX model is compatible with that observed in depressed patients (123).

Previous studies in our laboratory have shown that OBX induces a down-regulation of NMDA receptors. Following OBX, the hyperactivity induced by dizocilpine is markedly decreased. In keeping with this finding, within one week following OBX, [^{125}I]iodo-dizocilpine binding was decreased in the frontal and piriform cortices, and several limbic regions (124).

These different observations suggest that the NMDA receptors may be involved in the pathophysiology of depression and in the mechanism of action of antidepressant treatments.

As in our electrophysiological model, σ agonists are dose-dependently potentiating the NMDA response, we postulated that treatments with σ agonists by potentiating the effects of the down-regulated NMDA receptors produced by OBX, will compensate for this down regulation, thereby normalizing the response to dizocilpine and thus inducing smaller behavioral effects.

We investigated the effects of short- and long-term treatments with igmesine and L-678,384 (1-benzylspiro(1,2,3,4-tetrahydronaphthalene-1,4-piperidine). The OBX surgery increased the behavioral modifications on ambulatory distance and circling behavior induced by the acute administration of dizocilpine (200 mg/kg). A two-week treatment of OBX rats with low doses of igmesine and L-687,384 (125) (50-200 mg/kg/day) induced a reversal of this behavioral response to dizocilpine compared to OBX-saline treated rats (Figure 14-5). Interestingly, short-term (2 day) treatments with low doses did not induce any behavioral changes, whereas a short-term treatment with high doses (500-1000 mg/kg) of igmesine reversed behavioral modifications induced by OBX, whereas long-term treatments with the same doses did not (126). The dose-dependent relationship is very similar to what was observed in depressed patients (see above) and follows the bell-shaped dose-response curve previously mentioned. It could explain why short-term treatments with low doses or long-term treatments with high doses do not have any effect, if they induced plasma levels outside of the "agonist range." As effective antidepressant treatments are expected to reverse OBX-induced alterations, our results constitute another argument suggesting that σ agonists may possess antidepressant properties (126).

10. CONCLUSIONS

Both preclinical and clinical data, even if limited, strongly suggest that σ agonists could act as antidepressant treatments.

The effects of σ ligands and neurosteroids on the efficacy of the 5-HT neurotransmission and on the modulation of the NMDA response could explain how they exert their antidepressant properties. However, it appears clearly that, as a class, σ agonists, (even if we restrict this definition to σ_1 agonists), do not act in an homogenous way. As stated above, a compound efficient in clinical studies like igmesine or PRE-084 does not appear to modify the activity of 5-HT neurons. Conversely, in the NMDA model, if the majority of σ agonists potentiate selectively the NMDA response, some will also increase the response to quisqualate, or act as σ "inverse agonist" by reducing the response to NMDA in certain conditions. Moreover, compounds like progesterone will have an effect similar to (+)-pentazocine by increasing the firing activity of dorsal raphe 5-HT neurons, whereas in the dorsal hippocampus it acts as a σ antagonist by reversing the potentiation of the NMDA response induced by (+)-pentazocine.

The lack of effect of igmesine and PRE-084 may be due to the existence of several different σ_1 receptor subtypes, which has previously been suggested by data from our and many other laboratories. For example, we have previously shown that the potentiation induced by DTG and igmesine is mediated by a subtype of σ_1 receptor linked to a $G_{i/o}$ protein, whereas potentiation induced by (+)-pentazocine is mediated by another subtype of the σ_1 receptor not linked to a $G_{i/o}$ (127). Furthermore, in our NMDA model, following colchicine pretreatment, which destroys the mossy fibre system, the neuronal response induced by DTG and igmesine was abolished while (+)-pentazocine effect persisted, indicating the σ_1 receptor subtype mediating (+)-pentazocine's effect is located postsynaptically on pyramidal neurons while the σ_1 receptor subtype mediating DTG and igmesine's effects is located presynaptically (128). Furthermore, the potentiation of the NMDA response by (+)-pentazocine is reversed by naloxone, an opiate antagonist, while the potentiating effects of igmesine, BD737 [(+)-cis-N-methyl-N-[2-(3,4-dichlorophenyl)ethyl]-2-(1-pyrrolidinyl)cyclohexylamine] and L-687,384 were not (129). But even the existence of several subtypes of σ receptors would not readily explain why some neurosteroids like progesterone and testosterone could act as "agonists" in one model and "antagonists" in another. Obviously, we will have to take into account the intervention of other neurotransmitters like GABA to completely understand the effect of neurosteroids and possibly of σ ligands.

14. Major depression

The very recent creation of σ_1 knockout mice (130) and the future cloning of other subtypes of σ receptors will certainly be of a great help to understand these apparent discrepancies and thus obtain a reliable knowledge on the mechanisms of action of the different classes of σ ligands underlying their potential antidepressant properties.

Nonetheless, all the data accumulated up to now are certainly very encouraging and strongly support the notion that σ ligands could represent a totally new class of antidepressant. Their original mechanism of action could suggest a different clinical response and a better efficacy for the large number of depressed patients not responding adequately to classical antidepressants.

REFERENCES

1. Martin WR, Eades CG, Thompson JA, Huppler RE, Gilbert PE. The effects of morphine- and nalorphine-like drugs in the nondependent and morphine-dependent chronic spinal dog. J Pharmacol Exp Ther 1976, 197:517-532.
2. Quirion R, Bowen WD, Itzhak Y, Junien JL, Musacchio JM, Rothman RB, Su TP, Taylor DP. Classification of sigma binding sites: A proposal. In: *Multiple Sigma and PCP Receptor Ligands. Mechanisms for Neuromodulation and Neuroprotection?* In Kamenka JM, Domino EF (eds), pp. 927-933 (NPP Books, Ann Arbor, 1992).
3. Hanner M, Moebius FF, Flandorfer A, Knaus HG, Striessnig J, Kempner E, Glossmann H. Purification, molecular cloning, and expression of the mammalian σ_1-binding site. Proc Nat Acad Sci USA 1996, 93:8072-8077.
4. Largent BL, Gundlach AL, Snyder SH. Psychotomimetic opiate receptors labeled and visualized with (+)[^{3}H]3-(3-hydroxyphenyl)-N-(1-propyl) piperidine. J Neurosci 1984, 81:4983-4987.
5. Tam SW, Cook L. Sigma opiates and certain antipsychotic drugs mutually inhibit (+)-[^{3}H]SKF 10047 and [^{3}H]haloperidol binding in guinea pig brain membranes. Proc Natl Acad Sci USA 1984, 81:5618-5621.
6. Su TP. Evidence for σ opioid receptor: Binding of [^{3}H]SKF-10047 to etorphine-inaccessible sites in guinea-pig brain. J Pharmacol Exp Ther 1982, 223:284-290.
7. Su TP. HR 375: A potential antipsychotic drug that interacts with dopamine D_2 receptors and σ-receptors in the brain. Neurosci Lett 1986, 71:224-228.
8. Snyder SH, Largent BL. Receptor mechanisms in antipsychotic drug action: Focus on receptors. J Neuropsychiatry 1989, 1:7-15.
9. Cook L, Tam SW, Rochbach KW. Multiple σ and PCP receptor ligands: Mechanisms for neuromodulation and neuroprotection. In Kamenka JM, Domino EF (eds) (NPP Books, Ann Harbor, 1992).
10. Taylor DP, Dekleva J. Potential antipsychotic BMY 14802 selectively binds to σ sites. Drug Dev Res 1987, 11:65-70.
11. Taylor DP, Dekleva J. Sigma and phencyclidine like compounds as molecular probes in biology. In EF Domino, Kamenka JM (eds), pp. 345-355 (NPP Books, Ann Arbor, 1988).

12. Schmidt A, Lebel L, Koe BK, Seeger T, Heym J. Sertraline potently displaces (+)-[^{3}H]3-PPP binding to σ-sites in rat brain. Eur J Pharmacol 1989, 165:335-336.
13. Itzhak Y, Kassim CO. Clorgyline displays high affinity for σ-binding sites in C57BL/6 mouse brain. Eur J Pharmacol 1990, 176:107-108.
14. Bergeron R, Debonnel G, de Montigny C. Modification of the N-methyl-D-aspartate response by antidepressant σ receptor ligands. Eur J Pharmacol 1993, 240:319-323.
15. Roman FJ, Pascaud X, Martin B, Vauche D, Junien JL. JO-1784, a potent and selective ligand for rat and mouse brain σ sites. J Pharm Pharmacol 1990, 42:439-440.
16. Matsuno K, Nakazawa M, Okamoto K, Kawashima Y, Mita S. Binding properties of SA4503, a novel and selective $σ_1$ receptor agonist. Eur J Pharmacol 1996, 306:271-279.
17. Weber E, Sonders M, Quarum M, McLean S, Pou S, Keana JFW. 1,3-Di(2-[5-^{3}H]tolyl)guanidine: A selective ligand that labels σ-type receptors for psychotomimetic opiates and antipsychotic drugs. Proc Natl Acad Sci USA 1986, 83:8784-8788.
18. Steinfels GF, Alberici GP, Tam SW, Cook L. Biochemical, behavioral, and electrophysiologic actions of the selective σ receptor ligand (+)pentazocine. Neuropsychopharmacology 1988, 1:321-327.
19. Matsumoto RR, Bowen WD, Tom MA, Vo VN, Truong DD, De Costa BR. Characterization of two novel σ receptor ligands: Antidystonic effects in rats suggest σ receptor antagonism. Eur J Pharmacol 1995, 280:301-310.
20. Okuyama S, Imagawa Y, Ogawa S, Araki H, Ajima A, Tanaka M, Muramatsu M, Nakazato A, Yamaguchi K, Yoshida M. NE-100, a novel σ receptor ligand: In vivo tests. Life Sci 1993, 53:PL285-PL290.
21. Kobayashi T, Matsuno K, Nakata K, Mita S. Enhancement of acetylcholine release by SA4503, a novel $σ_1$ receptor agonist, in the rat brain. J Pharmacol Exp Ther 1996, 279:106-113.
22. Urani A, Roman FJ, Phan VL, Su TP, Maurice T. The antidepressant-like effect induced by $σ_1$-receptor agonists and neuroactive steroids in mice submitted to the forced swimming test. J Pharmacol Exp Ther 2001, 298:1269-1279.
23. Tottori K, Miwa T, Uwahodo Y, Yamada S, Nakai M, Oshiro Y, Kikuchi T, Altar CA. Antidepressant-like responses to the combined σ and 5-HT_{1A} receptor agonist OPC-14523. Neuropharmacology 2001, 41:976-988.
24. Skuza G, Rogoz Z. Effect of combined treatment with selective σ ligands and amantadine in the forced swimming test in rats. Pol J Pharmacol 2002, 54:699-702.
25. Urani A, Privat A, Maurice T. The modulation by neurosteroids of the scopolamine-induced learning impairment in mice involves an interaction with $σ_1$ receptors. Brain Res 1998, 799:64-77.
26. Kamei H, Noda Y, Nabeshima T. The psychological stress model using motor suppression. Nippon Yakurigaku Zasshi 1999, 113:113-120.
27. Noda Y, Kamei H, Nabeshima T. Sigma-receptor ligands and anti-stress actions. Nippon Yakurigaku Zasshi 1999, 114:43-49.
28. Sanchez C, Papp M. The selective $σ_2$ ligand Lu 28-179 has an antidepressant-like profile in the rat chronic mild stress model of depression. Behav Pharmacol 2000, 11:117-124.
29. Goldberg HL, Finnerty R. An open-label, variable-dose study of WY-45,030 (venlafexine) in depressed outpatients. Psychopharmacol Bull 1988, 24:198-199.
30. Schweizer E, Clary C, Weise C, Rickels K. An open-label, dose-finding study of WY-45,030, a novel bicyclic antidepressant. Psychopharmacol Bull 1988, 24:195-197.
31. Tollefson GD, Thompson RG, Heiligenstein JH, James S, Wong DT, Faries DE, Hyslop DL. Is the potent, dual uptake inhibition of norepinephrine and serotonin by LY248686 (duloxetine) associated with a more rapid and robust antidepressant response? ACNP Abstacts, 1993, 32.23.

32. Pande AC, Geneve J, Scherrer B. Igmesine, a novel σ ligand, has antidepressant properties. Int J Neuropsychopharmacol 1998; 1:S8-S9.
33. Boyd JH, Weissman MM. Epidemiology of affective disorders: a reexamination and future directions. Gen Psychiatry 1981, 38:1039-1046.
34. Weissman MM, Olfson M. Depression in women: implications for health care research. Science 1995, 269:799-801.
35. Nott PN, Franklin M, Armitage C, Gelder MG. Hormonal changes and mood in the puerperium. Br J Psychiatry 1976, 128:379-383.
36. Watson RD, Esler MD, Leonard P, Korner PI. Influence of variation in dietary sodium intake on biochemical indices of sympathetic activity in normal man. Clin Exp Pharmacol Physiol 1984, 11:163-170.
37. Fuchs AR, Fuchs F. Endocrinology of human parturition: a review. Br J Obstet Gynaecol 1984, 91:948-967.
38. Whiffen VE. The comparison of postpartum with non-postpartum depression: a rose by any other name. J Psychiatry Neurosci 1991, 16:160-165.
39. O'Hara MW, Schlechte JA, Lewis DA, Wright EJ. Prospective study of postpartum blues. Biologic and psychosocial factors. Arch Gen Psychiat 1991, 48:801-806.
40. Sipes SL, Malee MP. Endocrine disorders in pregnancy. Obstet Gynecol Clin North Am 1992, 19:655-677.
41. Glover V. Do biochemical factors play a part in postnatal depression? Prog Neuropsychopharmacol Biol Psychiatry 1992, 16:605-615.
42. Harris B, Lovett L, Newcombe RG, Read GF, Walkere R, Riad-Fahmy D. Maternity blues and major endocrine changes: cardiff puerperal mood and hormone study. Br Med J 1994, 308:949-953.
43. Pugh TF, Jerath BK, Smith WM, Reed RB. Rates of mental disease related to childbearing. New Engl J Med 1963, 268:1228.
44. Meltzer HY, Lowy MT. Psychopharmacology: The third generation of progress. In Meltzer HY (ed), pp. 513-526 (Raven Press, New York, 1987).
45. Siever LJ. Psychopharmacology The third generation of progress. Meltzer HY (ed), pp. 493-504 (Raven Press, New York, 1987).
46. Lapin IP, Oxenkrug GF. Intensification of the central serotonergic processes as a possible determinant of the thymoleptic effect. Lancet 1969, 18:132-136.
47. Coppen A. The biochemistry of affective disorders. Br J Psychiatry 1967, 113:1237-1264.
48. Blier P, de Montigny C. Current advances and trends in the treatment of depression. Trends Pharmacol Sci 1994, 15:220-226.
49. Owens MJ. Molecular and cellular mechanisms of antidepressant drugs. Depress Anxiety 1996, 4:153-159.
50. Bijak M. Monoamine modulation of the synaptic inhibition in the hippocampus. Acta Neurobiol Exp 1996, 56:385-395.
51. Hoyer D, Martin GR. Classification and nomenclature of 5-HT receptors: a comment on current issues. Behav Brain Res 1996, 73:263-268.
52. Nicholson VJ, Wieringa JH, Van Delf AML. Comparative pharmacology of mianserin, its main metabolites and 6-azamianserin. Naunyn Schmiedeberg Arch Pharmacol 1982, 319:48-53.
53. Chaput Y, de Montigny C, Blier P. Effects of a selective 5-HT reuptake blocker, citalopram, on the sensitivity of 5-HT autoreceptors: Electrophysiological studies in the rat. Naunyn Schmiedeberg Arch Pharmacol 1986, 333:342-348.

54. Beique JC, Blier P, de Montigny C, Debonnel G. Potentiation by (-)pindolol of the activation of postsynaptic 5-HT$_{1A}$ receptors induced by venlafaxine. Neuropsychopharmacology 2000, 23:294-306.
55. Aghajanian GK. Essays in neurochemistry and neuropharmacology. In Youdim MBH, Lovenberg W, Sharman DF, Lagnado JR (eds), pp. 1-32 (John Wiley and Sons, New-York, 1978).
56. Su TP, Wu XZ, Cone EJ, Shukla K, Gund TM, Dodge AL, Parish DW. Sigma compounds derived from phencyclidine: Identification of PRE-084, a new, selective σ ligand. J Pharmacol Exp Ther 1991, 259:543-550.
57. John CS, Vilner BJ, Bowen WD. Synthesis and characterization of [^{125}I]-N-(N-benzylpiperidin-4-yl)-4-iodobenzamide, a new σ receptor radiopharmaceutical: High-affinity binding to MCF-7 breast tumor cells. J Med Chem 1994, 37:1737-1739.
58. Bermack JE, Debonnel G. Modulation of serotonergic neurotransmission by short- and long-term treatments with σ ligands. Br J Pharmacol 2001, 134:691-699.
59. Chaput Y, Blier P, de Montigny C. *In vivo* electrophysiological evidence for the regulatory role of autoreceptors on serotonergic terminals. J Neurosci 1986, 6:2796-2801.
60. de Montigny C, Blier P, Chaput Y. Electrophysiologically-identified serotonin receptors in the rat CNS. Effect of antidepressant treatment. Neuropharmacology 1984, 23:1511-1520.
61. Blier P, de Montigny C. Short-term lithium administration enhances serotonergic neurotransmission: electrophysiological evidence in the rat CNS. Eur J Pharmacol 1985, 113:69-77.
62. Blier P, de Montigny C. Current advances and trends in the treatment of depression. Trends Pharmacol Sci 1994, 15:220-224.
63. Bermack J, Lavoie N, Debonnel G. Modulation of serotonergic neurotransmision by σ ligands in acute and chronic treatments in the rat: Electrophysiological studies in the rat. Soc Neurosci Abst 2000, 30:416.
64. Su TP, London ED, Jaffe JH. Steroid binding at s receptors suggests a link between endocrine, nervous and immune systems. Science 1988, 240:210-221.
65. McCann DJ, Weissman AD, Su TP. σ$_1$ and σ$_2$ sites in rat brain: Comparison of regional, ontogenetic, and subcellular patterns. Synapse 1994, 17:182-189.
66. Bethea CL, Gundlah C, Mirkes SJ. Ovarian steroid action in the serotonin neural system of macaques. Novartis Found Symp 2000, 230:112-130.
67. Paul SM, Purdy RH. Neuroactive steroids. FASEB J 1992, 6:2311-222X.
68. Weiland NG, Orchinik M. Specific subunit mRNAs of the GABAA receptor are regulated by progesterone in subfields of the hippocampus. Brain Res Mol Brain Res 1995, 32:271-278.
69. Baulieu EE. Neurosteroids: a new function in the brain. Biol Cell 1991, 71:3-10.
70. Wu FS, Gibbs TT, Farb DH. Pregnenolone sulfate: a positive allosteric modulator at the N-methyl-D-aspartate receptor. Mol Pharmacol 1991, 40:333-336.
71. Bethea CL, Mirkes SJ, Shively CA, Adams MR. Steroid regulation of tryptophan hydroxylase protein in the dorsal raphe of macaques. Biol Psychiatry 2000, 47:562-576.
72. Pecins-Thompson M, Bethea CL. Ovarian steroid regulation of serotonin-1A autoreceptor messenger RNA expression in the dorsal raphe of rhesus macaques. Neuroscience 1999, 89:267-277.
73. McQueen JK, Wilson H, Fink G. Estradiol-17 β increases serotonin transporter (SERT) mRNA levels and the density of SERT-binding sites in female rat brain. Brain Res Mol Brain Res 1997, 45:13-23.

74. Pecins-Thompson M, Brown NA, Bethea CL. Regulation of serotonin re-uptake transporter mRNA expression by ovarian steroids in rhesus macaques. Mol Brain Res 1998, 53:120-129.
75. Lakoski JM. Cellular electrophysiological approaches to the central regulation of female reproductive aging. Alan R. Liss Inc. (ed.), pp. 209-220, (Neural Control of Reproductive Function, 1989).
76. Raap DK, DonCarlos L, Garcia F, Muma NA, Wolf WA, Battaglia G, Van De Kar LD. Estrogen desensitizes 5-HT$_{1A}$ receptors and reduces levels of G(z), G(i1) and G(i3) proteins in the hypothalamus. Neuropharmacology 2000, 39:1823-1832.
77. Cyr M, Landry M, Di Paolo T. Modulation by estrogen-receptor directed drugs of 5-hydroxytryptamine-2A receptors in rat brain. Neuropsychopharmacology 2000, 23:69-78.
78. Summer BE, Fink G. Estrogen increases the density of 5-hydroxytryptamine(2A) receptors in cerebral cortex and nucleus accumbens in the female rat. J Steroid Biochem Mol Biol 1995, 54:15-20.
79. Wetzel CH, Hermann B, Behl C, Pestel E, Rammes G, Zieglgansger W, Holsboer F, Rupprecht R. Functional antagonism of gonadal steroids at the 5-hydroxytryptamine type 3 receptor. Mol Endocrinol 1998, 12:1441-1451.
80. Buster JE. Gestational changes insteroid hormone biosynthesis, secertion, metabolism and action. Clin Perinatol 1983, 10:527-552.
81. Pepe GJ, Rothchild I. A comparative study of serum progesterone levels in pregnancy and in various types of pseudopregnancy in the rat. Endocrinology 1974, 95:275-279.
82. Alves SE, Weiland NG, Hayashi S, McEwen BS. Immunocytochemical localization of nuclear estrogen receptors and progestin receptors within the rat dorsal raphe nucleus. J Comp Neurol 1998, 391:322-334.
83. Zhang L, Ma W, Barker JL, Rubinow DR. Sex differences in expression of serotonin receptors (subtypes 1A and 2A) in rat brain: A possible role of testosterone. Neuroscience 1999, 94:251-259.
84. Klink R, Robichaud M, Debonnel G. Gender and gonadal status modulation of dorsal raphe nucleus serotonergic neurons: I Effects of gender and pregnancy. Neuropharmacology 2002, 43:1119-1128.
85. Robichaud M, Debonnel G. Neurosteroid modulation of dorsal raphe nucleus serotonergic neurons. Soc Neurosci Abst 2000, 30:369.
86. Robichaud M, Klink R, Debonnel G. Gender and gonadal status modulation of dorsal raphe nucleus serotonergic neurons: II Regulatory mechanisms. Neuropharmacology 2002, 43:1129-1138.
87. Debonnel G, Monnet FP, de Montigny C. Sigma ligands potentiate NMDA-induced hippocampal neuron activation. Soc Neurosci Abst 1990, 16:396.11.
88. Monnet FP, Debonnel G, de Montigny C. *In vivo* electrophysiological evidence for a selective modulation of N-methyl-D-aspartate-induced neuronal activation in rat CA$_3$ dorsal hippocampus by σ ligands. J Pharmacol Exp Ther 1992, 261:123-130.
89. Bergeron R, de Montigny C, Debonnel G. Potentiation of neuronal NMDA response induced by dehydroepiandrosterone and its suppression by progesterone: effects mediated via σ receptors. J Neurosci 1996, 16:1193-1202.
90. Bergeron R, Debonnel G, de Montigny C. Biphasic effects on NMDA response of two antidepressants with high affinity for σ sites. Soc Neurosci Abst 1992, 18:16.9.
91. Debonnel G, Bergeron R, de Montigny C. Electrophysiological evidence for the existence of sub-types of σ receptors in the rat dorsal hippocampus. I. Antagonistic effects of high doses of sertraline, clorgyline and L-687-384. Soc Neurosci Abst 1992, 18:16.10.

92. Bergeron R, de Montigny C, Debonnel G. Biphasic effects of σ ligands on the neuronal response to N-methyl-D-aspartate. Naunyn Schmiedeberg Arch Pharmacol 1995, 351:252-260.
93. Bergeron R, Debonnel G. Effects of low and high doses of selective σ ligands: further evidence suggesting the existence of different subtypes of σ receptors. Psychopharmacology 1997, 129:215-224.
94. Maurice T, Hiramatsu M, Kameyama T, Hasegawa T, Nabeshima T. Behavioral evidence for a modulating role of σ ligands in memory processes. I. Attenuation of dizocilpine (MK-801)-induced amnesia. Brain Res 1994, 647:44-56.
95. Monnet FP, Debonnel G, de Montigny C. Neuropeptide-Y selectively potentiates N-methyl-D-aspartate-induced neuronal activation. Eur J Pharmacol 1990, 182:207-208.
96. Monnet FP, Debonnel G, Junien JL, de Montigny C. N-methyl-D-aspartate-induced neuronal activation is selectively modulated by σ receptors. Eur J Pharmacol 1990, 179:441-445.
97. Monnet FP, Debonnel G, de Montigny C. Electrophysiological evidence for the existence of subtypes of σ receptors in the rat dorsal hippocampus. II. Effect of lesioning the mossy fiber system. Soc Neurosci Abst 1992, 18:16.11.
98. Martin WJ, Roth JS, Walker JM. The effects of σ compounds on both NMDA- and non NMDA-mediated neuronal activity in rat hippocampus. Soc Neurosci Abst 1992, 18:16.6.
99. Bergeron R, Debonnel G, de Montigny C. Modification of the N-methyl-D-aspartate response by antidepressant σ receptor ligands. Eur J Pharmacol 1993, 240:319-323.
100. Walker JM, Hunter WS. Role of σ receptors in motor and limbic system function. Neuropsychopharmacology 1994, 10:837s.
101. Chaki S, Tanaka M, Muramatsu M, Otomo S. NE-100, a novel potent σ ligand, preferentially binds to σ_1 binding sites in guinea pig brain. Eur J Pharmacol 1994, 251:R1-R2.
102. Debonnel G, Bergeron R, de Montigny C. Potentiation by dehydroepiandrosterone of the neuronal response to N-methyl-D-aspartate in the CA_3 region of the rat dorsal hippocampus: An effect mediated via σ receptors. J Endocrinol 1996, 150:S33-S42.
103. Bergeron R, de Montigny C, Debonnel G. Potentiation of neuronal NMDA response induced by dehydroepiandrosterone and its suppression by progesterone: Effects mediated via σ receptors. J Neurosci 1996, 16:1193-1202.
104. Leander JD. Tricyclic antidepressants block N-methyl-D-aspartic acid-induced lethality in mice. Br J Pharmacol 1989, 96:256-258.
105. Watanabe Y, Saito H, Abe K. Tricyclic antidepressants block NMDA receptor-mediated synaptic responses and induction of long-term potentiation in rat hippocampal slices. Neuropharmacology 1993, 32:479-486.
106. Maj J, Rogóz Z, Skuza G. Antidepressant drugs increase the locomotor hyperactivity induced by MK-801 in rats. J Neural Transm 1991, 85:169-179.
107. Maj J, Rogóz Z, Skuza G, Sowinska H. The effect of antidepressant drugs on the locomotor hyperactivity induced by MK-801, a non-competitive NMDA receptor antagonist. Neuropharmacology 1992, 31:685-691.
108. Kitamura Y, Zhao X-H, Takei M, Yonemitsu O, Nomura Y. Effects of antidepressants on the glutamatergic system in mouse brain. Neurochem Int 1991, 19:247-253.
109. Reynolds IJ, Miller RJ. [^{3}H]MK801 binding to the NMDA receptor/ionophore complex is regulated by divalent cations: evidence for multiple regulatory sites. Eur J Pharmacol 1988, 151:103-112.

110. Sills MA, Loo PS. Tricyclic antidepressants and dextromethorphan bind with higher affinity to the phencyclidine receptor in the absence of magnesium and L-glutamate. Mol Pharmacol 1989, 36:160-165.
111. Löscher W, Hönack D. The behavioural effects of MK-801 in rats: Involvement of dopaminergic, serotonergic and noradrenergic systems. Eur J Pharmacol 1992, 215:199-208.
112. Abreu P, Sugden D. Characterization of binding sites for [^{3}H]-DTG, a selective σ receptor ligand, in the sheep pineal gland. Biochem Biophys Res Comm 1990, 171:875-881.
113. Trullas R, Skolnick P. Functional antagonists at the NMDA receptor complex exhibit antidepressant actions. Eur J Pharmacol 1990, 185:1-10.
114. Jesberger JA, Richardson JS. Brain output dysregulation induced by olfactory bulbectomy: an approximation in the rat of major depressive disorder in humans? Int J Neurosci 1988, 38:241-265.
115. Lumia AR, Teicher MH, Salchli F, Ayers E, Possidente B. Olfactory bulbectomy as a model for agitated hyposerotonergic depression. Brain Res 1992, 587:181-185.
116. Kelly JP, Wrynn AS, Leonard BE. The olfactory bulbectomized rat as a model of depression: An update. Pharmac Ther 1997, 74:299-316.
117. Grecksch G, Zhou D, Franke C, Schroder U, Sabel B, Becker A, Heuther G. Influence of olfactory bulbectomy and subsequent imipramine treatment on 5-hydroxytryptaminergic presynapses in the rat frontal cortex: behavioural correlates. Br J Pharmacol 1997, 122:1725-1731.
118. Leonard BE, Tuite M. Anatomical, physiological and behavioral aspects of olfactory bulbectomy in the rat. Int Rev Neurobiol 1981, 22:251-286.
119. Sakurada T, Kisara K. Effects of p-chlorophenylalanine (p-CPA) on sleep in olfactory bulb lesioned rats. Jpn J Pharmacol 1977, 27:389-395.
120. Sakurada T, Shima K, Tadano T, Sakurada S, Kisara K. Sleep-wakefulness rhythms in the olfactory bulb lesioned rats. Jpn J Pharmacol 1976, 26:97P.
121. Van Riezen H, Leonard BE. Effects of psychotropic drugs on the behavior and neurochemistry of olfactory bulbectomized rats. Pharmac Ther 1990, 47:21-34.
122. Watanabe A, Tohyama Y, Nguyen KO, Hasegawa S, Debonnel G, Diksic M. Regional brain serotonin synthesis is increased in the olfactory bulbectomy rat model of depression: an autoradiographic study. J Neurochem 2003, 85:469-475.
123. Butler J, Tannian M, Leonard BE. The chronic effects of desipramine and sertraline on platelet and synaptosomal 5HT uptake in olfactory bulbectomised rats. Prog Neuro-Psychopharmacol Biol Psychiat 1988, 12:585-594.
124. Beauchemin V, Lavoie N, Dennis T. Quantitative autoradiographic studies of the effects of olfactory bulbectomy in the rat brain: N-methyl-D-aspartate receptors. IUPHAR Abst. XII, P.13:2.11. 1994.
1. Middlemiss DN, Billington DC, Chambers M, Huston PH, Knight AK, Russell M, Thorn L, Tricklebank MD, Wong EHF. L-687,384 is a potent, selective ligand at the central σ recognition site. Br J Pharmacol 1991, 102:153.
125. Bermack J, Lavoie N, Dryver E, Debonnel G. Effects of σ ligands on NMDA receptor function in the bulbectomy model of depression: a behavioral study in the rat. Int J Neuropsychopharmacol 2002, 5:53-62.
126. Monnet FP, Debonnel G, Bergeron R, Gronier B, de Montigny C. The effects of σ ligands and of neuropeptide Y on N-methyl-D-aspartate-induced neuronal activation of CA_3 dorsal hippocampus neurons are differentially affected by pertussis toxin. Br J Pharmacol 1994, 112:709-715.

127. Debonnel G, Bergeron R, Monnet FP, de Montigny C. Differential effects of σ ligands on the N-methyl-D-aspartate response in the CA1 and CA3 regions of the dorsal hippocampus: Effect of mossy fiber lesioning. Neuroscience 1996, 71:977-987.
128. Couture S, Debonnel G. Some of the effects of the selective σ ligand (+)pentazocine are mediated via a naloxone-sensitive receptor. Synapse 2001, 39:323-331.
129. Langa F, Codony X, Tovar V, Lavado A, Gimenez E, Cozar P, Cantero M, Dordal A, Hernandez E, Perez R, Monroy X, Zamanillo D, Guitart X, Montoliu L. Generation and phenotypic analysis of σ receptor type I (σ_1) knockout mice. Eur J Neurosci 2003, 18:2188-2196.

Corresponding author: *Dr. Guy Debonnel, Mailing address: McGill University, Department of Psychiatry, 1033 Pine Avenue West, Montréal, Québec, Canada H3A 1A1, Phone: (514) 398-7310, FAX: (514) 398-4866, Electronic mail address: guy.debonnel@mcgill.ca*

Chapter 15

σ RECEPTORS AND DRUG ABUSE

Yun Liu, Yongxin Yu, Jamaluddin Shaikh, Buddy Pouw, AnTawan Daniels, Guang-Di Chen and Rae R. Matsumoto
Department of Pharmaceutical Sciences, University of Oklahoma Health Sciences Center, Oklahoma City, OK 73190, USA and Department of Pharmacology, University of Mississippi, Oxford, MS 38677, USA

1. INTRODUCTION

Drug abuse is a serious health and societal problem in industrialized and developing countries. Current treatments for drug abuse are limited and there is a need to develop new and effective pharmacological approaches to address this problem.

Many drugs of abuse interact with σ receptors, providing a logical target for medication development efforts. Drugs of abuse with significant affinities for σ receptors include some opiates, cocaine, amphetamines and phencyclidine (PCP). In addition, the actions of other abused substances, such as alcohol and nicotine, can be modulated by σ receptor ligands, even if the abused substances themselves do not bind to these receptors.

The following sections describe the manner in which σ receptors influence the actions of drugs of abuse. In addition, these sections summarize the ability of σ receptor ligands to alter the actions of drugs of abuse, and suggest their potential as medications for the treatment of drug abuse. Finally, directions for future research are provided.

2. CLASSICAL DOPAMINE HYPOTHESIS: LIMITATIONS AND ALTERNATE MEDICATION DEVELOPMENT TARGETS

For more than a decade, research on drug abuse has focused on the dopamine hypothesis (1). According to this hypothesis, "addiction" can be partially explained by the actions of drugs of abuse on reward pathways. Although the initial targets of drugs of abuse involve different receptors and transporters, they share a final common action of increasing dopamine neurotransmission in the brain reward system (2-4). Therefore, for over 15 years, intensive drug development efforts and research in the drug abuse field have focused on the dopamine system (5-7). However, no current medication robustly blocks drug reward or substantially relieves dependence in humans by targeting dopaminergic mechanisms (5). In recent years, medication development efforts which have focused on dopamine-related mechanisms have been further challenged by the introduction of transgenic mice in drug abuse research (8-10). While most studies have shown altered responding to drugs of abuse in these knockout animals, the compounds are still capable of producing rewarding or reinforcing effects. Together, these findings suggest that drug abuse research and medication development efforts must look beyond dopamine and its systems (8-12).

In recent years, research has begun focusing on cellular and molecular adaptations in the brain reward system that is produced by drugs of abuse (13,14). Post-receptor processes such as alterations in intracellular signaling mechanisms and gene expression have become the subject of intense research (15,16). Addiction research has also begun to overlap with the learning and memory and stress fields (17-20). With the exploration of potential new medication development targets, σ receptors emerged as a promising new avenue for drug abuse research (21,22).

3. MECHANISMS THROUGH WHICH σ RECEPTORS INFLUENCE THE ACTIONS OF DRUGS OF ABUSE

The mechanisms through which σ receptors influence the actions of drugs of abuse are diverse. These mechanisms can be divided into three categories, which are summarized below.

3.1 Direct binding of drugs of abuse to σ receptors

Many drugs of abuse interact with σ receptors at physiologically relevant concentrations. Receptor binding studies, which are summarized in Table 15-1, demonstrate that the following drugs of abuse can directly bind to σ receptors: cocaine, methamphetamine, methylenedioxymethamphetamine (MDMA), and phencyclidine. Many of these abused substances have preferential affinity for σ_1 receptors, which is the subtype that has been the primary focus of many recent studies and reviews (21,22). However, it is noteworthy that phencyclidine exhibits preferential affinity for σ_2 receptors (23), as do the (-)-isomers of opiate benzomorphans and morphinans (24). The ability of diverse classes of drugs of abuse to interact with σ receptors suggests that they can serve as a common medication development target to combat an array of abused substances (22).

3.2 σ Receptor-mediated modulation of other neurotransmitter systems

In addition to σ receptors, most drugs of abuse also interact with other receptors and transporters for classical neurotransmitter systems, such as dopamine (3,25), serotonin (26,27) and glutamate (28), to influence brain physiology and behaviors. Dopamine systems are accepted as important mediators of the actions of psychomotor stimulants, such as cocaine and methamphetamine. In addition, glutamate function is affected by drugs of abuse such as phencyclidine through NMDA receptors. Serotonin function is also altered by drugs of abuse such as MDMA. Interestingly, σ ligands can modulate the actions of all three of these neurotransmitter systems through a variety of mechanisms (28-31).

As described in earlier chapters, σ receptors are found in the plasma membrane of neurons, and also intracellularly. The ability of these receptors to translocate and interact with other proteins suggests diverse and varied mechanisms through which they can modulate the actions of drugs of abuse through classical neurotransmitter systems (28,31). Drugs of abuse can bind to σ receptors that are located on dopaminergic, serotonergic or glutamatergic neurons to affect their function. The activation of σ receptors by drugs of abuse may elicit alterations in ion channel function (32,33), second messenger signaling and/or trafficking of receptors for classical neurotransmitter systems (34-39). In addition to modulating the activity of dopaminergic, serotonergic and/or glutamatergic neurotransmission through post-synaptic mechanisms, the interaction of drugs of abuse with σ receptors may also alter pre-synaptic mechanisms that are involved in the synthesis and release of classical neurotransmitters (23,40).

3.3 σ Receptor mediated alterations in gene expression

Recent studies have shown that σ ligands can alter gene expression, including preventing changes that are induced by drugs of abuse (21,41-44,93). To date, these σ receptor-induced changes in gene expression have, for the most part, focused on immediate early genes, particularly those of the fos family of transcription factors (42-44). These changes are significant because many investigators have hypothesized that increases in immediate early gene expression by drugs of abuse may be the initial step by which these drugs alter the expression of late genes to produce enduring effects on nervous system function (45-51). The ability of σ receptor ligands to alter immediate early gene expression may therefore contribute to the transitional mechanisms which underlie changes in behavior over time when an organism responds to acute vs. repeated drug exposures (45-51). Moreover, the ability of σ antagonists to prevent psychostimulant-induced gene expression may help explain the ability of this class of compounds to combat an array of behaviors, both acute and subchronic, that are produced by drugs of abuse (52-54,93).

4. REPRESENTATIVE ABUSED DRUGS

The following sections summarize our current knowledge about interactions of representative drugs of abuse with σ receptors. In addition, these sections provide compelling evidence for the ability of σ receptor ligands to prevent a range of negative behaviors produced by drugs of abuse. In most cases, it appears that antagonism of σ_1 receptors is the most effective strategy for combating the actions of drugs of abuse (21,55). However, there are two notable exceptions. First, phencyclidine, which has preferential affinity for σ_2 receptors appears less responsive to σ_1 antagonists than many of the other abused substances described herein. Second, nicotine, which does not bind directly to σ receptors, appears to be attenuated by σ_1 agonists instead of antagonists. Due to the diverse mechanisms of actions of most abused substances and the flexibility with which σ receptors can affect neuronal function, putative mechanisms through which σ ligands combat the actions of each representative drug of abuse are described in each section.

15. Drug abuse

Table 15-1. Affinities (K_i in nM) of drugs of abuse for σ receptors

Drug of Abuse	Radioligand	Tissue	σ Subtype	Affinity (nM)	Reference
(−)-Cocaine	[³H]haloperidol	Rat cerebellum	σ_1	6700 ± 300	58
(−)-Cocaine	[³H](+)-pentazocine	Rat brain without cerebellum	σ_1	4635 ± 1146	93
(−)-Cocaine	[³H]DTG plus cold[a]	Rat brain without cerebellum	σ_2	39,606 ± 5580	93
(+)-Cocaine	[³H]haloperidol	Rat cerebellum	σ_1	25,900 ± 2400	58
(+)-Cocaine	[³H](+)-pentazocine	Rat brain without cerebellum	σ_1	24,393 ± 1206	93
(+)-Cocaine	[³H]DTG plus cold[a]	Rat brain without cerebellum	σ_2	63,050 ± 10031	93
Cocaine	[³H]DTG	Guinea pig brain	σ_1 and σ_2	>10,000	107
Cocaine	[³H]DTG	Rat brain without cerebellum	σ_1 and σ_2	7708 ± 1512	93
Cocaine	[³H](+)-pentazocine	Rat brain without cerebellum	σ_1	2909 ± 192	76
Cocaine	[³H]DTG plus cold[a]	Rat brain without cerebellum	σ_2	29,175 ± 12,945	76
Cocaine	[³H](+)-pentazocine	Mouse brain	σ_1	2280 ± 180	76
Cocaine	[³H]DTG plus cold[a]	Mouse brain	σ_2	30,950 ± 3520	76
Cocaine	[³H](+)-pentazocine	Rat heart	σ_1	14,655 ± 515[b]	93
Cocaine	[³H]DTG plus cold[a]	Rat heart	σ_2	29,800 ± 4200[b]	93
Cocaine	[³H](+)-pentazocine	Mouse heart	σ_1	5240 ± 1020	72
(−)-Norcocaine	[³H]haloperidol	Rat cerebellum	σ_1	13600 ± 1300	58
(−)-Norcocaine	[³H](+)-pentazocine	Rat brain without cerebellum	σ_1	2403 ± 234	93
(−)-Norcocaine	[³H]DTG plus cold[a]	Rat brain without cerebellum	σ_2	33,073 ± 9467	93
(−)-Norcocaine	[³H](+)-pentazocine	Mouse brain	σ_1	20,370 ± 2410	93
(−)-Norcocaine	[³H]DTG plus cold[a]	Mouse brain	σ_2	21,050 ± 3560	93

Table 15-1. Affinities (K_i in nM) of drugs of abuse for σ receptors (cont.)

Drug of Abuse	Radioligand	Tissue	σ Subtype	Affinity (nM)	Reference
(−)-Norcocaine	[³H](+)-pentazocine	Mouse heart	σ_1	6160 ± 800	93
(−)-Norcocaine	[³H](+)-pentazocine	Rat heart	σ_1	28,355 ± 1800[b]	93
(−)-Norcocaine	[³H]DTG plus cold[a]	Rat heart	σ_2	495 ± 276[b]	93
(−)-Cocaethylene	[³H](+)-pentazocine	Rat brain without cerebellum	σ_1	7827 ± 643	93
(−)-Cocaethylene	[³H]DTG plus cold[a]	Rat brain without cerebellum	σ_2	20,516 ± 541	93
(−)-Cocaethylene	[³H](+)-pentazocine	Rat heart	σ_1	30,900 ± 700[b]	93
(−)-Cocaethylene	[³H]DTG plus cold[a]	Rat heart	σ_2	7100 ± 3900[b]	93
(+)-Amphetamine	[³H]haloperidol	Rat cerebellum	σ_1	12,700 ± 1500	58
(+)-Amphetamine	[³H](+)-pentazocine	Rat brain without cerebellum	σ_1	24,810 ± 1329	93
(+)-Amphetamine	[³H]DTG plus cold[a]	Rat brain without cerebellum	σ_2	>100,000	93
S(+)-Methamphetamine	[³H](+)-pentazocine	Rat brain without cerebellum	σ_1	2161 ± 252	93
S(+)-Methamphetamine	[³H]DTG plus cold[a]	Rat brain without cerebellum	σ_2	46,673 ± 10,344	93
S(+)-Methamphetamine	[³H]DTG	Rat brain without cerebellum	σ_1 and σ_2	7927 ± 710	93
(±)-MDMA	[³H](+)-pentazocine	Rat brain without cerebellum	σ_1	3057 ± 45	93
(±)-MDMA	[³H]DTG plus cold[a]	Rat brain without cerebellum	σ_2	8889 ± 500	93
Morphine	[³H]DTG	Guinea pig brain	σ_1	>10,000	107
Morphine	[³H](+)-pentazocine	Rat brain without cerebellum	σ_1	>10,000	93
Morphine	[³H]DTG plus cold[a]	Rat brain without cerebellum	σ_2	>10,000	93
(+)-Morphine	[³H](+)-pentazocine	Rat brain without cerebellum	σ_1	24,390 ± 7215	93
(+)-Morphine	[³H]DTG plus cold[a]	Rat brain without cerebellum	σ_2	>100,000	93

15. Drug abuse

Table 15-1. Affinities (K_i in nM) of drugs of abuse for σ receptors (cont.)

Drug of Abuse	Radioligand	Tissue	σ Subtype	Affinity (nM)	Reference
Apomorphine	[^{3}H](+)-pentazocine	Guinea pig brain	σ_1	>10,000	98
Phencyclidine	[^{3}H]DTG	Guinea pig brain	σ_1 and σ_2	1050 ± 106[b]	107
Phencyclidine	[^{3}H]DTG	Guinea pig brain	σ_1 and σ_2	1020 ± 103	108
Phencyclidine	[^{3}H](+)-pentazocine	Guinea pig brain	σ_1	2871 ± 735[b]	98
Phencyclidine	[^{3}H](+)-pentazocine	Guinea pig brain	σ_1	1090 ± 490	109
Phencyclidine	[^{3}H](+)-pentazocine	Rat brain without cerebellum	σ_1	2754 ± 504	93
Phencyclidine	[^{3}H]DTG plus cold[a]	Rat brain without cerebellum	σ_2	425 ± 36	93

[a]σ_2 assays were conducted in the presence of cold (+)-pentazocine to block binding to σ_1 sites. [b]EC_{50} values (nM); all other values in K_i (nM).

DTG = di-o-tolylguanidine; MDMA = methylenedioxymethamphetamine

4.1 Cocaine

Psychomotor stimulants are a group of drugs that produce wakefulness, arousal and stimulatory behaviors. Among these drugs, cocaine is one of the most abused by humans and self-administered by animals. According to the 2001 Household Survey on Drug Abuse, during 2000, there were approximately 926,000 new cocaine users in the United States. The average age of those who first used cocaine during the year was 20 years. The 2001 Drug Abuse Warning Network survey showed that cocaine is the most frequently reported drug in emergency department visits (76 visits per 100,000 population). Although cocaine abuse is a serious health and societal problem, there are currently no effective medications to aid in breaking the cycle of abuse. Of the many medication development targets being pursued, σ receptors are among the most promising.

The hypothesis for an involvement of σ receptors in the effects of cocaine is supported by a number of facts. First, σ receptors are unique proteins located in many areas of the body, including the brain and heart, which are target organs for the addictive and toxic actions of cocaine (56,57). The ability of cocaine to bind to σ receptors was first reported in 1988 by Kuhar and coworkers (58). Since then, several additional laboratories have confirmed that cocaine has micromolar affinity for σ receptors (see Table 15-1) and these concentrations can be achieved in the body *in vivo* (59).

In addition to binding directly with σ receptors in organ systems that mediate the toxic and addictive actions of cocaine, σ ligands can modulate downstream neurotransmitter and effector systems which are traditionally believed to mediate the actions of cocaine. For example, σ ligands have been reported to modulate neurotransmitter systems such as dopamine, norepinephrine and serotonin, and NMDA receptors that contribute to the psychomotor and toxic effects of cocaine (28,39,60-69). σ Receptors have also been linked to the modulation or production of intracellular second messengers such as cGMP, inositol phosphates, protein kinases and calcium which are involved in transducing neurotransmitter-triggered signaling cascades, and which may contribute to the long-term actions of cocaine in addition to its immediate effects (34-39).

Numerous studies have reported on the ability of σ receptor antagonists to attenuate a variety of behaviors that are elicited by acute administration of cocaine. The historic σ receptor antagonist BMY 14802 (α-(4-fluorophenyl)-4-(5-fluoro-2-pyrimidinyl)-1-piperazine-butanol) and numerous novel σ receptor antagonists which are analogs of BD1008 (N-[2-(3,4-dichlorophenyl)ethyl]-N-methyl-2-(1-pyrrolidinyl)ethylamine) attenuate the acute toxic (convulsions, lethality) and/or stimulant (locomotor activity) effects of cocaine (70-75). In contrast, σ receptor agonists such as DTG (1,3-di-o-tolylguandidine), BD1031 (3R-1-[2-(3,4-dichlorophenyl)ethyl]-1,4-diazabicyclo [4.3.2]nonane) and BD1052 (N-[2-(3,4-dichlorophenyl) ethyl]-N-allyl-2-(1-pyrrolidinyl)ethylamine) exacerbate the convulsions and locomotor effects of cocaine (70-72,74,75).

It is also interesting that studies of rimcazole, a σ ligand which has high affinity for the dopamine transporter, and its analogs showed a significant correlation between the potencies of the rimcazole analogs to attenuate cocaine-induced convulsions and their binding affinities for σ_1 receptors, but not for dopamine transporters. This suggests that σ_1 receptors may be more promising medication development targets than dopamine transporters for some cocaine-induced behaviors, such as convulsions (76).

Numerous studies have also reported on the ability of putative σ receptor antagonists to attenuate behaviors that are elicited by repeated exposure to cocaine. The σ receptor antagonists BD1047 (N-[2-(3,4-dichlorophenyl) ethyl]-N-methyl-2-(dimethylamino)ethylamine) and NE-100 (N,N-dipropyl-2-[4-methoxy-3-(2-phenylethoxy)phenyl]), for example, have no rewarding or aversive properties by themselves, but significantly attenuate cocaine-induced reward as measured using the conditioned place preference paradigm (77-79). A number of σ receptor antagonists, such as BMY 14802, NPC 16377 (6-[6-(4-hydroxypiperidinyl)hexyloxy]-3-methyl-flavone), rimcazole and SR 31747A (N-cyclohexyl-N-ethyl-3-(3-chloro-4-cyclohexylphenyl)propen-2-ylamine), furthermore significantly attenuate the development of cocaine-induced locomotor sensitization. These findings suggest that σ receptor antagonists prevent neural adaptations that occur upon repeated exposure to cocaine in addition to blocking its acute psychomotor effects (56,70-72,74-76).

Although most of the σ receptor antagonists described above exhibit a high degree of selectivity for σ receptors compared to other receptors and proteins, it is difficult to conclusively prove the selectivity of the novel compounds. Therefore, to address the selectivity issue, new molecular biological tools that have become available since the sequencing of the σ receptor gene have been invaluable. Antisense oligonucleotides are short, single strands of DNA comprised of bases that are complementary to a segment of RNA that is unique to a gene of interest, in this case the gene for σ receptors. Hybridization of the antisense oligo to the RNA of interest interferes with the translation of the protein, resulting in a reduction in protein numbers, that is knockdown of σ receptors. In recent studies, two different antisense oligonucleotides targeting σ_1 receptors were shown to significantly attenuate cocaine-induced convulsions and locomotor activity, under conditions where there is about a 40% reduction in the number of σ receptors in the brain (70,72). Control, sense and mismatch oligonucleotides, did not have an effect. Administration of antisense oligonucleotides targeting σ_1 receptors also significantly attenuated cocaine-induced reward as measured using the conditioned place preference paradigm. These studies suggest that interfering with the access of cocaine to σ_1 receptors is sufficient to prevent these actions of cocaine, which further strengthens the interpretation that antagonism of σ receptor conveys anti-cocaine actions (70,72).

Since neuroactive steroids have been proposed as putative endogenous ligands for σ receptors, their effects on cocaine-induced behaviors is revealing. Previous studies have suggested that the σ-active neurosteroids DHEA (dehydroepiandrosterone) and pregnenolone act as agonists at σ receptors, whereas progesterone acts as a σ receptor antagonist. Alone, DHEA, pregnenolone or progesterone has no effect on conditioned place preference. However, all three steroids modulate cocaine-induced conditioned place preference. Progesterone, which is a putative σ receptor antagonist, attenuated cocaine-induced conditioned place preference similar to the effects produced by selective σ receptor antagonists and antisense oligonucleotides. In contrast, DHEA and pregnenolone, which act as putative σ receptor agonists, potentiated cocaine-induced conditioned place preference similar to the selective σ receptor agonist igmesine. These results suggest that σ-active neurosteroids can influence cocaine-induced reward (16,80-83).

Together, these results firmly indicate that σ receptors, particularly the σ_1 subtype, mediate some of the toxic and behavioral effects of cocaine and that σ receptor antagonists represent a previously unrecognized class of potential pharmacotherapeutic agents for treating cocaine abuse.

4.2 Amphetamines: methamphetamine and methylenedioxymethamphetamine

The amphetamine analogs, methamphetamine and MDMA, together with amphetamine are collectively referred to as amphetamines. They are all potent psychomotor stimulants. The main stimulant action of amphetamines is produced by increases in the synaptic activity of the neurotransmitters dopamine and norepinephrine. Recently studies have demonstrated that numerous amphetamine analogs, including methamphetamine and MDMA, have significant affinities for σ receptors (Table 15-1). However, amphetamine itself does not appear to interact with σ receptors directly at physiologically relevant concentrations.

The hypothesis for an involvement of σ receptors in the effects of amphetamines is supported by numerous functional studies. The σ receptor ligands, trishomocubanes, have been reported to significantly enhance amphetamine-stimulated dopamine release in a concentration-dependent manner. The enhancement produced by these compounds is fully reversed by the selective σ_2 receptor antagonists, Lu28-179 and BIMU-8, and the non-subtype selective antagonist BD1008 (40).

In addition, the putative σ receptor antagonist BMY 14802 reversed the amphetamine-induced elevation of neostriatal extracellular ascorbate levels, which has been shown to modulate neuronal function, without altering basal levels. BMY 14802 also elevated basal levels of extracellular 3,4-dihydroxyphenylalanine (DOPAC), a major dopamine metabolite, and reversed the amphetamine-induced decline (40,84).

At the behavioral level, σ receptor antagonists such as S14905 and BMY 14802 prevent the locomotor stimulation induced by amphetamine. Since amphetamine does not interact directly with σ receptors, any antagonism of its effects must be indirect (84,85).

The behavioral effects of methamphetamine can also be attenuated by σ ligands, particularly those that are associated with antagonist actions. Pretreatment of mice with the σ receptor antagonists BD1047 or BD1063 significantly attenuates the acute locomotor stimulant effect of methamphetamine (86). In addition, co-administration of the putative σ receptor antagonists, rimcazole and NE-100, can block methamphetamine-induced anticipation (84,87,88). The σ receptor antagonist BMY 14802, as well as the selective σ_1 receptor ligand MS-377, have also been reported to prevent the behavioral sensitization produced by repeated methamphetamine exposures (88-90).

Interestingly, it has been reported that repeated exposure of rats to methamphetamine results in a significant up-regulation of the σ_1 binding sites labeled with $[^3H](+)$-pentazocine in the frontal cortex and substantia nigra. This up-regulation of σ receptors in motor regions upon repeated exposure to methamphetamine may contribute to the development of sensitization observed in the behavioral studies (91,92).

Recently, knockdown studies performed in our lab demonstrate that antisense oligonucleotides targeting σ_1 receptors significantly attenuate methamphetamine-induced locomotor activity, under conditions where there is about a 40% reduction in the number of σ receptors in the brain. Control, sense and mismatch oligonucleotides, do not have an effect. These studies demonstrate that interfering with the access of methamphetamine to σ_1 receptors alone is sufficient to produce a significant anti-methamphetamine outcome (86).

Although few studies have investigated the association between MDMA and σ receptors, recent studies in our lab demonstrate that antagonism of σ receptors using the σ_1 preferring antagonist BD1063 significantly attenuates the locomotor stimulatory effects of MDMA. Additional studies are needed to fully characterize the nature of this interaction.

Together, the studies suggest that, like cocaine, antagonism of σ receptors, particularly the σ_1 subtype, prevents many effects that are induced by amphetamine derivatives.

4.3 Phencyclidine

Phencyclidine (PCP) was developed in the 1950s as an intravenous anesthetic. Its use in humans was discontinued in 1965 because patients often became agitated, delusional and irrational while recovering from its anesthetic effects. Although the unique behavioral effects of PCP appear to be mediated at the PCP binding site that is located within the NMDA-receptor-gated ion channel, many other mechanisms are thought to contribute to its effects since multiple neurotransmitter systems (dopamine, norepinephine, serotonin, glutamate and opioid receptors) are affected by it in the brain. The ability of PCP to interact with σ receptors is one of the earliest reports that related σ receptors to drugs of abuse. In fact, there are many similarities between σ receptors and PCP binding sites, and some σ ligands are derived from PCP (94-97).

Abuse of PCP can produce unwanted neural complications including schizophrenia-like psychosis and memory deficits. It is therefore noteworthy that σ ligands have been reported to modulate these PCP-induced effects. One caveat with the existing studies, however, is that most of them manipulated activity at σ receptors using ligands with preferential affinity for σ_1 receptors. Of the two major σ receptor subtypes, the σ_1 receptor appears to be the one with the greatest influence on most drugs of abuse. However, it is worth remembering that PCP is somewhat unique in its preferential affinity for the σ_2 subtype, instead of σ_1 (98).

Nevertheless, alterations in σ receptor function have been shown to influence a number of PCP-induced effects. Selective σ receptor ligands such as MS-377 and NE-100 can antagonize PCP-induced motor behaviors and attentional deficits in a variety of animal models (99,100). PCP-induced release of dopamine and serotonin in the rat brain can also be reversed by σ receptor ligands, such as MS-377 (23,101). Recent studies have further shown that σ receptors are involved in PCP-induced c-fos and hsp70 gene expression, which has been related to PCP-induced schizophrenia-like behavior (41,42,102).

In addition, σ receptors have also been shown to influence PCP-induced deficits in learning and memory. The selective σ_1 receptor agonists, SA4503 and (+)-pentazocine, significantly counteract the PCP-induced impairment of latent learning. The ameliorating effects of SA4503 and (+)-pentazocine were antagonized by NE-100, a selective σ_1 receptor antagonist (103).

Thus far, only a few studies have examined the role of σ receptors in PCP-induced behaviors which have implications for drug abuse. Although it has been reported that σ receptor ligands have little effect on the discriminative stimulus properties of PCP, it should be noted that the compounds used in this study preferentially target the σ_1 subtype, whereas PCP acts predominantly through the σ_2 subtype. The results from many early studies are also difficult to interpret because the compounds used in them (eg. (+)-SKF-10,047) interact with both σ and NMDA receptors. Therefore, systematic studies are needed to determine whether σ receptors contribute to the rewarding and reinforcing properties of PCP.

Together, the studies demonstrate that σ receptors can be targeted to ameliorate a number of PCP-induced behavioral effects. However, additional studies are needed to fully characterize the influence of the different σ receptor subtypes, particularly the role of σ_2 receptors, in these behaviors.

4.4 Alcohol

Alcohol is a fermentation product of yeast with profound effects on the body. It is one of the most commonly abused substances worldwide. In the United States, over half of the adult population has a family history of alcoholism or drinking problems, with nearly 14 million adults meeting the medical diagnostic criteria for alcohol abuse or alcoholism. Alcohol is the third leading cause of lifestyle-related deaths in the United States each year, with tobacco being the first and activity patterns second. Alcohol has diverse and complex effects on the nervous system. It can affect a multitude of neurotransmitter systems including GABA, glutamate, dopamine, norepinephrine, serotonin, opioid, and acetylcholine. However, the affinity of alcohol for σ receptors is unknown.

Recently, the ability of σ receptor antagonists to attenuate several behavioral manifestations of alcohol was reported. The σ receptor antagonist BD1047 was shown to significantly attenuate ethanol-induced locomotor activity, conditioned place preference and conditioned taste aversion. Although the mechanism underlying this effect has yet to be defined, it is thought to involve interactions with neural substrates and systems that are common to many abused substances, such as those already described (104).

4.5 Nicotine

Nicotine is the primary component in tobacco that acts on the brain. It is recognized as one of the most frequently used addictive drugs. According to the 1999 National Household Survey on Drug Abuse, approximately 64.6 million Americans were current users of tobacco products. In addition, each day in the United States more than 2,000 people under the age of 18 began daily smoking. As such, nicotine is often considered one of the most serious public health threats that currently exist.

Nicotine does not have significant affinity for σ receptors. However, there is compelling evidence that σ ligands can attenuate nicotine-induced behaviors. In particular, it has been shown that the σ receptor agonist, SA4503 (1-(3,4-dimethyoxyphenethyl)-4-(3-phenylpropyl)piperazine), attenuates the development of the rewarding effects of nicotine in the conditioned place preference model (105). In addition, buprorion, which is an effective pharmacotherapy for smoking cessation in humans, has significant affinity for σ receptors and most likely acts as an agonist at these proteins (106).

The mechanisms that are responsible for the ability of σ receptor ligands to alter responsiveness to nicotine have yet to be fully elucidated. However, several lines of evidence suggest that they involve noncompetitive allosteric interactions between σ and nicotinic receptors. Nicotine has been reported to accelerate the association of binding of the σ_1 selective radioligand [^{3}H](+)-pentazocine to σ receptors, while nicotine itself does not have significant affinity for σ receptors. σ Ligands have also been reported to noncompetitively inhibit nicotine-stimulated catecholamine release and intracellular calcium increases. The potencies of σ ligands for modulating these nicotine-induced effects are significantly correlated with their binding to σ_1 receptors. Therefore, it appears that σ_1 receptors and nicotinic receptors are associated with one another and activation of each of these receptors can modulate the activity of the other (105,106). It is thus conceivable that administration of a σ_1 receptor agonist to an organism exposed to nicotine will facilitate the binding of the σ_1 agonist to its receptor. This in turn would be expected to negatively modulate the effects of nicotine, such as catecholamine release in reward pathways, thereby conveying an anti-nicotine effect.

5. FUTURE DIRECTIONS

The ability of many drugs of abuse to interact with σ receptors suggests that these proteins may represent a common mechanism that can be targeted for medication development efforts. Most of the studies to date have focused on cocaine, and the ability of antagonism of σ receptors, particularly the σ_1 subtype, to attenuate cocaine-induced behaviors. In addition, other studies have shown that selective targeting of σ receptors can ameliorate the actions of abused substances besides cocaine. Therefore, there are still many fertile areas of research that need to be addressed in future studies. First, the involvement of the σ_2 subtype, as compared to σ_1, has been less studied, and this subtype may have an important role for certain classes of abused drugs, such as PCP and benzomorphans. Second, the behavioral consequences of σ ligands themselves, and when combined with drugs of abuse other than cocaine, is an area in which additional studies are needed. Third, although the medications development potential of σ receptor antagonists appears promising, very little is known about the cellular and molecular mechanisms that enable their protective effects against both the acute exposure to drugs of abuse, and also their ability to block longer lasting changes that may be involved in the transition between acute, occasional drug use and the loss of behavioral control over repeated drug taking that is critical for understanding drug addiction. Together, the data indicate that novel pharmacotherapies that target σ receptors represent a potential new way of treating drug abuse.

ACKNOWLEDGEMENTS

Much of the work described herein was supported by grants from the National Institute on Drug Abuse (DA11979, DA13978).

REFERENCES

1. Ritz MC, Kuhar MJ. Psychostimulant drugs and a dopamine hypothesis regarding addiction: update on recent research. Biochem Soc Symp 1993, 59:51-64.
2. Kilts CD. Imaging the roles of the amygdala in drug addiction. Psychopharmacol Bull 2001, 35:84-94.
3. Volkow ND, Fowler JS, Wang GJ, Goldstein RZ. Role of dopamine, the frontal cortex and memory circuits in drug addiction: insight from imaging studies. Neurobiol Learn Mem 2002, 78:610-624.

4. Whitlow CT, Liguori A, Livengood LB, Hart SL, Mussat-Whitlow BJ, Lamborn CM, Laurienti PJ, Porrino LJ. Long-term heavy marijuana users make costly decisions on a gambling task. Drug Alcohol Depend 2004, 76:107-111.
5. Carroll FI, Howell LL, Kuhar MJ. Pharmacotherapies for treatment of cocaine abuse: preclinical aspects. J Med Chem 199, 42:2721-2736.
6. Nader MA, Grant KA, Davies HM, Mach RH, Childers SR. The reinforcing and discriminative stimulus effects of the novel cocaine analog 2beta-propanoyl-3beta-(4-tolyl)-tropane in rhesus monkeys. J Pharmacol Exp Ther 1997, 280:541-550.
7. Villemagne VL, Rothman RB, Yokoi F, Rice KC, Matecka D, Dannals RF, Wong DF. Doses of GBR12909 that suppress cocaine self-administration in non-human primates substantially occupy dopamine transporters as measured by [^{11}C] WIN35,428 PET scans. Synapse 1999, 32:44-50.
8. Caine SB. Cocaine abuse: hard knocks for the dopamine hypothesis? Nat Neurosci 1998, 1:90-92.
9. Rocha BA, Fumagalli F, Gainetdinov RR, Jones SR, Ator R, Giros B, Miller GW, Caron MG. Cocaine self-administration in dopamine-transporter knockout mice. Nat Neurosci 1998, 1:132-137.
10. Sora I, Wichems C, Takahashi N, Li XF, Zeng Z, Revay R, Lesch KP, Murphy DL, Uhl GR. Cocaine reward models: conditioned place preference can be established in dopamine- and in serotonin-transporter knockout mice. Proc Natl Acad Sci USA 1998, 95:7699-7704.
11. Rocha BA. Stimulant and reinforcing effects of cocaine in monoamine transporter knockout mice. Eur J Pharmacol 2003, 479:107-115.
12. Caine SB, Koob GF. Effects of mesolimbic dopamine depletion on responding maintained by cocaine and food. J Exp Anal Behav 1994, 61:213-221.
13. Hall FS, Li XF, Sora I, Xu F, Caron M, Lesch KP, Murphy DL, Uhl GR. Cocaine mechanisms: enhanced cocaine, fluoxetine and nisoxetine place preferences following monoamine transporter deletions. Neuroscience 2002, 115:153-161.
14. Sora I, Hall FS, Andrews AM, Itokawa M, Li XF, Wei HB, Wichems C, Lesch KP, Murphy DL, Uhl GR. Molecular mechanisms of cocaine reward: combined dopamine and serotonin transporter knockouts eliminate cocaine place preference. Proc Natl Acad Sci USA 2001, 98:5300-5305.
15. Su TP, Chen TJ, Hwang SJ, Chou LF, Fan AP, Chen YC. Utilization of psychotropic drugs in Taiwan: an overview of outpatient sector in 2000. Zhonghua Yi Xue Za Zhi (Taipei) 2002, 65:378-391.
16. Su TP, Hayashi T. Understanding the molecular mechanism of sigma-1 receptors: towards a hypothesis that sigma-1 receptors are intracellular amplifiers for signal transduction. Curr Med Chem 2003, 10:2073-2080.
17. Everitt BJ, Dickinson A, Robbins TW. The neuropsychological basis of addictive behaviour. Brain Res. Brain Res. Rev. 2001, 36:129-138.
18. Fasano S, Brambilla R. Cellular mechanisms of striatum-dependent behavioral plasticity and drug addiction. Curr Mol Med 2002, 2:649-665.
19. Hyman SE, Malenka RC. Addiction and the brain: the neurobiology of compulsion and its persistence. Nat Rev Neurosci 2001, 2:695-703.
20. Kelley AE. Memory and addiction: shared neural circuitry and molecular mechanisms. Neuron 2004, 44:161-179.
21. Matsumoto RR, Liu Y, Lerner M, Howard EW, Brackett DJ. Sigma receptors: potential medications development target for anti-cocaine agents. Eur J Pharmacol 2003, 469:1-12.

22. Maurice T, Martin-Fardon R, Romieu P, Matsumoto RR. Sigma$_1$ (σ_1) receptor antagonists represent a new strategy against cocaine addiction and toxicity. Neurosci Biobehav Rev 2002, 26:499-527.
23. Ault DT, Werling LL. Phencyclidine and dizocilpine modulate dopamine release from rat nucleus accumbens via sigma receptors. Eur J Pharmacol 1999, 386:145-153.
24. Vilner BJ, de Costa BR, Bowen WD. Cytotoxic effects of sigma ligands: sigma receptor-mediated alterations in cellular morphology and viability. J Neurosci 1995, 15:117-134.
25. Letchworth SR, Nader MA, Smith HR, Friedman DP, Porrino LJ. Progression of changes in dopamine transporter binding site density as a result of cocaine self-administration in rhesus monkeys. J Neurosci 2001, 21:2799-2807.
26. O'Dell LE, Kreifeldt MJ, George FR, Ritz MC. The role of serotonin$_2$ receptors in mediating cocaine-induced convulsions. Pharmacol Biochem Behav 2000, 65:677-681.
27. Porrino LJ, Ritz MC, Goodman NL, Sharpe LG, Kuhar MJ, Goldberg SR. Differential effects of the pharmacological manipulation of serotonin systems on cocaine and amphetamine self-administration in rats. Life Sci 1989, 45:1529-1535.
28. Gronier B, Debonnel G. Involvement of sigma receptors in the modulation of the glutamatergic/NMDA neurotransmission in the dopaminergic systems. Eur J Pharmacol 1999, 368:183-196.
29. Bergeron R, Debonnel G, De Montigny C. Modification of the N-methyl-D-aspartate response by antidepressant sigma receptor ligands. Eur J Pharmacol 1993, 240:319-323.
30. Gewirtz GR, Gorman JM, Volavka J, Macaluso J, Gribkoff G, Taylor DP, Borison R. BMY 14802, a sigma receptor ligand for the treatment of schizophrenia. Neuropsychopharmacology 1994, 10:37-40.
31. Matos FF, Korpinen C, Yocca FD. 5-HT1A receptor agonist effects of BMY-14802 on serotonin release in dorsal raphe and hippocampus. Eur J Pharmacol 1996, 317:49-54.
32. Aydar E, Palmer CP, Klyachko VA, Jackson MB. The sigma receptor as a ligand-regulated auxiliary potassium channel subunit. Neuron 2002, 34:399-410.
33. Wilke RA, Lupardus PJ, Grandy DK, Rubinstein M, Low MJ, Jackson MB. K$^+$ channel modulation in rodent neurohypophysial nerve terminals by sigma receptors and not by dopamine receptors. J Physiol 1999, 517:391-406.
34. Derbez AE, Mody RM, Werling LL. Sigma$_2$-receptor regulation of dopamine transporter via activation of protein kinase C. J Pharmacol Exp Ther 2002, 301:306-314.
35. Mamiya T, Noda Y, Noda A, Hiramatsu M, Karasawa K, Kameyama T, Furukawa S, Yamada K, Nabeshima T. Effects of sigma receptor agonists on the impairment of spontaneous alternation behavior and decrease of cyclic GMP level induced by nitric oxide synthase inhibitors in mice. Neuropharmacology 2000, 39:2391-2398.
36. Novakova M, Ela C, Bowen WD, Hasin Y, Eilam Y. Highly selective sigma receptor ligands elevate inositol 1,4,5-trisphosphate production in rat cardiac myocytes. Eur J Pharmacol 1998, 353:315-327.
37. Nuwayhid SJ, Werling LL. Sigma$_1$ receptor agonist-mediated regulation of N-methyl-D-aspartate-stimulated [^{3}H]dopamine release is dependent upon protein kinase C. J Pharmacol Exp Ther 2003, 304:364-369.
38. Vilner BJ, Bowen WD. Modulation of cellular calcium by sigma-2 receptors: release from intracellular stores in human SK-N-SH neuroblastoma cells. J Pharmacol Exp Ther 2000, 292:900-911.
39. Yamamoto H, Yamamoto T, Sagi N, Klenerova V, Goji K, Kawai N, Baba A, Takamori E, Moroji T. Sigma ligands indirectly modulate the NMDA receptor-ion channel complex on intact neuronal cells via sigma$_1$ site. J Neurosci 1995, 15:731-736.

40. Liu X, Nuwayhid S, Christie MJ, Kassiou M, Werling LL. Trishomocubanes: novel sigma-receptor ligands modulate amphetamine-stimulated [^{3}H]dopamine release. Eur J Pharmacol 2001, 422:39-45.
41. Sharp JW, Williams DS. Effects of sigma ligands on the ability of rimcazole to inhibit PCP hsp70 induction. Brain Res Bull 1996, 39:359-366.
42. Sharp JW. Phencyclidine (PCP) acts at sigma sites to induce c-fos gene expression. Brain Res 1997, 758:51-58.
43. Guitart X, Farre AJ. The effect of E-5842, a sigma receptor ligand and potential atypical antipsychotic, on Fos expression in rat forebrain. Eur J Pharmacol 1998, 363:127-130.
44. Yanahashi S, Hashimoto K, Hattori K, Yuasa S, Iyo M. Role of NMDA receptor subtypes in the induction of catalepsy and increase in Fos protein expression after administration of haloperidol. Brain Res 2004, 1011, 84-93.
45. Daunais JB, Nader MA, Porrino LJ. Long-term cocaine self-administration decreases striatal preproenkephalin mRNA in rhesus monkeys. Pharmacol Biochem Behav 1997, 57:471-475.
46. Freeman WM, Nader MA, Nader SH, Robertson DJ, Gioia L, Mitchell SM, Daunais JB, Porrino LJ, Friedman DP, Vrana KE. Chronic cocaine-mediated changes in non-human primate nucleus accumbens gene expression. J Neurochem 2001, 77:542-549.
47. He X, Rosenfeld MG. Mechanisms of complex transcriptional regulation: implications for brain development. Neuron 1991, 7:183-196.
48. Hope BT. Cocaine and the AP-1 transcription factor complex. Ann NY Acad Sci 1998, 844:1-6.
49. Robertson HA, Paul ML, Moratalla R, Graybiel AM. Expression of the immediate early gene c-fos in basal ganglia: induction by dopaminergic drugs. Can J Neurol Sci 1991, 18:380-383.
50. Sheng M, McFadden G, Greenberg ME. Membrane depolarization and calcium induce c-fos transcription via phosphorylation of transcription factor CREB. Neuron 1990, 4:571-582.
51. Zachor DA, Moore JF, Brezausek CM, Theibert AB, Percy AK. Cocaine inhibition of neuronal differentiation in NGF-induced PC12 cells is independent of ras signaling. Int J Dev Neurosci 2000, 18:765-772.
52. Berridge M. Second messenger dualism in neuromodulation and memory. Nature 1986, 323:294-295.
53. Goelet P, Castellucci VF, Schacher S, Kandel ER. The long and the short of long-term memory--a molecular framework. Nature 1986, 322:419-422.
54. Morgan JI, Curran T. Stimulus-transcription coupling in neurons: role of cellular immediate-early genes. Trends Neurosci 1989, 12:459-462.
55. Maurice T, Su TP, Privat A. Sigma$_1$ (σ_1) receptor agonists and neurosteroids attenuate B25-35-amyloid peptide-induced amnesia in mice through a common mechanism. Neuroscience 1998, 83:413-428.
56. Matsumoto RR, Mack AL. (+/-)-SM 21 attenuates the convulsive and locomotor stimulatory effects of cocaine in mice. Eur J Pharmacol 2001, 417:R1-R2.
57. Novakova M, Ela C, Barg J, Vogel E, Hasin Y, Eilam Y. Inotropic action of sigma receptor ligands in isolated cardiac myocytes from adult rats. Eur J Pharmacol 1995, 286:19-30.
58. Sharkey J, Glen KA, Wolfe S, Kuhar MJ. Cocaine binding at sigma receptors. Eur J Pharmacol 1988, 149:171-174.
59. Ritz MC, George FR. Cocaine-induced seizures and lethality appear to be associated with distinct central nervous system binding sites. J Pharmacol Exp Ther 1993, 264:1333-1343.

60. Brackett RL, Pouw B, Blyden JF, Nour M, Matsumoto RR. Prevention of cocaine-induced convulsions and lethality in mice: effectiveness of targeting different sites on the NMDA receptor complex. Neuropharmacology 2000, 39:407-418.
61. Gonzalez-Alvear GM, Werling LL. Regulation of [3H]dopamine release from rat striatal slices by sigma receptor ligands. J Pharmacol Exp Ther 1994, 271:212-219.
62. Gonzalez-Alvear GM, Thompson-Montgomery D, Deben SE, Werling LL. Functional and binding properties of sigma receptors in rat cerebellum. J Neurochem 1995, 65:2509-2516.
63. Gonzalez-Alvear GM, Werling LL. Sigma receptor regulation of norepinephrine release from rat hippocampal slices. Brain Res 1995, 673:61-69.
64. Gonzalez GM, Werling LL. Release of [^{3}H]dopamine from guinea pig striatal slices is modulated by sigma$_1$ receptor agonists. Naunyn Schmiedebergs Arch Pharmacol 1997, 356:455-461.
65. Ishihara K, Sasa M. Modulation of neuronal activities in the central nervous system via sigma receptors. Nihon Shinkei Seishin Yakurigaku Zasshi 2002, 22:23-30.
66. Kobayashi T, Matsuno K, Murai M, Mita S. Sigma 1 receptor subtype is involved in the facilitation of cortical dopaminergic transmission in the rat brain. Neurochem Res 1997, 22:1105-1109.
67. Matsumoto RR, Brackett RL, Kanthasamy AG. Novel NMDA/glycine site antagonists attenuate cocaine-induced behavioral toxicity. Eur J Pharmacol 1997, 338:233-242.
68. Monnet FP, Mahe V, Robel P, Baulieu EE. Neurosteroids, via sigma receptors, modulate the [^{3}H]norepinephrine release evoked by N-methyl-D-aspartate in the rat hippocampus. Proc Natl Acad Sci USA 1995, 92:3774-3778.
69. Pouw B, Nour M, Matsumoto RR. Effects of AMPA/kainate glutamate receptor antagonists on cocaine-induced convulsions and lethality in mice. Eur J Pharmacol 1999, 386:181-186.
70. Matsumoto RR, McCracken KA, Friedman MJ, Pouw B, de Costa BR, Bowen WD. Conformationally restricted analogs of BD1008 and an antisense oligodeoxynucleotide targeting σ$_1$ receptors produce anti-cocaine effects in mice. Eur J Pharmacol 2001, 419:163-174.
71. Matsumoto RR, McCracken, KA, Pouw B, Miller J, Bowen WD, Williams W, de Costa BR. N-alkyl substituted analogs of the σ receptor ligand BD1008 and traditional σ receptor ligands affect cocaine-induced convulsions and lethality in mice. Eur J Pharmacol 2001, 411:261-273.
72. Matsumoto RR, McCracken KA, Pouw B, Zhang Y, Bowen WD. Involvement of sigma receptors in the behavioral effects of cocaine: evidence from novel ligands and antisense oligodeoxynucleotides. Neuropharmacology 2002, 42:1043-1055.
73. Matsumoto RR, Gilmore DL, Pouw B, Bowen WD, Williams W, Kausar A, Coop A. Novel analogs of the σ receptor ligand BD1008 attenuate cocaine-induced toxicity in mice. Eur J Pharmacol 2004, 492:21-26.
74. McCracken KA, Bowen WD, de Costa BR, Matsumoto RR. Two novel σ receptor ligands, BD1047 and LR172, attenuate cocaine-induced toxicity and locomotor activity. Eur J Pharmacol 1999, 370:225-232.
75. McCracken KA, Bowen WD, Matsumoto RR. Novel σ receptor ligands attenuate the locomotor stimulatory effects of cocaine. Eur J Pharmacol 1999, 365:35-38.
76. Matsumoto RR, Hewett KL, Pouw B, Bowen WD, Husbands SM, Cao JJ, Newman AH. Rimcazole analogs attenuate the convulsive effects of cocaine: correlation with binding to sigma receptors rather than dopamine transporters. Neuropharmacology 2001, 41:878-886.

77. Romieu P, Martin-Fardon R, Maurice T. Involvement of the sigma1 receptor in the cocaine-induced conditioned place preference. Neuroreport 2000, 11:2885-2888.
78. Romieu P, Phan VL, Martin-Fardon R, Maurice T. Involvement of the sigma$_1$ receptor in cocaine-induced conditioned place preference: possible dependence on dopamine uptake blockade. Neuropsychopharmacology 2002, 26:444-455.
79. Romieu P, Meunier J, Garcia D, Zozime N, Martin-Fardon R, Bowen WD, Maurice T. The sigma$_1$ (σ_1) receptor activation is a key step for the reactivation of cocaine conditioned place preference by drug priming. Psychopharmacology 2004, 175:154-162.
80. Hayashi T, Su TP. Regulating ankyrin dynamics: Roles of sigma-1 receptors. Proc Natl Acad Sci USA 2001, 98:491-496.
81. Hayashi T, Su TP. Sigma-1 receptors (sigma$_1$ binding sites) form raft-like microdomains and target lipid droplets on the endoplasmic reticulum: roles in endoplasmic reticulum lipid compartmentalization and export. J Pharmacol Exp Ther 2003, 306:718-725.
82. Romieu P, Martin-Fardon R, Bowen WD, Maurice T. Sigma$_1$ receptor-related neuroactive steroids modulate cocaine-induced reward. J Neurosci 2003, 23:3572-3576.
83. Su TP, Hayashi T. Cocaine affects the dynamics of cytoskeletal proteins via sigma$_1$ receptors. Trends Pharmacol Sci 2001, 22:456-458.
84. Wang Z, Haracz JL, Rebec GV. BMY-14802, a sigma ligand and potential antipsychotic drug, reverses amphetamine-induced changes in neostriatal single-unit activity in freely moving rats. Synapse 1992, 12:312-321.
85. Hascoet M, Bourin M, Payeur R, Lombet A, Peglion JL. Sigma ligand S14905 and locomotor activity in mice. Eur Neuropsychopharmacol 1995, 5:481-489.
86. Nguyen EC, McCracken KA, Liu Y, Pouw B, Matsumoto RR. Involvement of sigma (σ) receptors in the acute actions of methamphetamine: Receptor binding and behavioral studies. Neuropharmacology 2005, In press.
87. Shibata S, Ono M, Fukuhara N, Watanabe S. Involvement of dopamine, N-methyl-D-aspartate and sigma receptor mechanisms in methamphetamine-induced anticipatory activity rhythm in rats. J Pharmacol Exp Ther 1995, 274:688-694.
88. Takahashi S, Miwa T, Horikomi K. Involvement of sigma$_1$ receptors in methamphetamine-induced behavioral sensitization in rats. Neurosci Lett 2000, 289:21-24.
89. Ujike H, Okumura K, Zushi Y, Akiyama K, Otsuki S. Persistent supersensitivity of sigma receptors develops during repeated methamphetamine treatment. Eur J Pharmacol 1992, 211:323-328.
90. Ujike H, Kanzaki A, Okumura K, Akiyama K, Otsuki S. Sigma (σ) antagonist BMY 14802 prevents methamphetamine-induced sensitization. Life Sci 1992, 50:L129-L134.
91. Itzhak Y. Modulation of the PCP/NMDA receptor complex and sigma binding sites by psychostimulants. Neurotoxicol Teratol 1994, 16:363-368.
92. Stefanski R, Justinova Z, Hayashi T, Takebayashi M, Goldberg SR, Su TP. Sigma$_1$ receptor upregulation after chronic methamphetamine self-administration in rats: a study with yoked controls. Psychopharmacology 2004, 175:68-75.
93. Liu Y, Chen G-D, Lerner MR, Brackett DJ, Matsumoto RR. Cocaine up-regulates fra-2 and σ-1 receptor gene and protein expression in brain regions involved in addiction and reward. J Pharmacol Exp Ther 2005, 314:770-779.
94. Su TP, Wu XZ, Cone EJ, et al. Sigma compounds derived from phencyclidine: identification of PRE-084, a new, selective sigma ligand. J Pharmacol Exp Ther 1991, 259:543-550.
95. Gundlach AL, Largent BL, Snyder SH. Characterization of phencyclidine and sigma receptor-binding sites in brain. NIDA Res Monogr 1986, 64:1-13.

96. Sircar R, Zukin SR. Characterization of specific sigma opiate/phencyclidine (PCP)-binding sites in the human brain. Life Sci 1983, 33 (Suppl 1):259-262.
97. Sircar R, Nichtenhauser R, Ieni JR, Zukin SR. Characterization and autoradiographic visualization of (+)-[^{3}H]SKF10,047 binding in rat and mouse brain: further evidence for phencyclidine/"sigma opiate" receptor commonality. J Pharmacol Exp Ther 1986, 237:681-688.
98. de Costa BR, Bowen WD, Hellewell SB, Walker JM, Thurkauf A, Jacobson AE, Rice KC. Synthesis and evaluation of optically pure [^{3}H]-(+)-pentazocine, a highly potent and selective radioligand for sigma receptors. FEBS Lett 1989, 251:53-58.
99. Okuyama S, Ogawa S, Nakazato A, Tomizawa K. Effect of NE-100, a novel sigma receptor ligand, on phencyclidine- induced delayed cognitive dysfunction in rats. Neurosci Lett 1995, 189:60-62.
100. Takahashi S, Takagi K, Horikomi K. Effects of a novel, selective, sigma1-ligand, MS-377, on phencyclidine-induced behaviour. Naunyn Schmiedebergs Arch Pharmacol 2001, 364:81-86.
101. Takahashi S, Horikomi K, Kato T. MS-377, a novel selective sigma$_1$ receptor ligand, reverses phencyclidine-induced release of dopamine and serotonin in rat brain. Eur J Pharmacol 2001, 427:211-219.
102. Sharp JW. PCP and ketamine inhibit non-NMDA glutamate receptor mediated hsp70 induction. Brain Res 1996, 728:215-224.
103. Noda A, Noda Y, Kamei H, Ichihara K, Mamiya T, Nagai T, Sugiura S, Furukawa H, Nabeshima T. Phencyclidine impairs latent learning in mice: interaction between glutamatergic systems and sigma$_1$ receptors. Neuropsychopharmacology 2001, 24:451-460.
104. Maurice T, Casalino M, Lacroix M, Romieu P. Involvement of the sigma 1 receptor in the motivational effects of ethanol in mice. Pharmacol Biochem Behav 2003, 74:869-876.
105. Horan B, Gardner EL, Dewey SL, Brodie JD, Ashby CR Jr. The selective sigma$_1$ receptor agonist, 1-(3,4-dimethoxyphenethyl)-4-(phenylpropyl)piperazine (SA4503), blocks the acquisition of the conditioned place preference response to (-)-nicotine in rats. Eur J Pharmacol 2001, 426:R1-R2.
106. Paul IA. Sigma receptors modulate nicotinic receptor function in adrenal chromaffin cells. FASEB J 2003, 7:1171-1178.
107. Weber E, Sonders M, Quarum M, McLean S, Pou S, Keana JFW. 1,3-Di(2-[5-^{3}H]tolyl)guanidine: a selective ligand that labels sigma-type receptors for psychotomimetic opiates and antipsychotic drugs. Proc Natl Acad Sci USA 1986, 83:8784-8788.
108. Kavanaugh MP, Tester BC, Scherz MW, Keana JF, Weber E. Identification of the binding subunit of the sigma-type opiate receptor by photoaffinity labeling with 1-(4-azido-2-methyl[6-3H]phenyl)-3-(2-methyl[4,6-3H]phenyl)guanidine. Proc Natl Acad Sci USA 1988, 85:2844-2848.
109. de Costa BR, Radesca L, Di Paolo L, Bowen WD. Synthesis, characterization, and biological evaluation of a novel class of N-(arylethyl)-N-alkyl-2-(1-pyrrolidinyl)ethylamines: structural requirements and binding affinity at the sigma receptor. J Med Chem 1992, 35:38-47.

Corresponding author: *Dr. Rae R. Matsumoto, Mailing address: University of Mississippi, School of Pharmacy, Department of Pharmacology, 303*

Faser Hall, University, MS 38677, USA, Phone: (662) 915-1466, Fax: (662) 915-5148, Electronic mail address: rmatsumo@olemiss.edu

Chapter 16

σ_1 RECEPTORS AND THE MODULATION OF OPIATE ANALGESICS

Gavril W. Pasternak
Laboratory of Molecular Neuropharmacology, Department of Neurology, Memorial Sloan-Kettering Cancer Center, New York, NY 10021, USA

1. INTRODUCTION

Opiate analgesics are widely used for the management of pain. However, it is common practice to administer a number of other drugs in conjunction with the opiates in the hope of facilitating analgesia and minimizing side effects (1,2). These agents cover a wide range of classes, from antihistamines to antidepressants. Some have analgesic actions alone, but many do not. Rather, they have been suggested to enhance the actions of the opiates on pain without a corresponding increase in side effects. The ability to regulate opioid analgesia through non-opioid modulatory systems is quite intriguing from a number of perspectives. First, these systems may help explain the utility of the adjuvant drugs. Second, some of these systems have been shown to modulate the analgesic actions of opiates but not their side effects. Thus, it may be possible to increase analgesic responses without increasing side effects, thereby increasing the therapeutic index of the drug. Despite their overall effectiveness, opioid activity can vary markedly among patients. Similar observations have been made in animals. Among strains of mice, for example, the activity of opiates such as morphine can vary many-fold in standardized assays of nociception. Clearly, there are factors that influence the sensitivity of the animals to opiate analgesics. A variety of systems capable of modulating opioid actions have been implicated, including cholecystokinin (CCK), Tyr-W-MIF-1 and neuropeptide FF (3-9). Our laboratory has focused upon the role of σ_1 receptors in the regulation of opioid analgesia (10-20). σ Systems have been widely explored by many investigators. Furthermore many of the drugs widely used clinically also have σ actions. Thus, the σ system has unique characteristics that may prove valuable in the clinical management of pain.

2. BEHAVIOR

2.1 (+)-Pentazocine and haloperidol effects on opiate analgesia and gastrointestinal transit

Investigations on the role of σ_1 receptors began in our laboratory with studies looking at the opiate pentazocine (10,12-14). Pentazocine is widely used clinically as an opiate analgesic and is provided as a racemate of both the (+)- and (-)-isomers. Like many other opiates, the (-)isomer has high affinity for opioid receptors, particularly μ and κ binding sites. In contrast, (+)-pentazocine does not label opioid receptors with high affinity, but it does display high affinity for σ_1 receptors. To more fully understand the actions of pentazocine, we examined the actions of the two pentazocine stereoisomers independently.

(-)-Pentazocine was an effective analgesic, acting predominantly through κ receptors. (+)-Pentazocine alone had no analgesic actions. However, when coadministered with an active opiate analgesic, (+)-pentazocine effectively lowered the analgesic responses. This is well illustrated with morphine (Figure 16-1A). Here, both suprapsinal and spinal morphine analgesia is effectively reduced by (+)-pentazocine. Additional studies confirmed the dose-dependent manner of this blockade. The ability of (+)-pentazocine to modulate morphine actions was limited to analgesia. When we explored the effects of (+)-pentazocine on morphine's inhibition of gastrointestinal (GI) transit, we saw no change (Figure 16-1B). Morphine decreased the transit of a charcoal meal in a dose-dependent manner. Unlike the analgesic response, however, (+)-pentazocine did not alter the decrease in transit produced by morphine. Thus, (+)-pentazocine modulates analgesia, but not GI transit despite the fact that both actions were mediated through morphine.

16. Modulation of opiate analgesics

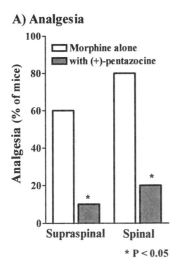
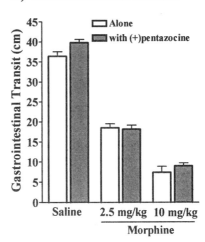

Figure 16-1. Effects of (+)-pentazocine on morphine analgesia and the inhibition of GI transit in mice. 1) Groups of mice received either morphine alone to elicit supraspinal analgesia (1.2 μg, i.c.v.) or spinal analgesia (600 ng, i.t.) alone or with (+)-pentazocine (10 mg/kg, s.c.). B) Groups of mice received either saline or the stated doses of morphine alone or with (+)-pentazocine at a dose (5 mg/kg, s.c.) that reduces systemic morphine analgesia by approximately 50%. Data from the literature (12).

(+)-Pentazocine is a well established σ_1 ligand. We therefore attempted to see whether the actions of pentazocine could be reversed by haloperidol, a known antagonist of both σ_1 and dopamine D_2 receptors (Figure 16-2). (+)-Pentazocine lowered the response to morphine, as expected. Haloperidol alone potentiated the actions of morphine alone. More important, haloperidol enhanced the response to morphine administered with (+)-pentazocine. Indeed, the response was the same as that seen with morphine and haloperidol without (+)-pentazocine. Thus, haloperidol completely reversed the actions of (+)-pentazocine. Since haloperidol can act at either σ_1 or D_2 receptors, we also examined (-)-sulpiride, a selective D_2 antagonist. Unlike haloperidol, (-)-sulpiride was ineffective in this model. Furthermore, haloperidol, but not (-)-sulpiride, reversed the actions of (+)-pentazocine. This selective reversal by haloperidol argued strongly for a σ effect, a conclusion that has been subsequently confirmed using an antisense paradigm (see below). Additional studies found similar effects of (+)-pentazocine and halopderidol on μ, δ κ_1, and κ_3 opiate analgesics. Indeed, the potentiation of κ analgesia was far more pronounced than that seen with either μ or δ drugs.

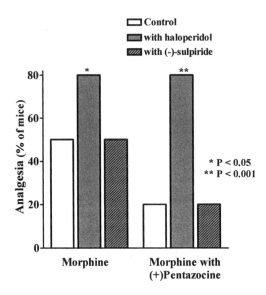

Figure 16-2. Effect of haloperidol on morphine analgesia in mice. Groups of mice received morphine and either saline, haloperidol, or (-)-sulpiride. Data from the literature (12).

It is important to note the difference between analgesia and GI transit. The ability of haloperidol to increase the response to opiates alone implied a tonic σ_1 activity in the mice. However, morphine-induced inhibition of GI transit and lethality are insensitive to σ_1 modulation. Thus, the increased analgesic response with haloperidol may be associated with an increased therapeutic index as well.

2.2 Role of σ_1 receptors in strain differences in opioid analgesia

Patients show a wide range of sensitivity to opiate drugs. Similarly, strains of mice also reveal differing sensitivities to opiates, particularly κ drugs (Table 16-1). One possibility for these different responses might be due to differing levels of tonic σ_1 activity. We therefore compared the effects of two κ opiates and haloperidol on CD1 and BALB/c mice (Table 16-1). CD1 mice are approximately 3-fold more sensitive to the analgesic actions of the κ_1 drug U50,488H than BALB/c mice. Haloperidol enhanced the analgesic response in both strains of mice, but the increase was far greater in the BALB/c mice. Indeed, in the presence of haloperidol the marked difference in potency of U50,488H was lost. Both strains showed virtually identical ED_{50}s. Similarly, the BALB/c animals were far less sensitive to the κ_3 analgesic naloxone benzoylhydrazone (NalBzoH). Coadministration of haloperidol eliminated the strain difference and enhanced their potency significantly.

The actions of σ drugs on opiates were not limited to mice. Rats displayed similar effects (13). (+)-Pentazocine reduced the analgesic actions of all classes of opiates. Like mice, rats also appeared to have a tonic level of σ_1 activity as evidenced by the ability of haloperidol to potentiate analgesia (Figure 16-3). Although haloperidol was inactive alone in the tail flick assay, haloperidol potentiated the analgesic actions of morphine (Figure 16-3A), U50,488H (Figure 16-3B) and NalBzoH (Figure 16-3C) in rats. The inability of the D_2 antagonist (-)-sulpiride to influence the analgesic actions suggests that haloperidol was acting through σ_1 receptors and not D_2 receptors in this model.

Together these studies indicate that σ_1 receptors modulate opioid analgesic responses, but not those inhibiting GI transit. Furthermore, the σ_1 system appears to be tonically active since the σ_1 antagonist haloperidol alone enhanced the analgesic actions of a series of opiate drugs.

Table 16-1. Effect of haloperidol on the analgesic potency of opiates

Opiate	Strain	Haloperidol		Control		Ratio
		ED_{50}	Confidence Limits	ED_{50}	Confidence Limits	
Morphine	CD1	4.5 mg/kg	(3.4, 6.1)	2.3 mg/kg	(1.7, 3.2)	2
DPDPE	CD1	312 ng	(213, 489)	103 ng	(64.5, 159)	3
U50,488H	CD1	4.8 mg/kg	(3.2)			
	BALB/c	16.9 mg/kg	(11.0, 30.4)	1.9 mg/kg	(1.2, 3.0)	8.9
NalBzoH	CD1	55.3 mg/kg	(38.7, 84)	21.4 mg/kg	(14.9, 31)	2.6
	BALB/c	10% at 100 mg/kg		23.2 mg/kg	(14.5, 38)	

Analgesia was assessed in the radiant heat tail flick assay after the indicated drugs were given s.c., with the exception of DPDPE ([D-Pen^2, D-Pen^5]enkephalin) which was given intracerebroventricularly. NalBzoH (Naloxone benzoylhydrazone) elicited only 10% analgesia in the BALB-C mice at the highest dose tests, 100 mg/kg. Higher doses could not be tested due to solubility problems. Data is from the literature (12).

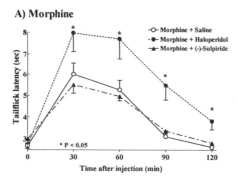

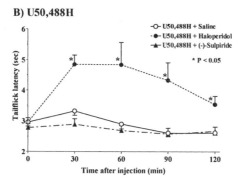

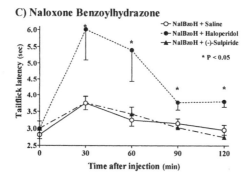

Figure 16-3. Effects of haloperidol on opioid analgesia in rats. Groups of rats received a) the μ opiate morphine, b) the κ_1 opiate U50,488H, or C) the κ_3 opiate naloxone benzoylhydrazone alone or in conjunction with either haloperidol of (-)-sulpiride. Data are from the literature (13).

3. MOLECULAR BIOLOGY OF σ_1 RECEPTORS

The pharmacological studies described above strongly imply a modulatory a role for σ_1 receptors in the actions of opioid analgesic mechanisms. However, the lack of selectivity of some of the drugs, particularly antagonists such as haloperidol, still left many questions unresolved. To definitively answer the role of σ_1 receptors in opioid analgesia, it is necessary to address these issues at a molecular level. The cloning of σ_1 receptors (21-23) offered major advances in our understanding of this particular receptor. Our own group cloned the σ receptor from both the mouse and rat (18,19), enabling us to explore its functions in our standard animal models.

3.1 Cloning the σ_1 receptor

Structurally, the receptor shows no homology to traditional G protein coupled receptors, such as the opioid receptors. It is a small protein of approximately 28 kD. Structural analysis suggests the possibility of two transmembrane domains. When expressed, the density of both the mouse and the rat receptors was markedly increased in cell lines and the binding showed the selectively that would be expected for a σ_1 receptor (18,19). Initially, it was difficult to assess the binding characteristics of the transfected protein. Although the levels vary, σ binding sites are present in virtually all cell lines (15,24-27). Thus, simply measuring the binding in membranes leaves open the question of whether or not the labeling corresponds to endogenously expressed or transfected receptors. Even the elevated levels of binding in the transfected cells might simply reflect modulation of expression of the endogenous receptor. To overcome this problem, we also examined binding in purified, transfected receptors. To obtain the purified receptors, we added a hemagluttinin (HA) tag at the amino terminus that enabled us to immunoprecipitate the transfected receptor. Western blots showed that the HA-tagged receptor was present in the cell membranes from the transfected cells and when the membranes were solubilized with CHAPS the immunoreactivity was found in the supernatant (18). Binding studies on these fractions confirmed the binding in the immunoprecipitated fraction (Figure 16-4). [^{3}H](+)-Pentazocine binding was readily seen in the cell membranes and in the crude supernatant. Virtually all the binding was immunoprecipitated, leaving very little in the treated supernatant. As a control, we examined nontransfected CHO cells treated in the same way. The binding in the cell membranes was far less, as anticipated since these control cells only showed endogenous expression of the receptor (i.e. they had not been transfected). Like the transfected cells, the endogenously expressed receptor were solubilized and were present in the crude supernatant, but they were not brought down with immunoprecipitation. All the endogenous binding remained in the treated supernatant. Thus, the binding in these HA-tagged receptors could be assessed independently of the endogenously expressed receptors by immunoprecipitation. Binding to these immunopreci-pitated receptors showed the selectivity expected for σ_1 receptors. The receptors had high affinity for pentazocine (1.8 nM) with a far lower affinity for its (-)-isomer (26.2 nM) (Table 16-2). Haloperidol retained high affinity for this binding site as well, with a K_i of 2.5 nM, whereas DTG displayed a far lower affinity. Thus, the cloned mouse σ_1 receptor clearly corresponded in these binding assays to the classical σ_1 receptor.

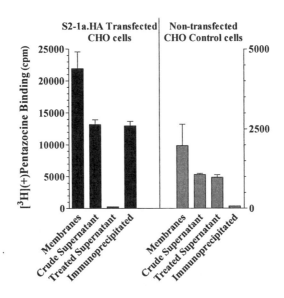

Figure 16-4. Purification of hemagluttinin (HA)-tagged σ_1 recepotrs. [^{3}H](+)-Pentazocine binding was assayed in the different membrane fractions of HA-S2-1a-transfected CHO cells (left axis) or CHO control cells (right axis). Note the different levels of binding. Only specific binding, the mean + S.E.M. of triplicate determinations, is presented. Lane A: Membrane fractions before solubiliation; Lane B: Solubilized crude supernatants before immunoprecipitation; Lane C: Solubilized treated supernatants after immoniprecipiration; Lane D: Immunoprecipitated receptors. Adapted from the literature (18).

Table 16-2. Binding characteristics of the HA-tagged/immunoprecipitated murine σ_1 receptor

Drug	K_i value (nM)
(+)-Pentazocine	1.8 ± 0.3
(-)-Pentazocine	26.2 ± 0.6
Haloperidol	2.5 ± 0.4
Di-o-tolylguanidine (DTG)	115 ± 13

The hemagluttinin (HA)-tagged σ_1 receptor was transfected into CHO cells and immunoprecipitated. Binding values are from the literature and are the mean ± S.E.M. of at least three determinations. Data is from the literature (18).

3.2 Functional assessment of the receptor

Having cloned the receptor and confirming its identity biochemically, we next sought to explore its functional significance. The actions of the protein *in vivo* can be best defined molecularly. Two major approaches are possible: knockout mice or antisense approaches. The knockout model has a number of advantages, but it is lengthy and with the wide-spread expression of the σ_1 receptor, we felt that there was a high likelihood that the gene disruption would be lethal. Therefore, we utilized an antisense technique. Antisense approaches have their own major advantages. By treating adults, it is possible to avoid compensatory mechanisms and developmental changes induced by the lack of the gene. Experimentally it is far simpler. Antisense probes are administered into the CNS over a period of a few days and behavior can then be tested. Of course, it is necessary to confirm that the antisense treatment effectively down regulated the expression of the protein. Unlike knockout models, antisense approaches do not completely eliminate the targeted mRNA and the down regulation of the receptor may be modest. Using antisense targeting the mouse σ_1 receptor that we cloned, we were able to down regulate σ_1 mRNA by approximately 65% with antisense treatment (18). A mismatch antisense control had no effects on the mRNA levels. We next explored the effects of this antisense treatment on opioid analgesia (Figure 16-5). We examined the μ opiate morphine, the κ_1 drug U05,488H and the κ_3 drug naloxone benzoylhydrazone (NalBzoH) given systemically, as well as the δ drug [D-Pen^2,D-Pen^5]enkephalin (DPDPE) given intracerebro-ventricularly. Down regulation of the σ_1 receptor with the antisense shifted the dose-response of each drug to the left, revealing significant increases in their analgesic potencies (Table 16-3). Of the four drugs tested, the shift for NalBzoH was the greatest, almost 5-fold. The others also were shifted, but only approx-imately 3-fold. This evidence strongly implied a role for σ receptors in opiate analgesia, confirming the earlier traditional studies with a far greater selectivity.

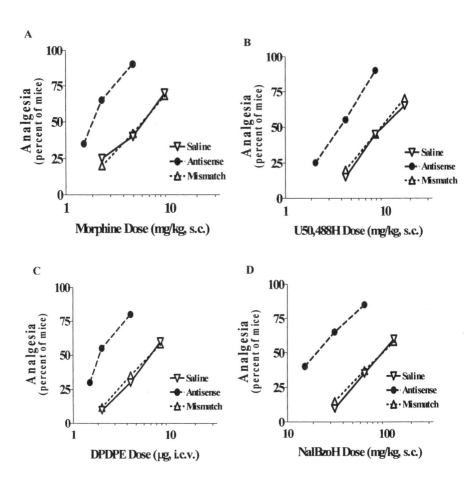

Figure 16-5. Effects of σ_1 receptor antisense treatment on opiate analgesia. Groups of mice (n=20) received an antisense or mismatch probe or saline and then were tested with the indicated doses of A) the μ opiate morphine, B) the κ_1 opiate U50,488H, C) the δ opiate [D-Pen^2, D-Pen^5]enkephalin (DPDPE), or D) the κ_3 opiate naloxone benzoylhydrazone (NalBzoH). Adapted from the literature (18).

16. Modulation of opiate analgesics

Table 16-3. Effects of σ_1 antisense on opiate analgesia in mice

	Saline	Mismatch	Antisense	Shift
Morphine (mg/kg, s.c.)	5.3	5.5	1.86	2.8
	(3.5, 7.9)	(3.8, 7.8)	(1.4, 2.5)	
U50,488H (mg/kg, s.c.)	10	9.7	3.4	2.9
	(6.7, 15)	(7.1, 13.3)	(2.5, 4.6)	
DPDPE (μg, i.c.v.)	6.1	6.4	2.1	2.9
	(4.1, 9.1)	(4.3, 9.5)	(1.6, 2.7)	
NalBzoH (mg/kg, s.c.)	96	91	19.7	4.9
	(65, 142)	(57, 146)	(12, 32)	

Values in the table represent ED_{50} (95% confidence limits). After treating groups of mice with either saline, a mismatch probe or an antisense probe, dose-response curves were performed with the indicated drugs and ED_{50} values and 95% confidence limits determined. Data from the literature (18). DPDPE = [D-Pen^2, D-Pen^5]enkephalin; NalBzoH = naloxone benzoylhydrazone.

The ability of haloperidol to reverse the actions of (+)-pentazocine strongly implied a role of σ_1 receptors. However, haloperidol also is a potent D_2 antagonist. The inactivity of (-)-sulpiride in the earlier studies argued against a D_2 site of action for haloperidol, but studies from groups have clearly implicated dopamine systems in the modulation of opioid actions (28-31). We therefore explored the role of dopamine D_2 receptors in opioid action using a knockout strategy (32). Here, we found that the D_2 receptor knockouts were modestly more sensitive to a range of opiate analgesics, suggesting that D_2 systems do interact with opioid ones. However, these mice may differ from the ones examined earlier, as shown by the ability of (-)-sulpiride to increase opioid analgesia in the wildtype mice. Thus, the modulation of opioid analgesia by D_2 receptors may be limited to subpopulations of mice. Most important, however, the σ_1 actions described earlier remained intact in the D_2 knockout animals. (+)-Pentazocine still blocked opioid analgesia in the D_2 knockout mice as effectively as in controls and this blockade was reversed by haloperidol (Figure 16-6A). Furthermore, haloperidol alone potentiated opioid analgesia in the D_2 knockout animals as effectively as in wildtype controls (Figure 16-6B). Clearly, the σ_1 system is important and independent of D_2 receptors.

4. CONCLUSION

σ Receptors remain an interesting area of investigation. They are biochemically quite unique and their biochemical mechanism of action is still unclear. Indeed, it is not even known whether or not they interact directly with opioid receptors or modulate pathways downstream from the opioids. However pharmacologically, they appear to be quite important, as demonstrated their profound ability to modulate opiate analgesia. It is particularly important to note that their increase in opiate potency appears to be limited to analgesia and not side effects such as the inhibition of gastrointestinal transit and lethality, a possible measure of respiratory depression. Furthermore, differing levels of their tonic activity among strains of mice appears to be important in determining the sensitivity of the animals to the opiates, particularly κ drugs. Obviously, these observations need to be explored clinically. It will be interesting to see how this field unfolds and whether or not drugs acting through σ systems may prove valuable in the clinical management of pain.

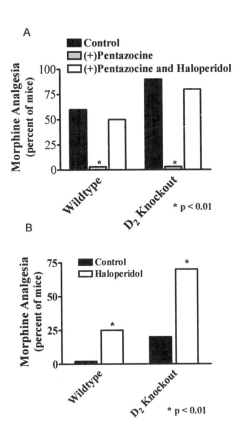

Figure 16-6. (+)-Pentazocine and haloperidol modulation of morphine analgesia in wildtype and D2 receptor knockout mice. A) Groups of mice received morphine (5 mg/kg, s.c.) and either (+)-pentazocine (3 mg/kg, s.c.) alone, or (+)-pentazocine (3 mg/kg, s.c.) with haloperidol (0.1 mg/kg, s.c.). Analgesia was assessed 30 min later. B) Groups of mice received morphine (2 mg/kg, s.c.) alone or with haloperidol (0.1 mg/kg, s.c.).

ACKNOWLEDGEMENTS

The work described was supported, in part, by research grants (DA02615, DA7242 and DA6241) and a Senior Scientist Award (DA00220) from the National Institute on Drug Abuse and a Core Grant (CA08748) from the National Cancer Institute.

REFERENCES

1. Payne R, Pasternak GW. Pain. In: Johnston MV, Macdonald RL, Young AB. (Eds.), Principles of drug therapy in neurology, F.A. Davis, Philadelphia, 1992, pp. 268-301.
2. Payne R, Pasternak GW. Pharmacology of pain treatment. In: Johnston,M.V., MacDonald R, Young AB. (Eds.), Contemporay Neurolog Series: Scientific Basis of Neurologic Drug Therapy, Davis, Philadelphia, 1992, pp. 268-301.
3. Faris PL, Komisaruk BR, Watkins LR, Mayer DJ. Evidence for the neuropeptide cholecystokinin as an antagonist of opiate analgesia. Science 1983, 219:310-312.
4. Watkins LR, Kinscheck IB, Mayer DJ. Potentiation of opiate analgesia and apparent reversal of morphine tolerance by proglumide. Science 1984, 224:395-396.
5. Noble F, Derrien M, Roques BP. Modulation of opioid antinociception by CCK at the supraspinal level: Evidence of regulatory mechanisms between CCK and enkephalin systems in the control of pain. Br J Pharmacol 1993, 109:1064-1070.
6. Lucas GD, Hoffmann O, Alster P, Wiesenfeld-Hallin Z. Extracellular cholecystokinin levels in the rat spinal cord following chronic morphine exposure: an in vivo microdialysis study. Brain Res 1999, 821:79-86.
7. Devillers JP, Boisserie F, Laulin JP, Larcher A, Simonnet G. Simultaneous activation of spinal antiopioid system (neuropeptide FF) and pain facilitatory circuitry by stimulation of opioid receptors in rats. Brain Res 1995, 700:173-181.
8. Harrison, LM, Kastin AJ, Zadina JE. Opiate tolerance and dependence: Receptors, G-proteins, and antiopiates. Peptides 1998, 19:1603-1630.
9. Malin DH, Lake JR, Arcangeli KR, Deshotel KD, Hausam DD, Witherspoon WE, Carter VA, Yang H-YT, Pal B, Burgess K. Subcutaneous injection of an analog of neuropeptide FF precipitates morphine abstinence syndrome. Life Sci 1993, 53:PL261-PL266.
10. Chien C-C, Pasternak GW. Functional antagonism of morphine analgesia by (+)-pentazocine: Evidence for an anti-opioid σ_1 system. Eur J Pharmacol 1993, 250:R7-R8.
11. Pasternak GW. Anti-opioid activity of σ_1 systems. Regul Pept 1994, 54:219-220.
12. Chien CC, Pasternak GW. Selective antagonism of opioid analgesia by a σ system. J Pharmacol Exp Ther 1994, 271:1583-1590.
13. Chien CC, Pasternak GW. Sigma antagonists potentiate opioid analgesia in rats. Neurosci Lett 1995, 190:137-139.
14. Chien CC, Pasternak GW. (-)-Pentazocine analgesia in mice: Interactions with a σ receptor system. Eur J Pharmacol 1995, 294:303-308.
15. Ryan-Moro J, Chien C-C, Standifer KM, Pasternak GW. Sigma$_1$ binding in a human neuroblastoma cell line. Neurochem Res 1996, 21:1309-1314.
16. King MA, Pan Y-X, Mei J, Chang A, Xu J, Pasternak GW. Enhanced kappa opioid analgesia by antisense targeting the σ_1 receptor. Eur J Pharmacol 1997, 331:R5-R7.

17. Chien CC, Carroll FI, Brown GP, Pan Y-X, Bowen W, Pasternak GW. Synthesis and characterization of $[^{125}I]3'(-)$-iodopentazocine, a selective σ_1 receptor ligand. Eur J Pharmacol 1997, 321:361-368.
18. Pan, YX, Mei JF, Xu J, Wan BL, Zuckerman A, Pasternak GW. Cloning and characterization of a σ_1 receptor. J Neurochem 1998, 70:2279-2285.
19. Mei J, Pasternak GW. Molecular cloning and pharmacological characterization of the rat σ_1 receptor. Biochem Pharmacol 2001, 62 349-355.
20. Mei JF, Pasternak GW. Sigma$_1$ receptor modulation of opioid analgesia in the mouse. J Pharmacol Exp Ther 2002, 300:1070-1074.
21. Hanner M, Moebius FF, Flandorfer A, Knaus HG, Striessnig J, Kempner E, Glossmann H. Purification, molecular cloning, and expression of the mammalian σ_1-binding site. Proc Natl Acad Sci USA 1996, 93:8072-8077.
22. Kekuda R, Prasad PD, Fei Y-J, Leibach FH, Ganapathy V, Fei YJ. Cloning and functional expression of the human type 1 σ receptor (hσR1). Biochem Biophys Res Commun 1996, 229:553-558.
23. Seth P, Leibach FH, Ganapathy V. Cloning and structural analysis of the cDNA and the gene encoding the murine type 1 σ receptor. Biochem Biophys Res Commun 1997, 241:535-540.
24. Garza HH Jr, Mayo S, Bowen WD, DeCosta BR, Carr DJJ. Characterization of a (+)-azidophenazocine-sensitive σ receptor on splenic lymphocytes. J Immunol 1993, 151:4672-4680.
25. John CS, Bowen WD, Varma VM, McAfee JG, Moody TW. Sigma receptors are expressed in human non-small cell lung carcinoma. Life Sci 1995, 56:2385-2392.
26. Kushner L, Zukin SR, Zukin RS. Characterization of opioid, δ, and phencyclidine receptors in the neuroblastoma-brain hybrid cell line NCB-20. Mol Pharmacol 1988, 34:689-694.
27. Largent BL, Gundlach AL, Snyder SH. σ Receptors on NCB-20 hybrid neurotumor cells labeled with (+) [^{3}H]SKF 10,047 and (+) [^{3}H]3-PPP. Eur J Pharmacol 1986, 124:183-187.
28. Rooney KF, Sewell RD. Evaluation of selective actions of dopamine D-1 and D-2 receptor agonists and antagonists on opioid antinociception. Eur J Pharmacol 1989, 168:329-336.
29. Michael-Titus A, Bousselmame R, Costentin J. Stimulation of dopamine D2 receptors induces an analgesia involving an opioidergic but non enkephalinergic link. Eur J Pharmacol 1990, 187:201-207.
30. Noble F, Cox BM. The role of dopaminergic systems in opioid receptor desensitization in nucleus accumbens and caudate putamen of rat after chronic morphine treatment. J Pharmacol Exp Ther 1997, 283:557-565.
31. Unterwald EM, Cuntapay M. Dopamine-opioid interactions in the rat striatum: a modulatory role for dopamine D_1 receptors in δ opioid receptor-mediated signal transduction. Neuropharmacology 2000, 39:372-381.
32. King MA, Bradshaw S, Chang AH, Pintar JE, Pasternak GW. Potentiation of opioid analgesia in dopamine$_2$ receptor knockout mice: Evidence for a tonically active anti-opioid system. J Neurosci 2001, 21:7788-7792.

Corresponding author: *Dr. Gavril Pasternak, Mailing address: Memorial Sloan Kettering Cancer Center, Department of Neurology, 1275 York Avenue, New York, NY 10021, USA, Phone: (212) 639-7046 Fax: (212) 794-4332, Electronic mail address: pasterng@mskcc.org*

Chapter 17

σ RECEPTORS IN THE IMMUNE SYSTEM: IMPLICATIONS FOR POTENTIAL THERAPEUTIC INTERVENTION – AN OVERVIEW

Sylvaine Galiegue, Hubert Vidal and Pierre Casellas
Sanofi~Aventis, 371 rue du Professeur Joseph Blayac, F-34184 Montpellier cedex 04, France

1. INTRODUCTION

σ Receptors are defined as non-opiate, non-dopaminergic and non-phencyclidine binding sites that interact with several chemically unrelated antipsychotic, neuroprotective and/or immunoregulatory agents. These include haloperidol, guanidine derivatives such as 1,3-di-o-tolylguanidine (DTG), phenylpiperidines such as 3-(3-hydroxyphenyl)-N-(1-propyl)piperidine (3-PPP), benzomorphans such as pentazocine and SKF-10,047, N-cyclohexyl-N-ethyl-3-(3-chloro-4-cyclohexylphenyl)propen-2-ylamine (SR 31747A) and derived molecules, which are devoid of central activity, and to a lesser degree steroids such as progesterone. There are at least two σ receptor subtypes, classified according to their pharmacological binding properties, i.e. σ_1 and σ_2 (1). The σ_1 receptor shows high affinity for benzomorphans, stereoselectivity for (+)-isomers, and is usually labeled with the radioligand $[^3H](+)$-pentazocine, while the σ_2 receptor exhibits a lower affinity for $[^3H](+)$-pentazocine, a reverse stereoselectivity, and is usually probed with DTG which binds to σ_1 and σ_2 with almost equal affinity (30 nM). The σ_1 receptor has been completely sequenced in different species (2-5) while σ_2 has not yet been molecularly characterized. An additional binding site in mammalian brain with high affinity for phenylaminotetralins and low affinity for most σ ligands was originally proposed as a σ_3 site, but was later shown to represent a subtype of histamine receptor (6). Based on binding properties and sequence homologies, a sterol isomerase emopamil binding protein (EBP) and its related SR 31747A-binding protein (SR-BP)2 are also considered to be members of the σ ligand binding site family. Altogether these proteins are thought to mediate the broad pharmacological activities of σ ligands.

The literature on σ receptors in immune functions is reviewed here. Our aim was first to summarize the expression pattern of σ receptors in the immune system, focusing on σ_1 and σ_2 receptor subtypes and second to present the immunomodulatory activities of σ ligands. In the last section of this review, we particularly pointed on SR 31747A and its back-up SSR125329 [(Z)-3-(4-adamantan-2-yl-3,5-dichlorophenyl)allyl]cyclohexyl-ethylamine], two σ ligands that entered into clinical trials. Specifically, we emphasize results which suggest that specific σ ligands could be used therapeutically to modulate immune functions through a σ receptor-mediated process with possible clinical benefit for the treatment of immune pathologies.

2. EXPRESSION OF σ RECEPTORS IN THE IMMUNE SYSTEM

2.1 Tissular expression

A decade ago, many studies reported the tissular expression pattern of σ receptors. These receptors are widely expressed throughout the body. They have been described in a variety of tissues including distinct regions of the central nervous system (CNS) (7-10) and peripheral organs such as testis, ovary (11), adrenal gland (11,12), vas deferens (13), gastrointestinal tract (14), liver (15), heart and kidney (16). σ Receptor expression in the immune system has not yet been exhaustively characterized at the mRNA or protein level. Nevertheless, a few studies have unequivocally demonstrated the presence of σ receptors in different immune tissues. Their presence was first reported by De Souza et al. in rat spleen (17), by Su et al. in guinea pig spleen (18) and by Wolfe et al. in human peripheral blood leukocytes (19) using [^{3}H]DTG and [^{3}H]haloperidol as σ ligands. Their localization in the immune system was subsequently confirmed using the selective radioligands [^{3}H](+)-pentazocine and [^{3}H](+)-3-PPP on mouse and rat T-enriched lymphocytes and B-enriched lymphocytes isolated from the spleen. The binding of [^{3}H](+)-pentazocine is saturable with T and B lymphocyte sites having similar K_d values of 401 ± 85 nM and 302 ± 46 nM, respectively

(20,21,22). σ Receptors were also detected in thymocytes with saturable high (K_{d1} 277 ± 92 nM) and low (K_{d2} 2.5 ± 1.2 µM) affinity sites for [^{3}H](+)-pentazocine (20,23).

All of the abovementioned results were obtained in binding or autoradiography studies using radiolabeled ligands on homogenates or tissue sections. Given this, they rarely addressed the identification of the receptor subtypes that are expressed in different cell populations of the immune system and did not compare their expression levels. Considering that the biochemical and pharmacological profiles of σ_1 and σ_2 differ markedly, the characterization of each receptor subtype expression pattern is an important question. In this context, subsequent studies using Northern blot probes or different selective ligands that can discriminate between σ_1 and σ_2 concomitantly give valuable information. Thus, using [^{3}H](+)-pentazocine in the presence of 100 µM naltrexone, Wolfe and coworkers demonstrated that σ_1 receptors were ubiquitously expressed in the spleen but most concentrated in the T cell zones of the white pulp (24). Using [^{3}H]DTG under σ_2 receptor-selective conditions, specific binding sites for σ_2 were observed in splenic homogenates, isolated splenocytes, and T and B cell lines (25). Cloning of human (3), mouse (4) and rat (5) cDNA encoding the σ_1 receptor made it possible to characterize the expression of this subtype at the mRNA level. Different groups unambiguously demonstrated high levels of the σ_1 receptor mRNA in thymus, spleen, placenta, lymph node, bone marrow, peripheral leukocytes, fetal spleen and fetal thymus (2,26).

Ganapathy and coworkers characterized the receptor subtype expressed in the human T lymphocyte Jurkat cells using pharmacological and molecular biology approaches (27). They demonstrated that these Th1 CD4+ T cells functionally express the σ_1 receptor subtype. They cloned the corresponding human σ_1 receptor cDNA using RT-PCR and specific primers and performed binding studies on the receptor expressed in a heterologous system (i.e. MCF-7 cells which express very low levels of σ_1 receptors). In addition, during the course of their analysis, they obtained evidence of an alternatively spliced form of the σ_1 receptor. This receptor variant resulted from the deletion of the third exon, which causes the deletion of 31 amino acids. The corresponding protein was inactive with respect to σ ligand binding. To a functional point of view, one can speculate that this inactive variant may interfere with σ_1 receptor functions. To date, the occurrence of an inactive splice variant of σ_1 receptor has been demonstrated neither in natural immune cells nor other cell types yet. Further studies are required to clarify this issue.

Immunological studies of σ receptor expression in the immune system using specific antibodies are sparse. Jbilo and coworkers produced a monoclonal anti-σ_1 antibody and characterized the subcellular distribution of

σ_1 in the THP1 human promonocytic cell line using confocal microscopy (2). They evidenced that σ_1 receptors are expressed in association with the nuclear envelope and endoplasmic reticulum and noted that this localization was also observed in a variety of different cell types including isolated lymphocytes, monocytes and macrophages (2). Using the same antibody, Dussossoy and coworkers observed similar localization in the T lymphocytic Jurkat and Ichikawa, in promyeloblastic leukemia HL60 and in monocytic U937 cell lines. In addition, they hypothesized, on the basis of electron microscopy experiments and identification of hydrophobic portions of the protein, that the σ_1 C-terminal part faces the endoplasmic reticulum and nuclear envelope lumens (28). Finally, they showed that σ_1 receptors are delocalized during the cell cycle at the mitosis step when the nuclear membranes disappear, in phytohemagglutinin-activated T lymphocytes (28).

2.2 Modulation of σ receptor expression

To our knowledge, no quantitative modulation of σ receptor expression in immune cells has been reported so far. It is noteworthy, however, that the promoter of the human σ_1 receptor gene contains consensus sequences for a variety of cytokine responsive factors besides a SP1 binding site, consensus sequences for the liver-specific transcription factor nuclear factor-1/L and for the xenobiotic responsive factor called the arylhydrocarbon receptor (29). These cytokine responsive elements are NF-GMa, NF-GMb, NF-κB, and IL-6RE. An NF-GMb site was also present in the murine promoter (4). The presence of various cytokine responsive elements in the promoter region suggests that cytokines control σ_1 receptor expression and that σ_1 receptor expression may vary along with different cytokine levels.

Concerning σ_2 receptors, several studies have reported that the receptor is overexpressed in tumor cells as compared with their normal counterparts. They include breast, neural, lung, prostate and melanoma tumors (30,31). σ_2 Receptors are also highly expressed on THP-1 leukemia cells, which are of hematopoietic origin (30).

3. σ RECEPTORS, A PUTATIVE LINK BETWEEN NEUROENDOCRINE AND IMMUNE SYSTEMS

The expression of σ_1 receptors in the immune system is well established. Considering σ_2, information is sparse. However, they also support the expression of σ_2 in some immune cells. Both σ_1 and σ_2 have already been reported to co-exist in rat liver (32) and brain (33). Thus, the two proteins may also co-exist in immune cells. Even though a minor contribution by σ_2 cannot be ruled out and given the well established expression of the σ_1 subtype in the immune system, σ_1 receptors are thought to mediate the effects of specific σ ligands on immune functions. As a putative link between the nervous and immune systems, it was initially suggested that σ binding sites mediate some aspects of steroid-induced alterations in immune functions. Su and coworkers tested the interaction of σ receptors with 20 representative gonadal and adrenal steroids (18) and demonstrated that certain steroids, particularly progesterone, inhibited σ receptor binding in homogenates of brain and spleen, as measured in competition binding experiments with [^{3}H]SKF-10,047 and [^{3}H]haloperidol, respectively. Similar potencies were obtained in brain and spleen homogenates, with K_is ranging from 200-300 nM for progesterone to 5 µM for 11-β-hydroxyprogesterone. Interestingly except for testosterone, all steroids previously reported as active in anti-inflammatory tests also displaced σ ligands while steroids inactive in the σ binding tests do not show anti-inflammatory activity. Given the binding of steroids to σ ligands together with their ability to cause changes in mood and psychological parameters, Su and coworkers proposed that σ receptors might constitute a link between the endocrine, immune and central nervous systems (34).

This property supports a broad spectrum of therapeutic applications for σ ligands. There are two potential key applications, regarding the expression of σ receptors in immune cells: 1) these receptors may be used as peripheral markers of CNS disorders, and 2) σ ligands may have σ receptor-mediated immunomodulatory properties, which would provide a basis for specific therapeutic interventions in the treatment of inflammatory conditions.

4. σ RECEPTORS IN IMMUNE CELLS AS PERIPHERAL MARKERS OF CNS DISORDERS

An aspect of σ receptors expressed at the periphery may be their use as peripheral biological markers for CNS disorders. One example is schizophrenia. Sequence analysis of the type 1 σ receptor gene in patients suffering from schizophrenia revealed a significant association between two polymorphisms and schizophrenia (35). In that study, genomic DNA was prepared from peripheral whole blood and the two polymorphisms were GC-241-240TT polymorphism in the 5'-flanking region and a missense polymorphism of a substitution from A to C predicting an alternation from Glutamine to Proline in codon 2 (Gln2Pro). Given that autoimmune mechanisms may play a role in the pathophysiology of schizophrenia (36), together with the immunomodulatory activity of σ ligands (see below), then human peripheral blood leukocytes represent a useful and obtainable peripheral marker for the study and/or diagnosis of this CNS defect.

5. IMMUNOMODULATORY ACTIVITIES OF σ LIGANDS

5.1 Pharmacological studies

The first indication of a role of σ receptors in immune functions came from the identification of phencyclidine (PCP) site of action in human peripheral blood leukocytes (PBL) (19). PCP binds with high affinity to both PCP and σ receptors. Given this, the absence of PCP receptors together with the high density of σ receptors in human PBL supported the hypothesis that σ receptors mediate PCP immunosuppressive activities. These activities include the inhibition of basal [^{3}H]thymidine and [^{3}H]2-deoxy-d-glucose uptake, mitogen-driven antibody production and lipopolysaccharide-driven interleukin-1 production (37). In addition, PCP analogs were found to cause a reduction in the resting potential of splenocytes and reduced the ability of these cells to depolarize and elevate their intracellular calcium levels in response to stimulation with concanavalin A (ConA) (38). Later, (+)-pentazocine, which has higher affinity and selectivity for σ receptor than PCP was shown to depress cellular and humoral immune responses *in vitro* at micromolar concentrations as shown by the suppression of mitogen-induced polyclonal Ig production (20,39). Using 14 σ ligands differing in their binding potency, Liu and coworkers demonstrated a positive correlation between the pharmacology of σ_1 receptors (K_i value at σ_1 receptors) and the ability of drugs to suppress ConA-induced splenocyte proliferation (EC_{50} in proliferative assay) (25). The slope of the correlation line was 0.98 when considering the most σ_1 selective test compounds (seven molecules) and 0.86 when considering the remaining seven tested compounds (25). Interestingly in that experiment, none of the tested compounds were less potent in suppressing proliferation than predicted on the basis of their relative binding potencies at σ_1 sites. In addition, (+)-pentazocine dose dependently reduced natural killer cell cytotoxic activity both *in vitro* and *in vivo* (80% inhibition measured at 50 mg/kg) while 10^{-9} M of DTG enhanced natural killer activity *in vitro* (20). The fact that (-)-pentazocine was ineffective in this test supported a σ_1 receptor-mediated process (40). Finally, (+)-azidophenazocine, (10^{-5} to 10^{-9} M) which selectively binds to the σ_1 receptor, dose dependently inhibited ConA-induced production of interferon by splenocytes (21).

5.2 Putative mechanisms of action

The mechanism by which σ ligands produce their immunosuppressant effects is unknown. Different scenarios have been proposed. An inhibitory action of σ receptors on T cell phosphoinositide turnover in splenocytes was speculated since this action was previously observed in rat brain (33). An alternative mechanism may involve the σ_1 receptor coupling to G proteins, as proposed by Itzhak and coworkers, although no evidence of this has been reported in other studies (41). Given the localization of the protein on the endoplasmic reticulum, a role of σ_1 receptors in regulating intracellular calcium was suggested. Hayashi and coworkers demonstrated that nanomolar concentrations of σ ligands affected calcium signaling by modulating intracellular calcium concentrations through a σ receptor-mediated process as evidenced using antisense oligonucleotides (42). Later, this group also demonstrated that σ_1 receptors anchor ankyrins to the endoplasmic reticulum membrane and form a trimeric complex with ankyrin B and inositol 1,4,5-triphosphate receptors, thus controlling the intracellular calcium homeostasis (43). One putative mechanism may also involve potassium channels. They are present on T lymphocytes, and they control membrane potential, calcium influx, lymphokine production, and proliferation of T lymphocytes (44). Potassium channels are activated during delivery of lethal hits by cytotoxic T lymphocytes, and during stimulation with mitogens such as phytohemagglutinin and succinyl concanavalin A (45,46). σ Receptors were recently shown to interact physically with voltage-gated potassium channels (47) and σ ligands are known to block potassium channels in the CNS and in tumor cells (48), likely through this protein-protein interaction. Since blockade of these channels can suppress lymphocyte proliferative responses (49), one can speculate that σ receptors may regulate immune responses through potassium channels. Further studies are warranted to address this important issue.

6. CLINICAL APPLICATIONS OF σ LIGANDS IN IMMUNE DISORDERS, TWO EXAMPLES: SR 31747A AND SSR125329

The only two σ ligands that have entered clinical trials with the specific indication to treat inflammatory conditions or autoimmune pathologies are SR 31747A and its back-up SSR125329. They are the archetypal members of a σ ligand subfamily, showing specific pharmacological and binding properties. They are peripheral σ ligands, which are devoid of central activity and exhibit highly potent activities regarding the control of inflammatory conditions.

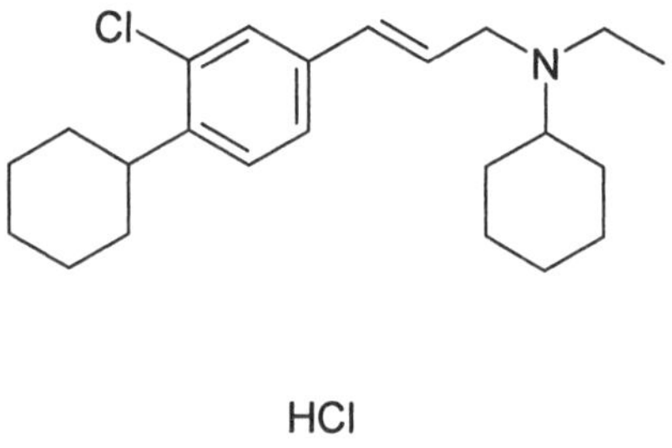

Figure 17-1. Chemical structure of SR 31747A

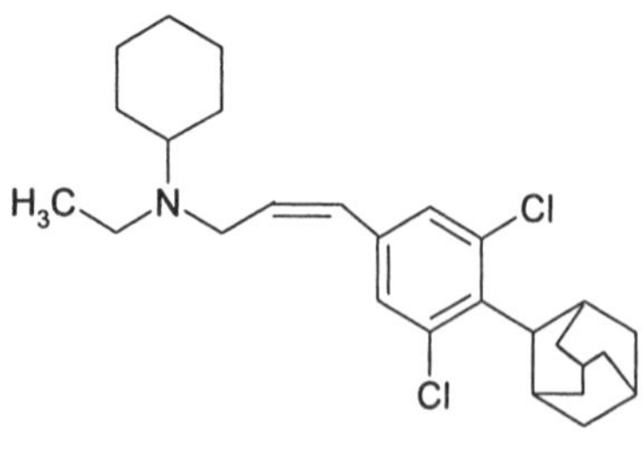

Figure 17-2. Chemical structure of SSR125329

Table 17-1. Distribution and characterization of SR 31747 binding sites on human leukocytes

Cell	K_d (nM)	B_{max} (sites/cell)
Granulocytes	2.0 ± 1.4	719,510 ± 445,179
NK Cells	2.2 ± 0.7	197,883 ± 67,412
T8 Lymphocytes	5.7 ± 3.4	161,998 ± 98,126
T4 Lymphocytes	14.7 ± 9.5	494,242 ± 393,410
B Lymphocytes	1.8 ± 0.6	119,540 ± 27,778
PBL	2.4 ± 0.3	210,483 ± 20,593

For displacement studies, 100 µl of [^{3}H]SR 31747A in PBS without Ca^{2+} or Mg^{2+} complemented with 0.1% bovine serum albumin were incubated with 200 µl of PBS-BSA buffer containing the different substances to be tested. 200 µl of cell suspension containing 10^6 cells in PBS without BSA was incubated for 2h at 2°C. Non-specific binding was determined in the presence of SR 31747A, 10^{-5} M. The cell-bound radioligand was separated from the free form by filtration on GF/B filters soaked with PEI 0.5%. Radioactivity was counted after washing with PBS (22). NK, natural killer; PBL, peripheral blood leukocytes.

6.1 Pharmacological properties

SR 31747A, N-cyclohexyl-N-ethyl-3-(3-chloro-4-cyclohexylphenyl) propen-2-ylamine hydrochloride (Figure 17-1), and its back-up SSR125329, ([(Z)-3-(4-adamantan-2-yl-3,5-dichlorophenyl)allyl]cyclohexylethylamine) (Figure 17-2), are atypical peripheral σ ligands. First, SR 31747A is one of the most potent molecules characterized so far in its capacity to compete with a variety of σ ligands (22). This molecule was found to interact with σ sites on rat spleen membranes, human PBL and purified subpopulations of human mononuclear cells (granulocytes, NK cells, T4, T8 and B lymphocytes) (see Table 17-1). In addition, SR 31747A displaced [^{3}H](+)-pentazocine, [^{3}H]DTG, [^{3}H](+)-3-PPP binding to σ sites in rat spleen membranes with high efficacy, IC_{50} = 1.3, 8 and 8 nM, respectively. Interestingly, the molecule was not reciprocally displaced by these ligands, suggesting an allosteric modulation of the σ subtypes by SR 31747A. SSR125329 has similar binding properties and shows even higher affinities than SR 31747A (50).

6.2 The family of SR 31747A binding proteins

One original feature of the SR 31747A subfamily concerns its binding protein repertoire. Historically, the identification of the molecular entity that binds SR 31747A was performed using a radiolabeled chemical probe in membrane preparations of the human T leukemia Ichikawa cells. This led to the purification and characterization of the SR 31747A binding protein (SR-BP), whose sequence is identical to $σ_1$ (2,3). Later, besides $σ_1$, binding

Table 17-2. SR 31747A binding characteristics

	σ_1 Yeast [^{3}H]SR 31747A	σ_2 Spleen/Rat [^{3}H]DTG	EBP Yeast [^{3}H]SR 31747A	SR-BP2 Yeast [^{3}H]SR 31747A
SR 31747A	1.7 ± 0.4	45 ± 7	1.5 ± 0.1	35 ± 3
Tamoxifen	511 ± 50	>1000	3.6 ± 0.4	15 ± 2
Haloperidol	5.3 ± 0.5	43 ± 2	>1000	>1000
(+)-Pentazocine	18 ± 2	38 ± 3	>1000	>1000
DTG	>1000	>1000	>1000	>1000

Binding studies were performed on membrane extracts from cells or yeast expressing σ_1, σ_2, EBP or SR-BP2. IC$_{50}$ values are expressed in nM.

Table 17-3. SSR125329 binding characteristics

	σ_1 Brain/Guinea pig [^{3}H]SR 31747A	σ_2 Spleen/Rat [^{3}H]DTG	EBP Yeast [^{3}H]SR 31747A	SR-BP2 Yeast [^{3}H]SR 31747A
SSR125329	0.4 ± 0.04	25 ± 2	0.29 ± 0.02	300 ± 20
SR 31747A	4.2 ± 0.2	45 ± 7	1.5 ± 0.1	35 ± 3
Tamoxifen	0.8 ± 0.1	43 ± 2	>1000	>1000
Haloperidol	4.3 ± 0.2	2900 ± 500	>1000	>1000
(+)-Pentazocine	62 ± 3	75 ± 10	>1000	>1000
DTG	38 ± 6	38 ± 3	>1000	>1000

Binding studies were performed on membrane extracts from cells or yeast expressing σ_1, σ_2, EBP or SR-BP2. IC$_{50}$ values are expressed in nM.

studies indicated that SR 31747A and its relative SSR125329 also bind σ_2 (51,50), and the human sterol isomerase (52,53), also called the emopamil binding protein (EBP). Finally, a fourth binding site was identified by sequence homology search, i.e. SR-BP2, which has 42% homology with EBP at the protein level but whose function is not yet known (64). SR 31747A and SSR125329 bind these four proteins with nanomolar affinities (see Tables 17-2 and 17-3). SR 31747A is the first molecule described so far with such potency for these four binding sites. Indeed, (+)-pentazocine, (-)-pentazocine, haloperidol, and DTG, which have nanomolar affinity for σ_1 and σ_2 have little affinity for EBP and SR-BP2.

Both EBP and SR-BP2 may be considered as new members of the σ receptor family and they form an original subfamily of σ binding sites with high affinity for SR 31747A and related compounds. EBP is a sterol C8-C7 isomerase. By contrast, no sterol isomerase activity of SR-BP2 has been detected (64). EBP is the mammalian counterpart of ERG2, the yeast C8-C7 isomerase of the ergosterol biosynthetic pathway, which is the only protein showing substantial homology with σ_1 (30% identity). ERG2 and σ_1 share identical transmembrane topologies, with two stretches of hydrophobic residues possibly involved in substrate binding and an N-terminal membrane anchor. In addition, ERG2 displayed nanomolar affinities for SR 31747A and other σ ligands such as opipramol, ifenprodil, amiodarone and emopamil (52,54). Based on these sequences, structural and binding property homologies, Moebius and coworkers proposed that the yeast ERG2 represents an ancestral member of the σ receptor family (54). In contrast with ERG2, EBP and its relative SR-BP2 do not show sequence or structural homology with σ_1. EBP and SR-BP2 have four transmembrane domains and cytoplasmic N- and C-terminal ends. Within a cell, EBP, SR-BP2 and σ_1 are colocalized at the endoplasmic reticulum and with the outer and inner membranes of the nuclear envelope (28,64). In humans, EBP and SR-BP2 are ubiquitously expressed and most abundant in liver, lung, and kidney. The two proteins are also found in immune organs. SR-BP2 was detected in several hematopoietic cell lines and the Burkitt's lymphoma cell line Raji shows high expression levels (64). Like σ_1 and ERG2, EBP binds the σ ligand, ifenprodil, with high affinity (55), but SR-BP2 does not. The antiestrogen tamoxifen binds the two proteins and is highly efficient at inhibiting SR 31747A binding (55). The binding of SR 31747A to EBP (56) blocked its sterol isomerase activity leading to sterol biosynthesis blockade. Given the role of sterol biosynthesis in the control of cell division and immune responses, together with the σ_1 mediated immunomodulatory activities, the pharmacological binding profiles of SR 31747A and SSR125329, which contrast with those of other σ ligands, would confer to the molecules powerful activities at the periphery regarding immune responses and cell proliferation.

6.3 Immunomodulatory activities of SR 31747A

SR 31747A was able to modulate several cellular immune responses *in vitro* and *in vivo* with potencies dramatically higher than that of other already described σ agonists.

In vitro, SR 31747A exerts a concentration and time dependent inhibition of proliferative response to mitogens (either ConA, allogeneic stimulation or phorbol-12-myristate-13-acetate (PMA) plus IL-2) on mouse and human lymphocytes (57). This suppressive effect occurs over a concentration range (IC_{50} = 20 nM) correlated with the pharmacological profile of the molecule, supporting a receptor-mediated process. Reversible inhibition progressively appeared over time and became very efficient after a 96-hour incubation. The observation that the release of IL-2, which occurs during the G1 phase of the cell cycle, was not affected by SR 31747A, together with the long incubation time period needed to observe this effect supported the fact that unlike cyclosporin A, SR 31747A does not alter early cellular events. Instead, the molecule interferes at steps further along the pathway leading to proliferation, most probably during the S phase. The effect of SR 31747A on T cell proliferation was blocked by the competitive ligand (+)-pentazocine, further arguing for a σ_1-mediated process (57).

In vivo treatment with SR 31747A selectively blocks immune responses in various animal models. SR 31747A prevented graft-versus-host disease in a manner quite similar to prednisone and azathioprine. SR 31747A decreased both splenomegaly (ED_{50} = 20 mg/kg, i.p.) and concomitant splenic cellular modifications induced by graft-versus-host reaction in mice, which include infiltration of macrophages and polymorphonuclear and T lymphocyte activation (57,23). SR 31747A also inhibited the delayed-type hypersensitivity granuloma formation induced by methylated-albumin, a model which measures the chronic aspects of an immune-mediated inflammatory disease, such as the formation of inflammatory tissue resulting from cell proliferation and edema fluid production. The molecule inhibited granuloma formation by 90% at 75 mg/kg (i.p.) compared to 75% with cyclosporin A at 100 mg/kg. In a different model, SR 31747A was shown to block *in vivo* lipopolysaccharide (LPS)-induced production of IL-1, IL-6 and tumor necrosis factor (TNF)α in a dose dependent manner (ED_{50} = 2 mg/kg, i.p.). Related to this inhibitory activity, SR 31747A also improved the survival rate by 34% after induction of lethal endotoxinic shock by injection of a high dose of LPS (15 mg/kg); this protective effect was similar to that obtained with a neutralizing anti-TNF antibody (58). To enlarge the analysis of SR 31747A immunomodulatory activities *in vivo*, Bourrie and coworkers studied the effect of a treatment with SR 31747A on *Staphylococcus enterotoxin B* (SEB)-induced lethality in mice. They observed that the drug had a potent protective effect (ED_{50} = 20 mg/kg, i.p.), which was only observed when SR 31747A was administered before SEB challenge (59,60). The authors showed that this protective effect stems from the inhibition of the systemic SEB-induced release of IL-2, IL-4, GM-CSF, IL-6 and TNFα and the concomitant stimulation of IL-10 production. Precisely SEB-induced IL-2, IL-4, GM-CSF, IL-6 and TNFα peaks were reduced by two- to four-fold in SR 31747A treated animals compared with mice treated with vehicle alone, while IL-10 release was induced by two-fold. These modulations at the protein level were also observed at the mRNA level (two- to five-fold change) and the dual effect was observed within a similar dose range. SSR125329 elicited the same dual effect when tested in similar conditions. By contrast, IL-10 secretion was only slightly increased above normal levels by cyclosporin, and dexamethasone did not consistently alter IL-10 secretion.

Importantly, the SR 31747A-induced effect on cytokine production was not observed in basal conditions, only in inflammatory situations. SR 31747A does not have a direct effect on the humoral response; in contrast to cyclosporin, up to 50 mg/kg SR 31747A does not inhibit the primary response to sheep red blood cells (T-dependent antigen). Moreover, SR 31747A does not inhibit the secondary antibody response of these cells (57). Altogether, these data support a selective site of action of the molecule on immune responses.

6.4 Applications for the treatment of inflammatory pathologies

Although [^{3}H]SR 31747A binding sites are present in the central nervous system, this compound and derived molecules are fully devoid of any central activity related to σ receptor functions (57). Given the dual immunomodulatory properties of the molecules, SR 31747A and its back-up molecule, SSR125329, may be interesting agents for the treatment of inflammatory conditions, septic shocks or autoimmune disorders. The dual immunomodulatory activity of SR 31747A and SSR125329 is particularly relevant considering the treatment of rheumatoid arthritis. Rheumatoid arthritis is a chronic, systemic, inflammatory disorder characterized by symmetrical polyarthritis leading to progressive joint destruction. This has been attributed to complex cellular interactions among acute and chronic inflammatory cells, chondrocytes, macrophages and other monocytic modulatory cells. TNFα plays a key role in the pathogenesis of rheumatoid arthritis in inflamed joints and exhibits a variety of complex inflammatory effects leading to joint destruction. By contrast, IL-10 acts as a powerful inhibitor of most mediators involved in this pathology (61). Current strategies aimed at inhibiting TNFα activity used neutralizing antibodies to the cytokine or soluble forms of its receptor. They have demonstrated substantial efficacy and are approved for rheumatoid arthritis therapy (62). However, such treatments may have some limitations since these proteins elicit an immune response that reduces their efficacy, thus prompting the need for concomitant treatment with immunosuppressive molecules (e.g. methotrexate). Alternatively, the use of molecules that inhibit TNFα synthesis is recent in rheumatoid arthritis and only a few products have reached the clinical phase of development. In this context, molecules that concomitantly inhibit TNFα synthesis and induce IL-10 expression at the inflammatory site may offer new prospects for treatment.

To test this hypothesis, SR 31747A was first evaluated against collagen-induced arthritis in mice, which is an experimental model of autoimmune arthritis. Precisely, in different sets of experiments, the molecule tested at 20 mg/kg, i.p. or 40 mg/kg, p.o. was able to decrease the incidence of the clinical signs, increase the period of outbreak of these signs, and finally reduce the extent of arthritis as evidenced by measuring paw volume. The molecule entered into clinical phase development and the first clinical phase IIa development was conducted in rheumatoid arthritis (Figure 17-3). Like SR 31747A, SSR125329 exhibited efficacy in different *in vivo* models. Particularly, SSR125329 was significantly active in reducing the severity of spontaneous rheumatoid-like disease in Mrl/lpr mice injected with Freund's adjuvant. Following a 21 day treatment at 40 mg/kg, i.p., the molecule dramatically reduced the pathological signs as shown by the reduction of footpad swelling and the low frequency of cartilage lesions (50). SSR125329 is in phase I testing and the targeted indications are rheumatoid arthritis and Crohn's disease (Figure 17-3).

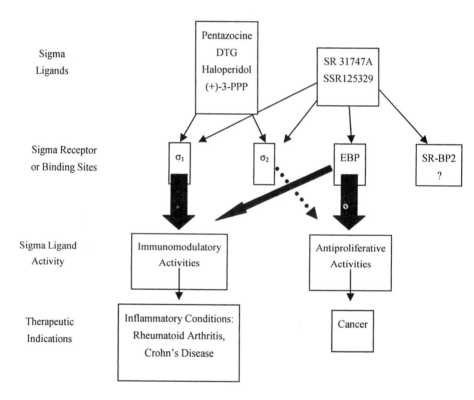

Figure 17-3. Summary scheme of contribution of peripheral σ ligand binding sites to the immunomodulatory activities of σ ligands and their modulation of cell proliferation, σ ligand binding properties and potential therapeutic applications are indicated. Prevailing contributions are indicated by large, solid arrows. The functions of σ_2 and SR-BP2 are not known with certainty. Current evidence suggests that σ_2 receptors, which are not yet cloned, have antiproliferative effects and mediate apoptosis.

17. Immune system 365

Besides its immunomodulatory activities, SR 31747A was shown to inhibit cell proliferation in yeast and in several human breast and prostate cancer cell lines both *in vitro* and *in vivo* (51,56,63; Figure 17-3). *In vitro*, nanomolar concentrations of SR 31747A were shown to inhibit the proliferation of breast cancer MCF-7 cells, MCF-7 derived cells, MDA MB 231 and BT20 cells. This inhibitory effect was reproduced in both hormono-responsive (LNCaP) and unresponsive (DU-145 and PC3) prostate cancer cell lines. *In vivo*, using the mouse xenograft model, SR 31747A treatment was shown to significantly reduce tumor development induced by the inoculation of human breast or prostatic cancer cell lines in nude mice. No toxicity was noted. Finally, *in vivo*, SR 31747A was shown to prevent the escape of breast cancer cells under long-term tamoxifen treatment, and to synergize with flutamide (51). The nanomolar efficacy of SR 31747A in inhibiting cell proliferation *in vitro* is in favor of a receptor-mediated event, with one mechanism depending on SR 31747A binding on EBP, which blocks cholesterol synthesis, thereby stopping cell proliferation (51,55; Figure 17-3). A global transcriptional analysis aimed at identifying gene modulations induced by the drug in breast and prostate cell lines *in vitro* revealed that a treatment with the molecule also inhibited the expression of genes regulating DNA replication and cell cycle progression and specifically inhibited the expression of three key enzymes of the nucleotide synthesis pathway, i.e. dihydrofolate reductase, thymidylate synthase and thymidine kinase. In addition, a classification based on transcriptional signatures, which reflect a molecule's mode of action, showed that SR 31747A does not belong to any previously characterized class of standard anticancer drugs. This study suggested that the antitumoral effect of SR 31747A resulted from an original mechanism of action (65). Given these properties, the potential of SR 31747A in prostate cancer is currently being explored.

7. CONCLUSIONS

In conclusion, the presence of σ receptors in the immune system and the ability of σ ligands to modulate some immune responses are well established. Numerous data indicate that the σ_1 receptor plays a prominent role in mediating these immunomodulatory activities. SR 31747A and its back-up, SSR125329, are members of the σ ligand family, showing highly potent immunomodulatory and anti-inflammatory activities in different models. To date, these molecules are the only σ ligands tested in the clinic for the treatment of inflammatory conditions. These compounds are distinct from other classical σ ligands in terms of binding properties. Precisely, they bind in the periphery not only to σ_1 and σ_2 receptors, but also to the sterol isomerase EBP, and its relative, SR-BP2. The latter two proteins are considered as peripheral σ ligand binding sites, and the binding of SR 31747A to EBP results in inhibition of sterol isomerase activity, leading to blockade of cell proliferation. The contribution of the σ_2 receptor, which has not yet been characterized at the molecular level, is less clear. However, current evidence suggests that σ_2 receptors are involved in induction of apoptosis in various types of cells (66) and may therefore also regulate cell proliferation (see review by Bowen in this volume). Given these properties, σ ligands and their binding sites are attractive targets to define novel therapeutic strategies useful for the treatment of immune disorders or cancer (Figure 17-3). Further studies would be warranted to accurately characterize members of the σ receptor family, their mechanism(s) of action and their contributions to peripheral σ ligand activities.

REFERENCES

1. Quirion R, Bowen WD, Itzhak Y, Junien JL, Musacchio JM, Rothman RB, Su TP, Tam SW, Taylor DP. A proposal for the classification of sigma binding sites. Trends Pharmacol Sci 1992, 13:85-86.
2. Jbilo O, Vidal H, Paul R, De Nys N, Bensaid M, Silve S, Carayon P, Davi D, Galiegue S, Bourrie B, Guillemot JC, Ferrara P, Losion G, Maffrand JP, Le Fur G, Casellas P. Purification and characterization of the human SR 31747A-binding protein. A nuclear membrane protein related to yeast sterol isomerase. J Biol Chem 1997, 272:27107-27115.
3. Kekuda R, Prasad PD, Fei YJ, Leibach FH, Ganapathy V. Cloning and functional expression of the human type 1 sigma receptor (hSigmaR1). Biochem Biophys Res Commun 1996, 229:553-558.
4. Seth P, Leibach FH, Ganapathy V. Cloning and structural analysis of the cDNA and the gene encoding the murine type 1 sigma receptor. Biochem Biophys Res Commun 1997, 241:35-540.
5. Seth P, Fei YJ, Li HW, Huang W, Leibach FH, Ganapathy V. Cloning and functional characterization of a sigma receptor from rat brain. J Neurochem 1998, 70:922-931.
6. Booth RG, Owens CE, Brown RL, Bucholtz EC, Lawler CP, Wyrick SD. Putative sigma$_3$ sites in mammalian brain have histamine H_1 receptor properties: evidence from ligand binding and distribution studies with the novel H_1 radioligand [^{3}H]-(-)(trans-1-phenyl-3-aminotetralin. Brain Res. 1999; 837: 95-105.
7. Gundlach AL, Largent BL, Snyder SH. Autoradiographic localization of sigma receptor binding sites in guinea pig and rat central nervous system with (+)^{3}H-3-(3-hydroxyphenyl)-N-(1-propyl)piperidine. J Neurosci 1986, 6:1757-1770.
8. Stojilkovic SS, Dufau ML, Catt KJ. Opiate receptor subtypes in the rat hypothalamus and neurointermediate lobe. Endocrinology 1987, 121:384-394.
9. Gonzalez-Alvear GM, Thompson-Montgomery D, Deben SE, Werling LL. Functional and binding properties of sigma receptors in rat cerebellum. J Neurochem 1995, 65:2509-2516.

10. Gonzalez-Alvear GM, Werling LL. Sigma receptor regulation of norepinephrine release from rat hippocampal slices. Brain Res 1995, 673:61-69.
11. Wolfe SA Jr, Culp SG, De Souza EB. Sigma-receptors in endocrine organs: identification, characterization, and autoradiographic localization in rat pituitary, adrenal, testis, and ovary. Endocrinology 1989, 124:1160-1172.
12. Rogers C, Cecyre D, Lemaire S. Presence of sigma and phencyclidine (PCP)-like receptors in membrane preparations of bovine adrenal medulla. Biochem Pharmacol 1989, 38:2467-2472.
13. Su TP, Wu XZ. Guinea pig vas deferens contains sigma but not phencyclidine receptors. Neurosci Lett 1990, 108:341-345.
14. Roman F, Pascaud X, Chomette G, Bueno L, Junien JL. Autoradiographic localization of sigma opioid receptors in the gastrointestinal tract of the guinea pig. Gastroenterology 1989, 97:76-82.
15. Samovilova NN, Nagornaya LV, Vinogradov VA. (+)-[^{3}H]SK&F 10,047 binding sites in rat liver. Eur J Pharmacol 1988, 147:259-264.
16. Hellewell SB, Bruce A, Feinstein G, Orringer J, Williams W, Bowen WD. Rat liver and kidney contain high densities of sigma 1 and sigma 2 receptors: characterization by ligand binding and photoaffinity labeling. Eur J Pharmacol 1994, 268:9-18.
17. De Souza EB, Kulsakdinum C, Wolfe S, Battaglia G, Jaffe JH. Sigma receptors in human perpheral blood leukocytes and rat spleen identification and characterization. Neurosci Branch NIDA Addiction Res Baltimore 1987, 46:1128.
18. Su TP, London ED, Jaffe JH. Steroid binding at sigma receptors suggests a link between endocrine, nervous, and immune systems. Science 1988, 240:219-221.
19. Wolfe SA Jr, Kulsakdinun C, Battaglia G, Jaffe JH, De Souza EB. Initial identification and characterization of sigma receptors on human peripheral blood leukocytes. J Pharmacol Exp Ther 1988, 247:1114-1119.
20. Carr DJ, De Costa BR, Radesca L, Blalock JE. Functional assessment and partial characterization of [^{3}H](+)-pentazocine binding sites on cells of the immune system. J Neuroimmunol 1991, 35:153-166.
21. Garza HH Jr, Mayo S, Bowen WD, DeCosta BR, Carr DJ. Characterization of a (+)-azidophenazocine-sensitive sigma receptor on splenic lymphocytes. J Immunol 1993, 151:4672-4680.
22. Paul R, Lavastre S, Floutard D, Floutard R, Canat X, Casellas P, Le Fur G, Breliere JC. Allosteric modulation of peripheral sigma binding sites by a new selective ligand: SR 31747. J Neuroimmunol 1994, 52:183-192.
23. Carayon P, Petitpretre G, Bourrie B, Le Fur G, Casellas P. In vivo effects of a new immunosuppressive sigma ligand, SR 31747, on mouse thymus. Immunopharmacol Immunotoxicol 1996, 18:179-191.
24. Wolfe SA Jr, Ha BK, Whitlock BB, Saini P. Differential localization of three distinct binding sites for sigma receptor ligands in rat spleen. J Neuroimmunol 1997, 72:45-58.
25. Liu Y, Whitlock BB, Pultz JA, Wolfe SA Jr. Sigma-1 receptors modulate functional activity of rat splenocytes. J Neuroimmunol 1995, 59:143-154.
26. Mei J, Pasternak GW. Molecular cloning and pharmacological characterization of the rat sigma1 receptor. Biochem Pharmacol 2001, 62:349-355.
27. Ganapathy ME, Prasad PD, Huang W, Seth P, Leibach FH, Ganapathy V. Molecular and ligand-binding characterization of the sigma-receptor in the Jurkat human T lymphocyte cell line. J Pharmacol Exp Ther 1999, 289:251-260.
28. Dussossoy D, Carayon P, Belugou S, Feraut D, Bord A, Goubet C, Roque C, Vidal H, Combes T, Loison G, Casellas P. Colocalization of sterol isomerase and sigma(1)

receptor at endoplasmic reticulum and nuclear envelope level. Eur J Biochem 1999, 263:377-386.
29. Prasad PD, Li HW, Fei YJ, Ganapathy ME, Fujita T, Plumley LH, Yang-Feng TL, Leibach FH, Ganapathy V. Exon-intron structure, analysis of promoter region, and chromosomal localization of the human type 1 sigma receptor gene. J Neurochem 1998, 70:443-451.
30. Vilner BJ, John CS, Bowen WD. Sigma-1 and sigma-2 receptors are expressed in a wide variety of human and rodent tumor cell lines. Cancer Res 1995, 55:408-413.
31. Wheeler KT, Wang LM, Wallen CA, Childers SR, Cline JM, Keng PC, Mach RH. Sigma-2 receptors as a biomarker of proliferation in solid tumours. Br J Cancer 2000, 82:1223-1232.
32. Hellewell SB, Bowen WD. A sigma-like binding site in rat pheochromocytoma (PC12) cells: decreased affinity for (+)-benzomorphans and lower molecular weight suggest a different sigma receptor form from that of guinea pig brain. Brain Res 1990, 527:244-253.
33. Bowen WD, Walker JM, de Costa BR, Wu R, Tolentino PJ, Finn D, Rothman RB, Rice KC. Characterization of the enantiomers of cis-N-[2-(3,4-dichlorophenyl)ethyl]-N-methyl-2-(1- pyrrolidinyl)cyclohexylamine (BD737 and BD738): novel compounds with high affinity, selectivity and biological efficacy at sigma receptors. J Pharmacol Exp Ther 1992, 262:32-40.
34. Su TP. Sigma receptors. Putative links between nervous, endocrine and immune systems. Eur J Biochem 1991, 200:633-642.
35. Ishiguro H, Ohtsuki T, Toru M, Itokawa M, Aoki J, Shibuya H, Kurumaji A, Okubo Y, Iwawaki A, Ota K, Shimizu H, Hamaguchi H, Arinami T. Association between polymorphisms in the type 1 sigma receptor gene and schizophrenia. Neurosci Lett 1998, 257:45-48.
36. Ganguli R, Brar JS, Chengappa KN, Yang ZW, Nimgaonkar VL, Rabin BS. Autoimmunity in schizophrenia: a review of recent findings. Ann Med 1993, 25:489-496.
37. Khansari N, Whitten HD and Fudenberg HH. Phencyclidine-induced immunodepression. Science 1984, 225:76-78.
38. Dornand J, Kamenka JM, Bartegi A, Mani JC. PCP and analogs prevent the proliferative response of T lymphocytes by lowering IL2 production. An effect related to the blockade of mitogen-triggered enhancement of free cytosolic calcium concentration. Biochem Pharmacol 1987, 36:3929-3936.
39. Carr DJ, Mayo S, Woolley TW, DeCosta BR. Immunoregulatory properties of (+)-pentazocine and sigma ligands. Immunology 1992, 77:527-531.
40. Maity R, Mukherjee R, Skolnick P. Stereoselective inhibition of natural killer activity by the sigma ligand (+)-pentazocine. J Neuroimmunol 1996, 70:7-13.
41. Itzhak Y. Multiple affinity binding states of the sigma receptor: effect of GTP-binding protein-modifying agents. Mol Pharmacol 1989, 36:512-517.
42. Hayashi T, Maurice T, Su TP. Ca^{2+} signaling via sigma$_1$-receptors: novel regulatory mechanism affecting intracellular Ca^{2+} concentration. J Pharmacol Exp Ther 2000, 293:788-798.
43. Hayashi T, Su TP. Regulating ankyrin dynamics: Roles of sigma-1 receptors. Proc Natl Acad Sci USA 2001, 98:491-496.
44. Cahalan MD, Wulff H, Chandy KG. Molecular properties and physiological roles of ion channels in the immune system. J Clin Immunol 2001, 21:235-252.
45. Mahaut-Smith MP, Mason MJ. Ca^{2+}-activated K+ channels in rat thymic lymphocytes: activation by concanavalin A. J Physiol 1991, 439:513-528.

46. Sharma B. Inhibition of the generation of cytotoxic lymphocytes by potassium ion channel blockers. Immunology 1988, 65:101-105.
47. Aydar E, Palmer CP, Klyachko VA, Jackson MB. The sigma receptor as a ligand-regulated auxiliary potassium channel subunit. Neuron 2002, 34:399-410.
48. Wilke RA, Mehta RP, Lupardus PJ, Chen Y, Ruoho AE, Jackson MB. Sigma receptor photolabeling and sigma receptor-mediated modulation of potassium channels in tumor cells. J Biol Chem 1999, 274:18387-18392.
49. Koo GC, Blake JT, Talento A, Nguyen M, Lin S, Sirotina A, Shah K, Mulvany K, Hora D Jr, Cunningham P, Wunderler DL, McManus OB, Slaughter R, Bugianesi R, Felix J, Garcia M, Williamson J, Kaczorowski G, Sigal NH, Springer MS, Feeney W. Blockade of the voltage-gated potassium channel Kv1.3 inhibits immune responses in vivo. J Immunol 1997, 158:5120-5128.
50. Bourrie B, Bribes E, De Nys N, Esclangon M, Garcia L, Galiegue S, Lair P, Paul R, Thomas C, Vernieres JC, Casellas P. SSR125329A, a high affinity sigma receptor ligand with potent anti-inflammatory properties. Eur J Pharmacol 2002, 456:123-131.
51. Berthois Y, Bourrie B, Galiegue S, Vidal H, Carayon P, Martin PM, Casellas P. SR 31747A is a sigma receptor ligand exhibiting antitumoural activity both in vitro and in vivo. Br J Cancer 2003, 88:438-446.
52. Silve S, Dupuy PH, Labit-Lebouteiller C, Kaghad M, Chalon P, Rahier A, Taton M, Lupker J, Shire D, Loison G. Emopamil-binding protein, a mammalian protein that binds a series of structurally diverse neuroprotective agents, exhibits delta8-delta7 sterol isomerase activity in yeast. J Biol Chem 1996, 271:22434-22440.
53. Hanner M, Moebius FF, Weber F, Grabner M, Striessnig J, Glossmann H. Phenylalkylamine Ca^{2+} antagonist binding protein. Molecular cloning, tissue distribution, and heterologous expression. J Biol Chem 1995, 270.7551-7557.
54. Moebius FF, Striessnig J, Glossmann H. The mysteries of sigma receptors: new family members reveal a role in cholesterol synthesis. Trends Pharmacol Sci 1997, 18:67-70.
55. Paul R, Silve S, De Nys N, Dupuy PH, Bouteiller CL, Rosenfeld J, Ferrara P, Le Fur G, Casellas P, Loison G. Both the immunosuppressant SR31747 and the antiestrogen tamoxifen bind to an emopamil-insensitive site of mammalian Δ8-Δ7 sterol isomerase. J Pharmacol Exp Ther 1998, 285:1296-1302.
56. Labit-Le Bouteiller C, Jamme MF, David M, Silve S, Lanau C, Dhers C, Picard C, Rahier A, Taton M, Loison G, Caput D, Ferrara P, Lupker J. Antiproliferative effects of SR 31747A in animal cell lines are mediated by inhibition of cholesterol biosynthesis at the sterol isomerase step. Eur J Biochem 1998, 256:342-349.
57. Casellas P, Bourrie B, Canat X, Carayon P, Buisson I, Paul R, Breliere JC, Le Fur G. Immunopharmacological profile of SR 31747: in vitro and in vivo studies on humoral and cellular responses. J Neuroimmunol 1994, 52:193-203.
58. Derocq JM, Bourrie B, Segui M, Le Fur G, Casellas P. In vivo inhibition of endotoxin-induced pro-inflammatory cytokines production by the sigma ligand SR 31747. J Pharmacol Exp Ther 1995, 272:224-230.
59. Bourrie B, Bouaboula M, Benoit JM, Derocq JM, Esclangon M, Le Fur G, Casellas P. Enhancement of endotoxin-induced interleukin-10 production by SR 31747A, a sigma ligand. Eur J Immunol 1995, 25:2882-2887.
60. Bourrie B, Benoit JM, Derocq JM, Esclangon M, Thomas C, Le Fur G, Casellas P. A sigma ligand, SR 31747A, potently modulates Staphylococcal enterotoxin B-induced cytokine production in mice. Immunology 1996, 88:389-393.
61. Katsikis PD, Chu CQ, Brennan FM, Maini RN, Feldmann M. Immunoregulatory role of interleukin 10 in rheumatoid arthritis. J Exp Med 1994, 179:1517-1527.

62. Feldmann M, Maini RN. Anti-TNF alpha therapy of rheumatoid arthritis: what have we learned? Ann Rev Immunol 2001, 19:163-196.
63. Cinato E, Peleraux A, Silve S, Galiegue S, Dhers C, Picard C, Jbilo O, Loison G, Casellas P. A DNA microarray-based approach to elucidate the effects of the immunosuppressant SR 31747A on gene expression in Saccharomyces cerevisiae. Gene Expr 2002, 10:213-230.
64. Vidal H, Mondesert G, Galiègue S, Carrière D, Dupuy PH, Carayon P, Combes T, Bribes E, Simony-Lafontaine J, Kramar A, Loison , Casellas P. Identification and pharmacological characterization of SRBP-2 — a novel SR 31747A-binding protein. Cancer Res 2003, 63:4809-18.
65. Ferrini JB, Jbilo O, Pelereau A, Combes T, Vidal H, Galiegue S, Casellas P. Transcriptomic classification of antitumor agents: application to the analysis of the antitumoral effect of SR 31747A. Gene Expr 2003, 11:125-39.
66. Crawford KW, Bowen WD. Sigma-2 receptor agonists activate a novel apoptotic pathway and potentiate antineoplastic drugs in breast tumor cell lines. Cancer Res 2002, 62:313-322.

Corresponding author: *Dr. Pierre Casellas, Mailing address: Sanofi-Aventis, 371 rue du Professeur Joseph Blayac, F-34184 Montpellier cedex 04, France, Phone: (33) 4 67 10 67 10, Fax: (33) 4 67 10 67 10, Electronic mail address: pierre.casellas@sanofi-aventis.com*

Chapter 18

σ RECEPTORS AND GASTROINTESTINAL FUNCTION

Francois J. Roman[1], Maria Chovet[1] and Lionel Bueno[2]
[1]Pfizer, 3-9 rue de La Loge, BP 100, 94265, Fresnes, France; and [2]Institut National de la Recherche Agronomique, 180 Chemin de Tournefeuille, 31932 Toulouse, Cedex, France

1. INTRODUCTION

The initial observation that σ receptors are located in the gastrointestinal tract of the guinea pig (1), and the discrete distribution that was found with a particularly dense distribution of σ binding sites in the mucosa and in the submucosal plexus has driven a lot of interest for investigating the role of σ receptors in gastrointestinal function. σ Ligands have been found able to stimulate mucosal alkaline secretion (2) and this effect has been demonstrated to correlate with their protective effects on experimental duodenal ulcers (3).

Other studies (4,5) measured the effects of σ ligands on short-circuit currents (I_{sc}) applied to mouse jejunum preparations and demonstrated a net proabsorptive effect by inhibiting I_{sc}. The study of σ ligands *in vivo* demonstrated potent intestinal antisecretory properties reflected by potent antidiarrheal effects in various models of toxigenic diarrhea. Igmesine (JO 1784; (+)-cinnamyl-1-phenyl-1-N-methyl-N-cyclopropylene) and JO 2871 [(E)-3-(1-cyclopropyl-methyl-2-azinanyl)-1-(3,4-dichlorophenyl)-1-propene] have shown antidiarrheal activity in *S. enteriditis* lipopolysaccharide (LPS)-, heat stable *E. coli* enterotoxin (*E. coli*-Sta)- and *C. difficile* toxin-induced diarrhea in mice (6,7). Turvill and colleagues (8) showed that igmesine inhibits cholera toxin and *E. coli* enterotoxin (*E. coli*-Sta)-induced jejunal secretion in rats.

σ Ligands *per se* have no or mild effects on gastrointestinal motility. However, they may influence affected gastrointestinal motility under stress conditions. At low doses administered by intracerebroventricular (i.c.v.) injection, σ ligands block the excitatory effects of stress and corticotropin-releasing hormone (CRH) on colonic motility (9).

2. LOCALIZATION OF σ RECEPTORS IN THE GASTROINTESTINAL TRACT

The presence of σ receptors was investigated in the guinea pig gastrointestinal tract using an autoradiographic approach, examining the binding of [^{3}H](+)-SKF-10,047 to tissue sections prepared from esophagus, stomach, duodenum, ileum and colon. The incubation conditions were essentially those described by Herkenham and Pert (10) and modified by Largent et al. (11). The results showed that the distribution of σ binding sites was heterogeneous (Table 18-1).

In most tissues, σ receptors were not clearly detected in smooth muscle layers, whereas mucosal and submucosal areas showed high concentrations of dark silver grains. These results were unexpected, as in a previous work, specific binding sites had been characterized in homogenates of myenteric plexus (1). This discrepancy could be explained by the different sensitivity of the two techniques. Considering the results obtained *in vitro* with membrane homogenates, it can be noticed that the binding capacity, i.e. the B_{max} value, in the myenteric plexus was very low, about seven-fold lower than the value reported in the brain by Largent et al. (11). This could explain the low dark silver grain density found in the muscular layers of the ileum, a region used for the preparation of our membrane homogenates.

Table 18-1. Quantitative analysis of [^{3}H](+)-SKF-10,047 binding to the different components of gastrointestinal tissue sections taken from different levels of the guinea pig gastrointestinal tract

Region	[^{3}H](+)-SKF-10,047 Binding (nCi/mg tissue)			
	LM	CM	SMP	M
Fundus	1.35 ± 0.06	2.55 ± 0.12	4.5 ± 0.34	6.88 ± 0.61
Antrum	2.21 ± 0.13	1.80 ± 0.08	3.8 ± 0.60	1.31 ± 0.06
Pylorus	2.35 ± 0.13	2.14 ± 0.11	7.47 ± 1.73	2.46 ± 0.35
Duodenum	0.88 ± 0.03	1.55 ± 0.07	9.99 ± 1.51	6.62 ± 0.98
Ileum	1.33 ± 0.06	1.90 ± 0.14	5.13 ± 1.05	2.35 ± 0.27
Colon	1.45 ± 0.07	1.23 ± 0.07	3.85 ± 0.48	2.9 ± 0.2

The binding site concentration for each component has been expressed in nCi of ligand bound per mg of tissue and is the mean ± S.E.M. of at least 20 determinations from 3 or 4 animals. CM, circular muscle; LM, longitudinal muscle; M, mucos; SMP, submucosal plexus.

18. Gastrointestinal function

The finding that the greatest area of density of σ receptor sites occurs in the mucosal and more specifically in the submucosal regions may indicate a role of σ receptors in the regulation of mucosal function. In the fundus and in the duodenum, the pattern of distribution was different from other areas, as patchy grains were distributed throughout the mucosa. Because these organs are of great importance from a secretory point of view, these results led us to make the hypothesis that σ receptors play an important role in the control of gastric and intestinal endocrine or exocrine secretions, or both. It must be underlined at this point that a pattern of distribution very similar to the one reported here has been described for various peptides on the basis of immunohistochemical observations. Submucosal nerve cell bodies of guinea pig intestine have been shown to contain vasointestinal peptide, galanin, neuropeptide Y (12), cholecystokinin, substance P (13) and pancreatic polypeptide (14). Daniel et al. (15) have characterized somatostatin, gastrin-releasing peptide, and enkephalin like immunoreactive material in the canine intestinal submucous plexus. The same observation has been made by Keats et al. (16) in other species including rat, dog, marmoset, and human. In most cases (12,14,17), these neurons project to the mucosal epithelium or supply submucous or mucous arterioles. They control intestinal blood flow, mucosal water and electrolyte exchange, and endocrine cell function; the latter two functions are well correlated with σ sites localized in the near luminal area.

These results confirm the work of other authors (18,19) who had postulated the presence of σ receptors in the guinea pig ileum on the basis of (±)-SKF-10,047 effects in isolated organ preparations. However, SKF-10,047 has been shown (19,20) to interact with naloxone-sensitive opiate binding sites and the same experiments should be reproduced using (+)-SKF-10,047 as a pharmacologic probe for σ receptors. When comparing these results with those of Nishimura et al. (21) on autoradiographic localization of opiate subtype receptors in the guinea pig and rat gastrointestinal tract, it is worth noting that the μ-type receptor distribution evidenced by [^{3}H]naloxone and [^{3}H]dihydromorphine binding overlaps with the σ distribution in the corpus and the antrum, where both receptors display a high concentration in the mucosal and submucosal regions. In the other regions, the quantitative repartition is different, especially in the muscular layers, which show a high density of μ receptors in most species (21,22) in contrast to σ receptors.

This autoradiographic binding study provides information on the distribution and localization of σ binding sites, but these do not necessarily represent functional receptors. No classical neurotransmitter has been shown to bind to σ sites and the exact nature of the endogenous ligand has not yet been established. However, the determination of its chemical nature will be a fundamental step in understanding the role of σ receptors. The binding sites for [^{3}H](+)-SKF-10,047 found in the guinea pig intestinal tract could thus represent receptors for an endogenous peptide acting as a neurotransmitter or transmitter-related substance. If σ receptors are involved in the secretory processes of the gastrointestinal wall, the development of agonists or antagonists could lead to compounds acting on intestinal and gastric secretions potentially useful for the treatment of secretory disorders.

3. INTESTINAL MOTILITY

3.1 Gastrocolonic reflex

σ Binding sites are present both in the brain (23-25) and the periphery (25,26). A particularly dense distribution is found in the digestive tract (1,27), the function of which is unknown. Therefore, the purpose of this study was to evaluate the effects of σ ligands and phencyclidine (PCP) ligands on basal and feeding-stimulated motility of the colon where σ sites are widely spread in both muscular and mucous layers (28). In this study, (±)-SKF-10,047, (+)-SKF-10,047, 1,3-di-o-tolylguanidine (DTG) a more specific ligand for σ sites (29), and igmesine [(+)-cinnamyl-1-phenyl-1-N-methyl-N-cyclopropylene, Jouveinal Laboratoires, Fresnes, France] a compound with high affinity for σ receptors (30) were tested. To establish the σ specificity of these effects, PCP was also tested for rejecting a possible effect of these compounds through PCP sites on N-methyl-D-aspartate (NMDA) receptors (27,31).

An attempt was also made to establish potential mechanisms of action using several selective antagonists of the opiate, dopaminergic and adrenergic systems (9). Colonic motor contractions were recorded using a chronically implanted strain gauge on the wall of the viscus.

When injected by the intravenous (i.v.) route at doses of 0.1-1 mg/kg, (±)-SKF-10,047, (+)-SKF-10,047, and saline did not influence either the pattern of colonic contractions or the motility indexes in both the proximal and transverse colon ($P > 0.05$). Injections of DTG, igmesine, and PCP (0.1-1 mg/kg) also had no effect. In contrast, when administered before feeding at a dose of 1 mg/kg, (±)-SKF-10,047 stimulated both the frequency and amplitude of colonic contractions and significantly increased ($P < 0.05$) postprandial (0-4 hours) motility indexes in the proximal and transverse colon. Similarly, (+)-SKF-10,047, DTG and igmesine stimulated colonic postprandial motility and increased proximal colonic motility indexes in a dose-related manner from 0.1-1 mg/kg for igmesine (r = 0.89) and (+)-SKF-10,047 (r = 0.92), while for DTG the maximal effect was observed at a dose of 0.25 mg/kg. The highest stimulatory effects were observed for igmesine. In contrast, PCP administered i.v. at doses of 0.1-1 mg/kg did not significantly alter either proximal or transverse colonic motility indexes. When injected centrally at doses of 0.1-1 mg/kg, both (+)-SKF-10,047 and igmesine did not influence ($P > 0.05$) the 0-4 hour or 0-8 hour postprandial motor indexes at either site.

Haloperidol (0.5 mg/kg), which had no effect on feeding-stimulated colonic motility, reduced the enhancement of colonic motility after i.v. administration of 1 mg/kg of either (+)-SKF-10,047, DTG, or igmesine. In contrast, the dopamine D_2 receptor antagonist sulpiride (0.5 mg/kg) and the D_1 dopamine receptor antagonist (+)-SCH 23390 [(R)-(+)-7-chloro-8-hydroxy-3-methyl-1-phenyl-2,3,4,5-tetrahydro-1H-3-benzazepine] did not reduce this effect on postprandial colonic motility.

Finally, the opiate antagonist naltrexone (0.1 mg/kg, i.v.) had no effect on feeding-induced colonic motor activity. However, it did not prevent or reduce the potentiation of postprandial colonic motor activity induced by the σ agonists DTG, (+)-SKF-10,047, and igmesine. In contrast, the cholecystokinin (CCK) antagonist devazepide (0.01 mg/kg, i.v.) reduced the motor index corresponding to the postprandial state at this dosage and it also attenuated the stimulatory effects of the three σ agonists tested. A similar blockade of the potentiation of the colonic motor response to eating induced by DTG (0.25 mg/kg) and igmesine (1 mg/kg) was seen following previous treatment with the α-adrenergic antagonist prazosin (0.1 mg/kg, i.v.).

The mechanisms involved in the stimulatory effects of σ ligands on colonic motility depend on the digestive state and these data. This agrees with the hypothesis of a physiological role of endogenous CCK-8 in these stimulatory effects of σ ligands on colonic postprandial motor responses.

The noradrenergic system is also involved in the postprandial increase of colonic motility because prazosin, an α-adrenoreceptor antagonist, suppresses the stimulatory effects of σ agonists on colonic motility. In contrast to the peripheral level, an increase of central adrenergic outflow by i.c.v. administration of yohimbine stimulates colonic motility, supporting the hypothesis that σ agonists may act through an increase in central adrenergic outflow in response to the stimulation of peripheral σ receptors, an effect antagonized centrally by prazosin. In summary, the present results suggest that σ ligands may be useful drugs specifically for enhancing postprandial colonic motility in patients with the irritable bowel syndrome with abnormally low postcibal motility.

3.2 Stress and altered colonic motility

Stress is known to alter gut function (32,33). Although there is increasing evidence that in humans stress alters gastrointestinal motility (34) or gastric emptying (35), little is known about the effects of stress on colonic motility. Recordings of colonic motility in humans during stress sessions show mainly a stimulation (36,37). Centrally acting stimuli such as alarming interviews or mental performance tests are more often associated with colonic motor alterations (38,39) than physical stressors (37,40).

In animals, physical stress models have been used to evaluate the effects of stress on gastrointestinal motility. In addition, wrap-restraint stress has been shown to influence colonic transit in rats (41).

Conditioned emotional stress to receive electric foot-shocks increases the frequency of colonic spike bursts in rats and this effect is mediated through the central release of corticotropin releasing hormone (CRH) (42). In rats, it has been shown that both neuropeptide Y (NPY) (43) and σ ligands (9) such as (+)-SKF-10,047 and igmesine centrally administered are able to prevent the excitation effect of stress and CRH centrally administered on colonic motility. Colonic motility was recorded by electromyography using chronically implanted electrodes in the colonic wall.

Since the effects of σ ligands to enhance colonic motor responses are mediated through the central and/or peripheral release of CCK (see section 3.1.), we have tested if CCK_{BA} acts on stress and CRH-induced colonic hyperkinesia. We also evaluated if the effects of a σ ligand and NPY are mediated through the release of CCK_{BA} by the use of selective antagonists at each type of receptor, devazepide (L 364,718) (44) for CCK_A receptors and L 365,260 (3R-(+)-N-(2,3-dihydro-1-methyl-2-oxo-5-phenyl-1H-1,4-benzodiazepine-3-yl)-N'-(3-methyl-phenyl)urea (45) for CCK_B receptors.

During the control period, i.e. 30 min before any treatment, the number of spike bursts occurring on the colon was 7.0 ± 0.9/10 min. Intracerebroventricular injection of vehicle (2 µl of sterilized water) in rats maintained in their home cage did not change the colonic spike burst frequency. Emotional stress produced a significant increase in colonic spike burst frequency (20.8 ± 4.2 versus 8.7 ± 20/10 min for controls).

NPY injected i.c.v. at a dose of 0.15 µg/kg had no effect *per se* on colonic spike burst frequency, but abolished the excitatory effect of emotional stress. Similarly, igmesine (0.1 µg/kg) injected i.c.v. 30 min before emotional stress did not affect the frequency of colonic spike bursts but suppressed ($P < 0.01$) the hyperkinesia induced by emotional stress (Table 18-2). Injected subcutaneously (s.c.) at a dose of 1 mg/kg, 40 min before emotional stress, the σ receptor antagonist BMY 14802 (α-(4-fluorophenyl)-4-(5-fluoro-2-pyrimidinyl)-1-piperazine-butanol) had no effect on the basal frequency of colonic spike bursts or on the colonic hyperkinesia induced by emotional stress. However, BMY 14802 abolished the inhibitory effects of the σ agonist igmesine and NPY on emotional stress-increased colonic spike burst frequency.

Previous (15 min) treatment with the CCK_A antagonist devazepide at a dose of 0.1 and 1 µg/kg, i.c.v. did not alter the colonic spike bursts activity or the increase in frequency produced by emotional stress. However, devazepide at a dose as low as 0.1 µg/kg reduced the effect of both NPY (0.15 µg/kg) and igmesine (0.1 µg/kg) on emotional stress-induced increases in the frequency of spike bursts, and at a dose of 1 µg/kg devazepide suppressed the effect of both NPY and igmesine on emotional stress-induced increases in colonic spike burst activity. The CCK_B antagonist L 365,260, which had also no effect *per se* on colonic motility at any dose given i.c.v. (0.1-10 µg/kg), blocked the effects of NPY and reduced significantly ($P < 0.05$) those of igmesine on emotional stress-induced increases in colonic spike bursts only at a dose of 1 µg/kg. At the highest dose (10 µg/kg), L 365,260 also abolished the effects of igmesine (Table 18-2).

Injected s.c. at a dose of 1 mg/kg, 40 min prior to emotional stress, BMY 14802 considered as a σ antagonist had no effect on the basal frequency of colonic spike bursts nor on the colonic hyperkinesia induced by emotional stress. However, it abolished the antagonistic effects of igmesine and NPY on emotional stress-induced colonic hyperkinesia.

NPY, CCK and σ receptors are present in the hypothalamus in the same areas as corticotropin releasing factor (CRF) neurons. Those neurons project to the nucleus tractus solitarius (NTS) and dorsal vagal complex (DVC) from where sympathetic and parasympathetic fibers emerge to the periphery. CCK_B receptors are widely distributed in the brain while only few areas show a high density of CCK_A receptors. The NTS is one of the richest regions in CCK_A receptors (46). This perhaps could be a region where σ receptors, CCK neurons and a CRF descending pathway may interact.

Moreover, other results suggest a close relationship between σ receptors and dopamine neurons (47), but no consistent pharmacological evidence has been given even though local application of σ ligands inhibits the firing of spontaneously active dopaminergic neurons (48). Similarly, NPY has been shown to activate dopaminergic neurons in the same brain area (49) and activation of dopaminergic neurons in the central nervous system may increase the colonic motor response to stressful events.

In summary, the present investigation supports the hypothesis that σ ligands and NPY suppress emotional stress-induced alterations of colonic motility by an action involving σ and CCK_A receptors on pathways activated by the release of CRF. However, the level of these interactions remains to be determined.

Table 18-2. Influence of previous treatment with devazepide and L 365-260 on the effects of igmesine and neuropeptide Y (NPY) on emotional stress-induced increases in the frequency of colonic spike bursts in fasted rats (mean ± S.D.; n=10)

Substance	Dose (µg/kg)	Control		Igmesine (0.1 µg/kg, i.c.v.)		NPY (0.15 µg/kg, i.c.v.)	
		Basal	ES	Basal	ES	Basal	ES
Vehicle		8.7 ± 2.0	20.8 ± 4.2*	9.7 ± 2.9	11.8 ± 3.2	8.0 ± 1.9	9.3 ± 1.6
Devazepide	0.1	9.5 ± 1.7	19.9 ± 3.0*	9.8 ± 1.3	19.8 ± 2.1*‡	9.1 ± 3.0	19.4 ± 1.7*‡
	1	10.3 ± 2.9	19.1 ± 3.8*	10.0 ± 0.9	20.2 ± 2.9*‡	8.7 ± 2.5	17.7 ± 3.0*‡
L 365,260	0.1	9.6 ± 2.5	19.2 ± 2.6*	9.0 ± 2.4	10.5 ± 2.2	8.5 ± 1.8	10.7 ± 2.1
	1	9.2 ± 1.9	18.8 ± 3.1*	8.6 ± 1.9	14.3 ± 2.5*	9.0 ± 1.9	17.2 ± 3.6*‡
	10	9.0 ± 1.4	20.4 ± 4.1*	9.2 ± 2.8	19.7 ± 4.1*	8.7 ± 2.4	19.9 ± 2.9*‡

Compounds were given by intracerebroventricular (i.c.v.) injection and frequency of colonic contraction was measured over a period of 30 min before (basal) and during emotional stress (ES). Values significantly ($P < 0.05$) different from corresponding control (*) or vehicle (‡) values.

4. ULCERS

4.1 Duodenal and gastric ulceration

The effect of igmesine (30) was investigated on duodenal and gastric ulcerations (3). Within 18 hours of treatment, all rats receiving cysteamine alone developed a major lesion covering a small part of the antrum, the pylorus, and a major part of the duodenal mucosa. Ulcerations were of variable depth, some of them large and perforated. In this group, the index of ulceration varied from 14 to 99 mm^2 with a mean of 33.2 ± 3.6 mm^2. Igmesine and DTG (1-10 mg/kg, p.o.) induced a dose-related decrease in the index value with a median effective dose (ED$_{50}$) of 4.14 (2.2-7.8) and 4.71 (1.7-12.8) mg/kg, respectively. In the same experimental conditions, 30 mg/kg of JO 1783, the active enantiomer of igmesine, did not produce any significant decrease in the ulcer index (27.3 ± 3.9 mm^2; P > 0.05), whereas the positive controls, omeprazole (proton pump inhibitor), ranitidine (histamine H$_2$ antagonist), and pirenzepine (muscarinic M$_1$ antagonist), were also protective with an ED$_{50}$ of 4.6 (2.1-10.2), 11.3 (4.7-27.5), and 11.0 (3.8-32.0) mg/kg p.o., respectively.

Similar results were reported with KB-5492 (50), a compound with selective affinity for [^{3}H]DTG-labeled σ receptors (51). In these studies, KB-5492 (4-methoxyphenyl-4-(3,4,5-trimethoxybenzyl)-1-piperazine acetate monofumarate monohydrate) prevented cysteamine-induced duodenal ulcers in rats with a potency comparable to that of the H$_2$ antagonist cimetidine.

Table 18-2. Effects of igmesine, DTG, omeprazole, ranitidine, and pirenzepine on restraint stress, ethanol-, aspirin-, and taurocholate-induced gastric mucosal damage in rats

	ED$_{50}$ (mg/kg, p.o.)			
	Restraint stress	Ethanol	Aspirin	Taurocholate
Igmesine	35.0 (12-109)	25.2 (20-32)	55.4 (25-123)	30.4 (19-49)
DTG (% at 10 mg/kg)	-48	640	-40	-41
Omeprazole	7.9 (2.8-22)	11.3 (2.6-49)	2.45 (0.83-7.2)	13.6 (5.2-35)
Ranitidine	18.0 (8.3-39)	63.9 (31-131)	7.76 (1.8-34)	95.7 (64-145)
Pirenzepine	11.1 (3.8-32)	49.9 (30-82)	5.7 (1.6-21)	15.9 (5.6-49)

Aspirin ulcerations were evaluated 4 h after the oral (p.o.) administration of 80 mg/kg acetylsalilycilic acid. Gastritis was induced either by p.o. administration of 1 ml absolute ethanol or 320 mg/kg sodium taurocholate, and the ulcer index was determined 1 h later. Stress ulcerations were evaluated 4 h after restraining rats in a thermoregulated room at 17°C. Gastric lesions were visually scored after killing the animals and removal of the stomach. Drugs were administered orally 1 h before the ulcerating agent.

On gastric lesions produced by restraint stress, ethanol, aspirin and taurocholate, igmesine had a weaker protective activity with ED_{50}s 6- to 13-times higher than that observed against cysteamine-induced duodenal ulcerations (Table 18-3).

Such a difference was also displayed by the σ agonist DTG; however, it was not possible to obtain an ED_{50} with this compound because of toxic symptoms appearing at doses >10 mg/kg. With this 10 mg/kg dose of DTG, the protection was around 40% in each of the gastric lesion tests. The ED_{50} of omeprazole and to a lesser extent, ED_{50} of ranitidine and pirenzepine, on the gastric mucosal injuries and especially aspirin and restraint stress ulcers were close to and sometimes smaller than the ED_{50} calculated for cysteamine-induced duodenal ulcerations. In an attempt to link the protective effect of igmesine on cysteamine-induced duodenal ulceration with its affinity for σ receptors, its activity was estimated after a pretreatment with the σ receptor antagonists haloperidol (1 and 3 mg/kg) and BMY 14802 (30 mg/kg, p.o.). Both drugs had no significant direct effect on the cysteamine-induced duodenal ulcers while they abolished the protective effects of igmesine (10 mg/kg, p.o.). The effect of the CCK_A antagonist devazepide (30 mg/kg, p.o.) was also tested, and it also induced a complete inhibition of the ulcero-protective effect of igmesine.

The σ ligand KB-5492 was also reported to have a direct, but limited, protective effect on the gastric mucosa *in vivo* against the damage induced by both ethanol and acidified aspirin (50). However, KB-5492 does not seem to play a key role in the mechanism of gastric mucosal protection by a direct protective effect on the surface epithelial cells.

4.2 Gastric acid secretion

Because the protective effect of omeprazole, ranitidine, and pirenzepine against gastric and duodenal lesions is linked to their gastric antisecretory properties, the effect of igmesine and DTG on gastric acid secretion was determined in 4-hour pylorus-ligated Shay rats (52). Given by the intraduodenal route, both igmesine (1-30 mg/kg) and DTG (0.5-10 mg/kg) did not significantly modify free and total gastric concentration and output. In the same experimental conditions, omeprazole, pirenzepine, and ranitidine produced a potent and dose-related inhibition with an ED_{50} on free acid output of 2.35 (0.75-7.4), 14.5 (2.9-73), and 16.3 (5.7-46) mg/kg, respectively.

In a similar manner, KB-5492 did not affect cysteamine-induced gastric acid hypersecretion at doses required to prevent the duodenal ulcers, suggesting that KB-5492 prevents cysteamine-induced duodenal ulcers but not by inhibiting gastric acid hypersecretion. In contrast, the H_2 antagonist cimetidine inhibited cysteamine-induced gastric acid hypersecretion at antiulcer doses, suggesting that cimetidine prevents cysteamine-induced duodenal ulcers by inhibiting gastric acid hypersecretion (53).

4.3 Duodenal bicarbonate secretion

Bicarbonate secretion by the duodenal mucosa was measured by the method described by Flemström and Kivilaakso (54). Under basal conditions, bicarbonate was spontaneously secreted in the rat duodenum. Saline injection did not modify bicarbonate output, because the mean value recorded within the 30 minutes following and preceding the saline injection was 7.85 ± 0.07 versus 7.56 ± 0.09 µEq/cm/h. Bolus i.v. injection of igmesine induced a dose-related increase of bicarbonate output. The threshold dose was 0.25 mg/kg, and the maximal secretory rate was obtained after 1 mg/kg. Within the 10 min following the administration of 1 mg/kg igmesine, the bicarbonate output increased from a basal value of 7.6 ± 0.14 µEq/cm/h to a mean secretory plateau of 14.8 ± 1.3 µEq/cm/h ($P < 0.001$), which lasted during the 2 hours following the injection. Doubling this dose did not produce any further increase of the secretory rate. No significant change in the arterial blood pressure was produced by igmesine. Under the same experimental conditions, JO 1783, the active enantiomer of igmesine, induced a small and insignificant increase in duodenal secretion. As already reported, DTG displayed a very similar effect to igmesine, whatever the dosage used.

The stimulating effect of igmesine on duodenal bicarbonate secretion was inhibited by the σ antagonists haloperidol (0.5 mg/kg) and BMY 14802 (5 mg/kg) and also after bilateral truncal vagotomy or by treatment with the ganglionic blocking agent hexamethonium (1 mg/kg) or the neural blocking agent tetrodotoxin (5 µg/kg), but not by the muscarinic receptor antagonist atropine (0.25 mg/kg). It was not decreased by D_1 and D_2 dopamine antagonists, nor by α_1- or α_2-adrenolytic drugs. In the same way, the effect was not significantly modified by naloxone or indomethacin at doses known to affect opioid responses and prostaglandin synthesis, respectively. These results show that that the stimulatory effect of igmesine on bicarbonate secretion is caused by its affinity for σ receptors and suggest that it is driven through nonadrenergic, noncholinergic fibers of the vagus. The lack of any stimulation observed in vagotomized rats also strongly supports the participation of a loop starting from the periphery and going through the central nervous system.

KB-5492 was shown to increase dose-dependently duodenal bicarbonate secretion in normal anesthetized rats at doses required to prevent duodenal ulcers (50). In the same conditions, cimetidine at 200 mg/kg did not affect duodenal bicarbonate secretion.

Selective σ ligands like igmesine and KB-5492 protect rats from cysteamine-induced duodenal ulceration; this effect may be related to their potent stimulating activity on duodenal secretion. These properties are related to their σ binding affinity, suggesting that σ receptors play a role in the control of duodenal bicarbonate secretion and thus in the defense mechanisms of this mucosa.

5. INTESTINAL SECRETION

5.1 *In vitro* experiments

Autoradiographic techniques with [^{3}H](+)-SKF-10,047 have demonstrated the presence of σ sites in the mucosa and submucosal plexus of the guinea pig gastrointestinal tract (28), suggesting that σ receptors might play a role in the regulation of intestinal fluid in this species.

Whole segments of jejunum were taken from male, ICR mice and mounted as sheets in standard Ussing flux chambers and transmural short-circuit current (I_{sc}) was monitored. Concentration-effect curves were generated non-cumulatively by treating different segments of jejunum with graded concentrations of agonist (4,5).

Serosal addition of igmesine produced a concentration-related decrease in I_{sc}. The IC_{50} value (concentration producing 50% of the maximal effect for igmesine) was 5.21 (3.43-7.92) µM. The action of igmesine was abolished by pretreatment with tetrodotoxin (0.1 µM), a neural blocking agent. Haloperidol, which had no significant effect on I_{sc} up to concentrations of 1 µM, blocked the igmesine effect on I_{sc} in paired tissues taken from the same mouse. The effect of igmesine was not altered by the selective D_1 receptor antagonist SCH 23390 (0.1 µM), nor by the selective D_2 receptor antagonist sulpiride (1 µM), suggesting that the above haloperidol blockade was not mediated through dopamine receptors, as for duodenal alkaline secretion. Similarly, other σ ligands such as DTG, (-)-SKF-10,047 and (+)-SKF-10,047 inhibited the I_{SC} in isolated mouse segments (Table 18-4).

Table 18-3. Concentrations producing 50% inhibition (IC_{50}) of short-circuit current for σ ligands in mouse jejunum

Compound	IC_{50} (nM)	95% Confidence Limits
Igmesine	11.13	7.84-15.78
JO 1783	12.58	9.0-17.58
DTG	15.6	7.5-32.1
(-)-SKF-10,047	61.1	22.5-165.8
(+)-SKF-10,047	91.9	17.9-471.0

Data from ref. (5).

5.2 *In vivo* experiments

In animals, on isolated jejunal segments, interleukin (IL)-1β a cytokine released by macrophages and monocytes, is involved in diarrheal states associated with infections or inflammatory bowel diseases and food allergy reactions. IL-1β alters ion transport in the ileum and triggers net water secretion in the proximal colon. The antisecretory effect of loperamide, somatostatin, and igmesine on IL-1β-induced colonic hypersecretion has been compared in anesthetized rats (55).

After a 2 hour equilibration period, net water flux in the isolated loop was calculated from ^{14}C concentration, measured in the effluent collected at 15 min intervals over 3 hours. IL-1β (5 µg/kg) or vehicle (NaCl 0.9%) were administered intraperitoneally (i.p.) 3 hours after the start of infusion. IL-1β or saline administration was preceded (20 min) by i.p. administration of either loperamide (1 mg/kg), somatostatin (1 µg/kg), igmesine (1 mg/kg), or their vehicle.

Loperamide failed to modify the IL-1β secretory effect, while in contrast, somatostatin and igmesine significantly reduced the IL-1β-induced hypersecretion. Moreover, antisomatostatin reversed the antisecretory effect of igmesine on IL-1β-induced colonic hypersecretion.

Since vasoactive intestinal peptide (VIP) is involved in many toxin-induced intestinal secretions, another study tested the ability of igmesine to inhibit VIP-induced jejunal hypersecretion in rats (56). Jejunal loops were filled with saline (2 ml, 37°C) and jejunal secretion was stimulated by a 30 min intraarterial infusion of VIP. At the end of the VIP infusion, the loop was collected, measured, and weighed before and after fluid removal to determine water net flux (mg/cm). Intravenous (i.v.) bolus injections of igmesine or the somatostatin analog, octreotide, were performed 15 min before starting VIP infusion. The neural blocking agent tetrodotoxin at 5 µg/kg, the somatostatin antagonist cyclosomatostatin at 1 mg/kg, or the σ antagonist BMY 14802 at 1 mg/kg were given by i.v. route 5 min before igmesine or octreotide.

As for IL-1β-induced jujenal hypersecretion, the VIP response was blocked by igmesine in a BMY 14802-, tetrodotoxin-, and cyclosomatostatin-sensitive manner, suggesting an indirect action on the enterocyte through σ receptors, nerve, and somatostatin pathways, respectively. The octreotide response was BMY-insensitive indicating that activation of somatostatin receptors during the igmesine response is likely to be distal to the σ receptors.

Cholera toxin, and *E. coli* heat labile (LT) and heat stable (Sta) enterotoxins induce small intestinal secretion in part by activating enteric nerves. Igmesine is a novel σ receptor ligand that inhibits neurally-mediated secretion. The antisecretory potential of igmesine in cholera toxin-, LT-, and Sta-induced water and electrolyte secretion was tested using an *in vivo* rat model of jejunal perfusion.

After pretreatment with igmesine (0.03-10 mg/kg, i.v.), jejunal segments of anesthetized, adult male Wistar rats were incubated with cholera toxin (25 μg), LT and Sta toxins of *E. coli* (25 μg), or saline. Jejunal perfusion with a plasma electrolyte solution containing a non-absorbable marker was undertaken. In some cases, 200 μg/l Sta was added to the perfusate. After equilibration, net water and electrolyte movement was determined. In additional experiments, rats received igmesine, intravenously or intrajejunally, after exposure to cholera toxin.

Cholera toxin induced net water secretion that was inhibited by 1 mg/kg igmesine (median −120 versus −31 μl/min/g, $P < 0.001$). LT- and Sta-induced secretion were also inhibited by 1 mg/kg igmesine (-90 versus −56, $P < 0.03$; and −76 versus −29, $P < 0.01$, respectively). Igmesine also reduced established cholera toxin-induced secretions (8).

Despite the great difference in the mechanisms of action of cholera toxin, and LT and Sta from *E. coli*, the antisecretory efficacy of igmesine implies a common pathway for these enterotoxins, on which the σ ligand acts. This pathway is likely to involve components of the enteric nervous system. As σ ligands inhibit VIP-induced secretion and attenuate the actions of acetylcholine, modulation of the neuronal arc involved in the action of such toxins, may represent the common pathway by which igmesine acts.

In man, the influence of σ ligands and particularly of igmesine on basal and stimulated intestinal water secretion has been tested (57). Jejunal absorption of water and electrolytes was measured with a three-lumen open-segment perfusion method in 16 volunteers. A double-blind crossover intraluminal infusion of prostaglandin E_2 (PGE_2) was performed after oral administration of placebo or igmesine at two doses.

Eight volunteers (group 1) received successively one 25 mg capsule of igmesine (dose expressed as igmesine base) and one capsule of placebo 24 hours apart in randomized order according to the crossover principle; the eight volunteers in group 2 were similarly treated with two 100 mg capsules of igmesine and two capsules of placebo. After 90 min, the infused solution was replaced by the same solution plus 5 µmol/l PGE_2 to stimulate intestinal secretion, and 15 min samples were collected for 270 min (end of the intestinal perfusion).

PGE_2 induced net secretion of water and electrolytes ($P < 0.01$ versus basal conditions). The effect of PGE_2 on water and electrolytes was not changed by 25 mg of igmesine but was suppressed by 200 mg of igmesine. This effect lasted at least 3 hours after a single oral dose. Igmesine at a dose of 200 mg also produced a significant decrease in basal rates of water and electrolyte absorption.

The antisecretory effect of igmesine on PGE_2-induced secretion found in this study quite agrees with the animal data and one may speculate that igmesine may have antidiarrheal properties. However, igmesine also alters basal absorption in an opposite direction, suggesting that σ receptors may have complex effects on the human mucosa. This result is in agreement with recent results showing that the σ antagonist haloperidol increased net basal absorption of water and electrolytes in human subjects through an unknown mechanism (58).

5.3   Diarrhea

Several studies were performed in order to confirm that the antisecretory properties of igmesine and related compounds correspond to antidiarrheal properties. In mice, the effects of igmesine were tested on various models of toxigenic diarrhea (6). Diarrhea was induced by administration of *Salmonella enteriditis* LPS (15 mg/kg, i.v.) in male DBA2 mice, heat stable *E. coli* enterotoxin (*EC*, 70 µg, p.o.) in male NMRI mice, or *C. difficile* toxin (*CD*, 6 ng, p.o.) in male NMRI mice. Diarrhea was evaluated by measuring cumulative stool weight 120 min after LPS treatment or by determination of the fecal water content 120 min after *EC* or *CD* treatment. Igmesine (0.1-1 mg/kg) was administered by oral route 1 hour before induction of diarrhea. The somatostatin antagonist cyclosomatostatin (1 µg/kg, i.v.), the σ antagonist BMY 14802 (1 mg/kg, p.o.), and the opiate antagonist naloxone (0.3 mg/kg, s.c.) were given 5 min before igmesine. Gastric emptying and intestinal transit of a nutritive liquid meal (milk) labeled with a non absorbable marker (^{51}Cr) were evaluated in DBA2 mice by determination of the movement of the marker along the intestine using the Geometric Center method. Igmesine (0.1-10 mg/kg) was given by oral route 1 hour before the meal.

LPS significantly increased stool weight (289 ± 36 mg versus 58 ± 8 mg in control mice, $P < 0.001$) during 120 min. *EC* or *CD* increased the 120 min fecal water excretion by 299% and 349%. Igmesine inhibited in a dose-related manner the three models of diarrhea (ED_{50} 0.34, 1.72 and 1.45 mg/kg, p.o. on LPS-, *EC*- and *CD*-induced diarrhea, respectively). The antidiarrheal activity of igmesine (1 mg/kg) on LPS-induced diarrhea was not modified by naloxone (5.6%, n.s.) but was suppressed by BMY 14802 (-97.7%, $P < 0.001$) and by cyclosomatostatin (-70.8%, $P < 0.001$). Igmesine (0.1, 1, 10 mg/kg) affected neither gastric emptying (-6.7, 2.6%, -3.2%, respectively) nor intestinal transit (-3.0%, 13%, -5.5%, respectively).

Finally, igmesine displays potent antidiarrheal activity in models of toxigenic diarrhea without affecting gastrointestinal transit. The antidiarrheal activity of igmesine did not involve opiate receptors and these results suggest an indirect action on the enterocytes through σ receptor and somastostatin pathways.

Other σ ligands derived from igmesine were also demonstrated to have antidiarrheal properties on similar toxigenic diarrhea in mice (7). Indeed, oral JO 2871 as well as igmesine dose-dependently inhibited toxigenic diarrhea in all models. ED_{50} values obtained using JO 2871 (1-20 µg/kg) were more than 40 times lower than those obtained with igmesine. Given after *EC*, oral JO 2871 also inhibited diarrhea in a dose-dependent manner (ED_{50} 50 µg/kg). Both σ ligands were active by the intravenous route on LPS- and *EC*-induced stool weight increases. JO 2871 administered intracerebroventricularly failed to block this effect at any dose tested. Both BMY 14802 and cyclosomatostatin reversed the antidiarrheal effect of oral JO 2871.

Since these results suggest the involvement of a somatostatin pathway in the antisecretory mechanism of action of JO 2871, it has been hypothesized that JO 2871 acts peripherally on σ receptors, probably located within the gut on somatostatinergic neurons, triggering the release of somatostatin which in turn acts directly or indirectly on receptors located on enterocytes or on secretomotor neurons to alleviate toxin-induced intestinal secretion. The hypothesis of a direct and indirect mechanism involved in the action of somatostatin is supported by previous findings showing a parallel reduction in cholera toxin-induced net fluid secretion and in VIP release from the small intestine of the cat (59). Activation of σ receptors inhibits acetylcholine release (60) and favors noradrenaline release (61); such effects may contribute to its antisecretory action. However, the lack of an anti-transit effect of JO 2871 and igmesine is not in agreement with this possible mechanism of action.

6. OVERALL SUMMARY AND FUTURE DIRECTIONS

The properties reported in this chapter for σ ligands are in agreement with the localization of σ receptors that has been described in the gastrointestinal tract. The stimulating effects of σ ligands on mucosal alkaline secretion have been found to correlate with their protective effects on experimental duodenal ulcers. In the intestine, σ ligands have been shown to have potent antisecretory effects which are related to potent antidiarrheal efficacy in various animal models of diarrhea produced by toxins, including cholera toxin.

Besides these local regulations of intestinal function, regulatory activities of gut motility induced by stress have been evidenced. These effects involve more complex mechanisms and central regulation.

In summary, these results show that σ ligands display potent regulatory effects on gastrointestinal function. Of particular interest, the very potent antidiarrheal effect of σ ligands offers new therapeutic potential for the treatment of diarrheas for which the present available medication is not fully satisfactory and for which there is still a medical need. With a new mechanism of action, σ ligands could be a complementary therapeutic approach in situations where classical opioid treatment is proscribed (infants, elderly) or ineffective (cholera).

REFERENCES

1. Roman F, Pascaud X, Vauche D, Junien JL. Evidence for a non-opioid σ-binding site in the guinea-pig myenteric plexus. Life Sci 1988, 42:2217-2222.
2. Pascaud X, Defaux JP, Roze C, Junien JL. Effect of selective σ ligands on duodenal alkaline secretion in the rat. J Pharmacol Exp Ther 1990, 255:1354-1359.
3. Pascaud XB, Chovet M, Soulard P, Chevalier E, Roze C, Junien JL. Effects of a new σ ligand, JO 1784, on cysteamine ulcers and duodenal alkaline secretion in rats. Gastroenterology 1993, 104:427-434.
4. Riviere PJ, Pascaud X, Junien JL, Porreca F. Neuropeptide Y and JO 1784, a selective σ ligand, alter intestinal ion transport through a common, haloperidol-sensitive site. Eur J Pharmacol 1990, 187:557-559.
5. Riviere PJ, Rao RK, Pascaud X, Junien JL, Porreca F. Effects of neuropeptide Y, peptide YY and σ ligands on ion transport in mouse jejunum. J Pharmacol Exp Ther 1993, 264:1268-1274.
6. Theodorou V, Chovet M, Dassaud M. Antidiarrhoeal effects of igmesine on various models of toxigenic diarrhea. Gastroenterology 1999, 116:A868.
7. Theodorou V, Chovet M, Eutamene H, Fargeau H, Dassaud M, Toulouse M, Bihoreau C, Roman FJ, Bueno L. Antidiarrhoeal properties of a novel σ ligand (JO 2871) on toxigenic diarrhoea in mice: mechanisms of action. Gut 2002, 51:522-528.

8. Turvill JL, Kasapidis P, Farthing MJ. The σ ligand, igmesine, inhibits cholera toxin and Escherichia coli enterotoxin induced jejunal secretion in the rat. Gut 1999, 45:564-569.
9. Junien JL, Gue M, Bueno L. Neuropeptide Y and σ ligand (JO 1784) act through a Gi protein to block the psychological stress and corticotropin-releasing factor-induced colonic motor activation in rats. Neuropharmacology 1991, 30:1119-1124.
10. Herkenham M, Pert CB. Light microscopic localization of brain opiate receptors: a general autoradiographic method which preserves tissue quality. J Neurosci 1982, 2:1129-1149.
11. Largent BL, Gundlach AL, Snyder SH. Pharmacological and autoradiographic discrimination of σ and phencyclidine receptor binding sites in brain with (+)-[^{3}H]SKF 10,047, (+)-[^{3}H]-3-(3-hydroxyphenyl)-N-(1-propyl)piperidine and [^{3}H]-1-(1-(2-thienyl)cyclohexyl)piperidine. J Pharmacol Exp Ther 1986, 238:739-748.
12. Furness JB, Costa M, Rokaeus A, McDonald TJ, Brooks B. Galanin-immunoreactive neurons in the guinea-pig small intestine: their projections and relationships to other enteric neurons. Cell Tissue Res 1987, 250:607-615.
13. Furness JB, Costa M, Keast JR. Choline acetyltransferase- and peptide immunoreactivity of submucous neurons in the small intestine of the guinea-pig. Cell Tissue Res 1984, 237:329-336.
14. Furness JB, Costa M, Emson PC, Hakanson R, Moghimzadeh E, Sundler F, Taylor IL, Chance RE. Distribution, pathways and reactions to drug treatment of nerves with neuropeptide Y- and pancreatic polypeptide-like immunoreactivity in the guinea-pig digestive tract. Cell Tissue Res 1983, 234:71-92.
15. Daniel EE, Costa M, Furness JB, Keast JR. Peptide neurons in the canine small intestine. J Comp Neurol 1985, 237:227-238.
16. Keast JR, Furness JB, Costa M. Distribution of certain peptide-containing nerve fibres and endocrine cells in the gastrointestinal mucosa in five mammalian species. J Comp Neurol 1985, 236:403-422.
17. Keast JR, Furness JB, Costa M. Origins of peptide and norepinephrine nerves in the mucosa of the guinea pig small intestine. Gastroenterology 1984, 86:637-644.
18. Kromer W, Steigemann N, Shearman GT. Differential effects of SKF 10,047 (N-allylnormetazocine) on peristalsis and longitudinal muscle contractions of the isolated guinea-pig ileum. Naunyn Schmiedeberg Arch Pharmacol 1982, 321:218-222.
19. Aceto MD, May EL. Antinociceptive studies of the optical isomers of N-allylnormetazocine (SKF 10,047). Eur J Pharmacol 1983, 91:267-272.
20. Pert CB, Snyder SH, May EL. Opiate receptor interactions of benzomorphans in rat brain homogenates. J Pharmacol Exp Ther 1976, 196:316-322.
21. Nishimura E, Buchan AM, McIntosh CH. Autoradiographic localization of mu- and delta-type opioid receptors in the gastrointestinal tract of the rat and guinea pig. Gastroenterology 1986, 91:1084-1094.
22. Bitar KN, Makhlouf GM. Specific opiate receptors on isolated mammalian gastric smooth muscle cells. Nature 1982, 297:72-74.
23. Tam SW. Naloxone-inaccessible σ receptor in rat central nervous system. Proc Natl Acad Sci USA 1983, 80:6703-6707.
24. Gundlach AL, Largent BL, Snyder SH. Phencyclidine and σ opiate receptors in brain: biochemical and autoradiographical differentiation. Eur J Pharmacol 1985, 113:465-466.
25. Su TP, Schell SE, Ford-Rice FY, London ED. Correlation of inhibitory potencies of putative antagonists for σ receptors in brain and spleen. Eur J Pharmacol 1988, 148:467-470.

26. Wolfe SA Jr, Culp SG, De Souza EB. Sigma-receptors in endocrine organs: identification, characterization, and autoradiographic localization in rat pituitary, adrenal, testis, and ovary. Endocrinology 1989, 124:1160-1172.
27. Largent BL, Gundlach AL, Snyder SH. Psychotomimetic opiate receptors labeled and visualized with (+)-[^{3}H]3-(3-hydroxyphenyl)-N-(1-propyl)piperidine. Proc Natl Acad Sci USA 1984, 81:4983-4987.
28. Roman F, Pascaud X, Chomette G, Bueno L, Junien JL. Autoradiographic localization of σ opioid receptors in the gastrointestinal tract of the guinea pig. Gastroenterology 1989, 97:76-82.
29. Weber E, Sonders M, Quarum M, McLean S, Pou S, Keana JF. 1,3-Di(2-(5-^{3}H)tolyl)guanidine: a selective ligand that labels σ-type receptors for psychotomimetic opiates and antipsychotic drugs. Proc Natl Acad Sci USA 1986, 83:8784-8788.
30. Roman FJ, Pascaud X, Martin B, Vauche D, Junien JL. JO 1784, a potent and selective ligand for rat and mouse brain σ-sites. J Pharm Pharmacol 1990, 42:439-440.
31. Zukin SR, Brady KT, Slifer BL, Balster RL. Behavioral and biochemical stereoselectivity of σ opiate/PCP receptors. Brain Res 1984, 294:174-177.
32. Beaumont W. Experiments and observations on the gastric juice in physiology of digestion. In Osler W, ed. New York: Dover 1833. 1:280.
33. Almy TP. Experimental studies on the irritable colon. Am J Med 1951, 10:60-67.
34. Stanghellini V, Malagelada JR, Zinsmeister AR, Go VL, Kao PC. Stress-induced gastroduodenal motor disturbances in humans: possible humoral mechanisms. Gastroenterology 1983, 85:83-91.
35. Taché Y. Stress induced alterations of gastric emptying. In Bueno L, Collins S, Junien JL, eds. Stress and Digestive Motility. London: John Libbey 1989, p. 123-132.
36. Narducci F, Snape WJ, Jr., Battle WM, London RL, Cohen S. Increased colonic motility during exposure to a stressful situation. Dig Dis Sci 1985, 30:40-44.
37. Wangel A, Deller D. Intestinal motility in man II. Mechanisms of constipation and diarrhea with particular reference to the irritable colon syndrome. Gastroenterology 1964, 72:383-387.
38. Chaudhary NA, Truelove SC. The irritable colon syndrome. Q J Med 1962, 31:307-322.
39. Welgan P, Meshkinpour H, Beeler M. Effect of anger on colon motor and myoelectric activity in irritable bowel syndrome. Gastroenterology 1988, 94:1150-1156.
40. Frexinos J, Bueno L. Influence of stress on colonic myoelectrical activity in IBS patients. Gastroenterology 1987, 92:1396.
41. Williams CL, Peterson JM, Villar RG, Burks TF. Corticotropin-releasing factor directly mediates colonic responses to stress. Am J Physiol 1987, 253:G582-G586.
42. Gue M, Junien JL, Bueno L. Conditioned emotional response in rats enhances colonic motility through the central release of corticotropin-releasing factor. Gastroenterology 1991, 100:964-970.
43. Jimenez M, Bueno L. Inhibitory effects of neuropeptide Y (NPY) on CRF and stress-induced cecal motor response in rats. Life Sci 1990, 47:205-211.
44. Chang RS, Lotti VJ. Biochemical and pharmacological characterization of an extremely potent and selective nonpeptide cholecystokinin antagonist. Proc Natl Acad Sci USA 1986, 83:4923-4926.
45. Lotti VJ, Chang RS. A new potent and selective non-peptide gastrin antagonist and brain cholecystokinin receptor (CCK-B) ligand: L-365,260. Eur J Pharmacol 1989, 162:273-280.
46. Hill DR, Campbell NJ, Shaw TM, Woodruff GN. Autoradiographic localization and biochemical characterization of peripheral type CCK receptors in rat CNS using highly selective nonpeptide CCK antagonists. J Neurosci 1987, 7:2967-2976.

47. Steinfels GF, Tam SW. Selective σ receptor agonist and antagonist affect dopamine neuronal activity. Eur J Pharmacol 1989, 163:167-170.
48. Piontek JA, Wang RY. Acute and subchronic effects of Rimcazole (BW 234U), a potential antipsychotic drug, on A9 and A10 dopamine neurons in the rat. Life Sci 1986, 39:651-658.
49. Harfstrand A, Eneroth P, Agnati L, Fuxe K. Further studies on the effects of central administration of neuropeptide Y on neuroendocrine function in the male rat: relationship to hypothalamic catecholamines. Regul Pept 1987, 17:167-179.
50. Morimoto Y, Shimohara K, Tanaka K, Hara H, Sukamoto T. 4-Methoxyphenyl-4-(3,4,5-trimethoxybenzyl)-1-piperazineacetate monofumarate monohydrate (KB-5492), a new anti-ulcer agent with a selective affinity for the σ receptor, prevents cysteamine-induced duodenal ulcers in rats by a mechanism different from that of cimetidine. Jpn J Pharmacol 1994, 64:221-224.
51. Hara H, Tanaka K, Harada Y, Sukamoto T. Effect of KB-5492, a new selective σ ligand, on gastric lesions and alkaline secretion in rats. Jpn J Pharmacol 1993, 61 (Supp 1):194P.
52. Shay H, Sun P. Studies on gastric secretion. Gastroenterology 1954, 26:906-913.
53. Szabo S, Haith LR Jr, Reynolds ES. Pathogenesis of duodenal ulceration produced by cysteamine or propionitrile: influence of vagotomy, sympathectomy, histamine depletion, H-2 receptor antagonists and hormones. Dig Dis Sci 1979, 24:471-477.
54. Flemstrom G, Kivilaakso E. Demonstration of a pH gradient at the luminal surface of rat duodenum in vivo and its dependence on mucosal alkaline secretion. Gastroenterology 1983, 84:787-794.
55. Theodorou V, Chovet M, Fioramonti J, Bueno L. Comparative antisecretory effect of loperamide somatostatin and the σ ligand igmesine on IL-1b-induced colonic hypersecretion in rats. Gastroenterology 1998, 114:A423.
56. Fargeau PJ, Rivière M, Junien JL, Chovet M. Igmesine inhibition of VIP-induced jejunal hypersecretion in rats, involvement of endogenous somatostatin and nervous pathway. Neurogastroenterology 1998, 10:443.
57. Roze C, Bruley Des Varannes S, Shi G, Geneve J, Galmiche JP. Inhibition of prostaglandin-induced intestinal secretion by igmesine in healthy volunteers. Gastroenterology 1998, 115:591-596.
58. Roze C, Molis C, Xiaomei FC, Ropert A, Geneve J, Galmiche JP. Peptide YY inhibition of prostaglandin-induced intestinal secretion is haloperidol-sensitive in humans. Gastroenterology 1997, 112:1520-1528.
59. Eklund S, Sjoqvist A, Fahrenkrug J, Jodal M, Lundgren O. Somatostatin and methionine-enkephalin inhibit cholera toxin-induced jejunal net fluid secretion and release of vasoactive intestinal polypeptide in the cat in vivo. Acta Physiol Scand 1988, 133:551-557.
60. Campbell BG, Scherz MW, Keana JF, Weber E. Sigma receptors regulate contractions of the guinea pig ileum longitudinal muscle/myenteric plexus preparation elicited by both electrical stimulation and exogenous serotonin. J Neurosci 1989, 9:3380-3391.
61. Campbell BG, Bobker DH, Leslie FM, Mefford IN, Weber E. Both the σ receptor-specific ligand (+)3-PPP and the PCP receptor-specific ligand TCP act in the mouse vas deferens via augmentation of electrically evoked norepinephrine release. Eur J Pharmacol 1987, 138:447-449.

Corresponding author: *Dr. Francois J. Roman, Mailing address: Euroscreen, 47 rue Adrienne Bolland, Gosselies, B-6041, Belgium, Phone:*

32 71 348 502, Fax: 32 71 348 519, Electronic mail address: froman@euroscreen.be

Appendix A

CHEMICAL NAMES OF COMPOUNDS

Compound	Chemical Name
Sigma Ligands:	
AC927	N-phenethylpiperidine
BD737	(+)-cis-N-methyl-N-[2-(3,4-dichlorophenyl)ethyl]-2-(1-pyrrolidinyl)cyclohexylamine
BD1008	N-[2-(3,4-dichlorophenyl)ethyl]-N-methyl-2-(1-pyrrolidinyl)ethylamine
BD1031	3R-1-[2-(3,4-dichlorophenyl)ethyl]-1,4-diazabicyclo[4.3.2]nonane
BD1047	N-[2-(3,4-dichlorophenyl)ethyl]-N-methyl-2-(dimethylamino)ethylamine
BD1052	N-[2-(3,4-dichlorophenyl)ethyl]-N-allyl-2-(1-pyrrolidinyl)ethylamine
BD1063	1-[2-(3,4-dichlorophenyl)ethyl]-4-methylpiperazine
BMY 14802	α-(4-fluorophenyl)-4-(5-fluoro-2-pyrimidinyl)-1-piperazine-butanol
CB-64D	(+)-1R,5R-(E)-8-benzylidene-5-(3-hydroxyphenyl)-2-methylmorphan-7-one
CB-184	(+)–1R,5R-(E)-8-(3,4-dichlorobenzylidene)-5-(3-hydroxyphenyl)-2-methylmorphan-7-one
DHEA	dehydroepiandrosterone
DTG	1,3-di-o-tolylguanidine
DuP 734	1-(cyclopropylmethyl)-4-(2'-oxoethyl)piperidine
E-5842	4-[4-fluorophenyl]-1,2,3,6-tetrahydro-1-[4-[1,-2,4-triazol-1-il]butyl]pyridine
4-IBP	N-(N-benzylpipeeridin-4-yl)-4-iodobenzamide
IDAB	N-(2-diethylaminoethyl)-4-iodobenzamide
IPAB	N-[2-(1'-piperidinyl)ethyl]-4-iodobenzamide
IPEMP	N-[2-(4-iodophenyl)ethyl]-N-methyl-2-(1-piperidinyl)ethylamine
Igmesine (JO 1784)	(+)-cinnamyl-1-phenyl-1-N-methyl-N-cyclopropylene
JL-II-147	2-[N-[2-[1-pyrrolidinyl]ethyl]-N-methylamino]-6,7-dichlorotetralin
JO 2871	(E)-3-(1-cyclopropylmethyl-2-azinanyl)-1-(3,4-dichlorophenyl)-1-propene
KB-5492	4-methoxyphenyl-4-(3,4,5-trimethoxybenzyl)-1-piperazine acetate monofumarate monohydrate
L-687,384	1-benzylspiro(1,2,3,4-tetrahydronaphthalene-1,4-piperidine)
Lu 28-179	1'-[4-[1-(4-fluorophenyl)-1H-indol-3-yl]-1-butyl]spiro[isobenzofuran-1(3H),4'-piperidine

Compound	Chemical Name
MS-377	(R)-(+)-1-(4-chlorophenyl)-3-[4-(2-methoxyethyl)piperazin-1-yl]methyl-2-pyrrolidinone
NE-100	N,N-dipropyl-2-[4-methoxy-3-(2-phenylethoxy)phenyl]ethylamine
NPC 16377	6-[6-(4-hydroxypiperidinyl)hexyloxy]-3-methylflavone
OPC-14523	1-[3-[4-(3-chlorophenyl)-1-piperazinyl]propyl]-5-methoxy-3,4-dihydro-2-quinolinone monomethamesulfonate
Panamesine (EMD 57445)	(S)-(-)-[4-hydroxy-4-(3,4-benzodioxol-5-yl)piperidin-1-ylmethyl]-3-(4-methoxyphenyl)-oxazolidin-2-one
PIMBA	N-[2-(1'-piperidinyl)ethyl]-3]iodo-4-methoxybenzamide
3-PPP	3-(3-hydroxyphenyl)-N-(1-propyl)piperidine
PRE-084	2-(4-morpholino)ethyl-1-phenylcyclohexane-1-carboxylate
Rimcazole	9-[3-(cis-3,5-dimethyl-1-piperzinyl)propyl]carbazole
SA4503	1-(3,4-dimethoxyphenethyl)-4-(3-phenylpropyl)piperazine
SA5845	1-(4-2'-fluroethoxy-3-methoxyphenethyl)-4-(3-(3-fluorophenyl)propyl)piperazine
SKF-10,047	N-allylnormetazocine
SM-21	Tropanyl 2-(4-chlorophenoxy)butanoate
SR 31747A	N-cyclohexyl-N-ethyl-3-(3-chloro-4-cyclohexylphenyl)propen-2-ylamine
SSR 125329	(Z)-3-(4-adamantan-2-yl-3,5-dichlorophenyl)allyl]cyclohexylethylamine

Other Neurotransmitter Ligands:

AMPA	α-amino-3-hydroxy-5-methylisoxazole-4-propionate
Devazepide (L 364,718)	3S(-)-N-(2,3-dihydro-1-methyl-2-oxo-5-phenyl-1H-1,4-benzodiazepine-3-yl)-1H-indole-2-caboxamide
L 365,260	3R-(+)-N-(2,3-dihydro-1-methyl-2-oxo-5-phenyl-1H-1,4-benzodiazepine-3-yl)-N'-(3-methylphenyl)urea
LSD	lysergic acid diethylamide
8-OH-DPAT	(±)-8-hydroxy-2-dipropylaminotetralin
PYX-1	Ac-[3-(2,6-dichlorobenzyl)Tyr^{27}, D-Thr^{32}]NPY(27-36)amide
SCH 23390	(R)-(+)-7-chloro-8-hydroxy-3-methyl-1-phenyl-2,3,4,5-tetrahydro-1H-3-benzazepine

Second Messenger Ligands:

GF-109203x	2-[1-(3-dimethylaminopropyl)indol-3-yl]-3-(indol-3-yl)maleimide
Go-6976	12-(2-cyanoethyl)-6,7,12,13-tetrahydro-13-methyl-5-oxo-5H-indolo[2,3-a]pyrrolo[3,4-c]carbazole
H-7	1-(5-isoquinolinesulfonyl)-2-methylpiperazine
H-1004	N-(2-guanidinoethyl)-5-isoquinolinesulfonamide
PMA	phorbol-12-myristate-13-acetate
U-73,122	1-[6-[[(17β)-3-methoxyestra-1,3,5(10)-trien-17-yl]amino]hexyl]-1H-pyrrole-2,5-dione

Appendix B

LIST OF ENDOGENOUS COMPOUNDS THAT DO NOT BIND TO σ RECEPTORS

Neurotransmitter	Tissue	Radioligand	Reference
Acetylcholine	Guinea pig brain	[^{3}H]SKF-10,047 + etorphine	12
Aldosterone	Human cerebellum	[^{3}H]haloperidol	2
	Rat cerebellum	[^{3}H]haloperidol	2
Angiotensin II	Guinea pig brain	[^{3}H]DuP 734	5
	Rat brain	[^{3}H](+)-3PPP	10
Aspartate	Guinea pig brain	[^{3}H]dextromethorphan	9
Bradykinin	Guinea pig brain	[^{3}H]DuP 734	5
	Rat brain	[^{3}H](+)-3PPP	10
Cadaverine	Rat brain	[^{3}H](+)-3PPP	3
Cholecystokinin	Rat brain	[^{3}H](+)-3PPP	10
Cholesterol	Rat liver	[^{3}H]haloperidol	18
Choline	Guinea pig brain	[^{3}H]dextromethorphan	4
Corticosterone	Rat liver	[^{3}H]haloperidol	18
Corticotropin RF	Guinea pig brain	[^{3}H]DuP 734	5
Cortisol	Rat liver	[^{3}H]haloperidol	18
Creatine	Guinea pig brain	[^{3}H]DTG	17
Creatinine	Guinea pig brain	[^{3}H]DTG	17
Cyclic AMP	Rat liver	[^{3}H]progesterone	18
Cysteine	Guinea pig brain	[^{3}H]dextromethorphan	9
Desoxycorticosterone	Sheep pineal gland	[^{3}H]DTG	1
Dopamine	Guinea pig brain	[^{3}H]DTG	17
	Guinea pig brain	[^{3}H](+)-SKF-10,047	15
	Guinea pig brain	[^{3}H]SKF-10,047 + etorphine	12
	Guinea pig brain	[^{3}H]DuP 734	5
	Rat brain	[^{3}H](+)-3PPP	10
	Rat liver	[^{3}H](+)-SKF-10,047	11

Neurotransmitter	Tissue	Radioligand	Reference
Dopamine	Sheep pineal gland	[^{3}H]DTG	1
Dynorphin A	Guinea pig brain	[^{3}H]DTG	17
Dynorphin (1-8)	Guinea pig brain	[^{3}H](+)-SKF-10,047	14
Dynorphin (1-9)	Guinea pig brain	[^{3}H](+)-SKF-10,047	14
Dynorphin (1-13)	Guinea pig brain	[^{3}H](+)-SKF-10,047	14
	Guinea pig brain	[^{3}H]SKF-10,047 + etorphine	12
	Rat spinal cord	[^{3}H]EKC + naloxone	13
	Rat liver	[^{3}H](+)-SKF-10,047	11
α-Endorphin	Guinea pig brain	[^{3}H](+)-SKF-10,047	14
	Rat brain	[^{3}H](+)-3PPP	10
	Rat liver	[^{3}H](+)-SKF-10,047	11
β-Endorphin	Guinea pig brain	[^{3}H]DTG	17
	Guinea pig brain	[^{3}H]SKF-10,047 + etorphine	12
	Rat brain	[^{3}H](+)-3PPP	10
	Rat liver	[^{3}H](+)-SKF-10,047	11
γ-Endorphin	Guinea pig brain	[^{3}H]SKF-10,047 + etorphine	12
	Guinea pig brain	[^{3}H](+)-SKF-10,047	14
	Rat liver	[^{3}H](+)-SKF-10,047	11
Leu-Enkephalin	Guinea pig brain	[^{3}H]DTG	17
	Guinea pig brain	[^{3}H]dextromethorphan	4
	Guinea pig brain	[^{3}H](+)-SKF-10,047	15
	Guinea pig brain	[^{3}H]SKF-10,047 + etorphine	12
	Guinea pig brain	[^{3}H]DuP 734	5
	Rat brain	[^{3}H](+)-3PPP	10
	Rat spinal cord	[^{3}H]EKC + naloxone	13
	Rat liver	[^{3}H](+)-SKF-10,047	11
Met-Enkephalin	Guinea pig brain	[^{3}H]dextromethorphan	4
	Guinea pig brain	[^{3}H](+)-SKF-10,047	14
	Guinea pig brain	[^{3}H]SKF-10,047 + etorphine	12
	Rat liver	[^{3}H](+)-SKF-10,047	11
Epinephrine	Rat liver	[^{3}H](+)-SKF-10,047	11
	Guinea pig brain	[^{3}H]SKF-10,047 + etorphine	12
Estradiol-17β	Rat liver	[^{3}H]haloperidol	18
GABA	Guinea pig brain	[^{3}H]DTG	17
	Guinea pig brain	[^{3}H]dextromethorphan	4

B. Endogoneous compounds that do not bind σ receptors

Neurotransmitter	Tissue	Radioligand	Reference
	Guinea pig brain	[^{3}H]DuP 734	5
GABA	Rat brain	[^{3}H](+)-3PPP	10
	Rat cerebellum	[^{3}H]haloperidol	2
	Rat liver	[^{3}H]haloperidol	18
	Human cerebellum	[^{3}H]haloperidol	2
	Sheep pineal gland	[^{3}H]DTG	1
Glutamate	Guinea pig brain	[^{3}H]DTG	6
	Guinea pig brain	[^{3}H]SKF-10,047 + etorphine	12
	Guinea pig brain	[^{3}H]DuP 734	5
	Guinea pig brain	[^{3}H]dextromethorphan	4
	Guinea pig brain	[^{3}H](+)-pentazocine	7
Glycine	Guinea pig brain	[^{3}H]DTG	6
	Guinea pig brain	[^{3}H](+)-pentazocine	7
	Guinea pig brain	[^{3}H]DuP 734	5
	Guinea pig brain	[^{3}H]dextromethorphan	4
	Human cerebellum	[^{3}H]haloperidol	2
	Rat cerebellum	[^{3}H]haloperidol	2
Gpp(NH)p	Human cerebellum	[^{3}H]haloperidol	2
	Rat cerebellum	[^{3}H]haloperidol	2
GTP	Human cerebellum	[^{3}H]haloperidol	2
	Rat cerebellum	[^{3}H]haloperidol	2
	Rat liver	[^{3}H]progesterone	18
Histamine	Guinea pig brain	[^{3}H]SKF-10,047 + etorphine	12
	Guinea pig brain	[^{3}H]dextromethorphan	4
	Rat liver	[^{3}H](+)-SKF-10,047	11
Histidine	Guinea pig brain	[^{3}H]dextromethorphan	4
Hydra-head activator	Rat liver	[^{3}H](+)-SKF-10,047	11
Hydrocortisone	Human cerebellum	[^{3}H]haloperidol	2
	Rat cerebellum	[^{3}H]haloperidol	2
Liver cell GF	Rat liver	[^{3}H](+)-SKF-10,047	11
Melatonin	Guinea pig brain	[^{3}H]dextromethorphan	4
	Sheep pineal gland	[^{3}H]DTG	1
Neuropeptide Y	Guinea pig brain	[^{3}H]DTG	6
	Guinea pig brain	[^{3}H](+)-SKF-10,047	16
	Guinea pig brain	[^{3}H]DuP 734	5

Neurotransmitter	Tissue	Radioligand	Reference
	Rat brain	[^{3}H](+)-SKF-10,047	16
	Rat cortex	[^{3}H](+)-3PPP	8
Neurotensin	Rat brain	[^{3}H](+)-3PPP	10
	Rat liver	[^{3}H](+)-SKF-10,047	11
Neurotensin (8-13)	Guinea pig brain	[^{3}H]DuP 734	5
Norepinephrine	Guinea pig brain	[^{3}H]SKF-10,047 + etorphine	12
Norepinephrine cont.	Guinea pig brain	[^{3}H]DuP 734	5
	Rat liver	[^{3}H](+)-SKF-10,047	11
	Sheep pineal gland	[^{3}H]DTG	1
Octopamine	Guinea pig brain	[^{3}H]dextromethorphan	4
Prostaglandin E$_1$	Guinea pig brain	[^{3}H]dextromethorphan	4
Prostaglandin E$_2$	Guinea pig brain	[^{3}H]dextromethorphan	4
	Rat liver	[^{3}H]progesterone	18
Peptide YY	Guinea pig brain	[^{3}H](+)-SKF-10,047	16
	Rat brain	[^{3}H](+)-SKF-10,047	16
Progesterone	Sheep pineal gland	[^{3}H]DTG	1
Putrescine	Rat brain	[^{3}H](+)-3PPP	3
trans-Retinoic acid	Rat liver	[^{3}H]progesterone	18
trans-Retinol	Rat liver	[^{3}H]progesterone	18
Serotonin	Guinea pig brain	[^{3}H]DTG	17
	Guinea pig brain	[^{3}H]SKF-10,047 + etorphine	12
	Guinea pig brain	[^{3}H]DuP 734	5
	Guinea pig brain	[^{3}H]dextromethorphan	4
	Rat spinal cord	[^{3}H]EKC + naloxone	13
	Rat liver	[^{3}H](+)-SKF-10,047	11
	Sheep pineal gland	[^{3}H]DTG	1
Spermidine	Rat brain	[^{3}H](+)-3PPP	3
Spermine	Rat brain	[^{3}H](+)-3PPP	3
Substance P	Guinea pig brain	[^{3}H]DTG	6
	Guinea pig brain	[^{3}H](+)-pentazocine	7
Taurine	Guinea pig brain	[^{3}H]dextromethorphan	4, 9
Testosterone	Sheep pineal gland	[^{3}H]DTG	1
Thyronine	Rat liver	[^{3}H]progesterone	18

DTG = di-o-tolylguanidine; DuP 724 = 1-(cyclopropylmethyl)-4-(2'-oxoethyl)piperidine; EKC = ethylketocyclazocine; GABA = γ-aminobutyric acid; GF = growth factor; 3-PPP = 3-(3-hydroxyphenyl)-N-(1-propyl)piperidine; SKF-10,047 = N-allylnormetazocine

REFERENCES

1. Abreu P, Sugden D. Characterization of binding sites for [^{3}H]-DTG, a selective sigma receptor ligand, in the sheep pineal gland. Biochem Biophys Res Comm 1990, 171:875-881.
2. Barnes JM, Barnes NM, Barber PC, Champaneria S, Costall B, Hornsby CD, Ironside JW, Naylor RJ. Pharmacological comparison of the sigma recognition site labeled by [^{3}H]haloperidol in human and rat cerebellum. Naunyn-Schmiedeberg Arch Pharmacol 1992, 345:197-202.
3. Contreras PC, Bremer ME, Gray NM. Ifenprodil and SL 82.0715 potently inhibit binding of [^{3}H](+)-3-PPP to σ binding sites in rat brain. Neurosci Lett 1990, 116:190-193.
4. Craviso GL, Musacchio JM. High-affinity dextromethorphan binding sites in guinea pig brain. II. Competition experiments. Mol Pharmacol 1983, 23:629-640.
5. Culp SG, Rominger D, Tam SW, De Souza EB. [^{3}H]DuP 734 [1-(cyclopropylmethyl)-4-(2'-(4''-fluorophenyl)-2'-oxoethyl)-piperidine HBr]: A receptor ligand profile of a high-affinity novel sigma receptor ligand in guinea pig brain. J Pharmacol Exp Ther 1992, 263:1175-1187.
6. DeHaven-Hudkins DL, Fleissner LC. Competitive interactions at [^{3}H]1,3-(2-tolyl)guanidine (DTG)-defined sigma recognition sites in guinea pig brain. Life Sci 1992, 50:PL65-70.
7. DeHaven-Hudkins DL, Fleissner LC, Ford-Rice FY. Characterization of the binding of [^{3}H](+)-pentazocine to σ recognition sites in guinea pig brain. Eur J Pharmacol 1992, 227:371-378.
8. Jeanjean AP, Mestre M, Maloteaux J-M, Laduron PM. Is the σ$_2$ receptor in rat brain related to the K$^+$ channel of class III antiarrhythmic drugs? Eur J Pharmacol 1993, 241:111-116.
9. Klein M, Musacchio JM. High affinity dextromethorphan binding sites in guinea pig brain. Effect of sigma ligands and other agents. J Pharmacol Exp Ther 1989, 251:207-215.
10. Largent BL, Gundlach AL, Snyder SH. Psychotomimetic opiate receptors labeled and visualized with (+)-[^{3}H]3-(3-hydroxyphenyl)-N-(1-propyl)piperidine. Proc Natl Acad Sci USA 1984, 81:4983-4987.
11. Samovilova NN, Nagornaya LV, Vinogradov VA. (+)-[3H]SK&F 10,047 binding sites in rat liver. Eur J Pharmacol 1988, 147:259-264.
12. Su T-P. Evidence for sigma opioid receptor: Binding of [^{3}H]SKF-10047 to etorphine-inaccessible sites in guinea-pig brain. J Pharmacol Exp Ther 1982, 223:284-290.
13. Tam SW. Naloxone-inaccessible σ receptor in rat central nervous system. Proc Natl Acad Sci USA 1983, 80:6703-6707.
14. Tam SW. (+)-[^{3}H]SKF 10,047, (+)-[^{3}H]ethylketocyclazocine, μ, κ, δ and phencyclidine binding sites in guinea pig brain membranes. Eur J Pharmacol 1985, 109:33-41.
15. Tam SW, Cook L. σ Opiates and certain antipsychotic drugs mutually inhibit (+)-[^{3}H]SKF 10,047 and [^{3}H]haloperidol binding in guinea pig brain membranes. Proc Natl Acad Sci USA 81:5618-5621.
16. Tam SW, Mitchell KN. Neuropeptide Y and peptide YY do not bind to brain σ and phencyclidine binding sites. Eur J Pharmacol 1991, 193:121-122.
17. Weber E, Sonders M, Quarum M, McLean S, Pou S, Keana JFW. 1,3-Di(2-[5-^{3}H]tolyl)guanidine: A selective ligand that labels σ-type receptors for psychotomimetic opiates and antipsychotic drugs. Proc Natl Acad Sci USA 1986, 83:8784-8788.
18. Yamada M, Nighigami T, Nakasho K, Nishimoto Y, Miyaji H. Relationship between sigma-like site and progesterone-binding site of adult male liver microsomes. Hepatology 1994, 20:1271-1280.

Appendix C

LIST OF CELL LINES EXPRESSING σ RECEPTORS

Cell Line	Species	Radioligand or Probe	Subtype	Reference
PC12 pheochromocytoma	Rat	[^{3}H]DTG, [^{3}H](+)-3-PPP	σ_2	1, 2
C6 glioma	Rat	[^{3}H](+)-Pentazocine, [^{3}H]DTG	σ_1, σ_2	3-5
C6-BU-1 glioma	Rat	[^{3}H]DTG	σ_2	6
U-138MG glioblastoma	Human	[^{3}H](+)-Pentazocine, [^{3}H]DTG	σ_1, σ_2	5
NB41A3 neuroblastoma	Mouse	[^{3}H](+)-Pentazocine, [^{3}H]DTG	σ_1, σ_2	3, 5
N1E-115 neuroblastoma	Mouse	[^{3}H](+)-Pentazocine, [^{3}H]DTG	σ_1, σ_2	3, 5
S-20Y neuroblastoma	Mouse	[^{3}H](+)-Pentazocine, [^{3}H]DTG	σ_1, σ_2	3, 5
N18TG2 neuroblastoma	Mouse	[^{3}H]DTG	σ_1, σ_2	7
SK-N-SH neuroblastoma	Human	[^{3}H](+)-Pentazocine, [^{3}H]DTG	σ_1, σ_2	5, 8
SH-SY5Y neuroblastoma	Human	[^{3}H]Haloperidol, [^{3}H](+)-Pentazocine	σ_1	9, 10
BE(2)-C neuroblastoma	Human	[^{3}H](+)-Pentazocine	σ_1	11
NG108-15 neuroblastoma x glioma hybrid	Mouse x Rat	[^{3}H](+)-Pentazocine, [^{3}H]DTG	σ_1, σ_2	3, 5, 12
NCB-20 neuroblastoma x Chinese hamster brain hybrid	Mouse x Hamster	[^{3}H]SKF-10,047	σ_1 and low affinity site (probably σ_2)	13
9L brain tumor	Rat	[^{3}H]DTG	σ_2	14
MCF-7 breast adenocarcinoma	Human	[^{3}H]DTG	σ_2; no σ_1	5
T47D breast ductal carcinoma	Human	[^{3}H](+)-Pentazocine, [^{3}H]DTG	σ_1, σ_2	5
SKBr3 breast adenocarcinoma	Human	[^{3}H]DTG	σ_2	15
MCF-7/Adr breast adenocarcinoma	Human	[^{3}H]DTG	σ_2	15

Cell Line	Species	Radioligand or Probe	Subtype	Reference
Mammary adenocarcinoma line 66 (diploid)	Mouse	[^{3}H]DTG	σ_2	14, 16
Mammary adenocarcinoma line 67 (aneuploid)	Mouse	[^{3}H]DTG	σ_2	14
EMT-6 mammary carcinoma	Mouse	[^{3}H](+)-Pentazocine, [^{3}H]DTG	σ_1, σ_2	17
ThP-1 leukemia	Human	[^{3}H](+)-Pentazocine, [^{3}H]DTG	σ_1, σ_2	5
Jurkat T lymphocyte	Human	[^{3}H]Haloperidol, [^{3}H]Progesterone, [^{3}H]DTG	σ_1, σ_2	18, 19
NCI-H727 lung carcinoid	Human	[^{3}H](+)-Pentazocine, [^{3}H]DTG, [^{125}I]IPAB	σ_1, σ_2	5, 20
NCI-H1299 non-small cell lung carcinoma	Human	[^{125}I]IPAB	σ_1	20
NCI-H838 non-small cell lung adenocarcinoma	Human	[^{125}I]IPAB	σ_1	20
DMS-114 small cell lung carcinoma	Human	[^{125}I]Azidococaine	σ_1	21
A2058 melanoma	Human	[^{125}I]IPAB	σ_1	22
A375 melanoma	Human	[^{3}H](+)-Pentazocine, [^{3}H]DTG	σ_1, σ_2	5
LNCaP.FGC prostate carcinoma	Human	[^{3}H](+)-Pentazocine, [^{3}H]DTG	σ_1, σ_2	5
DU-145 prostate carcinoma	Human	[^{3}H](+)-Pentazocine, [^{3}H]DTG, [^{125}I]PIMBA	σ_1, σ_2	23
Placental choriocarcinoma (JAR)	Human	cDNA, [^{3}H]Haloperidol	σ_1	24
Caco-2 colorectal adenocarcinoma	Human	cDNA	σ_1	24, 25
HT-29 colorectal carcinoma	Human	cDNA	σ_1	24

Note: Only cell lines in which σ receptors were directly demonstrated by radioligand binding assays or by mRNA expression using cDNA probes are shown. Cell lines where the presence of σ receptors was indicated solely by functional assays are not shown here. Abbreviations: DTG = 1,3-di-o-tolylguanidine; IPAB = N-[2-(1'-piperidinyl)ethyl]-4-iodobenzamide; PIMBA = N-[2-(1'-piperidinyl)ethyl]-3]iodo-4-methoxybenzamide; 3-PPP = 3-(3-hydroxyphenyl)-N-(1-propyl)piperidine; SKF-10,047 = N-allylnormetazocine

REFERENCES

1. Hellewell SB, Bowen WD. A sigma-like binding site in rat pheochromocytoma (PC12) cells: Decreased affinity for (+)-benzomorphans and lower molecular weight suggest a different sigma receptor form from that in guinea pig brain. Brain Res 1990, 527:244-253.
2. Yang ZW, Paleos GA, Byrd JC. Expression of (+)-3-PPP binding sites in the PC12 pheochromocytoma cell line. Eur J Pharmacol 1989, 164:607-610.

3. Vilner BJ, Bowen, WD. Characterization of sigma-like binding sites of NB41A3, S-20Y, and N1E-115 neuroblastomas, C6 glioma, and NG108-15 neuroblastoma-glioma hybrid cells: Further evidence for sigma-2 receptors. In: *Multiple Sigma and PCP Receptor Ligands: Mechanisms for Neuromodulation and Neuroprotection?* J.-M. Kamenka and E.F. Domino, eds. NPP Books, Ann Arbor, MI, 1992, pp. 341-353.
4. Vilner BJ, de Costa BR, Bowen WD. Cytotoxic effects of sigma ligands: Sigma receptor-mediated alterations in cellular morphology and viability. J Neurosci 1995, 15:117-134.
5. Vilner BJ, John CS, Bowen, WD. Sigma-1 and sigma-2 receptors are expressed in a wide variety of human and rodent tumor cell lines. Cancer Res 1995, 55:408-413.
6. Georg A, Friedl A. Characterization of specific binding sites for [3H]-1,3-di-o-tolylguanidine (DTG) in the rat glioma cell line C6-BU-1. Glia 1992, 6:258-263.
7. Barg J, Thomas GE, Bem WT, Parnes MD, Ho AM, Belcheva MM, McHale RJ, McLachlan JA, Tolman KC, Johnson FE, Coscia, CJ. In vitro and in vivo expression of opioid and sigma receptors in rat C6 glioma and mouse N18TG2 neuroblastoma cells. J Neurochem 1994, 63:570-574.
8. Vilner BJ, Bowen WD. Modulation of cellular calcium by sigma-2 receptors: Release from intracellular stores in human SK-N-SH neuroblastoma cells. J Pharmacol Exp Ther 2000, 292:900-911.
9. Hong W, Nuwayhid SJ, Werling LL. Modulation of bradykinin-induced calcium changes in SH-SY5Y cells by neurosteroids and sigma receptor ligands via a shared mechanism. Synapse 2004, 54:102-110.
10. Hong W, Werling LL. Binding of sigma receptor ligands and their effects on muscarine-induced Ca^{2+} changes in SH-SY5Y cells. Eur J Pharmacol 2002, 436:35-45.
11. Ryan-Moro, J, Chien, CC, Standifer, KM, and Pasternak, GW. Neurochem Res 1996, 21:1309-1314.
12. Georg A, Friedl A. Identification and characterization of two sigma-like binding sites in the mouse neuroblastoma x rat glioma hybrid cell line NG108-15. J Pharmacol Exp Ther 1991, 259:479-483.
13. Wu XZ, Bell JA, Spivak CE, London ED, Su T-P. Electrophysiological and binding studies on intact NCB-20 cells suggest presence of a low affinity sigma receptor. J Pharmacol Exp Ther 1991, 257:351-359.
14. Al-Nabulsi I, Mach RH, Wang L-M, Wallen CA, Keng PC, Sten K, Childers SR, Wheeler KT. Effect of ploidy, recruitment, environmental factors, and tamoxifen treatment on the expression of sigma-2 receptors in proliferating and quiescent tumour cells. Brit J Cancer 1999, 81:925-933.
15. Crawford KW, Bowen WD. Sigma-2 receptor agonists activate a novel apoptotic pathway and potentiate antineoplastic drugs in breast tumor cell lines. Cancer Res 2002, 62:313-322.
16. Mach RH, Smith CR, Al-Nabulsi I, Whirrett BR, Childers SR, Wheeler KT. Sigma-2 receptors as potential biomarkers of proliferation in breast cancer. Cancer Res 1997, 57:156-161.
17. Colabufo NA, Berardi F, Contino M, Fazio F, Matarrese M, Moresco RM, Niso M, Perrone R, Tortorella V. Distribution of sigma receptors in EMT-6 cells: preliminary biological evaluation of PB167 and potential for in-vivo PET. J Pharm Pharmacol 2005, 57:1453-1459.
18. Ganapathy, ME, Prasad, PD, Huang, W, Seth, P, Leibach, FH, Ganapathy, V. Molecular and ligand-binding characterization of the sigma-receptor in the Jurkat human T lymphocyte cell line. J Pharmacol Exp Ther 1999, 289:251-260.
19. Dehaven-Hudkins, DL, Daubert, JD, Sawutz, DG, Tibero, L, Baine, Y. [^{3}H]1,3-(2-tolyl)guanidine binds to a sigma-2 receptor on Jurkat cell membranes, but sigma compounds fail to influence immunomodulatory events in human peripheral blood lymphocytes. Immunopharmacol 1996, 35:27-39.

20. John CS, Bowen WD, Varma VM, McAfee JG, Moody TW. Sigma receptors are expressed in human non-small cell lung carcinoma. Life Sci 1995, 56:2385-2392.
21. Wilke RA, Mehta RP, Lupardus PJ, Chen Y, Ruoho AE, Jackson MB. Sigma receptor photolabeling and sigma receptor-mediated modulation of potassium channels in tumor cells. J Biol Chem 1999, 274:18387-18392.
22. John CS, Bowen WD, Saga T, Kinuya S, Vilner BJ, Baumgold J, Paik CH, Reba RC, Neumann RD, Varma VM, McAfee JG. A malignant melanoma imaging agent: Synthesis, characterization, in vitro binding and biodistribution of iodine-125-(2-piperidinylaminoethyl)4-iodobenzamide. J. Nucl. Med. 1993, 34:2169-2175.
23. John CS, Vilner BJ, Geyer BC, Moody T, Bowen, WD. Targeting sigma receptor-binding benzamides as in vivo diagnostic and therapeutic agents for human prostate tumors. Cancer Res 1999, 59:4578-4583.
24. Kekuda, R, Prasad, PD, Fei, YJ, Leibach, FH, and Ganapathy V. Cloning and functional expression of the human type 1 sigma receptor (hSigmaR1). Biochem Biophys Res Commun 1996, 229:553-558.
25. Fujita T, Majikawa Y, Umehisa S, Okada N, Yamamoto A, Ganapathy V, Leibach FH. Sigma receptor ligand-induced up-regulation of H^+/peptide transporter PEPT1 in the human intestinal cell line Caco-2. Biochem Biophys Res Commun 1999, 261:242-246.

INDEX

α-amino-3-hydroxy-5-methylisoxazole-4-propionate. See AMPA
Acetylcholine, 200, 242–243
Actinomycin D, 226
Activator protein (AP), 109
Adrenal gland, 13, 153
Affinity labels, 45–46
  based on α ligand structural classes, 46–59
Aging, 260–262
Akt, 221
Alcohol, 327–328
Alkyl phenyl group, 93
Alzheimer's disease
  nontransgenic, 259–260
Amine analogs, secondary, 32f
Amino acid residues
  in C-terminal, 120–121
  in transmembrane domain, 115–120
Amino acid sequences, 104f, 105f
  deduced, 116f
Amino acid substitutions
  in transmembrane domain, 119t
Amnesia
  benzodiazepine diazepam, 255
  cholinergic systems and, 252–253
  nontransgenic Alzheimer's disease-related models, 259–260
  pharmarcological models of, 250t–251t, 257t
AMPA, 169, 300
Amphetamines, 324–326
AMPPcP, 135
Anatomical distribution, 11–14
  in nervous system, 11–12
  in peripheral organs, 12–14
Ankyrin, 195
Anti-amnesic properties, 250t–251t
  in pathological models of cognitive defects, 256–260
Antibodies
  immunohistochemical studies for, 155–156
Antidepressants, 296f
Antipsychotics
  activity of, 281f
  atypical, 275–276
  receptor ligands, 277f
  α ligands, 279–285
  typical, 275–276
Antisense treatment, 346f
AP. See Activator protein
aPKC. See Atypical PKC
Ar-C_5-N pharmacophore, 94
Ar-N_3. See Aryl nitrenes
Ar-X_5-N moiety, 92, 95
Aryl nitrenes, 60
Aryl ring-substituted compounds, 86
Arylacetamides, 217
Arylpropylamines, 33
Asp126, 121
ATP-independence of channel modulation, 136f
Atypical PKC (aPKC), 204
Autoradiography, 185, 186f
Azides
  reactive species formed from, 60f
Azidococaine
  structure of, 64f
Azido-DTG, 61–63
  structure of, 61f
Azido-phenazocine, 63–64
  structure of, 63f

Bad, 221
BALB/c, 340–341
Bcl-2, 221
BD737, 174–175, 176, 202, 208, 225, 306
BD1008, 28, 73f, 95, 178, 203, 322
BD1036, 246
BD1047, 35, 175, 176, 323, 327

BD1063, 244, 326
BDNF. *See* Brain-derived neurotrophic factor
Behavioral models
learning and memory, 239–241
Benzodiazepine diazepam, 255
Benzomorphans, 1, 25, 26f, 99, 184, 185, 266
binding affinities of, at PCP, 48t, 49t
irreversible ligands derived from, 57–59
phenylethylene diamines and, 29f
Benzylidine phenylmorphans, 32
Bicarbonate secretions, 381–382
Binding affinity, 278f
Binding studies
endogenous ligands and, 7–8
BMY 14802, 129, 202, 253, 254, 322, 323, 325, 376, 381
antipsychotic properties of, 280–281
BNIT, 58, 59
effect of, on binding parameters, 58t
structure of, 58f
Brain, 11
Brain-derived neurotrophic factor (BDNF), 283
Brainstem nuclei, 196
Brest tumors, 222
1-(4-bromo-acetylamido-2-methylphenyl)-3-(2-methylphenyl) guanidine. *See* DIGBA

C. difficile, 371, 385
Ca^{2+}
cell proliferation and, 219–220
homeostasis, 246–247
influx, 167
mobilization, 172–178, 263
Calcitonin gene-related peptide (CGRP), 254
Calcium channels
modulation of, 133
Cancer cells
potassium channel inhibition in, 132f
Cancer diagnosis, 223–225
CAPP. *See* Ceramide-activated protein phosphatase
3-(2-carboxypiperazine-4-yl)propyl-1-phosphonic acid. *See* CPP
Carcinoma, 178
Caspase independence, 224
Catecholaminergic, 168
Catecholaminergic neurotransmission, 202–209
dopamine, 202–209
CB-184, 218

CB-64D, 218
C57BL/6, 153, 261, 272
C8-C7 sterol isomerase, 102
CCK. *See* Cholecystokinin
CD1 mice, 340
cDNA probes, 102
Cell cultures
α ligands in, 217
Cell lines, 228
α receptor expression in, 216
Cell proliferation
calcium and, 219–220
ceramide and, 220–221
signaling mechanisms and, 219–223
sphingosylphosphorylcholine and, 221–223
Cell types, 14
Central nervous system (CNS), 151
immune cells and, 355–356
Ceramide, 220–221
Ceramide-activated protein phosphatase (CAPP), 220
cGMP, 171
CGRP. *See* Calcitonin gene-related peptide
Channel modulation, 145. *See also* Ion channels
ATP independent, 136f
G proteins and, 135f
ligand independent, 139f, 140–141
membrane delimitation of, 136f
in oocytes, 138f
reconstitution of, 137
Chemical names of compounds, 393t–394t
Chemo-sensitization, 225–226
Chemotherapy, 223–225
Cholecystokinin (CCK), 337, 373, 377
Cholera toxin, 384
Cholinergic neurotransmission, 200
Cholinergic receptors, 168
Cholinergic systems, 242–243
amnesia and, 252–253
Chromaffin cells, 174
(+)-cis-N-(4-isothiocyanatobenzyl)-N-normetazocine. *See* BNIT
Classical transmitter systems
regulation of, 198–209
Clorgyline, 295
Cocaine, 35–36, 322–324
Cognitive responses, 241–247
aging and, 260–262
cholinergic system modulation and, 242–243

GABAergic inhibitory responses and, 245–246
Coimmunoprecipitation, 142f, 143
CoMFA models, 38–39, 76
 Contour Graphs, 38f
Conventional PKC (cPCK), 176, 182, 183f, 185
 translocation of, 184f
Corticotropin-releasing factor (CRF), 377
Corticotropin-releasing hormone (CRH), 372, 376
cPCK. *See* Conventional PKC
CPP, 206
C-terminus, 106
 amino acid residues in, 120–121
Cyclazocine, 3
Cyclic adenosine monophosphate (cAMP), 170
Cyclic phenylphentylamines
 α receptor affinities for, 83f
Cyclohexyl compound 7, 96
Cyclosomatostatin, 385

DAT. *See* Dopamine transporters
Dehydroepiandrosterone. *See* DHEA
Depression, 295–296
 major, 297
 measurements of, 296
 NMDA receptors and, 303–305
 serotonin and, 297–298
D-erythro-sphingosine, 221
Desensitization
 cPKC and, 183f
 in motor brainstem, 180f
Desipramine, 260
Destrallorphan, 3
Detergent-resistant lipid rafts, 160f
Devazepide, 378t
Dextrallorphan, 62
DHEA, 199, 208, 237, 239, 244, 254, 261, 263, 264, 300, 324
Diacylglycerol (DAG), 204
Diarrhea, 385–386
DIGBA, 55
DIGIE, 55
DIGIT, 53f
 efffects of, 54t
DIGMF, 55
3,4-dihydroxyphenylalanine (DOPAC), 325
Di(methyl)guanidine (DMG), 52
Di-o-tolyl-guanidine. *See* DTG
Di(phenyl)guanidine (DPG), 52
Ditolyguanidine, 131
Dizocilpine, 304f
DMG. *See* Di(methyl)guanidine
DMS-114 cell lines, 65, 131, 134
DOPAC. *See* 3,4-dihydroxyphenylalanine
Dopamine, 77, 202–207, 278, 317
 drug abuse and, 316
 regulation of, release, 206f
Dopamine transporters (DAT), 206
 binding results at, 37t
 $α_1$ ligands and, 35–39
Dorsal raphe nuclei (DRN), 298
Dorsal vagal complex (DVC), 377
Doxorubicin, 226
DPDPE, 345
DPG. *See* Di(phenyl)guanidine
DRN. *See* Dorsal raphe nuclei
Drug abuse, 315, 316–318
 affinities of drugs in, 319t–321t
 alcohol, 327–328
 amphetamines, 324–326
 cocaine, 322–324
 direct binding in, 317
 dopamine and, 316
 nicotine, 328
 PCP, 326–327
DTG, 4, 52, 54, 196, 215, 252, 262–263, 298, 306, 374
 structure of, 53f
Dual probes, 35–39
Duodenal bicarbonate secretions, 381–382
DuP 734, 203, 207
Dup 734, 73f
DVC. *See* Dorsal vagal complex

E-5842, 199, 202
 antipsychotic properties of, 282–283
E. coli, 371, 385
EBP. *See* Emopamil binding protein
EDC, 120
Electrophiles, 45–46
Emopamil binding protein (EBP), 352, 360
Endogenous compounds, 395t–398t
Endogenous ligands, 7–10, 27
 evidence from binding studies, 7–8
 evidence from fractionation studies, 8–9
 evidence from physiological studies, 9
Endoplasmic reticulum (ER), 151, 155, 197
 in NG108-15 cells, 157f
 rafts, 161
Enhanced yellow fluorescent protein (EYFP), 158, 159, 160
ER. *See* Endoplasmic reticulum
ERG2, 103, 105

Estrogen, 301
1-ethyl-3-(3-dimethylaminopropyl). *See* EDC
Etoxadrol, 51t, 52
Eye, 14
EYFP. *See* Enhanced yellow fluorescent protein

FGF. *See* Fibroblast growth factor
Fibroblast growth factor (FGF), 283
Fluoxetine, 296
Fractionation studies
   endogenous ligands and, 8–9
   for subcellular distribution, 152–154

G protein coupled receptors (GPCRs), 114, 222
G protein coupling, 170–172
G proteins, 133, 134, 159
   activators, 134
   independent potassium channel modulation, 135f
$GABA_A$, 237, 300, 327
GABAergic inhibitory responses, 245–246
Gastric acid secretion, 380–381
Gastric mucosal damage, 379t
Gastrocolonic reflex, 374–375
Gastrointestinal tract, 13, 372–374
GATA-1, 109
GBR 12909, 35
GBR 12935, 36
Gene expression, 318
GF109203x, 205, 206
Gilligan model, 76–77
   regions in, 77f
Glennon/Ablordeppey model, 78–94
   initial, 82f
   revised, 96f
Glial cells, 168
Glu172, 121
Glutamate, 171, 276, 317
   release of, 199–200
Glutamatergic neurons, 166
Glutamatergic neurotransmission, 198–200
Glutamatergic receptors, 168
Go-6976, 176, 205
GPCRs. *See* G protein coupled receptors
Green fluorescent protein (GRP), 106, 129
GRP. *See* Green fluorescent protein
Guanidine
   irreversible ligands derived from, 52–55
GVIA, 173

H-1004, 181
Haloperidol, 30, 31f, 62, 63, 73f, 174–175, 201, 341t, 375
   binding, 120–121
   chemical structures of, 122f
   morphine analgesia and, 348f
   opiate analgesics and, 338–340
   opioid analgesia and, 342f
HeLa cells, 103
Hemagluttinin (HA), 343
   purification of, 344f
[3H](+)-3-(3-hydroxyphenyl)-N-(1-propyl)piperidine. *See* 3PPP
Hippocampus, 156f, 159
   pyramidal layer of, 196
   SKF-10,047 in, 177f
$[^3H]N_3DTG$, 61, 62
Hydropathy analysis, 129
Hydrophilic substituents, 39
3-hydroxyl group, 56
Hypoglossal neurons, 197
Hypoxic insults, 256–259

IAC. *See* Iodoazidococaine
Ibogaine, 33
   ligands based on, 33f
4-IBP, 274
IDAB, 228
Ifenprodil, 174–175
Igmesine, 169, 237, 296, 305, 306. *See* JO 1784
Iinflammatory pathologies, 363–365
Imaging studies, 10t
Immune disorders, 358–365
Immune system, 351–352
   CNS disorders and, 355–356
   neuroendocrine system and, 354–355
   pharmacological studies of, 356–357
   tissular expression in, 352–354
Immunoassays
   subcellular fractionation studies using, 154
Immunocytohistochemical studies
   for subcellular localization, 155–159
Immunofluorescence, 155
Inositol 1,4,5-triphosphate (IP_3), 160, 169, 195, 203
Intestinal motility, 374–378
   stress and, 376
Intestinal secretions, 382–387
   in vitro experiments, 382
   in vivo experiments, 383–385
Intracellular calcium mobilzation, 172–178
Intracellular dynamics, 159–161

Intracellular signaling
  historical perspective, 165–166
  phospholipases and, 179–187
  protein kinases and, 179–187
Iodoazidococaine (IAC), 64–65, 131
Ion channels, 127–128, 141–143. *See also*
    Channel modulation
  modulation of, 133
  reconstitution of modulation of, in
    Xenopus oocytes, 137
IP$_3$. *See* Inositol 1,4,5-triphosphate
Irreversible ligands, 45–46
  benzomorphan-derived, 57–59
  guanidine-derived, 52–55
  phenylpiperidine-derived, 56
  structures of, 57f
Isothiocyanato analogues, 35–36
1—3-(2-methylphenyl) guanidine. *See*
    DIGIE

JAR cells, 102
JJC 1-059, 37
JJC 2-010, 39
JO 1784, 169
JO 2871, 386

Ketocyclazocine, 1
Kidney, 13
Kv1.4, 140, 141, 143, 145
  coimmunoprecipitation of, 142f

L-687,384, 73f, 305
Largent model, 75
Learning, 239–241
  associative, 240
  complex, 240–241
  impairments, 253–255, 255–256
Leu-105, 120
Leu-106, 120
Ligand binding
  in wild-type and mutant α receptors,
    118–120
Ligand independent channel modulation,
    139f, 140–141
Limbic regions, 11
Lipid rafts, detergent-resistant, 160f
Liver, 9, 153
L-NAME, 254
Long-term depression (LTD), 243
Long-term potentiation (LTP), 243
LPS, 386
LR172, 35
LTD. *See* Long-term depression

LTP. *See* Long-term potentiation
Lu 28-179, 202, 207

Major depression, 297
Malignant melanoma, 227
Manallack model, 75
MAOI. *See* Monoamine oxidase inhibitors
MAP. *See* Mitogen-activated protein
MCF-7, 103, 218, 224
MDMA. *See*
    Methylenedioxymethamphetamine
Medicinal chemistry
  history of, 25–27
MEDLINE, 72
Membrane delimitation of channel
    modulation, 136f
Membrane topology, 104–107
Memory, 239–241
  hypoxic insults and, 256–259
  NMDA receptors and, 243–245
  working, 241
Metaphit, 49
  effects of, 48t
  structure of, 47f
Methamphetamine, 324–326
Methylenedioxymethamphetamine
    (MDMA), 324–326
1-(4-Methylfumaryl-amido-2-methylphenyl)-3
    -(2-methyl-phenyl) guanidine. *See*
    DIGMF
1-(2-methyl-4-isothiocyanatophenyl)-3-(2-me
    thylphenyl)guanidine. *See* DIGIT
Mitochondria, 197
Mitogen-activated protein (MAP), 220
MK-801, 51t, 129
Mnesic effects, 247–248
Monoamine oxidase inhibitors (MAOI), 295
Motor brainstem
  desensitization in, 180f
Mr 22,000, 64
Mr 27,000, 64
Mr 33,000, 64
Mr 57,000, 64
Mr 63,000, 62
Mr 65,000, 62
Mr 200,000, 64
Mr 150,000 protein complex, 62
MS-377, 326
  antipsychotic properties of, 284
Muscarinic receptors, 242–243
MVIIC, 173

(-N$_3$). *See* Azide

NaDodSo$_4$, 62
NADPH-cytochrome P450, 154, 158
NalBzoH, 341, 345
Naltrexone, 353
NAN-70, 94
NAN-190, 93
N-butyl-3-(3-hydroxyphenyl) piperidine, 56
NCAM. *See* Neuronal cell adhesion molecules
NE-100, 31, 35, 73f, 121, 244, 323, 326
  antipsychotic properties of, 283–284
  chemical structures of, 122f
  secondary amine analogs and, 32f
Nerve growth factor (NGF), 170
Nervous system
  anatomical distribution in, 11–12
NET. *See* Norepinephrine transporters
Neuroactive steroids, 238–239, 254, 324
Neuroblastoma, 217
Neuroendocrine areas, 12
Neuroendocrine system
  immune system and, 354–355
Neurohypophysial nerve terminals
  potassium current, 131f
Neuroleptics, 4
Neuronal cell adhesion molecules (NCAM), 161
Neuronal plasticity
  signal transduction and, 167–170
Neuropeptide FF, 337
Neuropeptide Y, 8, 204–205, 254, 373, 376
Neurosteroids, 8, 195, 300
Neurotransmission
  catecholaminergic, 202–209
  cholinergic, 200
  glutamatergic, 198–200
  opioidergic, 201
NF. *See* Nuclear factor
NF-GMb, 354
NG108, 159, 171, 197
NG108-15, 154, 158
  ER in, 157f
NGF. *See* Nerve growth factor
Nicotine, 328
Nicotinic receptors, 242–243
Nimodipine, 255–256
Nitrogen, 85
NMDA receptors, 2, 25, 46, 166, 168, 174, 175, 176, 177–178, 198, 199, 202, 203, 208, 237, 274, 276, 282, 294f, 306
  depression and, 303–305
  learning impairments and, 253–255
  memory and, 243–245

α receptors and, 302–303
N-methyl-D-aspartate receptors. *See* NMDA receptors
N,N-Dimethyl compound, 80
Nociceptin/orphanin FQ, 201
Non-invasive tumor imaging, 226–228
Noradrenergic system, 375
Norepinephrine, 207–209
Norepinephrine transporters (NET), 36
NPC 16377, 323
  antipsychotic properties of, 284
N-phenylaklyl-4-benzylpiperidines, 82
N-terminus, 106, 151
NTS. *See* Nucleus tractus solitarius
Nuclear factor (NF), 109
Nucleus tractus solitarius (NTS), 377

OBX. *See* Olfactory bulbectomy
OCH$_3$ group, 56
Octahydrobenzo[f]quinoline, 73f
Olfactory bulbectomy (OBX), 304, 305
Oligodendrocytes, 156
OPC-14523, 252
Opiate analgesics, 337–338
  haloperidol and, 338–340
  pentazocine and, 338–340
  strain differences in, 340–342
Opioid analgesia
  haloperidol and, 342f
Opioid receptors, 1
Opioidergic neurotransmission, 201

p53 mutations, 223–224
Panamesine
  antipsychotic properties of, 284
PBL. *See* Peripheral blood leukocytes
PCP. *See* Phencyclidine
Pentazocine, 3, 73f, 99, 113, 121, 129, 134, 153, 196, 201, 202
  chemical structures of, 122f
  morphine analgesia and, 348f
  opiate analgesics and, 338–340
  perfusion of, 186f
(+)-pentazocine, 159
Pentyl moiety, 90–91
Peptidergic nerve terminals, 130
Peripheral blood leukocytes (PBL), 356
Peripheral organs
  anatomical distribution in, 12–14
PET. *See* Positron emission tomography
P-glycoprotein expression, 225
Pharmacophore elements. *See* Pharmacophoric groups

Pharmacophoric groups, 71
Phencyclidine (PCP), 2, 3, 129, 274, 294, 326–327, 356, 374
   benzomorphan binding affinities at, 48t, 49t
   irreversible ligands derived from, 46–52
   structure of, 47f
   wash-resistant inhibition of, 51t, 56t
Phencyclidine-based ligands, 50f
Phenoxyalkyl amines, 29
Phenyl-A ring, 79
Phenylacetamides, 31
Phenylalkyl piperadine, 73f
Phenylalkyl piperazine, 73f
Phenylalkylamines
   simple, 34f
Phenyl-B region, 83
Phenylbuten-2-ylpiperazine moiety, 83
Phenylethylamine, 78
Phenylethylene diamine-based sigma ligands, 28f
Phenylethylene diamines
   benzomorphans and, 29f
Phenylmethanesulfonyl-fluoride. See PMSF
Phenylmorphans
   ligands based on, 33f
Phenylpentyl moiety, 78
Phenylpentylamines, 29, 30f
   nitrogen atom, 86
   α receptor affinities of, 81f
3-Phenylpiperidine
   irreversible ligands derived from, 56
Phenylpiperazine, 79
Phenylpiperidine, 79
Phorbol-12-myristate-13-acetate. See PMA
Phospholipase, 179–187
Phospholipase C (PLC), 159, 196, 283
   blockade of, 205
   Ca^{2+} dependent, 170
Phosphorylation, 185, 185f
Photoaffinity labels, 60–61
Physiological studies
   endogenous ligands and, 9
Physostigmine, 242
Phytohemagglutinin, 357
PIMBA, 226, 227
Piperazines, 36, 89
Piperazines ntirogen atoms, 84
Piperidines, 89
PKA. See Protein kinase A
PKC. See Protein kinase C
Placenta, 153
Plasma membrane, 197

PLC. See Phospholipase C
PLCβ, 179
PMA, 182, 204, 361
PMSF, 64
Positron emission tomography (PET), 226
Postsynaptic membrane, 168
Potassium channels, 145
   G protein independent modulation of, 135f
   inhibition of, in cancer cells, 132f
   inhibition of, in neurohypophysial nerve terminals, 131f
   modulation of, 129–132
PPHT, 130, 131
3PPP, 4, 26, 56, 129, 176, 351
PRE-079, 87
PRE-084, 171, 175, 248, 252
   effects of, 249f
Pregnancy, 299f, 300
Pregnenolone sulfate, 175, 239
Progesterone, 27, 100, 301
Pro-mnesic effects, 247–248
Promoter region, 108–109
Prostaglandin E2, 384
Protein kinase, 179–187
Protein kinase A (PKA), 181
Protein kinase C (PKC), 159, 169, 180–181, 186, 196
   atypical, 204
   conventional, 176, 182, 183f, 185
   inhibitors, 204
Protein kinase mediation, 135
Psychosis, 293–294
Psychotomimetic opioids, 4

Quaternary amines, 87

Racemic compounds, 81
Radioligand binding assays, 152–154
Reactive species
   from azide, 60f
Remoxipride
   antipsychotic properties of, 284
Reproductive organs, 13
Rimcazole, 35–38, 237, 323
   antipsychotic properties of, 280
RT-PCR, 101, 353

S. cerevisiae, 101
S. enteriditis, 371, 385
SA4503, 200, 252, 259, 295, 328
Scatchard plots
   of pentazocine binding, 117
SCH 23390, 281

Schizophrenia
  pathophysiology of, 274–275
  potential treatment for, 276–277
  α receptors and, 278–279
SDS-PAGE, 141
Ser-99, 120
Serotonin, 77, 275, 317
  depression and, 297–298
  α receptors and, 298–299
Serotonin transporters (SERT), 36
Sertraline, 294f, 295
SH 3-28, 37
SH-SY-5Y cells, 179, 198
SH-SY-5Y neuroblastoma, 217
"α enigma," 166–167
α ligands
  antipsychotic, 277f, 279–285
  in cell cultures, 217–218
  in immune disorders, 358–365
  immunomodulatory activities of, 356–357
  medicinal chemistry of, 72
  phenylethylene diamine-based, 28f
  photoaffinity labels based on, 61–65
  template structure used to align, 76f
α_1 ligands, 30–31
  cross-section of, 73f
  dopamine transporters and, 35–39
  perspective uses of, in cognitive indications, 262–264
  pro-mnesic effects of, 247–248
α_2 ligands, 32–33
α receptors, 26–27, 26f
  affinities of representative phenylpentylamines, 81f
  anatomical distribution of, 11–14
  cell types, 14
  characteristics of, 5t
  coimmunoprecipitation of, 142f
  cyclic phenylphentylamine derivatives and, 83t
  distribution of, 196
  drug abuse and, 316–318
  drug selectivity profile of compounds for, 3t
  expression of, in tumors and cell lines, 216
  functions of, 127
  historical perspective on, 1–5, 25–27, 237–238
  imaging studies of, 10t
  ion channels and, 141–143
  models after discovery of, 76–94
  models prior to, 75–76
  NMDA receptors and, 302–303
  psychosis and, 293–294
  research on, 74f
  schizophrenia and, 278–279
  serotonin and, 298–299
  splice variants, 6–7
  steroids and, 299–302
  topology of, 128f
  types of, 6–7
  wash-resistant inhibition of, 51t
α_1 receptors
  amino acid residues in transmembrane domain of, 115–120
  assessment of, 345–347
  binding results at, 37t
  chromosomal location of, 108
  cloning of, 100–104, 343–344
  cognitive responses and, 241–247
  dopamine release regulation and, 206f
  human, 102–104
  intracellular dynamics, 159–161
  ligand binding in wild-type, 118–120
  membrane topology of, 104–107
  molecular biology of, 343–347
  promoter region of, 108–109
  rodent, 104
  splice variants, 107–108
  Xenopus, 118
α selective agents, 28–29
α subtype selective agents, 30–34
Signal transduction
  neuronal plasticity and, 167–170
Signal transduction mechanisms, 133–137, 144f, 145–146
Signaling mechanisms
  cell proliferation and, 219–223
Single photon emission computed tomography (SPECT), 226, 227
SKBr3 breast tumor cells, 222
SKF-10,047, 1, 25, 64, 65, 113, 129, 130, 134, 152, 153, 166, 181, 182, 184, 185, 202, 327, 372, 373, 374
  forms of, 2f
  in hippocampal pyramidal neurons, 177f
SKF-525A, 90–91
SK-N-SH neuroblastoma cells, 175, 217, 219
Sodium dodecyl sulfate. *See* NaDodSo$_4$
SPC. *See* Sphingosylphosphorylcholine
SPECT. *See* Single photon emission computed tomography
Sphingosylphosphorylcholine (SPC), 221–223
Spleen, 13, 153
Splice variants, 6–7

Index 413

SR 31747A, 103, 114, 323, 352
  binding characteristics of, 360t
  binding sites, 359t
  chemical structure, 358f
  family of, 359–361
  immunomodulatory activities of, 361–363
  in inflammatory pathologies, 363–365
  pharmacological properties of, 359
SR-BP2, 360
SSR125329, 352
  binding characteristics of, 360t
  chemical structure, 358f
  in inflammatory pathologies, 363–365
  pharmacological properties of, 359
Staphylococcus enterotoxin B, 362
Steroid biosynthesis, 27
Steroids
  binding of, 92
  sigma receptors and, 299–302
Stress, 376–377
Subcellular distribution, 152–154
Subcellular fractionation studies
  using immunoassays, 154
Subcellular localization
  immunocytohistochemical studies for, 155–159
Substance P, 373
Succinyl concanavalin A, 357
Synaptic plasticity
  historical perspective, 165–166

Tacrine, 242
TCP, 129
Thioxanthenes, 83
Tissular expression, 352–354
TM. *See* Transmembrane
TNFα, 363
Toluene rings, 54

Translocation
  of cPCK, 184f
Transmembrane (TM), 114
  amino acid residues in, 115–120
  amino acid substitutions in, 119t
Triton X-100, 161
Tropane analogs, 34
Tumors
  drug resistant, 223
  non-invasive imaging of, 226–228
  p53 mutations, 223–224
  sigma receptor expression in, 216
Tyr-103, 121
Tyr-W-MIF-1, 337

U50,488, 91, 341
U-73,122, 179, 205
U-73,343, 179
U101958, 130
Ulcers, 379–382
  duodenal, 379–380
  gastric, 379–380

Voltage dependent Ca^{2+} channels (VDCC), 247
  blockers, 255–256
Voltage sensitive Ca^{2+} channels (VSCC), 173
  R-type, 173

Western blot, 185
Wild-type sigma receptors
  ligand binding in, 118–120
  in *Xenopus*, 118

Xenopus, 115, 129, 143, 145
  reconstitution of ion channel modulation in, 137
  wild-type sigma receptors in, 118

Printed in the United States of America